THE ROUTLEDGE HANDBOOK OF TRANSPORTATION

The *Routledge Handbook of Transportation* offers a current and comprehensive survey of transportation planning and engineering research. It provides a step-by-step introduction to research related to traffic engineering and control, transportation planning, and performance measurement and evaluation of transportation alternatives.

The *Handbook of Transportation* demonstrates models and methods for predicting travel and freight demand, planning future transportation networks, and developing traffic control systems. Readers will learn how to use various engineering concepts and approaches to make future transportation safer, more efficient, and more sustainable.

Edited by Dušan Teodorović and featuring 29 chapters from more than 50 leading global experts, with more than 200 illustrations, the *Routledge Handbook of Transportation* is designed as an invaluable resource for professionals and students in transportation planning and engineering.

Dušan Teodorović is Professor at the Faculty of Transport and Traffic Engineering, University of Belgrade, Serbia, and Professor Emeritus of the Virginia Polytechnic Institute and State University, USA. Dr. Teodorović received a Ph.D. degree in engineering from the University of Belgrade and has worldwide academic experience. His primary research interests are in operations research and computational intelligence applications in transportation engineering. Dr. Teodorović has been elected a member of the Serbian Academy of Sciences and Arts and the European Academy of Sciences and Arts.

THE ROUTLEDGE HANDBOOK OF TRANSPORTATION

Edited by
Dušan Teodorović

Routledge
Taylor & Francis Group
LONDON AND NEW YORK

First published 2016
by Routledge

2 Park Square, Milton Park, Abingdon, Oxfordshire OX14 4RN
711 Third Avenue, New York, NY 10017

Routledge is an imprint of the Taylor & Francis Group, an informa business

First issued in paperback 2017

Library of Congress Cataloging in Publication Data
The Routledge handbook of transportation/edited by Dusan Teodorovic.—
Edition 1st.
pages cm
1. Transportation. 2. Shippers' guides. 3. Traffic engineering—Handbooks,
manuals, etc. I. Teodorovic, D. (Dusan), 1951– editor. II. Title: Handbook
of transportation.
HE151.R68 2015
388—dc23
2014046452

ISBN: 978-1-138-79821-2 (hbk)
ISBN: 978-0-8153-8191-4 (pbk)

Typeset in Bembo
by Florence Production Ltd, Stoodleigh, Devon, UK

For Product Safety Concerns and Information please contact our
EU representative: GPSR@taylorandfrancis.com
Taylor & Francis Verlag GmbH, Kaufingerstraße 24,
80331 München, Germany.

This is for you, Ljiljana.

CONTENTS

FIGURES

TABLES

CONTRIBUTORS

Dušan Teodorović is Professor at the Faculty of Transport and Traffic Engineering, University of Belgrade, Serbia. His primary research interests are in operations research applications in transportation engineering. He is Professor Emeritus of Virginia Tech, USA, and an elected member of the Serbian Academy of Sciences and Arts.

Montasir Abbas is an Associate Professor in Transportation Infrastructure and Systems Engineering at Virginia Tech. He holds a Bachelor of Science from the University of Khartoum, Sudan (1993), a Master of Science from the University of Nebraska-Lincoln (1997), and a Doctorate of Philosophy from Purdue University (2001), all in Civil Engineering.

Riccardo Accorsi is an Assistant Professor at the Department of Industrial Engineering at the University of Bologna, Italy. He has a Ph.D. in Mechatronics and Industrial Systems from the University of Padua, Italy. His main fields of research are modeling and simulation applied to industry and supply chain context, with particular focus on warehousing systems and distribution networks.

Tsz-Chiu Au is Assistant Professor at Ulsan National Institute of Science and Technology. He received his Ph.D. in Computer Science from the University of Maryland, College Park, at which he was a member in the AI planning group and the Laboratory for Computational Cultural Dynamics. He joined the Department of Computer Science at the University of Texas at Austin as a postdoctoral fellow in 2008. His research interests include automated planning, multiagent systems, intelligent transportation systems, and robotics.

Maria Battarra completed a Ph.D. in Operational Research at the University of Bologna in March 2010. Since October 2011, Dr. Battarra is an Assistant Professor at the Mathematics Department of Southampton University. Her research interests lie in the area of heuristic and exact algorithms for routing and scheduling problems.

Robert L. Bertini, Ph.D, PE, is Associate Professor of Civil and Environmental Engineering at California Polytechnic State University, San Luis Obispo. He received a BS at Cal Poly San Luis Obispo, an MS at San Jose State University and a Ph.D. at the University of California at Berkeley, all in civil engineering.

Todd Borkowitz, PLA, LEED AP, is a registered landscape architect and accredited green building professional with over ten years' experience. He received a BLA in Landscape Architecture from Iowa State University and is currently completing his MURP in Urban and Regional Planning with a focus in transportation at Portland State University.

Partha Chakroborty is the Rajeeva and Sangeeta Lahri Chair Professor in the Department of Civil Engineering at the Indian Institute of Technology Kanpur, India. He is also the founder editor-in-chief of the journal *Transportation in Developing Economies*. He works in the areas of public transit operations and traffic flow theory.

T. Donna Chen is a Ph.D. candidate in transportation engineering at University of Texas at Austin with publications in transportation economics, electric vehicles and crash safety. A licensed professional engineer, Donna has worked in the consulting sector performing roadway design and traffic operation analyses.

Giovanni Circella is a Research Engineer at Georgia Institute of Technology and the University of California, Davis. His research interests include travel behavior, land use planning and modeling, travel demand modeling, discrete choice modeling, energy consumption, and policy analysis.

Alexandre G. de Barros is an Associate Professor of Transportation Engineering at the University of Calgary, Canada. His main expertise is on air transportation and airports, having worked for many years as a consultant in the field and served as the Director, Airport Infrastructure of the Brazilian National Civil Aviation Agency.

Branka Dimitrijević is an Associate Professor at the University of Belgrade, Faculty of Transport and Traffic Engineering, where she completed her Ph.D. in Location Analysis. Her scientific focus is on operations research and its applications in planning, modeling and exploration of transportation and logistics systems.

Matthew Downey is a master's degree student in the Department of Civil and Environmental Engineering. A native of Denver, Colorado, Matthew earned his BS in civil engineering from Iowa State University. He is currently studying the effects of a new variable speed limit system on traffic flow and safety on a Portland area freeway.

Praveen Edara is James Dowell Associate Professor of Engineering in the Civil Engineering department at University of Missouri. He teaches and conducts research in ITS, emergency evacuation, and alternative intersection designs. He previously worked for the Virginia Center for Transportation Innovation and Research and the Turner Fairbanks Highway Research Center.

Nick Foster, AICP, is a Senior Planner in the Boise office of Kittelson & Associates. He received a BA in Economics and Applied Physics at Whitworth University and an M.S. in Civil Engineering at Portland State University, where he developed a level-of-service model for protected bicycle lanes.

Grigorios Fountas is a graduate student of Civil Engineering at the National Technical University of Athens, majoring in Transportation. His research interests include transportation planning, operations research and transportation economics.

Mauro Gamberi is Associate Professor of Industrial Mechanical Plants at the University of Padova, Italy. He is author of more than 90 scientific publications dealing with manufacturing systems, industrial processes and new products development. He coordinates as senior expert engineer numerous projects and activities cofinanced by public and private institutions.

Bruce Golden is the France-Merrick Chair in Management Science in the Robert H. Smith School of Business. His research interests include vehicle routing, heuristic search, combinatorial optimization, networks, and healthcare analytics. Bruce has served as Editor-in-Chief of the *INFORMS Journal on Computing* and *Networks*. He has published extensively and is widely cited.

Konstantina Gkritza is an Associate Professor in the Lyles School of Civil Engineering, and the Department of Agricultural and Biological Engineering at Purdue University, Indiana. Dr. Gkritza's research experience and interests lie in the application of economic concepts and statistical methods for efficient, cost-effective, safe, and sustainable transportation systems.

Nigel G. Harris is a career railway planner who has worked on fares policy, network modeling, and service planning. Since 1995 he has managed The Railway Consultancy, and has undertaken projects on every continent except Antarctica. He is a visiting lecturer at the Universities of Newcastle, Birmingham and Sydney.

Xiaozheng He is a postdoctoral research associate in the NEXTRANS center, Purdue University. He earned his Ph.D. degree in transportation engineering at the University of Minnesota in 2010. His research covers a variety of topics including traffic assignment problems, location and allocation problem, traffic flow control, network disruptions, vehicle routing, and signal optimization.

John Hourdos is the Director of the Minnesota Traffic Observatory, an Assistant Professor at the Department of Civil, Environmental, and Geo-engineering, and a Research Scholar at the Centre for Transportation Studies at the University of Minnesota, Minneapolis, USA.

Milan Janić is Senior Researcher at the Department of Transport and Planning in the Faculty of Civil Engineering and Geosciences and the Department of Air Transport and Operations in the Faculty of Aerospace Engineering of Delft University of Technology, the Netherlands. He is also Research Professor at the Faculty of Traffic and Transport Engineering of the University of Belgrade, Serbia.

Matthew Karlaftis is a Professor of Transportation Planning at the National Technical University of Athens. He has acted as the editor of the ASCE *Journal of Transportation Engineering and Transportation Research Part C*. His research interests include transportation planning and operations, traffic planning, statistics, operations research and computational intelligence.

Konstantinos Kepaptsoglou is a Lecturer of Transportation Planning at the National Technical University of Athens. In 2011 he received the ASCE State-of-the-Art award in Civil Engineering Award. His research interests include transportation planning, public transportation operations and management, operations research, computational intelligence and transportation economics.

Jalil Kianfar, Ph.D., is Assistant Professor of Civil Engineering at Parks College of Engineering, Aviation, and Technology at Saint Louis University. Kianfar's area of expertise broadly covers three major fields: traffic operations, intelligent transportation systems and connected vehicles, and application of optimization and heuristic techniques in transportation engineering.

Kara M. Kockelman, E.P. Schoch Professor of Civil, Architectural and Environmental Engineering at the University of Texas at Austin, received the Berkeley Medal, an NSF CAREER Award, MIT's Technology Review Top 100 Innovators Award, a Google Research Award, the Regional Science Association International's Hewings Award, and ASCE's Bartholomew Award, Huber Award, and Laurie Prize. Dr. Kockelman's primary research interests include the statistical modeling of urban systems, the economic impacts of transport policy, crash occurrence and consequences, energy and climate issues, and transport policy-making.

Sirisha Kothuri, Ph.D., is a post-doctoral researcher at Portland State University. She received a BS from Osmania University, Hyderabad, India, an MS from Louisiana State University, and a Ph.D. from Portland State University, all in civil engineering. Her dissertation investigated methods to incorporate pedestrian considerations into traffic signal timing.

Henry Liu is a Professor in the Civil and Environmental Engineering Department of the University of Michigan, Ann Arbor. He is also a Research Professor in the University of Michigan Transportation Research Institute (UMTRI).

Riccardo Manzini is Professor of Logistics and Maintenance at the Department of Industrial Engineering at the University of Bologna. His academic research deals with planning, design and control of production systems with particular reference to logistics and operations. He is Director of the Food Supply Chain Center and the Warehousing Center at UniBo.

Timothy C. Matisziw is Associate Professor, Departments of Geography, Civil & Environmental Engineering, and Informatics Institute at the University of Missouri. His primary areas of interest and expertise include transportation systems, spatial optimization, geographic information systems, network modeling, and regional planning.

Christopher M. Monsere, Ph.D., P.E., is Associate Professor and Associate Chair of Civil and Environmental Engineering at Portland State University. He received a BCE from the University of Detroit Mercy, an MSCE and Ph.D. from Iowa State University. He chairs the TRB committee on Safety Data, Analysis and Evaluation and is a licensed professional engineer in the state of Oregon.

Alan T. Murray is Professor in the College of Computing and Informatics, Professor in the School of Public Health, and Director of the Center for Spatial Analytics and Geocomputation at Drexel University. He is editor of *International Regional Science Review*, associate editor for *Socio-Economic Planning Sciences* and *Annals of the Association of American Geographer*. His research and teaching interests include: geographic information science; health informatics; land use planning; urban, regional, and natural resource planning and development; quantitative methods; infrastructure and transportation systems; spatial optimization; location modeling; databases and data structures; spatial representation; and techniques to support interactive planning and decision making.

Ranko Nedeljković is Professor of Operations Research and Statistics. His areas of interest include: reliability, queueing, quality assurance and official statistics. Besides his academic career, he was the Director General of Statistical Office of Serbia and Montenegro. Since 2008 Ranko is the President of Statistical Society of Serbia.

Francesca Pagliara is an Assistant Professor at the University of Naples Federico II, where she teaches Transportation Systems Technologies. Her research focuses on both the theoretical aspects and experimental analysis in the fields of transportation, urban planning and spatial modeling.

Srinivas Peeta is a Professor of Civil Engineering at Purdue University, USA, and the Director of the NEXTRANS Center, the USDOT Region 5 Regional University Transportation Center. His interests broadly span transportation and infrastructure systems.

Harilaos N. Psaraftis is Professor of Transport Optimization at the Department of Transport at the Technical University of Denmark. He has also been a faculty member at the Massachusetts Institute of Technology, USA (1979–1989) and at the National Technical University of Athens, Greece (1989–2013), and has served as a CEO of the Piraeus Port Authority (1996–2002).

Hesham Rakha is the Samuel Reynolds Pritchard Professor of Engineering in the Charles E. Via, Jr. Department of Civil and Environmental Engineering, a Courtesy Professor in the Bradley Department of Electrical and Computer Engineering, and the Director of the Center for Sustainable Mobility at the Virginia Tech Transportation Institute. He received M.S. and Ph.D. degrees in Civil Engineering from the University of Minnesota, Twin Cities, in 2007 and 2010, respectively. He is currently a research associate at the NEXTRANS Center, Purdue University.

Alberto Regattieri is a Full Professor at the Department of Industrial Engineering at the University of Bologna. His current research interests include logistics, the optimal design of manufacturing and service systems, and the maintenance of industrial plants. He manages several research projects on these fields granted by public institutions and private companies.

James E. W. Roseborough received degrees from York University (BA, MA, and Ph.D. in Psychology). His research focuses on personality traits, cognitions, and emotions, and their contribution to risky and aggressive driving. Roseborough is a member of the Canadian Association of Road Safety Professionals.

Matteo Mario Savino holds a Ph.D. in Industrial Operations, obtained in 1999 at the Department of Production and Industrial Engineering at the University of Naples, Italy. He is currently Professor of Industrial Operations at the University of Sannio, Benevento, Italy. His research interests are centered on models and techniques for quality, environment and supply chain management. His research works appeared on several peer-reviewed international journals.

Paul Schonfeld is a Professor in the Department of Civil and Environmental Engineering at the University of Maryland, College Park. He has B.S. and M.S. degrees from MIT and a Ph.D. from U.C. Berkeley. He has considerable experience in analyzing urban public transportation systems, waterways, freight logistics, road networks and airports. Twenty of his Ph.D. students have received university faculty appointments.

Milica Šelmić is Assistant Professor at the Faculty of Transport and Traffic Engineering, Department for Operational Research in Transport. She completed her Ph.D. thesis in Location Theory. Her areas of interest also include: transportation networks design, routing, Artificial Intelligence in transport and traffic (fuzzy systems, neural networks, evolutionary computation), and swarm intelligence.

Peter Stone is an Alfred P. Sloan Research Fellow, Guggenheim Fellow, AAAI Fellow, Fulbright Scholar, and University Distinguished Teaching Professor in the Department of Computer Science at the University of Texas at Austin. He received his Ph.D. in Computer Science in 1998 from Carnegie Mellon University. Peter's research interests include machine learning, multiagent systems, robotics, and e-commerce. In 2007, he was awarded the prestigious IJCAI 2007 Computers and Thought award.

Eiichi Taniguchi is Professor at the Department of Urban Management, Kyoto University, Japan. His research focused on city logistics and urban freight transport planning. He has performed modeling city logistics to evaluate policy measures for efficient and environmentally friendly urban freight transport systems based on Intelligent Transport Systems.

Konstantinos Triantis is the John Lawrence Professor of Industrial Engineering and Operations Research in the Grado Department of Industrial and Systems Engineering, an adjunct Professor in the Via Department of Civil and Environmental Engineering and the co-Director of the System Performance Laboratory at Virginia Tech.

Gudmundur F. Ulfarsson is a Professor of Transportation Engineering and the Head of the Faculty of Civil and Environmental Engineering in the School of Engineering and Natural Sciences at the University of Iceland.

Vinod Vasudevan is an Assistant Professor of Civil Engineering at the Indian Institute of Technology Kanpur, India. He is active in various international professional bodies including TRB, ASCE, and ITE. His areas of interest include traffic safety, highway financing, and transport economics.

Xingyin Wang obtained his Bachelor's degree from the National University of Singapore in July, 2011. He is now a Ph.D. student at the University of Maryland, doing research on Vehicle Routing Problems under the supervision of Dr. Bruce Golden. His research interests include heuristic algorithms and worst-case analysis.

Edward Wasil is a Professor at American University in Washington, DC. He has published more than 90 technical articles in a variety of areas. His research interests focus on network optimization and applications of metaheuristics to optimization problems. Currently, he serves as feature article editor of the *INFORMS Journal on Computing* and associate editor of *INFOR: Canadian Journal of Operational Research and Information Processing*.

Christine M. Wickens holds a Ph.D. in social and personality psychology from York University in Toronto, Canada. She is currently a project scientist at the Centre for Addiction and Mental Health, Canada's pre-eminent mental health and addictions hospital, and has been appointed as an assistant professor at the Dalla Lana School of Public Health at the University of Toronto. She has published extensively in the field of driver safety including driver aggression, street racing, and alcohol- and drug-impaired driving.

David L. Wiesenthal received his doctoral degree in social psychology from the State University of New York at Buffalo and was a postdoctoral fellow at York University where he is currently a professor in the Department of Psychology. He is the coeditor of *Contemporary Issues in Road User Behavior and Traffic Safety* (2005) with Dwight Hennessy. He has been a visiting professor at the Hebrew University of Jerusalem and has lectured at the National Road and Traffic Research Institute (Linköping, Sweden), the University of Linköping, the University of Umeå, Lund University, and Stockholm University. He conducted traffic psychology workshops for the University of Costa Rica as well as the College of Psychologists in May 2014.

Shun Zhang is a master student in the Integrated BS/MS Program in Computer Science at the University of Texas at Austin. His research is supervised by Dr. Peter Stone. His research interests are autonomous intersection management, reinforcement learning and robotics. He received the department scholarship in 2013–2014 academic year in his undergraduate study at the University of Texas at Austin.

FOREWORD

Since the first organization of human societies and the division of labor, the transportation of persons and goods had an essential role in the progress of human civilization, contributing to and gaining from the technological and societal evolution. For thousands of years, transportation was taking place on surface paths and roads as well as on waterways. It is only in the last couple of centuries where railways were introduced, and even more recently when air transportation entered the scene. While for thousands of years transportation was almost exclusively serving trade and other professional activities, an initially timid, but eventually dramatic increase of recreational transportation needs has been observed over the last centuries. Today, transportation in developed countries accounts for some 6 percent of the GDP and some 13 percent of the family budget on average. On the other hand, the societal, environmental and personal cost of transportation has also grown enormously. There are 1.24 million road traffic deaths worldwide every year. Transport is the second largest sector in terms of emissions, releasing 22 percent of global CO_2 emissions in 2011. The fast emissions growth of the transport sector was mainly driven by emissions from the road sector, which increased by 52 percent since 1990, accounting for about three-quarters of transport emissions in 2011. Global transport fuel demand is expected to grow by nearly 40 percent by 2035. These statistics generate rightly questions about the rationality of the evolution of transportation, particularly in the last decades, the most severe of these questions concerning the usage of (typically one-person occupancy) cars in large metropolitan areas.

The interrelated evolution and continuous expansion of transportation means, infrastructure and needs would not have been possible without the parallel gathering, development and enhancement of skills, techniques and experiences, which eventually evolved in the interdisciplinary area of Transportation Science. Transportation is the main subject of study for myriads of students at corresponding divisions, departments and universities at all higher education levels, as well as the subject of intensive research activity at industrial firms, consultants, universities and research institutions around the globe. Transportation Science encompasses the design, planning, operation, assessment of all transportation modes, both separately and in appropriate multimodal combinations, and covers a broad variety of aspects, spanning from traffic flow theory to human factors and economics. The Transportation discipline has close ties and indeed overlaps with many other scientific disciplines, a non-exhaustive list comprising fluid mechanics, simulation, dynamic systems, optimization, metrology, automatic control, communications, computer science, psychology and economics. Thus, progress in each of these disciplines may be reflected in corresponding advancement of Transportation Science as well, and this is a first engine and source for innovation in the area. A second and more important motivation for innovation and improvements are the difficulties and problems appearing in transportation practice due to the steady demand and infrastructure expansion, shifting objectives, priorities and emerging needs. This calls for close cooperation of agencies responsible for the planning and operation of Transportation systems with agencies addressing Transportation research, something that, like in other disciplines as well, is not always achieved in the best possible way.

Transportation typically involves discrete entities, i.e. vehicles or particles. In some cases, e.g. in aspects of road traffic, it is reasonable and helpful to view the transportation system in an aggregate way, as a particular continuous fluid on traffic networks, while in other cases (e.g. a rail system), the significance of each transportation vehicle must be reflected in the analysis and considerations, leading to discrete mathematical representations (e.g. the scheduling task). When the density of transportation particles within a transportation infrastructure is low, planning and design may be static and focus mainly on safety. As the density of particles increases, network efficiency concerns must also be considered, and the related methods and tools have to consider dynamic aspects of the transportation process as well. Even higher concentration of vehicles may lead to congestion, i.e. a situation where the transportation infrastructure is degraded due to excessive queuing and spillback phenomena, leading to accordingly excessive delays, fuel consumption and environmental impact. As a matter of fact, increasing demand or exogenous events may lead to such situations at specific locations and time periods on all transportation modes. This calls for the development of appropriate methods and means to protect the infrastructure from the detrimental degradation. Intelligent Transportation Systems (ITS) is a relatively new domain, boosted by recent technological advances in communication, computing and control, aiming at optimizing the performance of transportation systems and mitigating congestion phenomena.

This *Handbook of Transportation* gathers contributions from virtually all areas and modes of Transportation, with focus on recent developments. The Editor, Professor Dušan Teodorović, has done an excellent job in the selection of subjects and experienced authors for each included subject. The volume will be valuable for students, researchers and engineers interested in particular transportation subjects, but it can also be consulted or read to gain a global overview of the many different aspects of the transportation domain and to discover, except for the differences, also many similarities among different subjects and modes. I believe that this work will have its own contribution to transportation education, research and, last but not least, to a more efficient, safe and environmentally friendly transportation system in the years to come.

Markos Papageorgiou

PREFACE

From the time when I was a small boy, I always loved to travel. My first trips happened on Sundays. I remember, often on Sundays, we were invited for a lunch with close relatives. At the end of the 1950s in Belgrade, where I grew up, there were many trolleybuses. Driving on a weekly lunch by trolleybus represented for me an indescribable pleasure. At the end of the ride and at the end of the lunch they served wonderful cakes. Many years later, my childhood dreams became a reality when I was a passenger on major world airports, or when I was driving through major cities worldwide. I was in traffic. When I was for the first time at the big airport I was fascinated by the fact that planes take off and land simultaneously at few runways, and that thousands of passengers and pieces of luggage unerringly find their way through the airport. As a passenger, I have always enjoyed when the bus arrives at the bus stop on schedule. As a driver, I have always been thankful to "good traffic lights" that helped me not to suffer from congestion. Many years later I became the Editor of the *Routledge Handbook of Transportation*. What a pleasure. The following is my story.

I cautiously selected topics that are, in my opinion, essential to transportation planning and traffic control. The majority of transportation engineering books focus on highway transportation. I have tried in this handbook to cover *all* transportation modes. In other words, transportation planning, management and control of traffic flows over modern urban, road, rail, inland water, maritime, and air transportation networks are the foci of this book. The book introduces, step by step, concepts related to traffic engineering and control, transportation planning, and performance measurement and evaluation of transportation alternatives. It brings systematic coverage of the whole transportation engineering area. The heart of this book is on engineering solutions to transportation problems. The book offers to readers various techniques related to transportation modeling, transportation planning and engineering design. The book also helps readers to use models and exact methods when predicting travel and freight demand, planning future transportation networks, and developing traffic control systems. Readers of the book can learn how to use various engineering concepts and approaches to make future transportation safer, more efficient, and "greener". It can serve for upper-level undergraduate and graduate courses, as well as for transportation and traffic engineers, urban planners, consultants, and government employees.

Chapter 1 contains an introductory explanation describing transportation science as a scientific discipline dealing with planning, design, operations, control, management, maintenance and rehabilitation of transportation systems, services and components. Chapter 2 describes the basics of traffic flow theory and its application to the modeling of traffic flow. The basics of the Traffic Network Modeling (OD matrix, traffic assignment, and network design problem) are described in Chapter 3. Chapter 4 describes the present tools of transportation that engineers have currently at their disposal in order to manage freeway traffic efficiently and safely. The fundamental control parameters and methods used in controlling isolated and coordinated systems, including fixed time control, traffic responsive, and adaptive control are given in Chapter 5. Chapter 6 provides an overview of the different ITS technologies that have made an impact on transportation. The protocol called Semi-Autonomous Intersection Management, which allows vehicles with partially autonomous features such as adaptive cruise control to enter an intersection from different directions simultaneously, is given in Chapter 7. The role that psychology can play in understanding and coping with aggressive roadway behaviors is described in Chapter 8. Chapters 9 and 10 provide an overview of the design assumptions and considerations for pedestrian and bicycle facilities, while Chapter 11 is devoted to traffic crash analysis and prevention. Chapter 12 gives an overview of the current state of practice in understanding the relationship between traveler behavior, system performance and traffic network disruptions. The basics of transportation planning are given in Chapter 13, while Chapter 14 is devoted to the complex relationships between land use and transportation. Chapter 15 studies how public transport networks and schedules can be developed so that they can adequately satisfy a large number of trip makers. The Vehicle Routing Problem (VRP) that models the distribution of goods or services to a number of customers by a fleet of vehicles from a central depot is described in the Chapter 16. An overview of the airport planning and design process is given in Chapter 17. Chapter 18 describes and analyses the components and operations of the Air Traffic Control (ATC) system. Railway planning and operations issues are given in Chapter 19, while Chapter 20 is devoted to the inland water transportation. The basics of maritime transportation are given in Chapter 21. Chapter 22 deals with logistics and supply-chain management as strategic discipline to support strategic infrastructure and operative decisions on freight transportation. Chapter 23 is devoted to city logistics. The study of complex relations between transportation, energy and environment is given in Chapter 24. Transportation economics issues are considered in Chapter 25. Chapters 26 and 27 are respectively devoted to queueing in transportation and location analysis in transportation. The concepts and methods that have originated in the efficiency measurement literature and have been applied to transportation systems and processes are described in Chapter 28. Chapter 29 provides an introduction to GIS and discusses its connection to transportation planning and analysis.

This *Handbook of Transportation* brings together 52 researchers and practitioners, acknowledged experts in the field from twelve countries (USA, Canada, Japan, India, South Korea, Serbia, Greece, Italy, Iceland, Netherlands, Denmark, and the UK), dealing with various transportation engineering problems. The book highlights transportation engineering problems in *all* modes of transportation. The primary goal of the book is to acquaint the reader with some of the innovative ideas, models, methods and tools related to transportation engineering.

I am very grateful to all colleagues, authors and reviewers for their enthusiastic work and engagement. I am especially thankful to Routledge who gave me the opportunity to be the Editor of this *Handbook*. Nicole Solano and Judith Newlin from Routledge were a great help.
I sincerely hope that this *Handbook* will advance engineering practice and teaching, and further motivate research in all areas of transportation science.

Dušan Teodorović
Faculty of Transport and Traffic Engineering
University of Belgrade, Serbia

1

INTRODUCTION

Dušan Teodorović

UNIVERSITY OF BELGRADE

For thousands of years, human society was characterized by the continuous migration of the population. Humans constantly migrated to new regions, establishing new settlements and creating states. Continuous human mobility through history significantly contributed to economic growth, and simultaneously changed societies, enabled interaction among various ethnical groups, and helped the exchange of various cultural values and activities.

The Silk Road was a network of roads that connected China with central Asia, northern India and the Roman Empire. Chinese silk and spices for centuries were transported to the Roman Empire. Caravans rested at a distance of about 20–25 miles, which enabled a ten-hour walk and hike during the day. The Romans built over 85,000 km of roads. The usual road width was about 6.5 m. The road network, built by hand, connected all Roman provinces. A network of modern roads in the twentieth century was rapidly constructed with the development of automotive technology. The network of roads in Western Europe and America are of a very high density. Thus, the total length of American roads is over 6 million km.

Before the advent of the railway, a large amount of goods were transported along rivers and canals. Inland water transportation played an extremely important role in the Industrial Revolution. For example, the Erie Canal, 363 miles long, which opened in 1825 and which connected the Great Lakes with the Atlantic Ocean, was very important in the development of America's economic activities and the colonization of uninhabited territory. The total length of the European waterways network is over 35,000 km. Approximately 17,000 km of this network are possible for sailing vessels of 1,000 tons. The waterways network is characterized by low transportation costs, low levels of air pollution and a high level of traffic safety.

Rail traffic began to develop rapidly in the first half of the nineteenth century, and quickly became the dominant mode of transportation for many corridors in the world. By the development of the automobile and air transportation, rail transportation was repressed in some parts of the world. Nowadays, the railway networks are well developed in China, Japan and Western Europe. The dense networks of suburban and metro lines characterize many Western European cities.

Air transportation underwent rapid development in the twentieth century. Brothers Orville and Wilbur Wright made the first controlled human flight on December 17, 1903. The first international commercial flights started in 1919.

In the modern world, the need to travel is generated by various human activities. People travel motivated by business, professional, cultural, or interpersonal reasons. Thousands of vehicles and passengers that move from one place to another form flows of cars on highways, bicycles on streets, peoples in metro stations and shopping malls, pedestrians on crossings, and aircraft on airports' taxiways and runways. Passenger and freight flows are the consequences of spatial interaction between various regions. Spatial characteristics (human activities and natural resources) create demand for transportation that is fulfilled by various transportation modes. There are complex relationships between transportation and urbanization, industrial and agricultural development, costs of goods, touristic development, etc.

In ancient as well as in modern times people developed various transportation systems (based on animal-drawn wheeled vehicles or on Boeing-747 aircraft). Vehicles guide ways, and terminals and control policy represent basic elements of every transportation system. How is it possible to plan, design and maintain good transportation systems and liveable human communities? Transportation engineering and transportation science deal with planning, design, operations, control, management, maintenance and rehabilitation of transportation systems, services and components. Within transportation science technology, science, social sciences and cultural heritage come together. Transportation science techniques have high influence on transportation system performance in terms of the level of service, capacity, safety, reliability, resource consumption, and the environment, economics and finance.

Transportation and traffic engineers are faced with many various characteristics of roads, vehicles, and control devices in solving engineering problems. Thus, for example, the path may be motorway, highway, urban arterial or rural road. At the same time, drivers participating in traffic flow could be very young, young, or old with various driving experience. Some of these drivers could be under the influence of alcohol, and the weather could be sunny, or rainy. The intersection could be signalized, or unsignalized.

Drivers in any city in the world enjoy a green wave, when a series of traffic lights are coordinated in some of the main streets. When designing synchronized bus schedules it is necessary to try to minimize the total waiting times of all passengers at transfer nodes in a transportation network. Transportation and traffic engineering are about the cycle-length calculation, and calculate how long each phase at the intersection will have the right of way. Transportation and traffic engineering are also about schedule synchronization in public transit, as well as numerous other real-life transportation problems. Transportation engineers and planners try to find answers to some of the following questions.

How do the characteristics of competitive routes influence route choice by drivers? How do travelers' characteristics influence route choice? How to design an airline network and determine flight frequencies? How to design the transit network in order to maximize the number of transported passengers, to minimize the total number of passenger transfers, and to minimize the total travel time of all served passengers? What are the best dial-a-ride systems that should serve the transportation needs of people in low-density areas, as well as the needs of the disabled and elderly persons? How to organize parcel pick-up and delivery service in urban areas? Transportation and traffic engineering are about all these and many similar questions.

In an early development of transportation engineering and science, the main research problems were related to guide way and vehicle construction rather than operations. Various control systems in transportation have been initially developed primarily to increase the level of traffic safety. In the later stages, researchers and engineers began with the development of control systems aimed at decreasing the level of traffic congestion.

Transportation science is a quantitative scientific discipline. Traffic and transportation engineering problems are very complex. The gained experience showed that the wrong decisions in the field of transportation are environmentally damaging and costly. Simultaneously, naive methodological approaches like "trial and error" methods are unreasonable and could also be costly. In other words, transportation analysis must be always rigorous, based on proven mathematical methods and concepts. Various operations research techniques have proven to be applicable in solving a variety of transportation engineering problems. These techniques have been successfully used to improve traffic safety, decrease travel times and passengers' waiting times, decrease vehicle delays, increase throughput, decrease transportation costs, increase operator's revenue, utilization of vehicles, and operator's market share. The first operations research application in transportation was published in 1954 (Edie 1954). John Glen Wardrop, an English transport analyst studied, in his seminal work (1952), route choice and network equilibrium problems and created the principles, today known as Wardrop's first and second principle. Wardrop's first principle states: "The journey times in all routes actually used are equal and less than those which would be experienced by a single vehicle on any unused route." The traffic flows in the network that satisfy this principle are called user equilibrium flows. Wardrop's second principle states: "At equilibrium the average journey time is minimum." These principles represent the fundament of the modern transportation science and are applied in urban traffic, as well as in air traffic control and telecommunications. The first mathematical model of network equilibrium was formulated by Beckmann et al., 1956. Significant theoretical contribution to the transportation science arena was also given by German mathematician Dietrich Braess who showed (1968) that adding extra capacity to a transportation network can in some cases reduce overall network performances. This phenomena known as the Braess Paradox is based on the fact that network users selfishly choose their routes through the network.

Traffic and transportation systems show a complex behavioral pattern. For example, urban or air traffic congestion are an emergent phenomena, which are often unpredictable and even occasionally counterintuitive. The more drivers choose a certain route, the lower the probability the "incoming" ones will do the same. The higher the congestion on a particular link, the less likely it is for an arriving driver to choose that link. Traffic congestion is a consequence of many decisions that different drivers make. Traffic and transportation systems are composed of decentralized individuals (pedestrians, drivers, passengers, dispatchers, operators, air traffic controllers, vehicles, vessels, aircraft) and each individual interacts with other individuals according to localized knowledge. Sometimes individuals cooperate and at other times they are in conflict. They interact, at the same time, with transportation infrastructure and environment. Through the aggregation of the individual interactions, the overall image of the transportation system emerges.

A number of complex transportation engineering problems are characterized by multiple conflicting criteria. For example, fleet size and total transportation costs conflict with the level of service provided to passengers. When matching transportation demand and supply, these conflict criteria and interests should certainly be taken into account.

A wide range of traffic and transportation engineering parameters are characterized by uncertainty, subjectivity, imprecision and ambiguity. Human operators, dispatchers, drivers and passengers use this subjective knowledge or linguistic information on a daily basis when making decisions.

Numerous complex transportation problems are characterized by time dependence. Flows in transportation systems vary over time. Every day, certain buses have a delay, some of the planned flights are cancelled, traffic incidents occur on highways, some links are completely congested, and meteorological conditions cause lower traffic flow speeds in particular locations. Traffic engineers and operators must be also able to find reasonable "good" solutions for the complex transportation problems caused by unpredictable events.

Modern transportation networks are very large. For example, there are more than 20,000 airports in the USA and more than 60,000 flights per day controlled by Air Traffic Control. The largest regional public transportation provider in the Western Hemisphere serves a region of approximately 15 million people spread over 5,000 square miles (13,000 km^2). This company moves more than 8 million passengers per day. Large transportation networks are also characterized by complex relationships between particular nodes and links, and they are frequently congested. Therefore, it is not easy to monitor and analyze them to find appropriate solutions for problems, and probably impossible to control the whole network in a centralized way.

Transportation engineering is also characterized by the existence of the gap between the theoretical achievements and implementations in the field. For example, in many countries in the world, road traffic control strategies are still essentially fixed-time signal control strategies that are unsuccessfully optimized. At the same time, modern traffic-responsive control systems that are installed in some countries are frequently immature and not properly tested. A similar situation also occurs with many freeway networks in the world. Advanced real-time ramp metering systems, driver information, and route guidance systems that are adequately implemented could highly decrease the total time spent in freeway networks.

It is difficult to predict the future of transportation. Within transportation systems we have been using many modern technological products like sensors, cameras and computers, air bags, antilock brakes, etc. What will be used 50 or 60 years from now? Sustainable transportation, gas-powered cars versus electric cars and hybrids, and fully automated vehicles are just a few issues that should be carefully studied and analyzed in years to come. One can expect that autonomous vehicle technology will create a new generation of transportation systems in the future.

The planning, design, and control of traffic and transportation systems are complex subjects. A wide range of traffic and transportation engineering parameters are uncertain, nonlinear and dynamic. Complex traffic and transportation problems call for the development of modern systems that merge knowledge, techniques, and methodologies from various scientific areas. There is no doubt that transportation engineering and transportation science will be a continuously challenging area of research for scientists and engineers.

References

Beckmann, McGuire, and Winsten, (1956) *Studies in the Economics of Transportation*, New Haven, NJ: Yale University Press.

Braess, D. (1968) Uber ein paradoxen der verkehrsplanung. *Unternehmensforschu*. 12, pp. 258–268.

Edie, L. (1954) Traffic delays at toll booths. *Journal of the Operations Research Society of America*, 2 (2): 107, May.

Wardrop, J.G. (1952) Some theoretical aspects of road traffic research. *Proceedings of the Institute of Civil Engineers, Part II*, pp. 325–378.

2

TRAFFIC FLOW THEORY

Hesham Rakha

VIRGINIA POLYTECHNIC INSTITUTE AND STATE UNIVERSITY

Overview

This chapter describes the basics of traffic flow theory and its application to the modeling of traffic flow. The chapter begins with a description of the modeling of a single vehicle considering the longitudinal forces acting on the vehicle and assuming a point mass object. Subsequently, the modeling of the interaction of a vehicle with the vehicle preceding it is described. Next, the modeling of a traffic stream of vehicles is presented. Subsequently, a description of the lateral movement of vehicles (i.e. lane-changing) is described. Finally, some concluding remarks are presented. It should be noted that given the space constraints, the various topics are covered at a basic level.

Single Vehicle Motion Modeling

In this section we describe the modeling of the longitudinal motion of a single vehicle considering only the longitudinal forces acting on the vehicle assuming the vehicle to be a point mass object.

Vehicle Dynamics Acceleration Modeling

The modeling of vehicle accelerations can be modeled using a) a vehicle kinematics model; b) a vehicle dynamics model; or c) a gear-shifting model. Here we briefly describe the modeling of vehicle motion using a point mass vehicle dynamics model (Rakha et al., 2001; Rakha and Lucic, 2002; Rakha et al., 2004).

Vehicle dynamics models compute the maximum vehicle acceleration levels from the resultant forces acting on the vehicle (mainly vehicle tractive force that is a function of the engine throttle input and the various resistance forces). The computation of the tractive force is computed as the minimum of the engine tractive force and the maximum force that can be sustained between the vehicle wheels and the roadway surface, as demonstrated in Equation 1. The resistance forces include the aerodynamic, rolling and grade resistance forces, as demonstrated in Equation 2.

$$F(t) = \min\left(3600 f_p \beta\eta_d \frac{P(t)}{v(t)}, m_{ta} g\mu\right) \tag{1}$$

$$R(t) = \frac{\rho}{25.91} C_d C_h A_f v^2(t) + mg\frac{C_{r0}}{1000}\left(c_{r1}v(t) + c_{r2}\right) + mgG(t) \tag{2}$$

Here, $F(t)$(N) is the vehicle tractive effort at any instant t; v is the vehicle velocity (km/h). Rakha and Lucic (2001) introduced the β factor in order to account for the gearshift impacts at low traveling speeds when trucks are accelerating. This factor is set to 1.0 for light duty vehicles (Rakha et al., 2004). The f_p factor models the driver throttle input level and ranges from 0.0 to 1.0. Other parameter definitions are: η_d is the driveline efficiency (unitless); $P(t)$ is the vehicle power (kW) at instant t; m_{ta} is the mass of the vehicle on the tractive axle (kg); g is the gravitational acceleration (9.8067 m/s^2) and μ is the coefficient of road adhesion or the coefficient of friction (unitless); ρ is the air density at sea level and a temperature of 15 °C (1.2256 kg/m^3); C_d is the vehicle drag coefficient (unitless), typically 0.30, as demonstrated in Table 2.1; C_h is the altitude correction factor equal to 1–0.000085h where h is the altitude in meters (unitless); A_f is the vehicle frontal area (m^2), typically 0.85 multiplied by the height and width of the vehicle; C_{r0} is a rolling resistance constant that varies as a function of the pavement type and condition, as demonstrated Table 2.2 (unitless); C_{r1} is the second rolling resistance constant (h/km) (Table 2.3); C_{r2} is the third rolling resistance constant (unitless) (Table 2.3); m is the total vehicle mass (kg); and $G(t)$ is the roadway grade at instant t (unitless).

Table 2.1 Typical vehicle drag coefficients

Vehicle Type	*Drag Coefficient*
Automobile	0.25–0.55
Bus	0.50–0.70
Tractor-trailer	0.60–1.30
Motorcycle	0.27–1.80

Table 2.2 Typical rolling resistance coefficient as a function of pavement type and condition

Pavement Type	*Pavement Condition*	C_{r0}
Concrete	Excellent	1.00
	Good	1.50
	Poor	2.00
Asphalt	Good	1.25
	Fair	1.75
	Poor	2.25

Table 2.3 Typical rolling resistance coefficients as a function of vehicle tires

Tire Type	C_{r1}	C_{r2}
Bias ply	0.0438	6.100
Radial	0.0328	4.575

The vehicle acceleration is calculated as the ratio of the residual force (difference between tractive and resistance forces) and the vehicle mass (i.e., $a = (F - R)/m$). The vehicle speed at $t + \Delta t$ is then computed using Euler's first-order approximation as demonstrated in Equation 3 and the position is computed using a Euler's second-order approximation, as demonstrated in Equation 4.

$$v(t+\Delta t) = v(t) + 3.6\frac{F(t)-R(t)}{m}\Delta t \ . \tag{3}$$

$$x(t+\Delta t) = x(t) + v(t)\Delta t + \frac{1}{2}\frac{F(t)-R(t)}{m}\Delta t^2 \ . \tag{4}$$

Vehicle Dynamics Deceleration Modeling

Vehicle deceleration can be modeled using the vehicle dynamics approach considering the distribution of the forces on the vehicle as

$$F_{bf\,\max} = \frac{\mu mg}{L}\left(l_r + h(\mu + f_{rl})\right), \text{ and} \tag{5}$$

$$F_{br\,\max} = \frac{\mu mg}{L}\left(l_f - h(\mu + f_{rl})\right) \ . \tag{6}$$

Here $F_{bf\,\max}$ is the maximum possible brake force on the front axle (N); $F_{br\,\max}$ is the maximum possible brake force on the front axle (N);

$$F_{rl} = \frac{C_{r0}}{1000}\left(C_{r1}v(t) + C_{r2}\right) \ ;$$

l_r is the distance between the vehicle center of gravity and the rear axle (m); l_f is the distance between the vehicle center of gravity and the front axle (m); L is the spacing between the front and rear axles (m); and h is the height of the vehicle center of gravity from the roadway surface (m).

Vehicle Powertrain Modeling

A more advanced vehicle model entails modeling the vehicle powertrain Rakha et al., 2012). This model starts with the driver throttle and brake pedal input, as illustrated in Figure 2.1 (step 1). Based on the driver's throttle input (f_t), the engine speed (ω_e) is computed using a simple regression model (step 2 or Equation 1) that was empirically developed but is being refined as described later in this chapter. In the case of a manual transmission system the gear selection (step 3 or gear selection modeling) is made directly using the engine speed. Alternatively, in the case of an automatic transmission system the torque converter (step 2a or torque converter modeling) is modeled, as illustrated in Figure 2.1. In particular, the torque converter calculates the torque converter output speed and torque and selects the appropriate gear. Subsequently, the engine speed is used to estimate vehicle torque and power (step 4). The engine power and torque is computed considering an upper bound parabolic function that was proposed by Ni and Henclewood (Ni and Henclewood 2008) (*Pmax(ω)*). The actual power available is

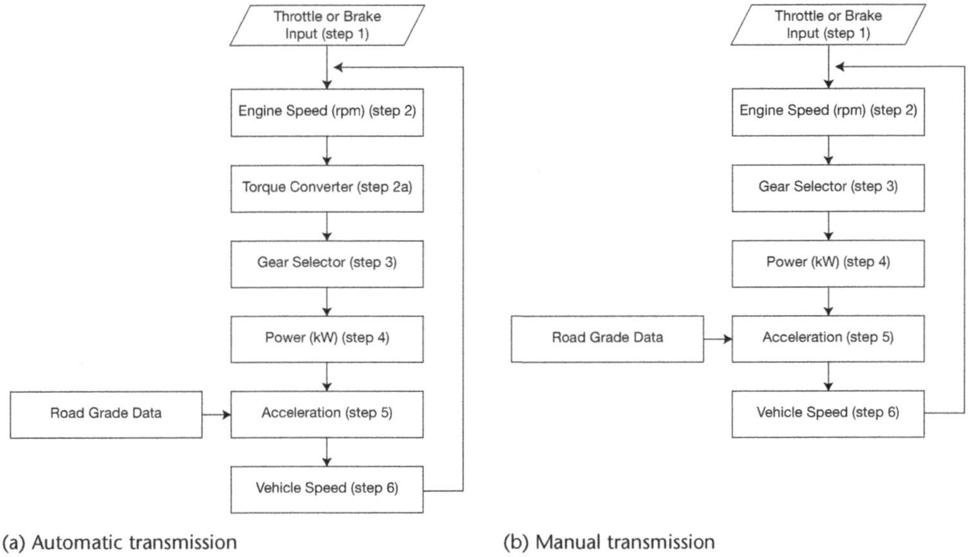

Figure 2.1 Proposed model structure

estimated as the proportion of the maximum power considering a linear relationship between throttle position and the proportion of maximum power available (i.e. $P(\omega) = f_p \times P_{\max}(\omega)$). The vehicle acceleration (step 5) is then computed considering a point mass vehicle dynamics model. The vehicle speed and position (step 6) are estimated by solving a second-order differential equation. The specifics of each of the components of the model are described in the following sections.

The model assumes that the powertrain is stiff and the transmission is modeled considering a gear ratio and efficiency. The model makes the following simplifying assumptions: (a) the clutch is assumed to be stiff (i.e. torque and angular speed remain the same before and after the clutch); (b) transmission inertia is disregarded (i.e. the model only considers a conversion ratio and efficiency of the transmission system); (c) the propeller shaft is assumed to be stiff; (d) the final powertrain is modeled considering an efficiency; and (e) the drive shafts are assumed to be stiff. Consequently, the overall efficiency is computed as power losses in the engine due to internal friction and other factors while the overall gear ratio is computed as the product of the transmission and final gear ratios.

Engine Modeling

Ni and Henclewood (2008) presented a model to empirically approximate the relationship between engine power, P_e (KW) and engine speed, ω_e (rpm) assuming a parabolic relationship between the engine torque, T_e(N-m) and engine speed. Using this assumption the relationship between the maximum engine power and speed can be written as

$$P_e(\omega_e) = \frac{P_{\max}}{2\omega_p^2}\left(3\omega_p - \omega_t\right)\omega_e - \frac{P_{\max}}{2\omega_p^2(\omega_p - \omega_t)}\left(\omega_e - \omega_t\right)^2 \omega_e \,, \tag{7}$$

where ω_t is the engine speed at peak torque (rpm). The engine torque, T_e(N-m) can be computed using Equation 8 (Guzzella and Sciarretta 2007).

$$T_e = 60{,}000 \frac{P_e}{2\pi w_e} \ . \tag{8}$$

Gear Selection Modeling

Once the powertrain speed is computed the next step is to evaluate the need to shift gears through the modeling of a transmission system. Wong (2001) indicates that the principal requirements for the transmission are: (a) to achieve the desired maximum speed with an appropriate engine; (b) to be able to start, fully loaded, in both forward and reverse directions on a steep gradient, typically 33 percent; and (c) to properly match the characteristics of the engine to achieve the desired operating fuel economy and acceleration characteristics.

A manual gear transmission usually consists of a clutch, a gearbox, a propeller shaft, and a drive axle with a differential (to allow relative rotation of the driven tires during turning maneuvers). The gearbox provides a number of gear reduction ratios ranging from 3 to 5 for passenger cars and 5 to 16 or more for commercial vehicles. The number of gear ratios is selected to provide the vehicle with the propulsive effort-speed characteristics as close to the ideal as possible in a cost-effective manner. The gear ratio for the highest gear is computed as (Wong, 2001):

$$\xi_n = \frac{n_{e1}(1-i)}{v_{\max}\xi_{ax}} \tag{9}$$

where ξ_n is the gear ratio of the highest gear in the gearbox for an n-speed vehicle; n_{e1} is the engine speed corresponding the maximum speed (about 10 percent higher than the speed at maximum power); r is the rolling radius of the tire; i is the tire slip (2–5 percent); $v_{\max}$ is the maximum desired speed; and ξ_{ax} is the gear ratio in the drive axle.

The lowest gear ratio is computed so that a vehicle can accelerate on a grade at a desired acceleration rate. The formula for computing this gear ratio varies depending on whether the vehicle is forward- versus rear-wheel drive. Once the highest and lowest gear ratios are determined, the remaining gear ratios are selected to establish the following relationship (Wong, 2001)

$$\frac{\xi_2}{\xi_1} = \frac{\xi_3}{\xi_2} = \ldots \frac{\xi_n}{\xi_{n-1}} = K_g \quad \text{where } K_g = \sqrt[n-1]{\xi_n/\xi_1} \ . \tag{10}$$

A model developed by Rakha et al. (2012) uses a simple algorithm that is solely based on the engine speed in shifting gears. The model assumes that drivers/transmission system makes gear shifts when the vehicle reaches the engine speed at peak torque, ω_t. Downshifts are made when the vehicle reaches an engine speed of 1,500 rpm based on typical values provided in the literature (1,000–2,000 rpm).

Longitudinal Vehicle Motion Modeling Considering Vehicle Interactions: A Microscopic Approach

Microscopic simulation software uses car-following models to capture the interaction of a vehicle and the preceding vehicle traveling in the same lane. The process of car following is modeled as an equation of motion for steady-state conditions plus a number of constraints that govern the behavior of vehicles while moving from one steady state to another (decelerating and accelerating). The first constraint governs the vehicle acceleration behavior, which is typically a function of the vehicle dynamics. The second and final constraint ensures that vehicles maintain a safe position relative to the lead vehicle in order to ensure asymptotic stability within the traffic stream.

Traffic stream motion can be formulated either as discrete entities or as a continuous compressible fluid. Microscopic car-following models characterize the relationship between a vehicle's desired speed and the spacing (s) between the lead and follower vehicles. Alternatively, another common representation of car-following behavior is to characterize the vehicle's desired acceleration and the speed differential between the lead and following vehicles. Specifically, Drew (1968) suggests

> the car-following laws are simplified descriptions of a very complicated response to the world of stimuli that confronts a driver. In fact, only two stimuli are considered: the relative speed between a vehicle and the one ahead and the spacing between the two vehicles. Obviously a driver considers more. Considerable realism can be achieved by including several vehicles ahead of and perhaps the vehicle immediately behind the driver. Most drivers are continually evaluating several gaps ahead and are extremely conscious of the proximity of the car behind.

Prior to describing the various car-following models the variables common across the models are first presented. These include the following: n is the index of the lead vehicle; $n+1$ is the index of the following vehicle; x, $\dot{x}$, $\ddot{x}$ are the vehicle position, velocity, and acceleration, respectively; $\ddot{x}_{max}$ is maximum acceleration level of the following vehicle; $\ddot{x}_{min}$ is the maximum deceleration level of the following vehicle; $\dot{x}_d$ is the desired velocity; Δ is the change in value for a given variable at a given time step between the leading and following vehicle; Δx_j is the vehicle spacing at a speed of zero (i.e. jam density spacing); t is the current time step; and τ is the perception reaction time.

Gaxis-Herman-Rothery (GHR) Model

The GHR type of model, also known as the GM model, involves research spanning from the late 1950s until the mid-1960s (Gazis et al. 1961). A discussion on the development of the General Motors models along with typical parameter values can be found in May (1990) among others. The fifth and final formulation of the model, sometimes referred to as the GM-5 model, generalizes the previous four forms and is seen in Equation 11.

$$\ddot{x}_{n+1}(t) = \left(\frac{\alpha[\dot{x}_{n+1}(t)]^z}{[\Delta x_{n \to n+1}(t-\tau)^l]} \right) \cdot [\Delta \dot{x}_{n \to n+1}(t-\tau)] \tag{11}$$

Here, α, z, l are model parameters that require calibration. The GHR model is particularly significant in the pantheon of car-following models as many other microscopic and macroscopic models can be represented as special cases of the GHR model. The first single-regime traffic stream model proposed by Greenshields in 1935 is one such case (Greenshields 1934). It can be expressed as a GHR model taking the z and l parameters to be 0 and 2, respectively. The CORSIM transportation simulation software operates based on the PITT model, which has been shown to be analogous to the Pipes car-following model (Rakha and Crowther 2002). In turn, the formulation of the Pipes model can be seen to be a special case of the GHR model with the z and l parameters set to zero.

Gipps Model

First published in 1981, Gipps developed a model as a response to the existing proliferation of GHR type models (Gipps 1981). The model seeks to mimic the behavior of real traffic, use parameters that correspond to characteristics of drivers and vehicles so that most can be assigned without calibration procedures, and provide reasonable results if the interval of speed and position calculations is similar to that of the reaction time. The model takes the form of Equation 12.

$$\dot{x}_{n+1}(t) = \min \begin{cases} \dot{x}_{n+1}(t-\tau) + 2.5\ddot{x}_{\max} \cdot \Delta t \cdot \left[1 - \frac{\dot{x}_{n+1}(t-\tau)}{\dot{x}_d}\right] \cdot \sqrt{0.025 + \frac{\dot{x}_{n+1}(t-\tau)}{\dot{x}_d}} \\ \ddot{x}_{d-\min} \cdot \Delta t + \sqrt{\left(\ddot{x}_{\min}^2 \cdot (\Delta t)^2 - \ddot{x}_{\min} \cdot A\right)} \end{cases} \quad (12)$$

where

$$A = 2[\Delta x_{n \to n+1}(t-\tau) - \Delta x_j] - [\dot{x}_{n+1}(t-\tau) \cdot \Delta t] - \frac{\dot{x}_n(t-\tau)^2}{\ddot{x}_{n-\min}}$$

The first component of the two-regime equation applies to traffic conditions with unconstrained flow, also known as steady-state flow conditions. The second component addresses conditions in constrained flow, where the velocity of the follower vehicle is based primarily on the time/space headway that must necessarily maintain a safe distance between it and the lead vehicle. Since the original publication of the Gipps model, a number of publications examining the calibration of the model have brought up critique in two areas: the feasible range for the expected maximum deceleration of the leading vehicle, and the relationship between the perception-reaction time and the time-step (Equations 13–15).

Wilson primarily examined the wave stability of the Gipps model, and specifically looked at cases where $\ddot{x}_{\min} > \ddot{x}_{\min,n}$ or $\ddot{x}_{\min} < \ddot{x}_{\min,n}$ (Equation 13). Wilson determined that where in some cases the anticipated deceleration of the leading vehicle is expected to be less than the desired maximum deceleration of the following vehicle, the resulting time-space wave phenomenon becomes unstable. If the original formulation of the Gipps model is used, the implication would be that any follower vehicle with a desired deceleration in excess of -3.0 m/s^2 would fall into the category of potentially being infeasible. For the purposes of the model calibration performed herein, Gipps's original formulation for the anticipated leading vehicle maximum deceleration is used.

Rakha et al. (2007) and Rakha and Wang (2009), building on the work of Wilson, developed a method to express the steady-state portion of the Gipps model on the macroscopic fundamental diagram. The speed-flow relationship can be computed using Equation 13 and subsequently, the perception-reaction time can be computed using Equation 14.

$$q = \frac{1000 \cdot \dot{x}_{n+1}}{\Delta x_j + \frac{1}{2.4} \cdot \tau \dot{x}_{n+1} + \frac{1}{25.92 \ddot{x}_{\min}} \cdot \left(1 - \frac{\ddot{x}_{\min}}{\ddot{x}_{\min,n}}\right) \cdot \dot{x}_{n+1}^2} \tag{13}$$

$$\tau_m = 2.4 \left(\frac{1000}{q_c} - \frac{\Delta x_j}{\dot{x}_c} - \frac{\dot{x}_c}{25.92 \ddot{x}_{\min}} \left[1 - \frac{\ddot{x}_{\min}}{\ddot{x}_{\min,n}} \right] \right) \tag{14}$$

Here, q is the traffic stream flow rate (veh/h/lane); q_c is the flow rate at capacity (veh/h/lane); $\dot{x}_c$ is the speed-at-capacity (m/s); and τ_m is the perception reaction time calculated from macroscopic measures (s).

Intelligent Driver Model

The Intelligent Driver Model (IDM) was developed as a high-fidelity model trying to reproduce traffic as realistically as possible, with a small number of parameters and realistic breaking reactions (Trieber and Kesting 2013). The authors of the IDM, as shown in Equation 15, cite the Gipps model as achieving these goals, although they attempt to overcome its limitations by introducing instabilities or hysteresis effects for vanishing fluctuations.

$$\ddot{x}_{n+1}(t) = \ddot{x}_{\max} \left\{ 1 - \left[\frac{\dot{x}_{n+1}(t)}{\dot{x}_d} \right]^{\delta} - \left[\frac{(\Delta x_j - l_n) + (\dot{x}_{n+1}(t) \cdot \tau_S) + \frac{\dot{x}_{n+1}(t)[\Delta x_{n \to n+1}(t)]}{2 \cdot \sqrt{\ddot{x}_{\max} \cdot |\ddot{x}_{\min}|}}}{\Delta x_{n \to n+1}(t) - l_n} \right] \right\} \tag{15}$$

Here, δ is the acceleration exponent; τ_s is the safe time headway; and l_n is the approximate length of vehicle n, taken to be 4.5 meters.

The IDM examines the maximum possible acceleration, and combines both acceleration and deceleration strategies into the formulation. The acceleration strategy takes the instantaneous ratio of the current velocity and desired velocity, and raises this value by the acceleration exponent, taken to be a default value of 4 in the original documentation. The deceleration strategy uses the desired spacing as a function of velocity of the following vehicle and the difference in velocity between the leading and following vehicles, and squares the result of this desired spacing divided by the current spacing. Effectively, this formulation balances the pull of the desired velocity against the push of the desired spacing.

Rakha-Pasumarthy-Adjerid Model

The Rakha–Pasumarthy–Adjerid (RPA) model is a simplified behavioral vehicle longitudinal motion model, currently incorporated in the INTEGRATION traffic simulation software. The RPA model is composed of a steady-state first-order model (also known as the fundamental diagram), collision avoidance constraints, and vehicle acceleration constraints. The vehicle acceleration and collision avoidance constraints revert the model from a first- to a second-order traffic stream model (Sangster and Rakha 2014), as will be discussed later and illustrated in Figure 2.2. The figure clearly demonstrates that the model output (light gray points) does not result in solutions solely on the fundamental diagram, as is the case with a first-order model,

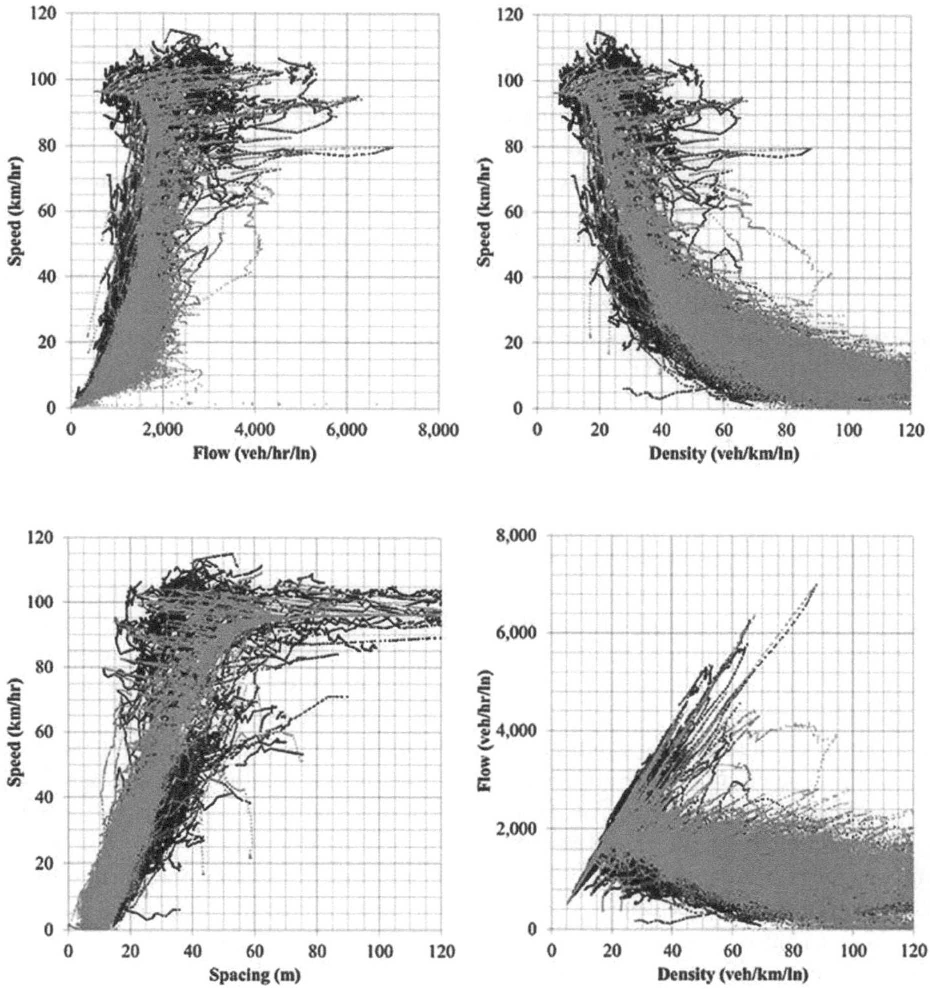

Figure 2.2 Modified RPA model predictions vs. empirical naturalistic data for microscopic speed, density, flow and spacing observations

Source: Sangster and Rakha 2014

and is consistent with empirical observations (black points). It should be noted that the data presented in Figure 2.2 were obtained from a naturalistic study where data were gathered using an on-board data acquisition system (DAS) over a period of one year. A more detailed description of the analysis and the modifications that were made to the RPA model are presented in the literature (Sangster and Rakha 2014).

The underlying steady-state fundamental diagram currently used in the RPA model is the Van Aerde formulation (Van Aerde 1995; Van Aerde and Rakha 1995; Rakha and Arafeh 2010) which combines the Pipes triangular and Greenshields parabolic fundamental diagrams into a single model, as will be described later. This model requires four parameters for calibration but offers the flexibility of having the speed-at-capacity independent of the free-flow speed. It is sufficient to state here that the model is a single-regime functional form by Van Aerde and Rakha (1995). The model computes the vehicle speed using Equation 16.

$$\ddot{x}_{n+1}(t) = \frac{= c_1 + c_3 \cdot \dot{x}_d + \Delta x_{n \to n+1}(t) - \sqrt{[c_1 + c_3 \cdot \dot{x}_d + \Delta x_{n \to n+1}(t)]^2 - 4c_3[\Delta x_{n \to n+1}(t) \cdot \dot{x}_d - c_1 \dot{x}_d - c_2]}}{2c_3} \quad (16)$$

Here, c_1, c_2, c_3 are model coefficients that require calibration and $\dot{x}_d$ is the driver desired speed (also known as the free-flow speed) for vehicle $n + 1$.

The model incorporates a variable power vehicle dynamics model that computes the vehicle's tractive effort, aerodynamic, rolling and grade-resistance forces, as described in detail and validated in the literature Rakha et al. 2001; Rakha and Lucic 2002).

The RPA model imposes limitations on the steady-state Van Aerde model to ensure that collisions do not occur under nonsteady-state conditions (e.g. a vehicle is approaching a slower vehicle ahead of it) and to account for vehicle dynamics constraints, as demonstrated by Park et al. (2011). The collision avoidance component is shown in Equation 17.

$$\dot{x}^2_{n+1}(t) = \sqrt{[\dot{x}_n(t)]^2 + 2\ddot{x}_{\min}[\Delta x_{n \to n+1}(t) - \Delta x_j]} \quad (17)$$

Here, $\ddot{x}_{\min}$ is the minimum acceleration (m/s^2) or maximum deceleration level. This can be as high as the coefficient of friction μ multiplied by the gravitational acceleration g (9.8067 m/s^2). This assumes a braking efficiency of 1.0.

The model then computes the speed as the minimum of the speeds computed using Equations 3 and 16 in addition to the minimum of Equations 3 and 17 when the speed of the vehicle ahead of it is lower than the subject's vehicle speed. These conditions ensure that the speed is the lowest speed subject to the acceleration, steady-state, and collision-avoidance constraints.

The separation of and distinction between the driver and vehicle requires the calibration of two sets of parameters. On the driver side the parameters include the driver's desired speed (aka free-flow speed), the headway at maximum discharge rate (saturation flow rate), the speed at the maximum discharge rate (speed-at-capacity), the spacing from the driver's vehicle front bumper to the rear bumper of the lead vehicle when completely stopped (jam density when you add the length of the vehicle), the driver throttle input, and the driver brake pedal input. The vehicle parameters are typically available from auto manufacturer websites. Figure 2.3 illustrates sample empirical driver throttle input together with model estimates of the driver throttle input and the resulting vehicle acceleration behavior.

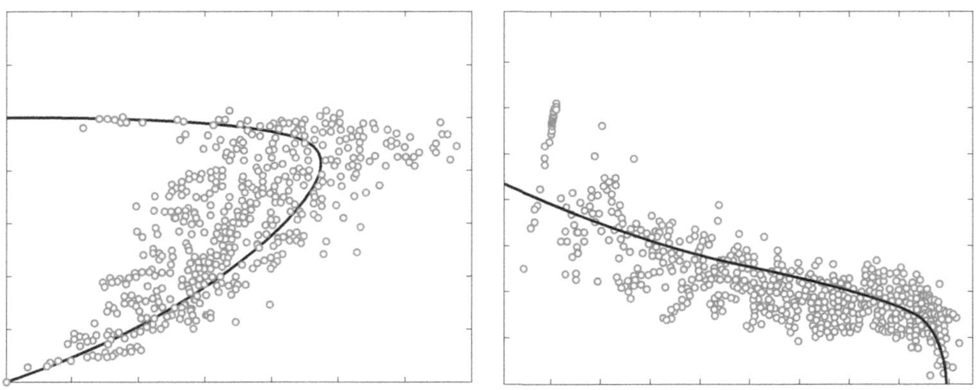

(a) Speed/desired speed versus driver throttle input (b) Acceleration versus speed

Figure 2.3 Sample empirical and model predicted vehicle accelerations

Longitudinal Traffic Stream Modeling: A Macroscopic Approach

Traffic stream models describe the motion of a traffic stream by approximating for the flow of a continuous compressible fluid. The traffic stream models relate three traffic stream variables, namely: the traffic stream flow rate, traffic stream density, and traffic stream space-mean-speed. Consequently, the macroscopic and microscopic traffic models are related to each other by the relationship between these microscopic and macroscopic parameters that feed into each of the respective models. Specifically, considering a relatively long analysis period (T_0), the duration of the analysis period can be approximated for the summation of the time headways (h) of successive vehicles within the analysis period, as demonstrated in Equation 18. Consequently, the flow rate can be expressed as the inverse of the average vehicle time headway. Similarly, the traffic stream density can be approximated for the inverse of the average vehicle spacing (s) for all vehicles within a section of roadway of length L, as demonstrated in Equation 19.

$$\begin{aligned} T_0 &\approx \sum_{i=1}^{r} h_i \\ q^{-1} &= \frac{T_0}{r} \approx \sum_{i=1}^{r} h_i \end{aligned} \tag{18}$$

$$\begin{aligned} L &\approx \sum_{i=1}^{p} s_i \\ k^{-1} &= \frac{L}{p} \approx \frac{1}{p}\sum_{i=1}^{p} s_i \end{aligned} \tag{19}$$

In modeling traffic flow behavior, it is typical to assume that all vehicles move at the same distances with respect to one another (same spacing) and with the same time-independent speed. Solutions to these models have different names depending on the literature source, but are commonly known as "homogenous," "equilibrium," or "stationary" states or, in some instances, "steady-state" model solutions. In this proposal we refer to these solutions as steady-state solutions.

With the exception to research conducted by Kerner (2004), the fundamental hypothesis of all traffic flow theories is the existence of a site-specific unique relationship between traffic stream flow and traffic stream density, commonly known as the fundamental diagram, or the traffic stream motion model. The assumption is that all steady-state model solutions lie on the fundamental diagram and thus are referred to as fundamental diagram approaches (Kerner 2004). Given that traffic stream space-mean speed can be related to the traffic stream flow and density, a unique speed-flow-density relationship is derived from the fundamental diagram for each roadway segment.

The basic hydrodynamic equation is given by

$$q = kv\ . \tag{20}$$

The dynamic flow continuity equation ensures that vehicle conservation is maintained and is written as

$$\frac{\partial k(x,t)}{\partial t} + \frac{\partial q(x,t)}{\partial x} = 0\ , \tag{21}$$

where $q(x,t)$ is the traffic stream flow rate over a time interval Δt at any instant t and location x along the roadway (veh/h) and $k(x,t)$ is the traffic stream density over a section of roadway Δx at instant t and location x. Equation 21 has two independent variables, k and q. Lighthill and Whitham (1955) and Richards (1956) independently hypothesized a relationship between traffic stream flow and density—the fundamental diagram. Consequently, Equation 21 was modified to derive

$$\frac{\partial k(x,t)}{\partial t} + c\frac{\partial k(x,t)}{\partial x} = 0 , \tag{22}$$

where c is the derivative of q with respect to k and represents the kinematic wave (KW) velocity. This relationship is a function of a single variable (density), and thus can be solved by discretizing the solution space over space and time considering initial and boundary conditions. It should be noted that the flow-density curve must be concave so that Equation 22 only renders solutions with deceleration shock waves as suggested by Whitham (1974). A concave flow-density relationship results in a concave car-following model (speed-spacing relationship).

Traffic Stream Models

A number of traffic stream models have been proposed in the literature. Here we present three models: the Greenshields model, the Pipes model and the Van Aerde model.

Greenshields Model

The first and most famous single-regime traffic stream model is the Greenshields model (1934) that was developed based on observations of speed-density measurements obtained from a photographic study (May 1990). Using these data, Greenshields concluded that the relationship between space-mean-speed and traffic density is linear. Using the linear speed-density relationship together with the basic traffic stream model, the flow-density relationship is represented as a parabolic relationship. The speed-at-capacity computed is found to be equal to half the free-speed and the density-at-capacity is equal to half the jam density. It should be noted that the calibration of the Greenshields model only requires the estimation of any two parameters of three parameters, namely, the free-flow speed (v_f), capacity (q_c), or jam density (k_j).

$$q = v_f\left(k - \frac{k^2}{k_j}\right) \tag{23}$$

Rakha demonstrated that the Greenshields model is a GHR (also known as GM-5) model (Equation 11) with the driver sensitivity speed exponent equal to 0 ($z = 0$) and the driver sensitivity spacing exponent equal to 2 ($l = 2$).

Pipes Model

Pipes (1953, 1967) proposed a linear car-following model (GHR model) constrained by the maximum facility free-flow speed (v_f). Rakha et al. (2004) demonstrated that the Pipes model is identical to a GHR model with the z and l parameters equal to zero in Equation 11. The Pipes model produces a triangular flow-density relationship (fundamental diagram). Specifically, using the uncongested relationship, Rakha and Crowther (2002) demonstrated that the speed-at-capacity for a Pipes model is equal to the free-flow speed and that the driver sensitivity factor can be computed as

$$c_3 = \frac{1}{q_c} - \frac{1}{k_j v_f} \tag{24}$$

This model requires the calibration of three parameters, namely: v_f, q_c, and k_j. The flow-density dual-regime relationship was derived by Rakha and Crowther (2002) as

$$q = \begin{cases} -kv_f - & k \leq \dfrac{q_c}{v_f} \\ \dfrac{q_c v_f}{k_j v_f - q_c} & k > \dfrac{q_c}{v_f} \end{cases} . \tag{25}$$

The congested regime (second regime) is a negative sloped linear function, where the slope of the relationship represents the wave velocity (w) and can be computed as

$$w = -\frac{q_c v_f}{k_j v_f - q_c} . \tag{26}$$

Equation 26 demonstrates that the wave velocity for a triangular flow-density relationship is dependent on the roadway capacity, its free-flow speed, and jam density.

Van Aerde Model

The functional form that is utilized in this study is the Van Aerde nonlinear functional form that was proposed by Van Aerde (1995) and Van Aerde and Rakha (1995), which was presented earlier in Equation 16.

As was demonstrated by Rakha and Crowther (2002) this functional form amalgamates the Greenshields and Pipes car-following models. This combination provides a linear increase in vehicle speed as the distance headway increases with a smooth transition from the congested to the uncongested regime. This combination provides a functional form with four degrees of freedom by allowing the speed-at-capacity (v_c) to differ from the free-flow speed (v_f), which is a common assumption in a number of models including the Pipes model, or half the free-flow speed, as is the case with Greenshields's model. Specifically, the first two parameters of Equation 16 provide the linear increase in the vehicle speed as a function of the distance headway, while the third parameter introduces curvature to the model and imposes a constraint on the vehicle's speed to ensure that it does not exceed the facility free-flow speed through the use of a continuous function. Demarchi (2002) demonstrates that by considering three boundary conditions the model constants can be computed as

$$c_1 = \frac{v_f}{k_j v_c^2}\left(2v_c - v_f\right) \; ; \; c_2 = \frac{v_f}{k_j v_c^2}\left(v_f - v_c\right)^2 \; ; \; c_3 = \frac{1}{q_c} - \frac{v_f}{k_j v_c^2} \tag{27}$$

Ignoring differences in vehicle speeds and spacings within a traffic stream and considering the direct inverse relationship between traffic stream density and traffic spacing, the speed-density relationship can be derived as

$$k = \frac{1}{c_1 + \frac{c_2}{v_f - v} + c_3 v}. \tag{28}$$

A more detailed description of the mathematical properties of this functional form can be found in the literature (Rakha 2009), as can a discussion of the rationale for its structure.

Of interest is the fact that Equation 28 reverts to Greenshields's model, when the speed-at-capacity and density-at-capacity are both set equal to half the free-flow speed and jam density, respectively (i.e. $v_c = v_f/2$ and $k_c = k_j/2$). Alternatively, setting $v_c = v_f$ results in the Pipes model (Rakha 2009). Furthermore, Rakha (2006) demonstrated that the wave speed at jam density (denoted as w_j) can be computed as

$$w_j = -\left[\left(\frac{k_j}{q_c} - \frac{v_f}{v_c^2}\right) + \frac{(v_f - v_c)^2}{v_f v_c^2}\right]^{-1}. \tag{29}$$

As was demonstrated earlier, the Van Aerde model reverts to the Pipes linear model when the speed-at-capacity is set equal to the free-flow speed. Consequently, it can be demonstrated that under this condition the wave speed of Equation 29 reverts to Equation 26. Furthermore, when $v_c = v_f/2$ and $k_c = k_j/2$ the wave speed at jam density is consistent with the Greenshields model estimates and is equal to $-v_f$ (Rakha 2009). A sample calibration of the Van Aerde model to empirical data is presented in Figure 2.4. The figure clearly demonstrates that the functional form is capable of fitting to empirically observed data.

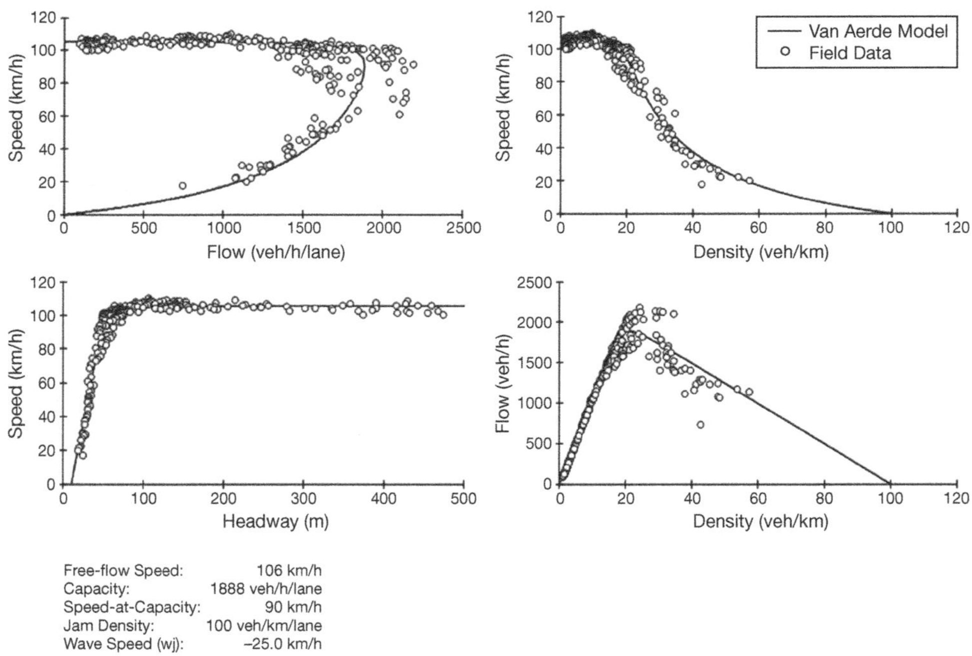

Figure 2.4 Van Aerde model fit to freeway data (Highway 401, Toronto, Canada)

Shockwaves

Stationary Bottlenecks

Lighthill and Whitham (1955) as well as Richards (1956) made the first successful attempts at such a description. They both demonstrated the existence of traffic shock waves and proposed a first theory of one-dimensional waves that could be applied to the prediction of highway traffic flow behavior. Equations 20 and 30 represent their model. The first equation defines the relation between volume, density, and speed that was developed from the application of fluid dynamics theory. Using Equation 20, Equation 30 was then developed to describe the speed at which a change in traffic characteristics, or shock wave, propagates along a roadway.

$$w = \frac{\Delta q}{\Delta k} \tag{30}$$

It should be noted that Equation 30 is a direct consequence of the mass conservation equation (Equation 21) that was presented earlier, because it ensures that the flow entering the wave equals the flow exiting the wave.

Moving Bottlenecks

A moving bottleneck is a phenomenon caused by a slow-moving vehicle within the traffic stream. An example illustration of a moving bottleneck was simulated in the INTEGRATION microscopic traffic simulation software (Van Aerde 1985; Van Aerde and Yagar 1988; Van Aerde and Rakha 2013), as illustrated in Figure 2.4. The simulation considers a single lane with a moving bottleneck (slow-moving truck). Given that vehicles are unable to overtake the truck, two waves are observed in the traffic stream, namely, a forward-forming wave at the moving bottleneck and a forward-recovery wave at the tail of the congestion upstream of the moving bottleneck. In this scenario all vehicles are modeled with no driver differences captured.

The interest in the study of moving bottlenecks started in the 1990s. Gazis and Herman (1992) were the first to attack the problem and to address some of its issues in 1992. Assuming that the moving obstruction impacts the vehicles that are behind it and aside it in the same way, they were able to study the formation and the evolution of the queue upstream of the slow vehicle. In order to overcome some of the limitations of the Gazis and Herman formulation, a more

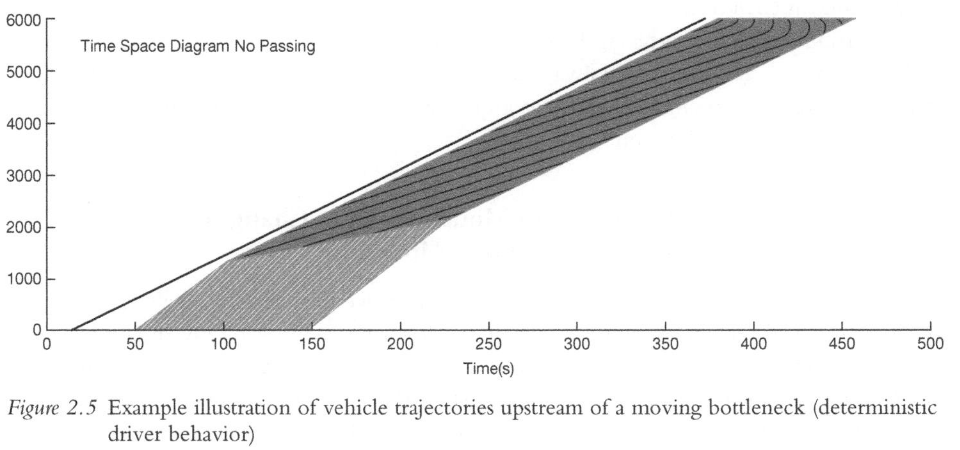

Figure 2.5 Example illustration of vehicle trajectories upstream of a moving bottleneck (deterministic driver behavior)

complete model based on the Lighthill-Whitham-Richards (LWR) theory was developed separately by Newell (1993, 1998) and by Lebacque et al. in 1998. A third model was proposed by Muñoz and Daganzo in 2002. Muñoz and Daganzo demonstrated that their model was more accurate in modeling moving bottlenecks at high speeds. Considering these three models, Leclercq et al.(2004) proposed a unified theory for the study of moving bottlenecks and demonstrated that the Gazis and Herman (1992) as well as the Daganzo and Muñoz (2002) models can be cast as special cases of the Newell and Lebacque models. Consequently, it was demonstrated that the latter model provides a complete framework for the modeling of moving bottlenecks. In a more recent effort, Kerner and Klenov (2010) proposed an alternative theory for the analysis of traffic breakdown at moving bottlenecks that is based on Kerner's three-phase traffic flow theory (Kerner 2009).

In addition to analytical models, numerical models were developed in an attempt to provide a solution to the moving bottleneck problem. Lebacque et al. proposed a numerical solution to the phenomenon in 1998; however, the developed model is limited to bottleneck speeds that are integer submultiples of the free-flow speed of the roadway. This solution was achieved by considering a moving reference that is moving at the same speed of the bottleneck, and thus, the moving bottleneck is transformed to a stationary bottleneck. Later, in 2002, another model was proposed by Giorgi et al. (2002). However, it includes the same limitation of the Lebacque model, as the bottleneck speed still has to be an integer submultiple of the free-flow speed. Finally, it is to be noted that a more complete model was developed by Daganzo and Laval (2003) in 2005, who considered an alternative approach that consisted of discretizing the bottleneck trajectories using step functions. In other words, they proposed to transform the problem into another problem whose solution is very close to the actual solution. This numerical model overcame the limitation that a certain relationship should exist between the bottleneck speed and the free-flow speed. However, the offered solution is still approximate.

Furthermore, the phenomenon was studied in the field by Giorgi et al.(2002) who conducted some experiments on an urban street with buses, and by Daganzo and Muñoz (2002) who introduced a slow vehicle in the traffic stream which allowed them to estimate the flow passing the bottleneck as a function of its speed, the free-flow speed and the traffic demand. They then highlighted deficiencies in existing models based on the results of their experiments, and proposed a new model that overcomes these limitations.

More recently, Fadhloun (2013) and Fadhloun et al. (in press) used simulated data generated from the INTEGRATION software to develop an analytical model that accounts for the underlying fundamental diagram in the computation of the passing flow. Furthermore, Fadhloun (2013) and Fadhloun et al. (in press) demonstrated that, unlike what was reported by Daganzo and Muñoz, the relationship is not always linear, even in the case in which the flow-density fundamental diagram is triangular, and that a quadratic convex relationship provides a better representation of the phenomenon.

Lateral Vehicle Movement Modeling: Lane-Changing and Gap Acceptance Modeling

Up to this point we have only discussed the longitudinal movement of vehicles along a roadway. Consequently, the lateral movement of vehicles across lanes will be briefly discussed in this section. The modeling of lane-changing logic entails capturing mandatory and discretionary lane changes. Mandatory lane-changing takes place when the current lane ceases to be a feasible option and the driver must shift to another lane in order to proceed along their path.

Discretionary lane-changing occurs when the driver selects a lane to maximize the perceived utility function. The lane-changing logic includes three actions: a) scanning the various lanes; b) deciding on making a lane change; c) executing a lane change. The execution of a lane change requires that the driver identify a feasible gap in the adjacent to make his/her maneuver. This is called gap acceptance modeling. The gap is deemed acceptable if it is of sufficient size depending on the vehicle's speed, distance to the point where the lane change should be completed, and the time spent searching for a gap. This section describes one perspective of modeling lateral vehicle movement and that is from the standpoint of the INTEGRATION software. The interested reader can review the vast breadth of literature on lane-changing modeling (Laval and Daganzo 2006; Bham 2007; Kesting et al. 2007; Toledo and Zohar 2007; Wang et al. 2007; Kim et al. 2008; Laval and Leclercq 2008; Tang et al. 2008; V. Ramanujam 2008; Zhang et al. 2008).

Modeling Discretionary Lane-Changing

Within the INTEGRATION software a determination whether to make a discretionary lane change is made based on the perceived speeds in the current lane and the adjacent lanes to the left and right every second. In addition, all lanes are scanned every five seconds in order to identify any potential gaps across multiple lanes (e.g. a high-occupancy lane). The INTEGRATION model considers a pre-specified bias for a vehicle to remain in the lane in which it is already traveling by adding an inertia factor to the vehicle's desired speed when computing its perceived speed. The use of such a bias factor reduces the number of unnecessary lane changes by increasing the attractiveness of the current lane, which is achieved using three factors, a relative speed factor, an absolute speed factor, and a maximum inertia threshold. Furthermore, the inertia factor increases as the vehicle length increases. The use of the vehicle length equivalency factor within the inertia factor ensures that trucks make fewer lane changes than passenger cars.

The INTEGRATION model incorporates a bias towards travel in specific lanes depending on the number of lanes on the roadway when vehicles travel outside the influence area of merge and diverge sections. Specifically, the model biases passenger cars to travel toward the middle lanes for roadways with three or more lanes. This bias is achieved by altering the perceived speed in a specific lane in an attempt to achieve field-observed traffic volume distributions across roadway lanes, as demonstrated from empirical data (Carter et al. 1999). In addition, the model biases trucks towards use of the shoulder lane.

Modeling Mandatory Lane Changes

In situations where a trip destination imposes a constraint on vehicle movement; for example, exiting vehicles at a ramp-freeway diverge section, mandatory lane changes are performed to ensure that vehicles maintain lane connectivity at the end of each link. This lane connectivity at any diverge or merge point can be computed internal to the model.

Prior to reaching the physical diverge point, the obligatory lane-changing logic in INTEGRATION assigns two imaginary boundaries upstream of the diverge gore. The first boundary, located farther upstream, is denoted as the "softwall," while the second boundary is denoted as the "hardwall" and is located closer to the physical diverge point. While the hardwall indicates the location where exiting vehicles are unable to proceed closer to the diverge section on the original lane and thus must abandon the lane due to lane discontinuity downstream, the softwall defines the location where the driver recognizes the need to change lanes in order to exit at a diverge section. The distance between these two boundaries represents a transition from the absolute discretionary nature to the absolute mandatory nature for the vehicle under consideration.

In order to ensure the smooth transition of flows from one lane to the next, the mandatory lane-changing logic within INTEGRATION has been made highly stochastic rather than purely deterministic. This is reflected by the significant variation in the locations of the softwalls and the hardwalls upstream of the diverge gore. Within INTEGRATION, the mean locations of the softwalls are at a distance of $100n$ times the jam density spacing upstream of the diverge gore, where n is the minimum number of lane changes required to complete the maneuver. Likewise, the mean locations of the hardwalls are at a distance of $10n$ times the jam density spacing back from the point of diverge, where n is as defined earlier. This implies that some very cautious drivers will strive to be in the rightmost lane a considerable distance upstream of the diverge gore.

A commonly observed phenomenon at merge and weaving sections is the movement of mainline vehicles from the shoulder lane to the middle and median lanes in order to avoid any interaction with merging vehicles. This behavior should be captured in any modeling tool. The bias should entice vehicles to use the desired lane at low volumes; however, it should have a minor effect at high volumes. As is the case with the soft and hardwalls, the location at which vehicles respond to the specified bias should be randomized in order to ensure that not all lane-changing occurs at the same location.

Concluding Remarks

This chapter has provided a brief overview of traffic flow theory as it pertains to the microscopic modeling of traffic. The interested reader is encouraged to read the various references in this chapter to acquire a deeper understanding of traffic flow theory.

References

Bham, G. H. (2007). Gap Acceptance Behavior in Mandatory Lane Changes under Congested and Uncongested Traffic on a Multi-lane Freeway. Transportation Research Board (TRB).

Carter, M., H. Rakha and M. V. Aerde (1999). "Variability of traffic-flow measures across freeway lanes." *Canadian Journal of Civil Engineering* 26(3): 270–281.

Daganzo, C. F. and J. Laval (2003). On the Numerical Treatment of Moving Bottlenecks. Institute of Transportation Studies, University of California, Berkeley, CA.

Demarchi, S. H. (2002). A New Formulation for Van Aerde's Speed-Flow-Density Relationship (in Portuguese). XVI Congresso De Pequisa e Ensino em Transportes, Natal, RN, Brazil.

Drew, D. R. (1968). *Traffic Flow Theory and Control.* New York, McGraw Hill.

Fadhloun, K. (2013). A Microscopic and Macroscopic Analysis of Moving Bottlenecks, Virginia Tech.

Fadhloun, K., H. Rakha and A. Loulizi (in press). "Impact of the fundamental diagram on the passing rate of moving bottlenecks." *Transportation Letters: The International Journal of Transportation Research.*

Gazis, D. and R. Herman (1992). "Moving and 'phantom' bottlenecks." *Transportation Science*: 223–229.

Gazis, D., R. Herman and R. Rothery (1961). "Nonlinear follow-the-lead models of traffic flow." *Operations Research* 9(4): 545–567.

Giorgi, F., L. Leclercq and J. B. Lesort (2002). Traffic Flow Model for Urban Traffic Analysis: Extensions of the LWR Model for Urban and Environmental Applications. 15th International Symposium on Transportation and Traffic Theory, Adelaide, Australia.

Gipps, P. G. (1981). "A behavioral car-following model for computer simulation." *Transportation Research 15B*: 105–111.

Greenshields, B. D. (1934). "A study of traffic capacity." Proceedings of the Highway Research Board 14: 448–477.

Guzzella, L. and A. Sciarretta (2007). *Vehicle Propulsion Systems: Introduction to Modeling and Optimization*. New York, Springer.

Kerner, B. S. (2004). *The Physics of Traffic*, Berlin, New York, Springer-Verlag.

Kerner, B. S. (2009). *Introduction to Modern Traffic Flow Theory and Control: The Long Road to Three-Phase Traffic Theory*. Berlin, Springer.

Kerner, B. S. and S. L. Klenov (2010). "A theory of traffic congestion at moving bottlenecks." *Journal of Physics A: Mathematical and Theoretical* 43 425101.

Kesting, A., M. Treiber and D. Helbing (2007). "General lane-changing model MOBIL for car-following models." *Transportation Research Record* (1999): 86–94.

Kim, J.-T., J. Kim and M. Chang (2008). "Lane-changing gap acceptance model for freeway merging in simulation." *Canadian Journal of Civil Engineering* 35(3): 301–311.

Laval, J. A. and C. F. Daganzo (2006). "Lane-changing in traffic streams." *Transportation Research Part B: Methodological* 40(3): 251–264.

Laval, J. A. and L. Leclercq (2008). "Microscopic modeling of the relaxation phenomenon using a macroscopic lane-changing model." *Transportation Research Part B: Methodological* 42(6): 511–522.

Lebacque, J. P., J. B. Lesort and F. Giorgi (1998). "Introducing buses into first-order traffic flow models." *Transportation Research Record* 1644: 70–79.

Leclercq, L., S. Chanut and J. B. Lesort (2004). "Moving bottlenecks in the LWR model: a unified theory." *Transportation Research Record* 1883: 3–13.

Lighthill, M. J. and G. B. Witham (1955). "On kinematic waves. I: Flood movement in long rivers, II. A theory of traffic flow on long crowded roads." Proceedings of the Royal Society of London A229: 281–345.

May, A. (1990). *Traffic Flow Fundamentals*, Prentice Hall, Upper Saddle River, NJ.

Muñoz, J. and C. F. Daganzo (2002). Moving Bottlenecks: A Theory Grounded on Experimental Observation. 15th International Symposium on Transportation and Traffic Theory, Adelaide, Australia.

Newell, G. (1993). A Moving Bottleneck. Institute of Transportation Studies, University of California, Berkeley, CA.

Newell, G. (1998). "A Moving Bottleneck." *Transportation Research: Part B* 32B(8): 531–537.

Ni, D. and D. Henclewood (2008). "Simple engine models for VII-enabled in-vehicle applications." *IEEE Transactions on Vehicular Technology* 57(5): 2695–2702.

Park, S., H. Rakha, R. Alfelor, C. Y. D. Yang and D. Krechmer (2011). "Empirical study of impact of icy roadway surface condition on driver car-following behavior." *Transportation Research Record: Journal of the Transportation Research Board* 2260(1): 140–151.

Pipes, L. A. (1953). "An operational analysis of traffic dynamics." *Journal of Applied Physics* 24: 274–287.

Pipes, L. A. (1967). "Car-following models and the fundamental diagram of road traffic." *Transportation Research 1* 21–29.

Rakha, H. A. (2006). Validation of Van Aerde's simplified steady-state car-following and traffic stream model. 85th Transportation Research Board Annual Meeting, Washington, DC, Transportation Research Board.

Rakha, H. (2009). "Validation of Van Aerde's Simplified Steady-state Car-following and Traffic Stream Model." *Transportation Letters: The International Journal of Transportation Research* 1(3): 227–244.

Rakha, H. and M. Arafeh (2010). "Calibrating steady-state traffic stream and car-following models using loop detector data." *Transportation Science* 44(2): 151–168.

Rakha, H. and B. Crowther (2002). "Comparison of Greenshields, Pipes, and Van Aerde car-following and traffic stream models." *Transportation Research Record: Journal of the Transportation Research Board* 1802(1): 248–262.

Rakha, H. and I. Lucic (2002). "Variable power vehicle dynamics model for estimating maximum truck acceleration levels." *Journal of Transportation Engineering* 128(5): 412–419.

Rakha, H., I. Lucic, S. H. Demarchi, J. R. Setti and M. V. Aerde (2001). "Vehicle dynamics model for predicting maximum truck acceleration levels." *Journal of Transportation Engineering* 127(5): 418–425.

Rakha, H., P. Pasumarthy and S. Adjerid (2004). Modeling Longitudinal Vehicle Motion: Issues and Proposed Solutions. Transport Science and Technology Congress, Athens, Greece.

Rakha, H., C. C. Pecker and H. B. B. Cybis (2007). "Calibration procedure for Gipps car-following model." *Transportation Research Record: Journal of the Transportation Research Board* 1999(1): 115–127.

Rakha, H., M. Snare and F. Dion (2004). "Vehicle dynamics model for estimating maximum light-duty vehicle acceleration levels." *Transportation Research Record: Journal of the Transportation Research Board* 1883(1): 40–49.

Rakha, H. and W. Wang (2009). "Procedure for Calibrating Gipps Car-Following Model." *Transportation Research Record: Journal of the Transportation Research Board* 2124(1): 113–124.

Rakha, H. and Y. Zhang (2004). "INTEGRATION 2.30 framework for modeling lane-changing behavior in weaving sections." *Transportation Research Record: Journal of the Transportation Research Board* 1883(1): 140–149.

Rakha, H. and Y. H. Zhang (2006). "Analytical procedures for estimating capacity of freeway weaving, merge, and diverge sections." *Journal of Transportation Engineering-Asce* 132(8): 618–628.

Rakha, H. A., K. Ahn, W. Faris and K. S. Moran (2012). "Simple vehicle powertrain model for modeling intelligent vehicle applications.", *IEEE Transactions on Intelligent Transportation* Systems 13(2): 770–780.

Ramanujam, V., C. Choudhury, M. Ben-Akiva (2008). An Extended Lane Changing Model to Capture Lane-Change Duration. 87th Transportation Research Board Annual Meeting (TRB).

Richards, P. I. (1956). "Shock waves on the highway." *Operations Research* 4: 42–51.

Sangster, J. and H. Rakha (2014). Enhancing and Calibrating Rakha-Pasumarthy-Adjerid Car-Following Model Using Naturalistic Driving Data. Submitted for Publication in the *International Journal of Transportation Science and Technology*.

Tang, T., H. Huang, S. C. Wong and R. Jiang (2008). "A car-following model with the anticipation effect of potential lane changing." *Acta Mechanica Sinica/Lixue Xuebao* 24(4): 399–407.

Toledo, T. and D. Zohar (2007). "Modeling duration of lane changes." *Transportation Research Record* (1999): 71–78.

Trieber, M. and A. Kesting (2013). Traffic Flow Dynamics: Data, Models and Simulation, Springer, Berlin.

Van Aerde, M. (1985). Modelling of Traffic Flows, Assignment and Queueing in Integrated Freeway/Traffic Signal Networks. Civil Engineering. Waterloo, Waterloo. Ph.D.

Van Aerde, M. (1995). "Single regime speed-flow-density relationship for congested and uncongested highways." Presented at the 74th TRB Annual Conference, Washington, DC, Paper No. 950802.

Van Aerde, M. and H. Rakha (1995). "Multivariate calibration of single regime speed-flow-density relationships." Proceedings of the 6th 1995 Vehicle Navigation and Information Systems Conference: 334–341.

Van Aerde, M. and H. Rakha (2013). INTEGRATION © Release 2.40 for Windows: User's Guide—Volume I: Fundamental Model Features. Blacksburg, M. Van Aerde & Assoc.

Van Aerde, M. and S. Yagar (1988). "Dynamic Integrated Freeway/Traffic Signal Networks: A Routeing-Based Modelling Approach." *Transportation Research* 22A(6): 445–453.

Wang, D.-H., L.-J. Zhou and W.-Q. Li (2007). Modeling lane changing behavior using fuzzy logic, Chengdu, China, American Society of Civil Engineers, Reston, VA 20191–4400.

Whitham, G. B. (1974). *Linear and Nonlinear Waves*, John Wiley & Sons, New York.

Wong, J. Y. (2001). *Theory of Ground Vehicles*, John Wiley & Sons, New York.

Zhang, Y. and H. Rakha (2005). "Systematic analysis of weaving section capacity." Transportation Research Board 84th Meeting.

Zhang, F., H.-Y. Xuan and Q.-X. Zhao (2008). "Lane changing model based on finite state automaton." Zhongguo Gonglu Xuebao/China *Journal of Highway and Transport* 21(3): 97–100.

3

TRAFFIC NETWORK MODELING

Srinivas Peeta,[1] *Henry Liu,*[2] *and Xiaozheng He*[1]

(1) PURDUE UNIVERSITY, (2) UNIVERSITY OF MICHIGAN

Overview

Traffic network modeling is a broad domain covering various problems that aid transportation design, planning, operations and policy. A key problem in traffic network modeling is to analyze the highway system performance.

The analysis of highway system performance is built upon traffic assignment, which is a process of allocating the given origin-destination (OD) demand to an analytical representation of the traffic network under some path choice principles. The traffic assignment process incorporates the demand (represented by the OD demand matrix) and supply (represented by the traffic network) characteristics that interact through a certain path choice mechanism. The outputs of this interaction are the link (path) flows and corresponding link (path) travel times.

Through this process, decision-makers are able to predict the paths taken by travelers and the resulting traffic volumes on the highway system. The outputs of the network modeling can be used to assess the deficiencies of the existing highway system and evaluate alternative proposals to improve it.

The rest of this chapter will discuss traffic network modeling, focusing on each component. The next section discusses the process to translate a highway system into a topological structure. Section 3 provides approaches for modeling the travel delay in a highway system. Section 4 describes how an OD demand matrix is obtained for traffic network modeling. Section 5 introduces traffic assignment in terms of the fundamental path decision principles and their extensions. Section 6 summarizes the network design problem to improve the existing highway system.

1. Network Representation

A highway traffic network is modeled as a directed graph consisting of a set of links (or arcs) to represent the roads carrying traffic flows, and a set of nodes (or vertexes) to represent the intersections. Nodes and links in the graph are indexed to identify them, and have associated attributes to represent their specific characteristics, such as capacities and travel times. This section focuses on the topological representation of the traffic network and associated attributes of nodes and links.

Intersection Representation

An intersection of roads in a highway traffic network is typically represented as a node in a graph. Figure 3.1a illustrates a typical representation of a four-legged intersection. Nodes that generate trips are called origins, while nodes that absorb trips are called destinations. A node can be an origin, a destination, both or neither.

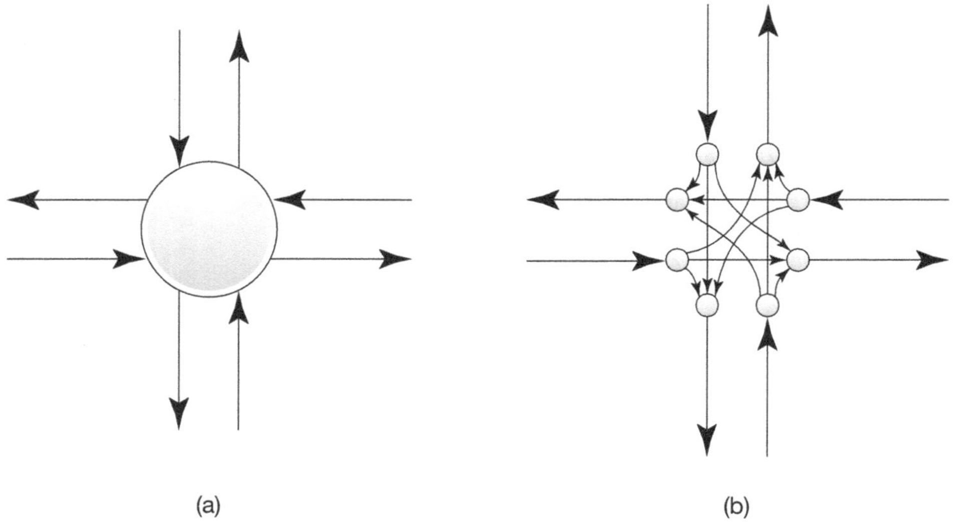

Figure 3.1 Intersection representation

Detailed movements at an intersection can be represented by a set of nodes connected by a set of links that represent the vehicular movement directions at the intersection. Figure 3.1b illustrates the representation of a four-legged intersection when detailed vehicular movements are considered.

In most traffic network problems, nodes have no associated attributes. In some specific problems, e.g., dynamic traffic assignment, nodes may have capacities that are associated with the control strategies imposed at intersections (Tampère et al., 2011).

Road Representation

A road in a highway traffic network is typically represented as a directed link whose direction is that of the traffic movement. A link is associated with several attributes, such as capacity, free flow speed (speed when the road is empty), flow, and travel time. Some link attributes, such as number of lanes and link length, can also be implicitly represented by other attributes such as link capacity and free flow travel time.

An important link attribute is the link performance function. It determines the link travel time given the link flow. Typically, the link travel time is illustrated as increasing with link flow due to congestion. A widely used link performance function is the Bureau of Public Roads (BPR) function:

$$t_a = t_a^0 \left[1 + \alpha \left(\frac{x_a}{c_a} \right)^\beta \right] \tag{1}$$

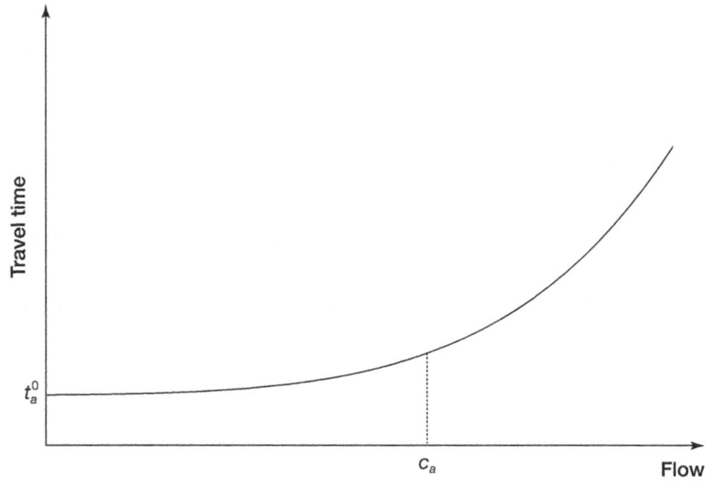

Figure 3.2 Bureau of Public Roads function

where t_a^0, x_a and c_a represent free flow travel time, link flow and capacity of link a, respectively. Parameters α and β determine the link travel time increase ratio. In general, $\alpha = 0.15$ and $\beta = 4$, indicating that link travel time increases in proportion to the fourth power of the volume-to-capacity ratio.

Figure 3.2 shows a plot of the BPR function, where link flow can exceed link capacity. However, consistent with the real world, the relationship between link travel time and link flow is more appropriately represented by a backward-bending curve as illustrated in Figure 3.3, where link flow decreases as congestion increases and link flow cannot exceed link capacity.

In Figure 3.3, as two different travel time values correspond to a link flow value, it is difficult to analytically represent this relationship in a simple form. Hence, BPR type link performance functions are used for modeling purposes when queuing in links is not a major concern.

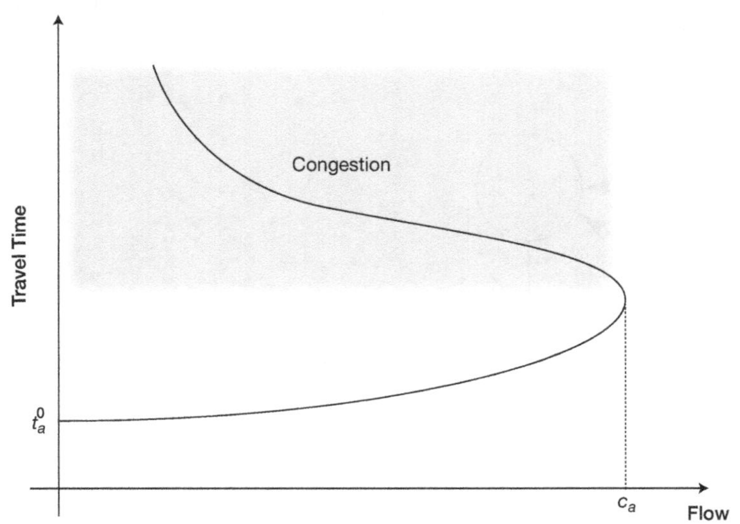

Figure 3.3 Relationship between link travel time and link flow

When modeling congestion is important, various approximation approaches (such as point queues, spatial queues, etc.) or simulation tools are used. We will discuss how to model travel delay under congestion in Section 2.

Topological Structure

The topological structure of a traffic network is established through the analytical representation of the relationships between links and nodes. There are two types of representation: node-link incidence matrix and node-node adjacency matrix.

The topological structure of a network can be fully described by the node-link incidence matrix, defined as:

$$A_{ia} = \begin{cases} 1 & \text{if link } a \text{ starts at node } i \\ -1 & \text{if link } a \text{ ends at node } i \\ 0 & \text{otherwise.} \end{cases}$$

where A_{ia} denotes the entry in *i*th row and *a*th column of the node-link incidence matrix.

For example, suppose we have a directed network with four nodes connected by four links as shown in Figure 3.4. Then, the link-node incidence matrix A describes the topological structure of the network.

$$\begin{array}{c} \\ A = \end{array} \begin{array}{c} \begin{matrix} a & b & c & d \end{matrix} \\ \begin{bmatrix} 1 & 1 & & \\ -1 & & 1 & \\ & -1 & -1 & 1 \\ & & & -1 \end{bmatrix} \end{array} \begin{matrix} \\ 1 \\ 2 \\ 3 \\ 4 \end{matrix}$$

Another network topology representation is the node-node adjacency matrix. Each element Bij in the matrix indicates whether node i is adjacent to node j. Thereby, the node-node adjacency matrix B also describes the topological structure of the network shown in Figure 3.4.

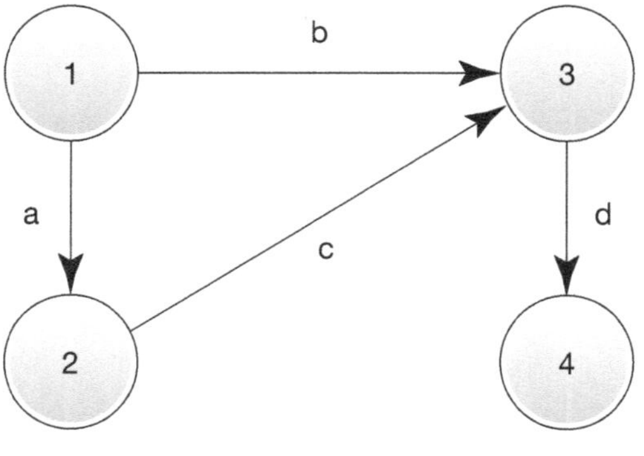

Figure 3.4 A directed network

$$B = \begin{array}{c} \\ \\ \\ \\ \end{array} \begin{array}{c} \begin{array}{cccc} 1 & 2 & 3 & 4 \end{array} \\ \left[\begin{array}{cccc} & 1 & 1 & \\ & & 1 & \\ & & & 1 \\ & & & \end{array}\right] \end{array} \begin{array}{c} \\ 1 \\ 2 \\ 3 \\ 4 \end{array}$$

Path Representation

A path in a highway traffic network is a sequence of links or nodes that connect an origin to a destination. Trips generated from an origin arrive at destinations through paths. Therefore, path flows represent the numbers of vehicles traveling on specific paths. The path flow representation allows the translation of OD demands into a link flow pattern.

One way to represent paths in a network is the link-path incidence matrix. Let Δ denote a link-path incidence matrix. Each entry of the matrix denotes whether a link lies on a path. That is, $\Delta_{ap}^{w} = 1$ if link a lies on path p connecting OD pair w, and $\Delta_{ap}^{w} = 0$ otherwise. Each row of the matrix corresponds to one of the links and each column of the matrix corresponds to one of the paths. Thereby, the 1s in the row for link a indicate which paths pass through that link, and the 1s in the column for path p indicate which links are on path p.

The link-path incidence matrix can translate path flows into link flows. Here, a path (or link) flow is the volume (in number of vehicles per hour) of traffic traversing along a particular path (or link). Denote x_a as the flow on link a and f_p as the flow on path p. As $\Delta_{ap}^{w} = 1$ when path p passes through link a and $\Delta_{ap}^{w} = 0$ when it does not, the total flow through link i can be computed as:

$$x_a = \sum_{w \in W} \sum_{p \in P_w} \Delta_{ap}^{w} f_p^{w}$$

where P_w represents the set of paths connecting OD pair w. The above equation shows that a link flow is the summation of flows on all paths that pass through that link. Let $\mathbf{x}$ be a vector of link flows and $\mathbf{f}$ be a vector of path flows. The above equation can be written as:

$$\mathbf{x} = \Delta \mathbf{f} \tag{2}$$

In addition, an OD-path incidence matrix, denoted by Φ, can be constructed to establish the relationship between path flows and OD demands. Each row of the matrix corresponds to one of the OD pairs and each column of the matrix corresponds to one of the paths. Then, Φ_{wp} if path p connects OD pair w, and Φ_{wp} otherwise. Therefore,

$$d_w = \sum_{p \in P} \Phi_{wp} f_p^{w}.$$

Let d be a vector of OD demands. Then, we have a succinct representatoion of the above equation in matrix form as:

$$\mathbf{d} = \Phi \mathbf{f}. \tag{3}$$

2. Modeling Travel Delay in Traffic Network

The BPR function (Equation 1) is unrealistic because it does not consider link interactions, capacity constraints and junction signals. To model the travel delay due to congestion and signals as well as link interactions and capacity constraints, various symmetric and asymmetric link performance functions were proposed in the literature (Nagurney, 1993).

A link performance function, including traffic signal delay, can be formulated by incorporating a signal delay function $D(x,\lambda)$, as:

$$t(x,\lambda) = t_0 + D(x,\lambda) \tag{4}$$

where x is the link flow, and λ is the green split. In particular, Webster's delay formula can be used to compute the signal delay:

$$D(x,\lambda) = \frac{x}{\lambda s(\lambda s - x)} \tag{5}$$

where s is the saturation flow rate. However, Webster's delay formula (5) only describes the undersaturated traffic condition, which may not be suitable when congestion occurs. For the saturated traffic condition, the Akcelik function, which includes a uniform delay term and an overflow delay term, can be used to compute the signal delay:

$$D(y,\lambda) = \frac{0.5L(1-\lambda)}{(1-\lambda y)} + 900T\left[y - 1 + \sqrt{(y-1)^2 + \frac{8(y-0.5)}{MT}}\right] \tag{6}$$

where L is signal cycle length, y is the volume-to-capacity ratio, T is the study duration of the flow period, and M is the capacity.

The above functions only model the average cost of traversing a link and cannot capture on-and-off signal change in reality. In addition, both delay functions (Equation 5) and (Equation 6) do not capture the delay due to the vehicular queue. To capture the travel delay due to either on-and-off signal change or vehicular queue, traffic flow models are required to formulate a link performance function. These traffic flow models can be categorized as whole link models, queuing models and continuum traffic flow models.

When a whole link model is used, a general time-dependent link performance function considering traffic flow dynamics can be expressed as a function of the number of vehicles on a link $x(\tau)$, inflow rate to a link $u(\tau)$ and outflow rate $v(\tau)$ (Carey, 2001):

$$t(\tau) = f\left(x(\tau), u(\tau), v(\tau)\right) \tag{7}$$

where $t(\tau)$ is the delay or the link traversal time of the vehicle entering a link at time τ. The simplest form of a delay function only depends on the traffic volume. Accordingly, the traffic flow dynamics can be written as:

$$\begin{cases} \dot{x}(\tau) = u(\tau) - v(\tau) \\ t(\tau) = f\left(x(\tau)\right) \\ v\left(\tau + t(\tau)\right) = u(\tau) / \left[1 + \dot{t}(\tau)\right]. \end{cases} \tag{8}$$

The link travel time function in (8) can be specified as a linear delay function:

$$f\left(x(\tau)\right) = \alpha + \beta x(\tau), \quad \alpha > 0, \beta > 0 \tag{9}$$

or a piece-wise linear function:

$$f\left(x(\tau)\right) = \begin{cases} \{\alpha, \text{ if } x(\tau) < \alpha / \beta \\ x(\tau) / \beta, \text{ otherwise} \end{cases} \tag{10}$$

or any smooth and convex functions which satisfy the following conditions:

$$\frac{f\left(x(\tau_2)\right) - f\left(x(\tau_1)\right)}{x(\tau_2) - x(\tau_1)} \cdot \frac{x(\tau_2) - x(\tau_1)}{\tau_2 - \tau_1} > -1, \forall \tau_2, \tau_1,$$

$$f''\left(x\right) \text{ is continuous,}$$

$$f\left(0\right) = \alpha$$

$$f'\left(x\right) > 0, f''\left(x\right) > 0 \text{ for all } x > 0, \text{ and}$$

$$f'\left(x\right) \to \beta \text{ as } x \to \infty .$$

When the point-queue model is adopted, the link performance function can be represented as the free-flow travel time plus the queuing delay:

$$x(\tau) = t_0 + \frac{q(\tau)}{M} \tag{11}$$

where $q(\tau)$ is the queue length at time τ and M is the capacity of the bottleneck. The point-queue model can be replaced by the spatial queue model or the double-queue model to compute the queue length.

If a continuum traffic flow model (e.g., LWR model) is used to describe traffic flow dynamics, then link travel time can be determined by a specified speed-density relationship, for instance, the modified Greenshields equation:

$$v(\tau) = v_0 + \left(v_f - v_0\right)\left(1 - \frac{k(\tau)}{k_j}\right)^{\alpha} \tag{12}$$

where $v(\tau)$ is the space-mean speed at time τ, v_0 is the minimum speed, v_f is the free-flow speed, $k(\tau)$ is the link density at time τ, k_j is the jam density, and α is a user-specific parameter. Then the link travel time can be computed by $t(\tau) = l/v(\tau)$ where l is link length.

3. OD Demand Matrix

An origin-destination (OD) demand matrix represents the demand aspects of the traffic network. Each element in the matrix denotes the travel desires from an origin to a destination. When an OD matrix is given, demand can be assigned onto the traffic network, and a link flow pattern will be produced.

Unfortunately, obtaining the OD demand matrix is not easy. In the classical four-step urban transportation planning model (Martin and McGuckin, 1998), the first three steps (trip generation, trip distribution, and mode split) are focused on determining the OD demand matrix. Traditional methods of estimating the OD matrix rely on large-scale surveys like home interview survey, roadside interview, and license plate method. These surveys are costly in terms of monetary expense, time, and manpower. Through surveys, data related to social and economic attributes of households are collected. After the socioeconomic data are collected, the trip generation process is performed to estimate the number of trips generated from each traffic analysis zone. Statistical regression is used to model the trip generation as a function of the socioeconomic attributes of households.

Based on the numbers of trips generated from traffic analysis zones, several methods can be used to assign the trips to different destinations, including the gravity, growth factor, and entropy-maximization models (Ortuzar and Willumsen, 2011). The gravity model is used extensively in the trip distribution process because it accounts for the attributes of the highway system and land-use characteristics. The gravity model for trip distribution is based on the concept that the attractiveness of a location declines with increasing distance (or travel time). The mathematical representation of the gravity model is as follows:

$$T_{ij} = K_i K_j T_i T_j f(C_{ij}) . \tag{13}$$

where T_{ij} denotes the total number of trips between origin i and destination j, T_i is the total number of trips departing from origin i, T_j is the total number of trips attracted to destination j, C_{ij} denotes the travel cost between origin i and destination j, K_i and K_j are balancing factors, and $f(.)$ is a distance decay function. The output of the trip distribution process is a trip table. The summation of row i of the trip table is the total number of trips generated from origin i, and the summation of row j of the trip table is the total number of trips attracted to destination j. Note that the trips can be completed using one or more of several modes (auto, transit, biking, walking, etc.).

To estimate the demand by mode, a mode split process is needed. Mode choice of highway network users can be modeled by performing discrete choice analysis (Ortuzar and Willumsen, 2011). Utility maximization and maximum likelihood estimation can be used to split trips to different travel modes. After the mode split process, the vehicular OD demand matrix is obtained which can be used in traffic assignment.

When traffic count information is available on the links, the OD demand matrix can be estimated through mathematical programming. A general optimization formulation for OD estimation using traffic counts can be represented as:

$$\min \gamma_1 F_1(T, \hat{T}) + \gamma_2 F_2(x, \hat{x}) , \tag{14}$$

subject to the link flows, x, as the result of a traffic assignment process. In the formulation, $\hat{T}$ is the target OD matrix, and $\hat{x}$ the observed traffic counts with F_1 and F_2 being distance measure functions. The objective function (equation 14) can be specified using maximum likelihood, generalized least squares, or Bayesian Inference (Cascetta and Russo, 1997).

4. Traffic Assignment

This section focuses on a core issue of traffic network modeling: how is the travel demand between different origins and destinations distributed over the network? The travel demand is loaded onto the network based on a principle by which travelers choose a path to complete their trips. The simplest path choice principle is all-or-nothing, where all travelers between an OD pair choose the same path (for example, based on the lowest travel cost). However, the all-or-nothing principle is limiting in that capacity constraints are ignored, implying that congestion is ignored.

The most widely used principles for traffic assignment are the user optimal and system optimal principles. They have been further extended by considering demand elasticity, perception errors, traffic dynamics, etc. This section will start with the user optimal and system optimal path decision principles, followed by their extensions known as elastic-demand traffic assignment, stochastic traffic assignment, and dynamic traffic assignment.

Fundamental Path Decision Principles

The path decision of traveler is impacted by many factors, such as travel time, fuel consumption, toll, comfort, convenience, etc. All these factors that influence the decision can be ascribed monetary values and summed to a generalized travel cost. As travel time is the most significant factor in traffic network modeling, it is typically used to represent the generalized travel cost.

Two fundamental path decision principles of highway network users were introduced by Wardrop (1952):

First principle: The journey times of all routes actually used are equal, and less than those which would be experienced by a single vehicle on any unused route.
Second principle: At equilibrium the average journey time is minimum.

The equilibrium characterized by Wardrop's first principle is called "user equilibrium" as it corresponds to the state in which travelers seek to minimize their own travel cost, whereas the equilibrium characterized by Wardrop's second principle is called "system optimum" as it corresponds to the state in which the total cost in the traffic network system is minimized.

The next few sections discuss how these two fundamental principles are rigorously formulated as mathematical models. A special example, known as the Braess paradox, is used to illustrate the difference between them.

Traffic Assignment Based on the User Optimal Principle

User optimal traffic assignment follows Wardrop's first principle. It is based on the assumptions that all travelers are homogeneous and fully rational, and they have perfect knowledge of travel costs. Travelers make path choice decisions based on average travel cost. The solution of user optimal traffic assignment corresponds to the Nash Equilibrium, namely, travelers selfishly compete to travel on the shortest paths and continue to modify their path choice until they minimize their travel cost. User equilibrium is reached when all travelers stop changing their paths and none can reduce travel cost by unilaterally changing path.

Beckmann et al. (1956) rigorously formulated user equilibrium as an equivalent constrained optimization problem. In the formulation, the objective function is equal to the sum of the integrals of the link cost functions, integrated between zero and the current link flows. Mathematically, it has the form:

$$\min_{x} z(\mathbf{x}) = \sum_{\alpha \in A} \int_{0}^{x_\alpha} t_\alpha(\omega)\, d\omega \tag{15}$$

subject to link flow definitional constraints (Equation 2), demand conservation constraint (Equation 3), and nonnegativity constraints f_p^w, for all $p \in P_w$. This nonlinear programming formulation can be applied to highway traffic networks with separable link cost functions, such as the BPR function (Equation 1).

The first-order optimality conditions for the optimal solution to the constrained optimization problem (Equation 15) are:

$$f_p^w \cdot \left(c_p^w - \pi^w\right) = 0, \quad \forall p \in P_w, w \in W \tag{16}$$

$$c_p^w - \pi^w \geq 0, \quad \forall p \in P_w, w \in W \tag{17}$$

$$\sum_{p \in P_w} \Phi_{wp} f_p^w = d_w \quad \forall w \in W \tag{18}$$

$$f_p^w \geq 0 \quad \forall p \in P_w, w \in W \tag{19}$$

$$\pi^w \geq 0, \; d^w \geq 0 \quad \forall w \in W \tag{20}$$

where $c_p^w = \sum_{a \in A} \Delta_{ap} t_a$ represents the path cost and π^w represents the minimal path cost for OD pair w. The mathematical formulation (Equations 16–20) is called the nonlinear complementarity problem (NCP).

Nagurney (2004) rephrased user equilibrium as: "user travel costs on used paths for each OD pair are equalized and minimal." Based on this statement, the user equilibrium flows must satisfy:

$$c_1^w = c_2^w = \ldots c_p^w = \pi^w \leq c_{p+1}^w = \ldots c_p^w \quad \forall p \in P_w, w \in W \tag{21}$$

where flows on paths 1, 2, . . ., p have positive values while flows on paths $p + 1$, . . ., P are zero. It has been shown that path flows satisfying conditions (Equation 21) must satisfy the NCP (Equations 16–20), and vice versa. Thus, the NCP (Equations 16–20) is also used as an alternative formulation for user optimal traffic assignment.

In addition, based on conditions (Equation 21), it can be shown that user equilibrium flows $\mathbf{f}^*$ must satisfy an inequality:

$$(\mathbf{f} - \mathbf{f}^*)^T \mathbf{c}(\mathbf{f}^*) \geq 0 \quad \forall \mathbf{f} \in \Omega \tag{22}$$

where Ω represents the feasible path flow set, i.e.,

$$\Omega = \left\{\mathbf{f} \mid \Phi\mathbf{f} = \mathbf{d}, \mathbf{f} \geq 0\right\}.$$

Inequality (Equation 22) is called the variational inequality problem (VIP). Both the NCP and VIP models are more general than the optimization formulation for user equilibrium, as they can be applied to traffic networks with non-separable link cost functions.

Traffic Assignment Based on the System Optimal Principle

The system optimal traffic assignment follows Wardrop's second principle. It addresses travelers' path choice decisions from the system perspective, namely, travelers choose their paths so as to minimize the total travel time in the highway system. Therefore, this principle may not be consistent with the more reasonable behavioral notion that travelers seek to minimize their individual path travel times.

In system optimal traffic assignment, travelers are assumed to make path choice decisions based on the marginal travel cost, instead of the average cost. Therefore, the mathematical models for user equilibrium can be revised by replacing the average cost with the marginal cost to formulate the system optimal state. Accordingly, the constrained optimization formulation for system optimal traffic assignment is:

$$\min_x z(\mathbf{x}) = \sum_{\alpha \in A} t_\alpha(x_\alpha) \cdot x_\alpha = \sum_{\alpha \in A} \int_0^{x_\alpha} \tilde{t}_\alpha(\omega)\, d\omega \tag{23}$$

subject to link flow definitional constraints (Equation 2), demand conservation constraint (Equation 3), and nonnegativity constraints $f_p^w \geq 0$ for all $p \in P_w$. In the formulation, $\tilde{t}_a$ represents the marginal travel time, i.e., $\tilde{t}_a(x_a) = t_a(x_a) + t'_a(x_a) \cdot x_a$.

Similarly, an NCP can be developed for system optimal traffic assignment as:

$$f_p^w \cdot \left(\tilde{c}_p^w - \tilde{\pi}^w\right) = 0, \quad \forall p \in P_w, w \in W \tag{24}$$

$$\tilde{c}_p^w - \tilde{\pi}^w \geq 0, \quad \forall p \in P_w, w \in W \tag{25}$$

$$\sum_{p \in P_w} \Phi_{wp} f_p^w = d_w \quad \forall w \in W \tag{26}$$

$$f_p^w \geq 0, \quad \forall p \in P_w, w \in W \tag{27}$$

$$\tilde{\pi}^w \geq 0, d \geq 0 \quad \forall w \in W \tag{28}$$

where $\tilde{c}_p^w = \sum_{a \in A} \Delta_{ap} \tilde{t}_a$ represents the marginal path cost and $\tilde{\pi}^w$ represents the minimal marginal path cost for OD pair w. And a VIP model for system optimal traffic assignment can be formulated as:

$$\left(\mathbf{f} - \mathbf{f}^*\right)^T \tilde{\mathbf{c}}(\mathbf{f}^*) \geq 0 \quad \forall \mathbf{f} \in \Omega \tag{29}$$

The Braess Paradox

This section presents the well-known Braess paradox (Braess, 1969; Braess et al., 2005) to illustrate the difference between the concepts of user equilibrium and system optimum. By intuition, we expect that building more roads can enhance the highway system performance. However, the Braess paradox shows a contradiction in which the addition of a link causes the increase of average travel cost.

The classical Braess network is shown in Figure 3.5, where link 1 (from node 1 to node 2), link 2 (from node 1 to node 3), link 3 (from node 2 to node 4) and link 4 (from node 3 to node 4) form the basic network (Figure 5(a)); while link 5 (from node 2 to node 3) is a new link to be added (Figure 5(b)). The link performance functions are listed beside the links. The total demand traveling from origin node 1 to destination node 4 is 6.

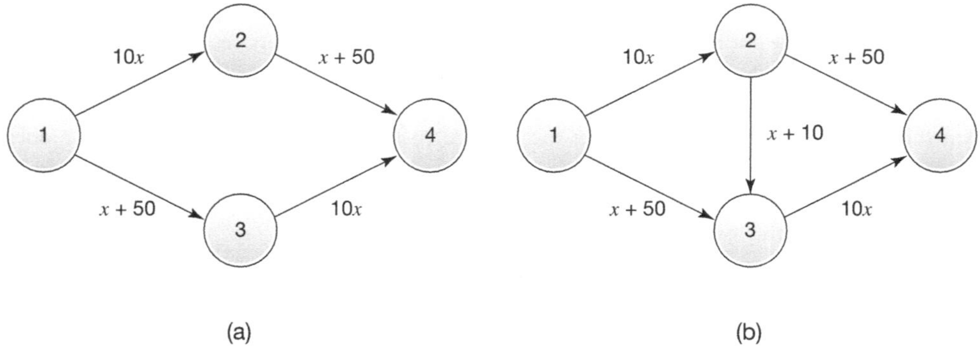

Figure 3.5 Braess' paradox network: (a) Before building a new road; (b) after building a new road

Before the new link is added, there are two alternative paths: path 1 consists of node sequence 1: 2–4 and path 2 consists of node sequence 1: 3–4. In user equilibrium, 3 units of flow would choose path 1 and the rest choose path 2, resulting in the path travel cost (10 × 3) + (3 + 50) = 83 time units. After link 5 is added, a new alternative path can be chosen by travelers. The new path consists of node sequence 1 . . . 2: 3–4. In the new user equilibrium, all three paths are used, each of which carries two units of flow. The resulting path travel cost is (10 × 4) + (2 × 50) = (10 × 4) + (2 + 10) + (10 × 4) = 92 time units. Thus, each traveler has a higher travel cost after the new road is built. It illustrates that building more roads may not enhance the highway system performance.

The reason for the occurrence of the Braess paradox is due to the selfishness in travelers' path choice behavior. Pas and Principio (1997) analyzed the relationship between the occurrence of the Braess paradox and the parameters in link performance functions and demand level. They showed that the Braess paradox occurs only when the travel demand is within a certain range.

User Equilibrium with Elastic Demand

The standard User Equilibrium (EU) network assignment problem can be viewed as a model of traveler path choice regardless of the congestion level. However, travelers may adjust not only their path but also trip destination, mode, and departure time to respond to congestion. The elastic demand equilibrium assignment problem can be viewed as a model to describe the choice of not to travel, in addition to path choices.

Elastic demand equilibrium models consider the case when the OD trip rate is influenced by the level of service on the network. As the congestion level increases, travelers may change their travel mode, shift the departure time, alter the trip destination, or even cancel the trip. Therefore, the trip rate can be assumed to be a monotonically decreasing function of the shortest travel time of an OD pair.

Let $D_w(\omega)$ denote the demand function for OD pair w. It determines the value of demand as $d_w = D_w(\pi_w)$. As $D_w(\omega)$ is a monotonically decreasing function, its inverse function $D_w^{-1}(\omega)$ exists. Thus, the shortest travel cost between OD pair w is $\pi_w = D_w^{-1}(d_w)$. Applying the inverse demand function $D_w^{-1}(\omega)$ to the nonlinear programming model (Equation 15), the mathematical programming formulation for user equilibrium with elastic demand is as follows:

$$\min_{x,d} z(\mathbf{x}) = \sum_{\alpha \in A} \int_0^{x_\alpha} t_\alpha(\omega)\, d\omega - \sum_{w \in W} \int_0^{d_w} D_w^{-1}(\omega)\, d\omega \tag{30}$$

subject to link flow definitional constraints (Equation 2), demand conservation constraint (Equation 3), and nonnegativity constraints

$$f_p^w \geq 0 \ \forall p \in P_w \ \text{ and } \ d_w \geq 0 \ \forall w \in W.$$

Stochastic Traffic Assignment

The standard user equilibrium network assignment problem relies on the assumption of perfect knowledge of travel cost. However, in the real world, users may perceive travel costs differently leading to variability in response, which entails the need to model travelers' perception errors.

Under stochastic traffic assignment, travelers choose their paths according to their perceived generalized costs. Utility maximization theory is used to model their path choice decision, where a rational traveler will choose the path with the minimum perceived generalized cost. The perceived generalized cost includes a deterministic component of travel cost and a random error term that captures variations in travel cost perception, tastes of individual trip making, as well as measurement errors. Therefore, the perceived travel cost (or disutility) can be formulated as:

$$C_p^w = \overline{c}_p^w + \varepsilon_p^w, \quad \forall p \in P_w, w \in W \tag{31}$$

where C_p^w represents the perceived path travel cost, $\overline{c}_p^w$ represents the deterministic (or measured) path travel cost, and ε_p^w represents the random error. Then, the probability that an individual driver will choose path p from the path set P is determined by:

$$\begin{aligned} \text{Prob}\left(p \mid P_w\right) &= \text{Prob}\left(C_p^w < C_q^w, \ \forall p \in P_w, q \neq p\right) \\ &= \text{Prob}\left(\varepsilon_p^w - \varepsilon_q^w < \overline{c}_q^w < \overline{c}_p^w, \ \forall p \in P_w, q \neq p\right). \end{aligned} \tag{32}$$

The joint cumulative distribution function of $\varepsilon_p^w - \varepsilon_q^w$ is evaluated at $\overline{c}_q^w - \overline{c}_p^w$.

If the distribution of ε_p is known, Equation 32 can be used to compute the probability of an individual's path choice. Two probability distributions are widely used for the error term ε_p. If ε_p follows the multinomial normal distribution, a probit model is obtained. However, the multinomial probit model does not have an analytically closed form. Solving the probit model relies on approaches such as the Monte Carlo simulation.

Another distribution used in stochastic traffic assignment is the Gumbel distribution, which has a cumulative density function as:

$$F(x) = \exp\left[-\exp\left(x\right)\right]. \tag{33}$$

Let $\zeta_p^w = -(1/\theta)\varepsilon_p^w$ be an independently and identically distributed Gumbel random variable, where θ represents a dispersion parameter related to the variance of the Gumbel random error term. Then, the path choice probability has a closed form:

$$\text{Prob}\left(p \mid P_w\right) = \frac{\exp(-\theta \overline{c}_p^w)}{\sum_{q \in P_w} \exp\left(-\theta \overline{c}_q^w\right)}. \tag{34}$$

For a given OD pair w, the corresponding flow on path $p \in P_w$ is determined by

$$f_p^w = d^w \cdot \text{Prob}\,(p \mid P_w).$$

As $\sum_p \text{Prob}\,(p \mid P_w) = 1$, it ensures that $\sum_p f_p^w = d^w$.

To identify a solution in stochastic traffic assignment, Daganzo and Sheffi (1977) introduced the Stochastic User Equilibrium (SUE) principle.

> *Stochastic User Equilibrium principle:* At stochastic user equilibrium, no traveler can improve his or her perceived travel time by unilaterally changing routes.

Based on the SUE principle, Fisk (1980) proposed a nonlinear programming model for SUE:

$$\min_x z(\mathbf{x}) = \sum_{\alpha \in A} \int_0^{x_\alpha} t_\alpha(\omega)\, d\omega + \frac{1}{\theta} \sum_{w \in W} \sum_{p \in P_w} f_p^w \ln f_p^w \tag{35}$$

subject to link flow definitional constraints (Equation 2), demand conservation constraint (Equation 3), and nonnegativity constraints, $f_p^w \geq 0$ for all $p \in P_w$. Sheffi and Powell (1982) proposed an unconstrained minimization program for the SUE problem as follows:

$$\min_x z(\mathbf{x}) = -\sum_{w \in W} d^w S_w\left[C^w(\mathbf{x})\right] + \sum_{a \in A} x_a t_a(x_a) - \sum_{a \in A} \int_0^{x_a} t_a(\omega)\, d\omega \tag{36}$$

where

$$S_w\left[C^w(\mathbf{x})\right] = E\left[\min_{w \in P_w}\left\{C_w^p \mid C^w(\mathbf{x})\right\}\right] \tag{37}$$

Equation 37 defines the expected perceived travel cost for OD pair w, which depends on the perceived costs of the entire path set. As the expectation cannot be evaluated readily, the Monte Carlo simulation is used for solving this unconstrained optimization problem.

Dynamic Traffic Assignment

All models discussed heretofore are based on the assumption that path flow is present simultaneously on all links of the path. These models, called static traffic assignment models, focus on representing average or steady-state conditions over an analysis time period that is long enough to allow all traffic flows to arrive at their destinations. However, these models cannot account for the variations of travel times and flows, because travel times and flows on links and paths are constant over the analysis period. Therefore, they are unable to analyze phenomena related to time-dependent traffic conditions, such as oversaturated traffic flow, queue spillback, dynamic routing, and peak spreading.

As active traffic management strategies require the accurate modeling of the within-day traffic dynamics and traveler adaptation to traffic dynamics, advanced traffic analysis tools are needed to predict network conditions and analyze network performance in both the planning and operational stages. Dynamic traffic assignment (DTA) (Peeta and Ziliaskopoulos, 2001) models can be used to aid decision-making for operational, construction, or demand management actions that are likely to induce a temporal or spatial traffic pattern shift. DTA models can interact with other traffic management or control systems, to provide travel mode, departure time, path, and other traffic information.

DTA follows the same process as static traffic assignment. However, DTA requires time-dependent OD matrices, and the time-dependent flow-density relationship (also known as fundamental diagram of traffic flow) needs to be specified for each link in the network. In DTA, travel demand enters the network over time, characterized by the time-dependent OD matrix. In addition, as the fundamental diagram is used in DTA models, travel time and flow on a link are more accurately associated with traffic density, such that the backward bending curve for travel time illustrated in Figure 3.3 can be captured.

The path decision principles used in DTA can be viewed as extensions of Wardrop's first and second principles and the SUE principle. Demand elasticity and perception errors can also be considered. For example, the path decision principle for user equilibrium DTA can be defined as:

> *User equilibrium DTA principle:* The routes chosen by those departing at the same time between the OD pair should have equal experienced travel time.

Modeling user equilibrium DTA is significantly more complex than modeling static traffic user equilibrium assignment as the temporal dimension needs to be included in the model. Analytical models have been developed by formulating user equilibrium DTA as a constrained optimization program, variational inequality system, and nonlinear complementarity problem. Constraints in these analytical approaches include flow conservation constraints, flow propagation constraints, boundary constraints, and nonnegativity constraints, all of which are time-dependent. The flow propagation constraints in analytical DUE approaches rest on various link performance models, such as point-queue model, whole link model, and cell transmission model. Another class of DTA models is the simulation-based DTA models, where the traffic flow modeling is represented using a traffic simulator. They circumvent some of the analytical modeling challenges associated with traffic flow modeling, and are hence more commonly used for DTA deployment in practice.

5. Network Design Problem

The highway network design problem seeks to optimize the improvement of a highway network with respect to a system objective while considering the path choice behavior of network users and deployment constraints. A typical network design problem includes four main components: objectives, decision variables, constraints, and underlying path choice behavior. A comprehensive review on highway network design problems is provided by Farahani et al. (2013)

A highway network design problem generally includes two objectives: (1) at the system level the objective is to optimize the overall system performance; (2) at the user level, the individual network users seek to minimize their travel costs. The decision-making at the system level will affect the responses of network users' path choices. If the decision-maker knows how network users respond to a decision he/she makes, then the network design problem is a Stackelberg game from a game theory perspective.

A Stackelberg game is usually formulated as a bi-level programming problem, whose applications in transportation have been reviewed by Clegg et al. (2001). The upper-level problem is formulated from the perspective of the leader (decision-maker). The upper-level problem has the following generic structure:

$$\min_{x \in X} F\left[x, y(x)\right] \tag{38}$$

subject to $G\left[x, y(x)\right] \le 0$, where $y(x)$ is a solution of the lower level (follower) optimization problem for any fixed x:

$$\min_{y \in Y} f\left(x, y\right) \tag{39}$$

subject to $g\left(x, y\right) \le 0$.

The structure of the bi-level programming problem typically does not maintain convexity even under the condition that both the upper and lower problems are convex. Therefore, it may have multiple local optima and is difficult to solve.

As the lower-level problem represents the path choice behavior of highway network users, it can be formulated as a fixed demand user equilibrium (Equation 15), user equilibrium with elastic demand (Equation 30), stochastic user equilibrium (Equation 35), or dynamic user equilibrium. If the lower-level program employs nonlinear complementarity (Equations 16–20) or variational inequality (Equation 22) to represent users' path choice behavior, then the network design problem is formulated as a mathematical program with equilibrium constraints (MPEC) and mathematical program with complementarity constraints (MPCC), respectively.

The upper-level problem represents the decision-maker's objective. It can be categorized as discrete or continuous network design problems, depending on whether or not the decision variables are integer. Discrete network design relates to the topology of the highway network; for example, building new roads or road closure. Continuous network design addresses the parameterization of the network, for example, capacity expansion, road pricing, and signal timings. Some network design problems contain both integer and real decision variables. Such mixed network design problems address both the topology and parameterization of the network, for example, adding new infrastructure and expanding capacity.

The main constraints in the upper-level program include budget constraints, resource availability constraints, capacity constraints, upper and lower bounds, etc. Objectives of the upper-level programming problem include minimizing total system travel time or construction cost, or maximizing social welfare, total revenue, the ratio of total revenue to total cost, or the network reserve capacity, etc.

References

Beckmann, M. J., McGuire, C. B., and Winsten, C. B. (1956). *Studies in the Economics of Transportation*, Yale University Press, New Haven, CT.

Braess, D. (1969). ber ein paradoxon aus der verkehrsplanung. *Unternehmensforschung*, 12, 258–268.

Braess, D., Nagurney, A., and Wakolbinger, T. (2005). On a paradox of traffic planning. *Transportation Science*, 39(4), 446–450.

Carey, M. (2001). Dynamic traffic assignment with more flexible modelling within links. *Networks and Spatial Economics*, 1(3–4), 349–375.

Cascetta, E. and Russo, F. (1997). Calibrating aggregate travel demand models with traffic counts: Estimators and statistical performance. *Transportation*, 24(3), 271–293.

Clegg, J., Smith, M., Xiang, Y., and Yarrow, R. (2001). Bilevel programming applied to optimising urban transportation. *Transportation Research Part B*, 35(1), 41–70.

Daganzo, C.F. and Sheffi, Y. (1977). On stochastic models of traffic assignment. *Transportation Science*, 11(3), 253–274.

Farahani, R.Z., Miandoabchi, E., Szeto, W.Y., and Rashidi, H. (2013). A review of urban transportation network design problems. *European Journal of Operational Research*, 229(2), 281–302.

Fisk, C. (1980). Some developments in equilibrium traffic assignment. *Transportation Research Part B*, 14(3), 243–255.

Martin,W.A. and McGuckin, N.A. (1998). *Travel Estimation Techniques for Urban Planning. NCHRP Report* 365, Transportation Research Board, Washington, DC.

Nagurney, A. (1993). *Network Economics: A Variational Inequality Approach*. Kluwer Academic Publishers, Boston, MA.

Nagurney, A. (2004). Spatial equilibration in transport networks. *Handbook of Transport Geography and Spatial Systems*, 583–608.

Ortuzar, J.de D. and Willumsen, L.G. (2011). *Modelling Transport*. Wiley (4th ed.), New York.

Pas, E. and Principio, S. (1997). Braess' paradox: some new insights. *Transportation Research Part B*, 31(3), 265–276.

Peeta, S. and Ziliaskopoulos, A. (2001). Foundations of dynamic traffic assignment: the past, the present and the future. *Networks and Spatial Economics*, 1(3/4), 233–266.

Sheffi, Y. and Powell, W.B. (1982). An algorithm for the equilibrium assignment problem with random link times. *Networks*, 12(2),191–207.

Tampère, C.M., Corthout, R., Cattrysse, D. and Immers, L.H. (2011). A generic class of first order node models for dynamic macroscopic simulation of traffic flows. *Transportation Research Part B*, 45(1), 289–309.

Wardrop, J.G. (1952). Some theoretical aspects of road traffic research. *Proceedings of the Institute of Civil Engineers*, Part II, pp. 325–378.

4

FREEWAY TRAFFIC CONTROL

John Hourdos

UNIVERSITY OF MINNESOTA

Introduction

The *Highway Capacity Manual* (HCM) (TRB 2000) defines a freeway as a divided highway with full control of access, and two or more lanes for the exclusive use of traffic in each direction. Freeways provide uninterrupted flow. "Uninterrupted" in this context is used to describe the type of facility, not the quality of the traffic flow at any given time. A freeway experiencing extreme congestion, for example, is still an uninterrupted-flow facility because the causes of congestion are internal. As society's need for transportation increases, keeping freeway flow uninterrupted is an everyday exercise in control and resource management.

Freeways require huge physical construction projects and, therefore, are one of the costliest of public works activities. For a while adding lanes (capacity) to the system was the obvious answer. In 2001 the Minnesota Department of Transportation (MnDOT) funded a project aimed at answering the big question "What would it take to build our way out of congestion in our city?" As described in Davis et al. 2001, for the modest in size metropolitan area of Minneapolis and St. Paul, MN, 1,146 lane miles would be needed to accommodate the demand predicted for the year 2020. Even a fraction of such expansion is impossible; therefore, effort must be directed in better managing the resources that are already available.

Freeway management is a collection of activities with the goal of minimizing congestion in a freeway network. As congestion is the result of demand exceeding capacity, these activities mainly aim at either increasing capacity or decreasing demand. This chapter will describe the present tools that transportation engineers currently have at their disposal in order to manage freeway traffic efficiently and safely.

Traffic Surveillance and Data Collection

Congestion on freeways can evolve very fast. Given that flows can reach upwards to 2,000 veh/h/ln, when a capacity-reducing incident happens queues can extend upstream with speeds of up to 20 mph and rapidly involve hundreds of travelers. Since their inception it was understood that freeway networks must be under constant surveillance to detect and respond to disturbances. Traffic Management Centers (TMC) were established and comprehensive surveillance and data collection systems were deployed. The majority, if not all of the freeway traffic control techniques described in this chapter depend on one or both of these systems.

Traffic surveillance systems are, as the word implies, comprised by cameras deployed along the roadway. Typically, these cameras are deployed every half-mile and are equipped with Pan Tilt Zoom (PTZ) features allowing the traffic operators to monitor the full length of the road. The simplest systems involve just manually controlled cameras but more sophisticated systems have been developed which allow the automatic detection of unwanted elements like debris, stalled vehicles, or inclement weather. In a typical TMC, operators are continuously scanning the cameras looking for incidents during the am and pm peak demand periods. Once a problem is detected, manual or automatic response strategies are implemented to minimize the impact and inform the road users.

Automated or traffic-responsive control strategies require real-time traffic information. Modern freeways are equipped with detectors providing a variety of information. In the simplest of cases, the data-collection system is comprised by standard inductive-loop traffic detectors embedded in the pavement or nonintrusive video or radar-based sensor mounted on top of the driving lanes. Such detectors usually provide counts of vehicles and time occupancy (the percentage of time the detector is occupied by a vehicle) over a predefined time interval. Such intervals can range from 10 seconds to 1 minute. More sophisticated traffic detection systems provide speed and density measurements as well as vehicle classification. The collected traffic measurements are archived, disseminated to the public through public and private portals, but most importantly they drive automated traffic control strategies like the ones described in the rest of this chapter. Recently, with the proliferation of GPS-equipped vehicles, probe data are also available which include travel times.

Freeway Traffic Control

Although better roads and freeways help move commuters, clearly there are limits to their effectiveness. As traffic continues to increase worldwide, transportation engineers are looking into innovative ways to ease congestion without widening roadways. Directly or indirectly controlling demand and supply (capacity) on freeways are the main weapons against an ever increasing congestion problem. There are various ways to exert such control and usually combinations of more than one method are implemented.

Controlling Demand

Since controlling directly the transportation needs of people is impossible, engineers can either control the flow of vehicles after they enter the system or provide information to road users to compel them to plan their trips more equitably. This section primarily focuses on the control of demand directed to freeways, with a short discussion of traveler information-based methods. Adjusting or metering the flow of vehicles entering the freeway is a method as old as the freeway system. However, expanding that control and coordinating it with traveler information systems and arterial control are the most recent ways of controlling demand.

Ramp Metering

Ramp metering improves traffic conditions by appropriately regulating inflow from on-ramps to the freeway mainstream. Ramp metering has been used since the 1950s with more than 4,000 ramp meters deployed. Although for many years studies have highlighted the benefits of

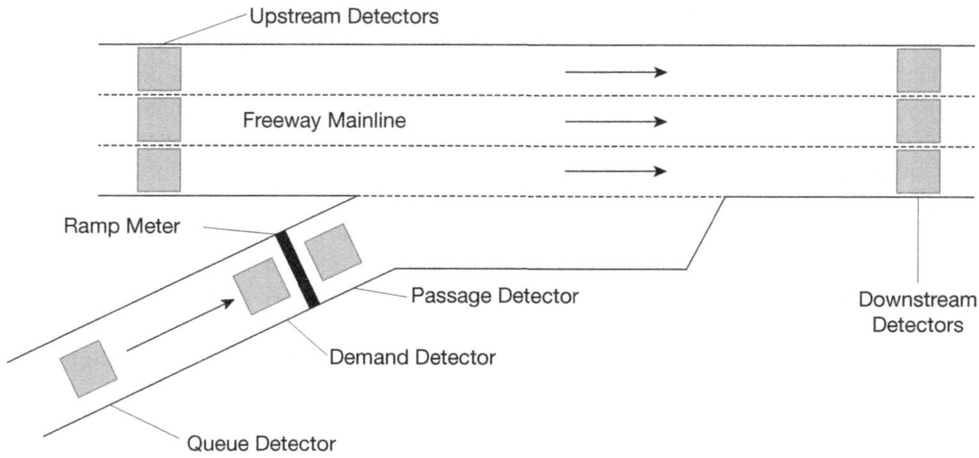

Figure 4.1 Metered freeway entrance ramp instrumentation

ramp metering, public perception has not always been positive. Doubts regarding the effectiveness of ramp metering in reducing freeway congestion prompted MnDOT to enact a six-week ramp metering "holiday" where all the meters in the Twin Cities freeway system were turned off. A before and after evaluation concluded that meters were responsible for a 22 percent reduction in freeway travel times, a 7 percent increase in freeway speeds, as well as a 26 percent reduction in crashes (Cambridge Systematics, 2001). Simulation studies during the same period showed similar results and estimated a 26 percent decrease in fuel consumption and emissions (Hourdakis and Michalopoulos 2002). A comprehensive accounting of reported benefits from ramp metering can be found in the *Ramp Management and Control Handbook* commissioned by the FHWA in 2006 (FHWA 2006).

The earliest and most simple method of metering the traffic at freeway entrance ramps is through pre-timed rate plans. This type of control depends on historical information and today is used in places where recurrent congestion is predictable. Another reason for using pre-timed ramp control is to break up the platoons of vehicles generated by upstream traffic lights to facilitate easier and safer merging into the mainline flow. Traffic-responsive ramp-metering strategies, as opposed to pre-timed, are based on real-time measurements from sensors installed in the freeway network and can be classified as isolated or coordinated. Figure 4.1 shows the typical traffic control setup on a metered freeway entrance ramp.

Isolated Ramp Control

Isolated ramp-metering strategies make use of measurements from the vicinity of a single ramp to control each entrance ramp independently of all other ramps in the network. Isolated ramp-control strategies can be divided into two major categories– Demand/Capacity and Occupancy-based. The Demand/Capacity strategies try to regulate the entrance ramp flow so the sum of that and the mainline flow from upstream does not exceed the capacity of a downstream section of the road. Occupancy-based strategies control metering rates based on the occupancy measured upstream of the entrance ramp. The following table presents a summary of the most notable isolated ramp-control strategies.

Table 4.1 Field-tested isolated ramp-control strategies

Control Algorithm	*Mainline Measurements*				*Ramp Measurements*
	Predefined Capacity/ Flow	*Occupancy*	*Speed*	*Flow*	
Demand-capacity	Yes	Yes	No	Y	None
Speed control	No	Yes	Yes	N	None
ALINEA	Yes	Yes	No	Y	Queue length, flow
ANCONA	No	Yes	Yes	N	Flow rate
Gap acceptance	No	Yes	Yes	N	Queue, vehicle presence

EXAMPLE OF ISOLATED RAMP CONTROL: ALINEA

ALINEA is the most widely used isolated ramp-control strategy. It was first applied to on entrance ramps of the Boulevard Peripherique in Paris (Papageorgiou et al., 1991). ALINEA is a feedback regulator that adjusts in real-time the metering rate to keep the occupancy downstream of the ramp at a specified value (critical occupancy). The strategy calculates the metering rate based on the following equation (also termed the feedback law):

$$\tau(t) = r(t-1) + K_R\left[\hat{o} - o_{\text{out}}(t-1)\right]$$

where:

$r(t)$ is the meter rate at time interval t
K_R is a constant regulator parameter (veh/hr)
$o_{\text{out}}(t-1)$is the last measured downstream occupancy value (%)
$\hat{o}$ is the desired value for the downstream occupancy (predefined).

The value $\hat{o} = o_{\text{cr}}$ is selected so the downstream flow comes close to capacity. This strategy reacts to excessive occupancies $o_{\text{out}}(t-1)$ only after the o_{cr}threshold value is exceeded. Recently, new extensions of the ALINEA algorithm have been developed in an effort to account for issues not sufficiently covered by the original strategy. These include FL-ALINEA which uses flow instead of occupancy, UP-ALINEA which takes into consideration the upstream instead of the downstream occupancy, and X-ALINEA/Q which takes into consideration the size of the queue on the ramp.

Coordinated Ramp Metering

Coordinated ramp-metering strategies make use of measurements from a wide area of the freeway to control all metered ramps within a particular corridor. In difference to isolated ramp metering strategies, coordinated ones try to optimize flow along a stretch of the freeway and spread the restriction in entrance flow over several ramps. Coordinated strategies may be more equitable and efficient than local ramp-metering strategies, particularly in the frequent case of restricted ramp-storage spaces. The downside is that such strategies require a considerably larger communication infrastructure with the controller centrally located transmitting rates to the ramps for implementation. Coordinated ramp-metering strategies can be divided into three major types (Zhang et al., 2001):

- *Cooperative strategies*: After computing the metering rate for each entrance ramp, further adjustment is done to avoid both congestion at the bottleneck and spillback at critical ramps. These strategies are still reactive to critical conditions and perform the adjustments in an ad hoc manner which can cause traffic instability. Examples of such strategies are the Helper and the Linked-Ramp algorithms.
- *Competitive strategies*: In these strategies, two sets of metering rates are computed based on both local and global traffic conditions; the most restrictive rate is selected for implementation. Examples of such strategies are the Compass, the Stratified, and the Swarm algorithms.
- *Integral strategies*: Strategies of this type have a clear control objective(s) that is explicitly or implicitly linked to the control action. The objective is usually travel time, or throughput of the entire corridor. They decide ramp metering rates through optimizing the objective function while considering system constraints, such as maximum allowable ramp queue, bottleneck capacity, and so forth. Two examples of integral strategies that have been implemented in the field are the Washington DOT Fuzzy Logic and METALINE which is an extension of ALINEA.

Table 4.2 Field-tested coordinated ramp-control strategies

Type	*Control Algorithm*	*Mainline Measurements*				*Ramp Measurements*
		Predefined Capacity/ Flow	*Occu-pancy*	*Speed*	*Flow*	
Cooperative	HELPER	Yes	Yes	No	No	Queue, vehicle presence
	Linked-ramp	Yes	Yes	No	Yes	Queue
Competitive	COMPASS	Yes	Yes	No	Yes	Queue
	ZONE	Yes	Yes	No	Yes	Exit flow
	Stratified zone metering	Yes	Yes	No	Yes	Queue, entrance and exit flow
	Bottleneck	Yes	Yes	No	Yes	Queue, entrance and exit flow
	SWARM	No	Yes	No	No	Queue
Integral	Fuzzy Logic	No	No	No	No	Queue
	METALINE	Yes	Yes	No	No	Flow rate
	Linear programming	Yes	Yes	Yes	Yes	Queue

Example of Coordinated Ramp Control: Stratified Zone Metering

In 2003, the Stratified Zone Metering (SZM) strategy replaced the ZONE metering strategy which was in operation in Minnesota freeways from 1969. The objective of the SZM is (a) regulating the zone inputs so that the total entering volumes do not exceed the zone capacity and (b) limiting ramp wait times below a predetermined value during the control period. To realize this, the Stratified Zone Metering strategy employs a hierarchical control structure and introduces two ramp state variables.

1. The freeway is divided into zones; zones of certain spatial attributes are grouped into layers.
   - The zone capacity is allocated among the ramp meters inside the zone, in proportion to the respective ramp demand. This yields the zone proposed release rate for each ramp within that zone.
   - A ramp meter may belong to several overlapping zones, each of which would give a zone proposed release rate. The most restrictive one constrains the final zone proposed release rate.
2. Two state variables, ramp demand and minimum release rate, are introduced for measuring real-time entrance ramp demand and adjusting metering rates:
   - The ramp demand is measured in real time by a queue detector, which is deployed on the upstream end of the ramp. This ensures that the zone proposed release rate varies in relation to the ramp demand.
   - The time-varying minimum release rate is determined from the actual ramp queue size. It is introduced for limiting the ramp wait time below the predetermined value. When the selected zone-proposed release rate is less than the ramp's minimum release rate, the latter will override the zone proposed rate.

SZM relies on real-time data, collected on a 30-second basis, to compute applicable metering rates. Such data include occupancy and volume; the occupancy data is directly used by the algorithm while 30-second volumes are first converted into an equivalent hourly flow rate and then smoothed to even out fluctuations and indicate trends.

In SZM, the acceptable range of metering rates is determined to be between 1,714 veh/h and 240 veh/h. The upper bound of this range (1,714 veh/h) is determined based on the consideration that 2.1 seconds is the minimum reasonable headway between vehicles entering the freeway mainline. The lower bound of acceptable metering rates is set to be 240 veh/h; this value is derived from empirical observations that 15 seconds is the maximum waiting time the first motorist in the ramp queue would tolerate before significant violations occur.

Ramp demand is defined as the hourly flow rate of the vehicles desiring to enter a ramp. Under normal conditions, ramp demand is obtained from direct measurements of the queue detector. When the ramp queue extends beyond the queue detector, the flow rate at the queue detector is no longer an accurate measurement of ramp demand. In such case, the measured flow rate is increased incrementally until the queue recedes from the queue detector. The selected 30-second increment step is 150 veh/h to result in a ramp demand of 1,714 veh/h in no more than 5 minutes (10 sampling periods).

The primary public concern regarding metering is excessive long wait times on ramps. Consequently, one of the most important control objectives of the SZM is to limit the wait times lower than the maximum allowed for each ramp. To realize this, a minimum release rate (R_{min}) is designed and applied to each metered ramp:

$$R_{min} = \frac{N}{T_{max}}$$

Where R_{min} is the minimum release rate, N is the queue size (in vehicles) and T_{max} is the predetermined maximum allowed wait time for the ramp, which in Minnesota is 120 seconds for freeway to freeway ramps and 240 seconds for local access ramps. These two thresholds are policy decision based on public input. The minimum release rate R_{min} ensures the last vehicle in the ramp queue would not have to wait longer than the maximum allowed wait time T_{max}.

ZONE FLOW BALANCE

Detector stations are groups of detectors on the mainline located approximately every half-mile. The older ZONE algorithm employs real bottleneck locations to divide the freeway mainline into metering zones. By contrast, in the SZM strategy, all detector stations are used as endpoints to identify all potential metering zones and zones are grouped into layers. A layer is defined as a sequence of successive zones comprising an equal number of stations. SZM is designed to handle zones from 0.5 mile to 3 miles in length; thereby, layers of up to six levels can be established. Specifically, the layer at level one contains the zones consisting of exactly two stations (0.5 mile in length) while the layer at level two contains the zones with three stations (1 mile in length). Following this pattern, the layer at level 6 would contain the zones consisting of as many as 7 stations (3 miles in length). Figure 4.2 illustrates this concept of zones and layers.

Besides limiting ramp wait time, the other important control objective of the SZM strategy is to keep the number of vehicles entering a zone (input) less than the ones leaving (output). The relationship of inputs and outputs within a given zone can be described as a zone inequality:

$$M + A + U \leq B + X + S$$

or

$$M \leq B + X + S - A - U$$

where

M is the total metered entrance ramp volume;
A is the measured upstream mainline volume;
U is the total measured non-metered entrance ramp volume;
X is the total measured off-ramp volume;
B is the downstream mainline capacity. It has traditionally been called the "bottleneck," although it does not represent an actual bottleneck in Stratified Ramp Control strategy. The value used for this term is the expected mainline capacity at that location;
S is the spare capacity which represents the storage space available within the zone.

The SZM coordinates and determines the metering rates based on a system level. To realize this, zones are processed sequentially in an ascending order of layer level. This means zones are processed one at a time from the first zone in layer 1 to the last zone in layer 6. For each individual zone, total ramp volume (*M*) is distributed over all metered ramps of the zone in proportion to their demands. This rule processing is done for all zones sequentially layer after layer. As zones overlap some ramps are assigned multiple release rates. Final ramp release rate is the minimum release rate *(rmin)* or the most restrictive of the assigned rates, whichever is higher.

Mainline Metering

A mainline meter throttles traffic flow from one segment of a highway to the next by directly metering the highway's traffic. While it can create congestion on the mainline upstream of the controlled section, it can help maintain uncongested flow on the mainline downstream. Such a scheme is typically implemented in specialized situations such as bridges and tunnels.

The longest example in operation of mainline metering in the US is westbound I-80 at the San Francisco-Oakland Bay Bridge (MacCalden 1984). This installation places the mainline

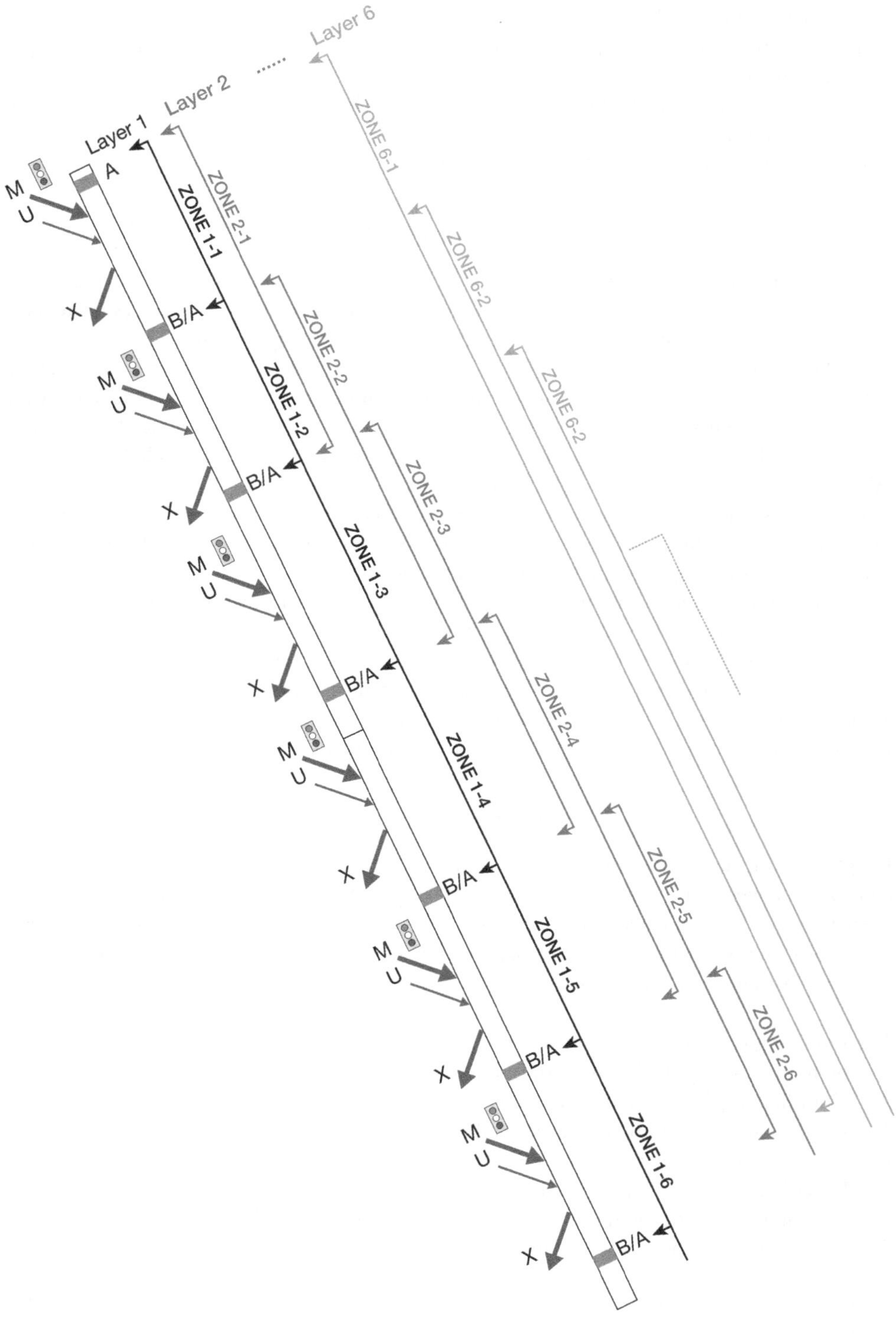

Figure 4.2 Zone–layer structure of stratified zone metering

meters just downstream of a 22-bay toll plaza. Westbound traffic approaching the San Francisco–Oakland Bay Bridge passes through the toll plaza and is then metered to narrow the 22 lanes of traffic into four lanes as efficiently as possible. HOV lanes allow HOVs to bypass the traffic queues.

Mainline metering has also been used on one lane at the westbound entrance to the Holland Tunnel in New York City (FHWA 2011a). A before-and-after study concluded that metering improved traffic volume throughput by approximately 7 percent.

Traveler Information Systems

Advanced Traveler Information Systems (ATIS) are aimed at providing users of the transportation system with more information with which to make decisions about route choices, estimate travel times, and avoid congestion. ATIS are a subtle way of controlling demand. The assumption is that when users know the traffic condition on all alternative routes they will better utilize the entire system's capacity. Such information can be travel times for popular origin-destination pairs, section speeds, and incidents generating nonrecurring congestion. Traditional ATIS involve telephone and Web-based roadway information for pre-trip support and Variable Message Signs (VMS) for en route. On July 21, 2000 the Federal Communications Commission designated "511" as the single traffic information telephone number to be made available to states and local jurisdictions across the country. Most, if not all US states currently operate 511 systems with many of these also supporting Web-based information.

In the age of communication, the vast amounts of traffic information TMCs collect from freeways across the country is available to the public and private sectors. Corporations like Google, Microsoft, INRIX and others have developed turn-by-turn navigation systems that take into account real-time traffic measurements and incident information. Virtually any owner of a smartphone can access traffic conditions on most metropolitan areas. Transportation engineers have yet to understand the effect this abundance of information has on traffic conditions.

Integrated Corridor Management

Integrated Corridor Management (ICM) is a recent attempt to extend the concept of freeway control to cover more than one roadway. Transportation corridors often contain unused capacity in the form of parallel routes, the nonpeak direction on freeways and arterials, single-occupant vehicles and transit services that could be leveraged to help reduce congestion. ICM is a combination of practices such as:

1. Providing comparative travel time to the public and operating agencies for the freeway, high occupancy vehicle lane, frontage roads, arterial streets, and transit lines.
2. Coordinating freeway ramp metering with arterial signal control.
3. Using simulation to predict travel conditions for improved incident response.
4. Improve transit operations with dedicated lanes and signal priority for buses.
5. Diverting traffic to a strategic arterial with improved signal control.
6. Shifting travelers to the transit system for major incidents on the freeway.
7. Implementing joint response plans among agencies.

ICM is still in its infancy with only two finished implementations in Dallas, TX (FHWA, 2011b) and San Diego, CA (FHWA, 2012). Six other corridors have been selected by FHWA as "Pioneer Sites" to implement ICM in the future. The ICM concept, especially due to the introduction of online traffic simulation, has great promise for managing congestion by harnessing the capacity of the network in a dynamic and proactive way.

Controlling Supply (Capacity)

Earlier sections discussed ways of controlling demand locally, from one facility to another, or globally through information and coordination. The other avenue for managing congestion is to control the supply, in the context of traffic that is the capacity of the roadway. Before the advent of ITS, roadway capacity was considered primarily a result of the road geometry and in extreme cases, weather. Today there are systems and methods that allow traffic managers to change the operational capacity of a freeway by closing lanes, imposing restrictions to the type of vehicles allowed to use certain lanes, or even dynamically changing the speed limit of a particular road section. The following sections describe these methods with some examples that have been implemented in the field.

Managed Lanes

A managed lane is a lane or set of lanes within a highway facility where operational strategies are proactively implemented and managed in response to changing conditions. Managed lanes have the following common elements, as described by the FHWA (Jacobs, 2013).

- The managed lane concept is typically a "freeway-within-a-freeway" where a set of lanes within the freeway cross-section is separated from the general-purpose lanes.
- The facility incorporates a high degree of operational flexibility so that over time, operations can be actively managed to respond to growth.
- The operation and demand on the facility is managed using a combination of tools and techniques in order to continuously achieve an optimal condition, such as free-flow speeds.

In the context of controlling capacity, the following management strategies have emerged:

- Eligibility: the lanes are managed by allowing certain vehicles or restricting others.
- Access: dynamic lane closures or shoulder use.
- Road classification: advisory or enforceable speed limits.

Dedicated Lanes

Either permanently or by time of day, certain freeway lanes are dedicated to only certain types of vehicles or numbers of occupants.

- *High-Occupancy Vehicle (HOV) lanes*: HOV lanes are dedicated to the exclusive use of HOVs, including buses, carpools, vanpools, or a combination thereof. Depending on the management strategy there are HOV, HOV+2, and HOV+3 allowing vehicles with a minimum of two, three, or four occupants respectively.
- *High-Occupancy Toll (HOT) Lanes*: HOT lanes are priced managed lanes that use both pricing and eligibility strategies, where HOVs are given free or discounted use while other vehicles are tolled. In the more advanced cases of HOT, the price is dynamically controlled to ensure a minimum level of service.

- *Express Toll Lanes (ETL)*: ETLs are priced managed lanes where all vehicles pay a toll to use them. Unlike HOT lanes, ETLs charge all vehicles except for some cases where buses and vanpools may be toll-exempt.
- *Express Lanes Without Toll*: All vehicles are allowed but access is limited during long stretches of the facility, minimizing turbulence in the flow of vehicles.
- *Bus Toll Lanes (BTL) and Truck-Only Toll* (TOT) *Lanes*: BTLs and TOT lanes are priced managed lanes that give priority to buses or trucks.

Dynamic Shoulder Lanes

Freeway shoulders are normally used for emergency stopping and for incident management. Hard Shoulder Running, as it is referred to in Europe, is an attempt to increase the capacity of the freeway by allowing the shoulder to be used as a driving lane. In the more recent implementations of the concept the shoulder use is dynamically managed through variable lane signs (see the example on p. 54). The shoulder of a particular freeway section has to fulfill a number of requirements regarding width, surface type, and continuity to be eligible. As described in the Scan Report (Mirshahi et al. 2007), temporary addition of a shoulder lane allows congested roadways to have higher throughput at reduced speeds. However, the key to dynamic shoulder lane use is that the segment must extend through the roadway bottleneck. If it does not extend beyond the bottleneck, traffic is simply fed at a greater rate into the segment that is already over capacity, thereby compounding the congestion.

Safety is generally the greatest concern when implementing shoulder-running strategies, since use of the roadway outside shoulder as a travel lane results in the loss of a continuous emergency refuge area for disabled vehicles. Depending on the length of the shoulder lane running section, European examples suggest that the provision of infrequent paved emergency refuge areas or pull-outs should be considered during analysis and design of the affected segment. Alternative refuge areas would be outside the shoulder area and would provide a designated place for stalled or disabled vehicles while allowing use of the hard shoulder as a travel lane. In Britain, for example, the alternative refuge areas are spaced every third or quarter of a mile. In the US implementation is mixed: a 1.5-mile section of I-35W in Minneapolis, MN includes a Priced Dynamic Shoulder Lane (PDSL), which is a shoulder open to HOT traffic during the AM peak period. In this case, no pull-outs have been implemented but the PDSL operation is paired with an Active Traffic Management system (see the following section) for rapid incident management.

Dynamic Junction Control

This strategy consists of dynamically allocating lane access on mainline and ramp lanes in interchange areas where high traffic volumes are present and the relative demand on the mainline and ramps change throughout the day. For off-ramp locations, this may consist of assigning lanes dynamically either for through movements, shared through-exit movements, or exit-only. For on-ramp locations, this may involve a dynamic lane reduction on the mainline upstream of a high volume entrance ramp, or might involve extended use of a shoulder lane as an acceleration lane for a two-lane entrance ramp which culminates in a lane drop. The volumes on the mainline lanes and ramps can be continuously monitored and lane access can be dynamically changed based on real-time or anticipated conditions.

Dynamic Speed Limits

This strategy adjusts speed limits based on real-time traffic, roadway, and/or weather conditions. Dynamic speed limits, also referred to as Speed Harmonization, can either be enforceable (regulatory) speed limits or recommended speed advisories, and they can be applied to an entire roadway segment or individual lanes. Monitoring real-time traffic conditions can be used to adjust the speed limits to meet an agency's goals/objectives for safety, mobility, or environmental impacts.

Active Traffic Management

The freeway control techniques described so far can be implemented independently or can be elements of a greater plan. Active Traffic Management (ATM) is a method of increasing peak capacity and smoothing traffic flows on busy major highways and combines a number of these techniques. ATM is a concept and its details differ on each implementation. To illustrate it, a recently implemented ATM system is presented in the following section.

ATM Example: Smart Lanes

Smart Lanes (MnDOT 2010) is the brand name of Minnesota's Active Traffic Management (ATM) system on I-35W and I-94. The ATM system was first deployed on an 18-mile section of I-35W in 2010 followed by a second implementation in 2012 on an 8-mile section of I-94 between downtown Minneapolis and downtown St. Paul. Traffic entering both freeways was already metered. The ATM system consists of individual electronic signs over each lane of traffic that provide real-time information to help drivers make informed decisions about their commute.

A series of overhead signs known as Intelligent Lane Control Signs (ILCS) above each lane of the two freeways are used to inform drivers of upcoming conditions or controls in place. Overhead signs are used to indicate which lanes are closed to access, blocked because of a crash or obstruction, and which adjacent lanes are impacted by such events. The ILCS are also used to post-advisory speed signs warning travelers to slow down in anticipation of stopped traffic ahead. Spacing of the signs is approximately half a mile apart but exact spacing is based on relations to bridges, existing signs, and sight lines. The managed lane control system serves four key purposes:

- To inform drivers when the left lanes are open to HOV and/or HOT lanes.
- To inform drivers of advisory speed limits (when necessary) in order to slow down traffic that is approaching stopped traffic (note: the practice at this time is that uniform speeds are posted across all lanes; MnDOT has not ruled out the idea of varying speeds in lanes and may opt to implement this in the future).
- To inform drivers of lanes closed.
- To inform drivers of hazards such as standing water or debris on the roadway and encouraging travelers to merge away from the hazardous lanes.

The managed lanes are controlled by traffic operators that have full authority to select from available messages, after events are verified. Figure 4.4 presents the ATM control concept through an example of signing for a crash blocking the right lane of a five lane section. Red "X" and Flashing Yellow Arrow are posted over the lane with the blockage and adjacent lanes (respectively) at the first upstream ILCS. Advanced warning messages are posted on all signs

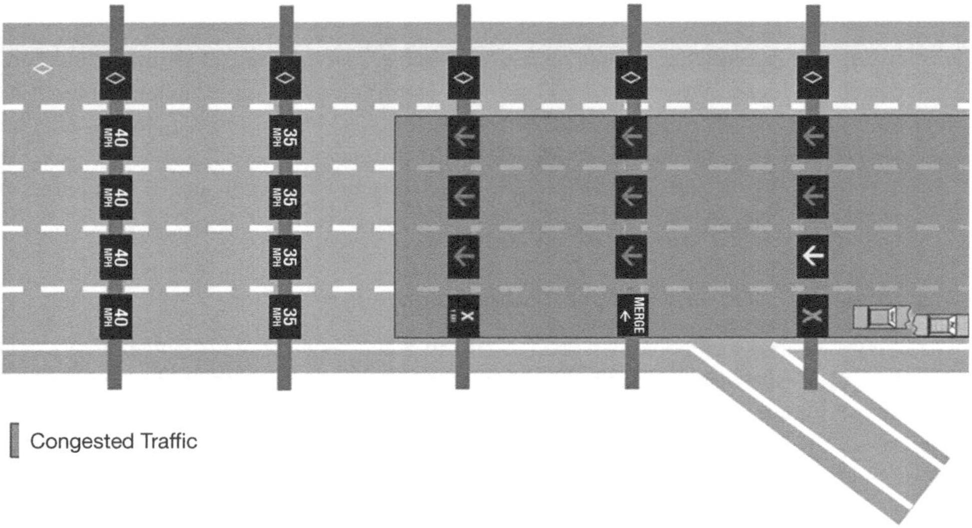

Figure 4.3 Display of intelligent lane control signals during an incident

located upstream of the blockage within a distance of at least 1 mile. As motorists approach a closed lane they encounter an Advanced Yellow "X" sign one mile before the closure. The next sign display is a Merge Left Arrow within a half-mile of the actual Red "X" sign over the lane that is closed. This is to ensure that travelers have advanced warning of the closure before reaching the incident. Further upstream, the signs display automated advisory speeds which change based on real-time traffic conditions. When display signs are posting either a Red "X" or flashing yellow arrow message, all other signs display a green arrow.

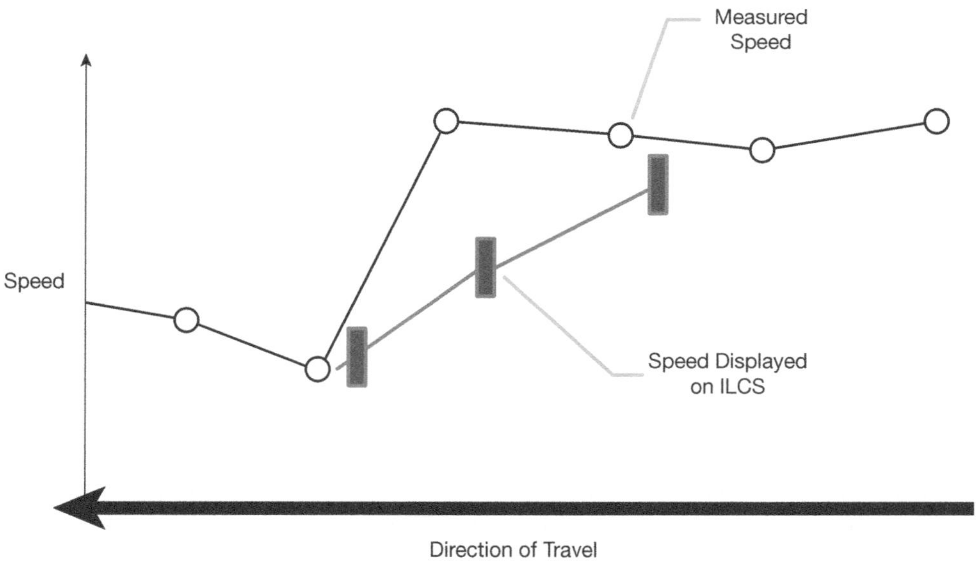

Figure 4.4 Display of advisory speed limits on DMS as it relates to freeway speeds

The selections of advisory Variable Speed Limits (VSL) to be posted are computed by an algorithm developed by MnDOT and the University of Minnesota—Duluth (Kwon et al. 2007 and Kwon et al. 2011). Operators have the option to override the calculated advisory speeds or to accept the recommendation and verify the posting of the message. The goal of the advisory VSL system is to mitigate shockwave propagation from a downstream bottleneck by gradually reducing speed levels of the incoming traffic flow. Figure 4.4 illustrates how speed data is collected through traffic sensors on the roadway at point locations shown as black circles on the chart. Without advisory VSL, vehicles approaching congested traffic are forced to change speeds within a very short distance leading to sudden stopping and possible rear-end collisions. The advisory speed limits, shown by the yellow boxes in the figure, are posted to allow for a more gradual deceleration between upstream free-flowing traffic and congested traffic.

As congestion levels develop, two or three sets of signs prior to the congestion display an advisory speed limit based on the algorithm depending on what the speed differential is between upstream and downstream traffic. Speeds are currently posted up to 11/2 miles upstream of the congestion.

Advisory speeds posted on the overhead signs change by no more than 5 mph with each change in speed, and can be updated every 30 seconds if traffic conditions warrant. The minimum advisory speed displayed is 30 mph and the maximum advisory speed displayed is 50 mph—5 less than the legal speed limit. If the current speeds on the roadway are below 30 mph, the signs go blank.

References

Cambridge Systematics (2001). MnDOT ramp metering study—Final report. Available at: www.dot.state.mn.us/rampmeter/finalreport.html. Accessed May 8, 2014.

Davis, G.A. & Sanderson, K. (2001). Building our way out of congestion? Highway capacity for the twin cities. Retrieved from the University of Minnesota Digital Conservancy. Available at: http://purl.umn.edu/746.

FHWA (2006). *Ramp Management and Control Handbook*. Available at: http://ops.fhwa.dot.gov/publications/ramp_mgmt_handbook/manual/manual/pdf/rm_handbook.pdf. Accessed February 23, 2015.

FHWA (2011a) Managed Lane chapter of the *Freeway Management and Operations Handbook*. Federal Highway Administration. Available at: http://ops.fhwa.dot.gov/freewaymgmt/publications/frwy_mgmt_handbook/revision/jan2011/mgdlaneschp8/sec8.htm. Accessed May 8, 2014.

FHWA (2011b) Integrated Corridor Management, ICM Pioneer Sites–Dallas, TX. Federal Highway Administration. Available at: www.its.dot.gov/icms/pioneer_dallas.htm. Accessed May 8, 2014.

FHWA (2012). Integrated Corridor Management, ICM Pioneer Sites–San Diego, CA. Federal Highway Administration. Available at: www.its.dot.gov/icms/pioneer_sdiego.htm Accessed May 8, 2014.

Highway Capacity Manual (2000). Transportation Research Board. National Research Council, Washington, DC, 113.

Hourdakis, J. & Michalopoulos, P. (2002). Evaluation of ramp control e_ectiveness in two twin cities freeways. Transportation Research Record. *Journal of the Transportation Research Board*, 1811, 21–29.

Jacobs (2013). *Managed Lanes and Ramp Metering Manual*. Prepared by Jacobs Engineering Group Inc. for the Nevada DOT. Available at: www.nevadadot.com/uploadedFiles/NDOT/About_NDOT/NDOT_Divisions/Planning/Safety_Engineering/Introduction%20and%20Policies%20(2).pdf. Accessed May 9, 2014.

Kwon, E., Brannan, D., Shouman, K., Isackson, C., & Arseneau, B. (2007). Development and Field Evaluation of Variable Advisory Speed Limit System for Work Zones. *Transportation Research Record: Journal of the Transportation Research Board*, No. 2015,Transportation Research Board of the National Academies, Washington, DC, pp. 12–18.

Kwon, E., Park, C., Lau, D., & Kary, B. (2011). Minnesota Variable Advisory Speed Limit System, Compendium of Papers, 2011 Annual Transportation Research Board Meeting, Washington, DC, January.

MacCalden, M.S. (1984). A Traffic Management System for the San Francisco–Oakland Bay Bridge, *ITE Journal*, 54(5): 46–51.

Mirshahi, M., Obenberger, J., Fuhs, C. A., Howard, C. E., Krammes, R. A.,. Kuhn, B. T, Mayhew, R. M., Moore, M.A., Sahebjam, K., Stone, C. J., & Yung, J. L. (2007). *Active Traffic Management: The Next Step in Congestion Management*. United States Federal Highway Administration. Office of International Programs. Washington, DC: US Department of Transportation.

MnDOT (2010). Smart lanes offer innovative approach to traffic management. MnDOT newsline. August 11. www.newsline.dot.state.mn.us/archive/10/aug/11.html. Accessed February 23, 2015.

Papageorgiou, M., Hadj-Salem, H., & Blosseville, J. M. (1991). ALINEA: A local feedback control law for on-ramp metering. Transportation Research Record (1320).

Zhang, M., Kim, T., Nie, X., Jin, W., Chu, L., & Recker, W. (2001). Evaluation of on-ramp control algorithms. California PATH Research Report UCB-ITS-PRR-2001–36.

5

ROAD TRAFFIC CONTROL

Montasir Abbas

VIRGINIA POLYTECHNIC INSTITUTE AND STATE UNIVERSITY

Introduction

One of the major issues in signal-timing practice is the lack of general guidelines that link theoretical, practical, and technological issues into one seamless process. Many of the available documents, as described below, either focus on pragmatic issues, or provide a general view of the issues involved. As many practitioners know, there is no comprehensive one document "textbook" in the signal-timing field. In this chapter, we will focus on providing a comprehensive overview of the issues involved in designing optimal control strategies and the underlining control methods. We will start with a description of the traffic state definition and estimation, followed by a description of the fundamental control parameters and methods used in controlling isolated and coordinated systems for normal and special circumstances, and finally, we follow with a description of system operation modes, including fixed time control, traffic responsive, and adaptive control.

Background and Related Literature

The National Operations Coalition "2007 Signal Report Card"[1] assigned an overall grade of "D" to signal infrastructure and operations across the USA. Although this grade is better than the 2005 Report Card grade of D–[2], it is still on the low side. The report pointed out that there are specific problems in signal timing and equipment operations and maintenance. The traffic engineering community recognizes these challenges and strives to improve traffic signal performance. For example, the Transportation Research Board (TRB) traffic signal systems committee highlighted current and future challenges in signal system operation, from a technical and management perspective, along with emerging issues and needs.[3]

Trying to find solutions, the Federal Highway Administration (FHWA) funded several research studies to address these issues, including the funding of the Signal Timing Manual (STM).[4] Most of these funded reports were either broad with general information in their coverage or focused on issues other than signal timing. For example, the FHWA Traffic Control Systems handbook provided general information about traffic control and management techniques.[5] The handbook provided useful information on isolated, arterial, network control concepts, special control at closely spaced intersections, preemptions and priority systems, detectors, controller hardware, and signal maintenance.

The FHWA historical primer also provides essential information on traffic signals as tools for improving safety and efficiency. Simple strategies, reasoning, operation, and maintenance behind traffic signals, along with funding for signal improvements, are included.[6] Cross-jurisdictional signal coordination issues were also recognized in the past and were addressed by another FHWA study. The Cross-Jurisdictional Signal Coordination Case Studies project surveyed several agencies across the country to identify the coordination issues.[7] Case studies were conducted in Pennsylvania, Maryland, Arizona, New York, and Colorado. The project concluded that cross-jurisdictional coordination is possible by establishing cooperation and communication among agencies.

The FHWA informational guide on signalized intersections[8] is another report that includes interesting information on road users, geometric design, traffic control and illumination design, project process, safety and operational analysis, along with a system-wide, intersection-wide, approach, and individual movement's treatments. This report, however, provides a general description of intersection components, without going into much detail about the methodology.

The FHWA also produced a toolbox to reduce red-light running in 2003 with background causation and engineering measures.[9] The report identified measures for preventing intentional red-light running, some of which are related to signal retiming. These measures include signal optimization, modifications to signal-cycle length, yellow-change interval, all-red clearance interval, and dilemma zone protection. The report also lists the advantages and disadvantages of changes in various signal timing parameters.

In addition to studies sponsored by the FHWA, some agencies and Department of Transportations (DOTs) developed their own manuals trying to address signal-timing issues. Some of these developed manuals providing insights related to local issues, while others also developed tools that are similar to commercially available timing tools. For example, the Texas Traffic Signal Operations Handbook[10] includes a useful review of some other state agency signal-timing practices, but it also invested resources in producing a Signal Timing Optimizer that optimizes signal control optimization parameters without going into more challenging issues such as, for example, oversaturation control strategies and traffic responsive control.

The National Cooperative Highway Research Program (NCHRP) also recognized the importance of signal-timing operational issues and funded some studies. NCHRP synthesis 307 provided a summary of practice using a system engineering approach for developing traffic signal system design and operation parameters.[11] NCHRP also funded projects that are applicable to specific issues, such as left-turn operation. NCHRP 493[12] provided information on traffic signal displays for protected/permissive left-turn control. The study took into consideration laboratory and field studies, and focused on both safety and efficiency principles.

More recently, the NCHRP funded projects that address more challenging issues that require state-of-the-art techniques. This includes the NCHRP project 3–90 "Operation of Traffic Signal Systems in Oversaturated Conditions" and 3–97 "Traffic Signal Analysis with Varying Demands and Capacities."

Understanding Traffic States and Optimal Control Strategies

Efficient operation of signalized intersections can be achieved through the optimal design of control strategies, be it in the form of a predefined set of optimal timing plans/schedules that assume repetitive traffic patterns or intelligent/adaptive modes of operation (i.e., adaptive

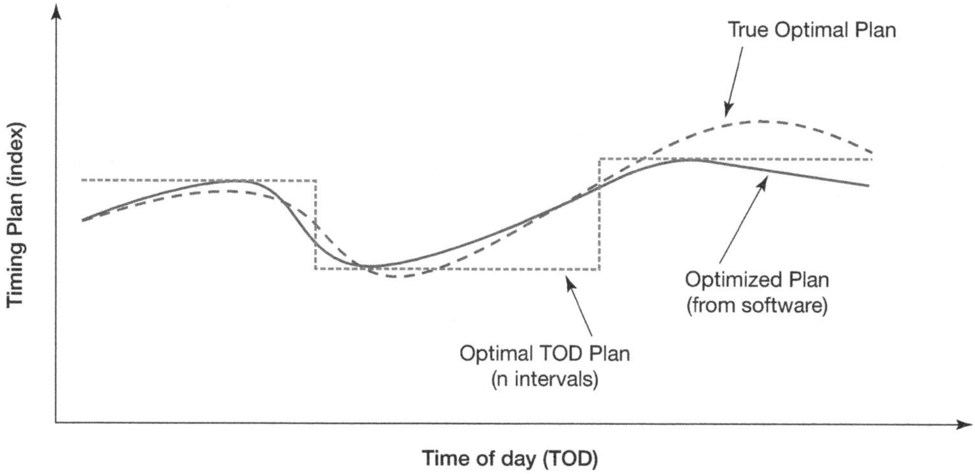

Figure 5.1 Conceptual illustration of optimal timing plans during the day

control, proactive control, traffic responsive control, etc.). In order to develop optimal timing plans, an accurate representation of traffic states is needed first. Once this accurate representation is achieved, an optimization process can be carried over to yield better signal performance in terms of the desired objective (delay, number of stops, or other objectives).

Timing plans for conventional control are typically optimized using commercial software packages such as Synchro, PASSER V or Transyt7-F.[13] These software packages typically produce "optimal" timing plans for a "snapshot" of traffic states in most cases (undersaturated conditions, simple network geometry, relatively condensed through-movement platoons, etc.). However, since the traffic state continuously changes during the day, a deviation from optimality as calculated for the design states is therefore expected. Traffic agencies will therefore need to invest in an adaptive control system that will continuously react (or better yet pro-act) to the changing traffic state. Otherwise, the engineer will have to solve an additional optimization problem to determine "n" number of timing plans and their optimal schedule during the day. Figure 5.1 shows a conceptual illustration of these concepts. The true "optimal/adaptive" and "optimized" timing plans are shown to smoothly change during the day, while an optimal set of "n" timing plans is chosen to operate the signal system over three periods in the figure (for illustration purposes only).

This conceptual understanding raises important issues that need to be recognized by the traffic agency, as summarized by the following questions:

1. What is the actual traffic state at any point in time? What are the important characteristics of the traffic states (i.e., volumes, travel times, critical routes, etc.) and how to obtain an accurate representation of the traffic state using today's technology?
2. How dynamic and variant is the traffic state during any given day? In other words, does the traffic state change in a predictable trend (where a time-of-day operation would be enough), or does it change in an unpredictable trend (where an adaptive mode of operation is needed)?
3. How would an optimal adaptive system work in response to existing and evolving traffic states? What heuristics does an adaptive system perform to be able to reach an "optimal" solution in real time? What are the impacts of such heuristics on system performance?

4. How to evaluate the performance of a given control strategy? What objectives need to be met and how can we map the performance measures to these objectives?

The importance of each one of these questions (and some more detailed related questions) depends on the agency goals, the traffic network, the traffic patterns, and the used control solution. The third and fourth points are better analyzed using detailed field data that log both the traffic state characteristics and the control actions performed by the evaluated system.

Traffic States Characterization

The importance of traffic state estimation accuracy is twofold:

1. The degree of optimality of a control strategy is dependent on the accuracy of the traffic state estimation. This estimation is either done by the traffic engineer when designing a control strategy or done directly by the control system if an adaptive system is used.
2. The understanding of system performance has to be related to the traffic state at the time of evaluation. This is critical to the understanding and prediction of future performance of an evaluated control system.

Traffic state in this context refers to the existing traffic pattern for all approaches at a certain point in time. Traffic state for an arterial network can be represented by a set of major external movements (arterial and cross street movements), a set of a microscopic-level origin-destination (MLOD) matrices (describing OD relationships between network nodes that could be used to extract critical route information, as opposed to planning-level OD matrices that are based on zone data which are not useful in this context), and a set of internal intersection movements as illustrated in Figure 5.2. The external movements need to be measured, possibly by system detectors or tube data, while the internal movements can be either measured or calculated.

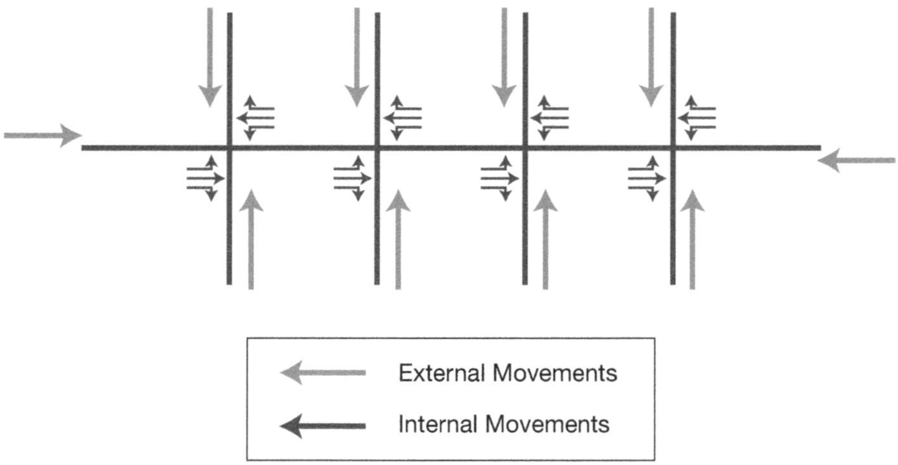

Figure 5.2 Depiction of an arterial traffic state

Traffic State Estimation

Many studies have been conducted in the field of traffic state estimation. The methods used range from multivariate time-series state space modeling of urban traffic flow[14] to accurate and robust freeway travel time prediction.[15] Some advanced algorithms can also be found in the literature for off-line estimation of route-level travel times based on time series of traffic-speed observations.[16] Data screening techniques such as the Kalman Filtering Technique (KFT) for dynamic traffic state estimation from probe data were also developed.[17] Some of these methods were also used in conjunction with real-time and adaptive control systems.[18–20] These methods were also implemented in the field to measure the reliability of estimation using artificial intelligence and Markov Chains (stochastic approaches).[21–25]

Probably one of the most important overlooked applications of traffic state estimation is done by practitioners during the development of timing plans. Section 7.2.5 of the current STM touches on this concept by suggesting to "plot the traffic volume as a function of time of day for the two or three most important intersections." However, as illustrated in Figure 5.1, the combination of significant changes at different intersection approaches might cause significant changes in traffic state, and therefore a need for different timing plans. The engineer's job starts by defining and determining the traffic states that will be used in designing the timing plans. Simple designs might resort to the "peak traffic, off peak traffic" definition of traffic states. However, even in these simplified circumstances, the engineer should consider changes in traffic patterns in all approaches when determining the cut-off for each state (i.e., the engineer should not use only one movement to define the traffic state).

A system-level characterization of traffic states, including a procedure for estimating means and variance should be used to determine the level of control needed to address the system needs. Initial research suggests that a volatile traffic state is an indicator of the need for adaptive traffic control, while low-traffic-state variance indicates the sufficiency of fixed-traffic control. There is not much literature on thresholds or guidelines for selecting the type of control based on this characterization. Some literature merely indicates that control methods require parameters which are largely based on traffic flows and quantify the benefit of using traffic-responsive or traffic-adaptive control.[26]

Basic Traffic Control Concepts

All traffic control systems share some basic control concepts and parameters. The distinguishing aspects between different control systems is the way they optimize these parameters and the additional advanced features that they can implement at a tactical level. In the following sections, we will describe some basic and advanced control parameters that form the basis of traffic control at isolated intersections. We will start with the basic traffic control parameters and describe when and how they can be relaxed in isolated operation and under which conditions.

Basic Traffic Control Parameters

The most basic traffic control parameters are the cycle, splits, and offset. Although these three parameters are technically well defined for coordinated systems, they can occur naturally if the control system was designed to respond to changes in traffic demand.

Figure 5.3 shows the basic phase numbering as defined by the USA National Electrical Manufacturers Association (NEMA). The phase numbers, 1 through 8, are shown in the figure to correspond to the eight main conflicting movements in the intersection. Not all numbers

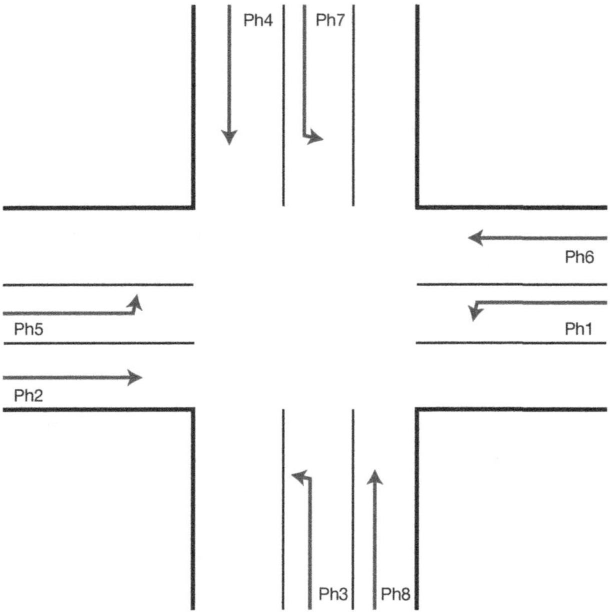

Figure 5.3 Phase numbering as defined by NEMA

have to exist at an intersection, nor do they have to follow the same sequence shown in the figure. However, doing so would avoid confusion and simplify the communication with other engineers and technicians.

Figure 5.4 shows a corresponding Ring Barrier Controller (RBC) configuration for the standard NEMA phase numbers shown above. There are multiple concepts illustrated in the figure that need to be examined closely. First, it should be noted that the figure shows two rings (the two rectangles with a series of green, yellow, and red indications) and two barriers (the two black vertical lines). The smaller rectangles shown with only green and yellow colors are the pedestrian phases. The main idea of the RBC configuration is that phases on different rings on the same side of the barrier can time simultaneously without causing a safety problem. For example, phase 2 on the first ring can time simultaneously with phase 5 or 6 that are shown in the ring below. Also, phase 1 on the upper ring can time simultaneously with either phase 5 or 6 in the lower ring. Another concept that is shown in the figure is that phase 2 can be configured to time before phase 1 (i.e., the left turn phase lags the through phase in this case). Looking at the same left side of the first barrier, phase 102 is shown to time simultaneously with phase 2 (i.e., pedestrians are shown the walk sign while phase 2 is timing). Phase 106 is shown to time simultaneously with phase 6. The pedestrian flashing "don't walk" is shown in the figure as the yellow pedestrian duration that is ending at the end of the all-red clearance interval for phase 2 in the figure.

The general rules for the RBC configuration is that the barriers cannot be broken. In other words, the green indication for the group of phases on the left side of each ring must end together. In the configuration above, phase 2 and phase 6 might end before their predefined time if the controller was configured to end the phase early if there was no demand on the phase (as will be described in more detail shortly). If we look at the phases on the right side of the first barrier, if there is not enough demand on phase 4, it might end earlier than anticipated,

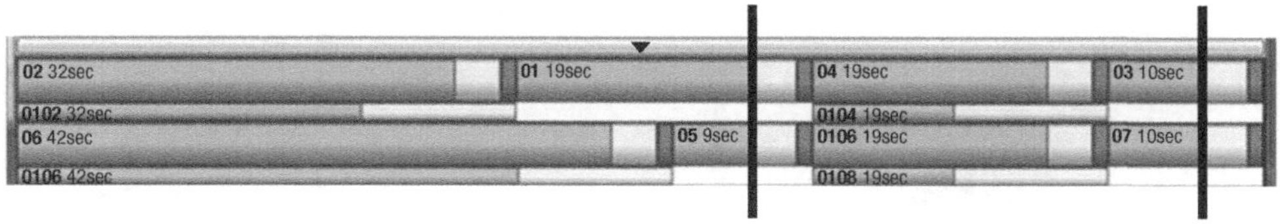

Figure 5.4 Example of a dual ring barrier configuration with standard NEMA phase numbering

resulting in a situation where phase 3 starts earlier and times simultaneously with phase 8. This possibility is what makes it very important to have only compatible phases on other rings at the same side of the barrier (because they might just happen to time with the phases on the other ring).

The RBC configuration determines the three basic control parameters: the duration of green+yellow+all-red assigned to a certain phase is called the phase split. The summation of all phase splits in one ring is the cycle length. The difference in time between the end of green for the coordinated phase and that of the same phase at the master intersection is referred to as the offset. Offsets can also be referenced to other intersections' reference points. In such cases, they are referred to as "relative offsets."

If there is one rule that forms the basic imbedded safety mechanism of the RBC configuration, it must be this: conflicting phases must be configured to be on the same ring, or on the other side of the barrier, guaranteeing that one phase cannot start until the end of the other.

Actuated Phase Operation

In the previous section, we mentioned that a phase can terminate earlier if the controller was configured to utilize that feature. In other words, if the controller was operated in an actuated mode. To be operated in the actuated mode, a phase has to have vehicle detection assigned to it. Vehicle detection is usually used either as stopbar detection, advance detection, or both.

Figure 5.5 shows a schematic of a three-legged intersection with stopbar detection. The stopbar detection is usually used to accommodate queues at the intersection, with the purpose of having the phase "gap-out" when vehicle queues clear the approach. A "gap-out" is a terminology used to mean that a predefined time gap has lapsed before any vehicle could register a call on the detector assigned to a given phase. In such case, the controller determines that it is time to end the phase. A "gap-out" is one of two methods a phase can be terminated in an actuated setting. The other method is called a "max-out," which refers to instances where the allocated time for a given phase is reached. In that case, the controller terminates the phase even if the demand on that phase still exists.

Detection Modes: Concepts and Applications

Optimal configuration of phase detection entails two main parameters: 1) detection length, and 2) passage time/vehicle extension time. The detection length and passage time (i.e., time without detection activation, after which the phase gaps out) work together to clear the queues at an intersection. In theory, a long detector does not need a passage time to clear the queue. However, a long detector is costly. Therefore, passage time is used to compensate for shorter detectors and should be configured correctly to provide an operation that is both adequate and snappy.

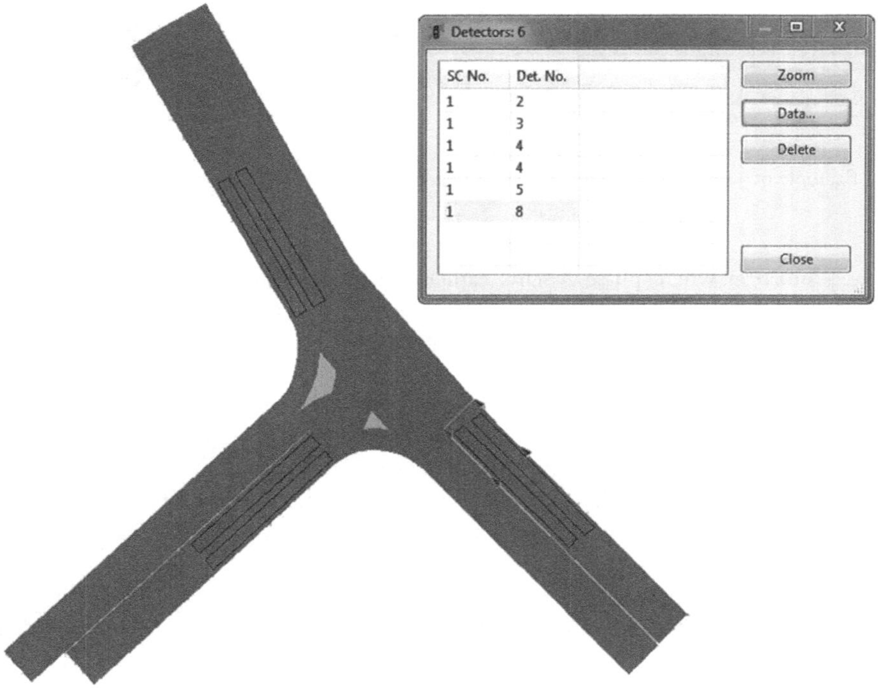

Figure 5.5 Stopbar detection for actuated control

Video Detection Implications

Video detection can be used instead of the standard loop detectors. Several video detection systems are available in the market. The quality of the available systems varies, and care should be taken in selecting and validating a selected system. Once configured, a major point that needs to be mentioned here is that having the flexibility of defining the video detection zones as long as needed negates the need for passage time at the stop bar. However, passage time at advanced detection is still needed to screen out dilemma zone vehicles and minimize the max out probability (which will be discussed in the operation and safety section).

Timing Plan Generation

As mentioned earlier, three basic control parameters are used in traditional coordinated systems. The reason we started with the description of these three parameters is the notion that they can be used as optimization parameters in an optimization setting. That being said, it should also be recognized that the general configuration of the controller requires a set of constraints to go with these parameters. For example, the summation of time in each ring should be equal. The summation of splits in each side of the barrier should be equal, etc.

If an intersection is operated in an isolated mode, then the cycle length becomes no longer constant. Also, the offset is no longer relevant. In such case, the splits are implemented at the controller using max timers.

Arterial Timing Tools

Several reports and papers in the literature provide guidelines on selecting software for use in the arterial signal-timing process.[27,28] The literature does not report any performance consistency of optimized signal timings with different simulation software. The selection of a timing software tends to be related to the agency familiarity with the software and their licensing agreements. Most of the timing software will produce near-optimum timing plans for a given snapshot of the traffic state.

Environmental Aspects of Signal Timing

The literature presents several approaches to signal-timing optimization, including the minimization of fuel consumption, CO_2 and NO_x emissions, as well as the PM10.[29] Some of the literature uses innovative methods to detect traffic congestion using emission data of individual road sections.[30] These methods are based on the fact that exhaust emissions are largely dependent on fleet speed, deceleration speed, and queuing time in idling mode.[31,32]

Another set of literature focuses on affecting driver behavior to reduce emissions.[33] The literature is also rich with papers that evaluate new technology to reduce emissions.[34,35] These comparisons include case studies on traffic-calming measures on both signals and roundabouts.[36–39] There is also a rich area of research that looks into evaluating the impact of alternative traffic control and management strategies on environmental factors using microscopic simulation.[40–42]

Optimizing timing plans based on environmental effect is still conducted at the research level and is not widely implemented in commercial signal timing software.

System Control Objectives and Algorithms

A wide variety of traffic signal control strategies for managing coordinated systems has been developed and proposed by various researchers over the years. These control systems mostly target the minimization of overall system delay and stops, maximization of throughput, and/or a combination of these objectives with other specific objectives. The following is an overview of signal control strategies grouped by application categories.

Queue Length Control

Managing queues in a coordinated system should be treated as a constraint, rather than an objective. As long as queues can be stored within their corresponding section of the road, they would constitute no issues with the optimization process. However, care should be taken that the calculated queues do not exceed the storage length, otherwise the whole operation regime would change, and the underlying delay and stop equations used in performance estimation would not be accurate.

The literature contains examples of research algorithms that utilize queue management. For example, Gordon[43] introduced the concept of storing queues at upstream links to avoid queue spill-back on shorter arterial links. Prosser and Dunne[44] presented a procedure that accounts for queue-blocking effects when determining movement capacities at closely spaced intersections. Messer extended the Prosser-Dunne model to calculate the effective green times and offset between closely spaced intersections. Queue management-control strategies are most relevant in oversaturated regimes and at diamond intersections.

Traffic Metering/Gating

The concept of "metering" refers to purposely impeding traffic input upstream of critical intersections to avoid the increase of demand to critical levels. Metering can be also applied at an area-wide level. The engineer must determine suitable locations on the network to apply metering, if needed. Links with enough storage capacity are the ones at which "gating" will occur. Metering and gating are used to basically "postpone" the arrival of excess demand to the following time periods.

Multi-Objective Design and Analysis

The multi-objective design of timing plans has seen an increase in interest in recent years. The concept refers to the fact that different control strategies can be superior in addressing a certain control objective, but not the other. For example, a timing plan with long cycle length can lead to a minimum number of vehicle stops, but can also lead to longer queues, which in turn can result in an increased delay and spill-backs. Also, a control strategy that aims to maximize throughput might intentionally induce delay on certain links and even on the overall network. A control strategy that is better in terms of minimizing delay cannot be considered better than another strategy that aims to increase throughput. If both control strategies are better in one aspect but not the other, they are said to fall on the Pareto front, or the surface on non-compromised solutions. The shape of the Pareto front itself provides useful information to the analyst, as it shows how much improvement in one objective can result in an inclination in another objective.

Control strategies designed using this concept can only be found at the research level at this time. Unfortunately, no commercial software currently utilizes multi-objective optimization.

Advanced Traffic Operations Concepts

Modern traffic controllers provide several advanced control features in addition to the basic signal control parameters. At the operational level, the engineer can use features such as overlaps to time some phases in parallel with the combination of other phases. For example, a right turn arrow can be displayed in conjunction with its adjacent through and the right side left turn movements.

Advanced Phase Assignments

As mentioned earlier, the standard NEMA phasing number scheme does not have to be strictly followed. There are special circumstances where different phase numbering can simplify certain operations. For example, split phasing (where all movements facing a direction start and end together, followed by all movements facing the opposite directions) can be easily achieved by assigning all movements to one phase on a ring (e.g., phase 1) and all opposite movement to the adjacent phase on the ring (e.g., phase 2). Peer-to-peer communication can also be used to activate certain phases at the downstream intersection based on actuations happening at the upstream intersection, and vice versa. Similar tricks are typically being used by competent engineers and technicians to overcome practical field challenges.

Diamond Interchanges Operation

Diamond intersection operation constitutes a special case from the perspective of traffic signal control. This type of facility usually requires control with special phasing sequences. Several authors represented and modeled control for signalized diamond interchanges. Koonce, Urbanik II, and Bullock used PASSER III along with HILS for optimizing and modeling diamond interchange operations.[45] Later papers also presented special signal control strategies for diamond interchange,[46,47] that used overlap phases with the timing that is guided by travel time between two signals. In the diamond intersection, control progression is usually provided for major movements, and one controller can control two intersections.

Preemption

Emergency vehicle preemption (PE) is an advanced signal-timing technique that, if needed at a particular intersection, and if properly set, can produce significant benefits in the system. However, during the design and implementation phase of PE, different constraints might be encountered. As described by the literature, those constraints might be, for example, high pedestrian volume or high number of turning vehicles.[48]

Beside those issues, previous research recognizes issues with the transition strategies to exit PE control,[49,50] or the implementation of PE on a coordinated arterial.[51] In addition to a normal type of PE, a different approach has to be taken with railroad PE. In this case, the operation effects, such as queue formation, might lead to significant safety issues (i.e., PE trap).[52,53] For the PE trap, some previous research provided the solution of using the gate-down confirmation mechanism.[54] Besides individual research papers, NCHRP synthesis 271—Traffic Signal Operations Near Highway-Rail Grade Crossings provided useful details on signal traffic operation near rail-grade crossings.[55] This synthesis report provided information on warning devices and systems, traffic signals only, and interconnection of signals and warning devices.

Transit Priority

Transit Signal Priority (TSP) is recognized as one of the advanced techniques in signal-timing design that deserves special consideration. As provided in the literature, there are different strategies and technologies available for Transit Signal Priority,[56,57] in addition to the two general types of implementation, Red Truncation and Green Extension. However, as funding for the deployment of TSP is limited, practitioners must be judicious in their choice of deployment sites.[58] Although TSP can be deployed in several options (e.g., centralized TSP, discrete TSP systems based on loop detection, and GPS or Adaptive Transit Signal Priority), practical recommendations need to be developed in accordance with NTCIP 1211: Signal Control and Prioritization (SCP) standard.[59,60] Many agencies, including the FHWA, recognized the importance of TSP, which in turn led to several research efforts. Reports such as Transit Signal Priority (TSP) planning and implementation handbook[61] and the report on the effectiveness of bus signal priority[62] provide extensive information on operational and design guidelines for TSP. Besides research reports, many individual research efforts were directed toward providing guidelines or developing TSP systems. For example, literature provided guidelines on some control strategies for TSP[63] or investigated the impacts of priority on service reliability.[64] On the other hand, some authors tried to develop advanced TSP control strategies by adding an intelligence component into the system.[65,66,67] Finally, literature presents available methods to improve the impacts of TSP through transit vehicle scheduling techniques.[68]

Programmable Logic and Phase Manipulations

Some controllers provide programmable logic features that can prove very useful in certain applications. These features allow the controller inputs and outputs to be based on a set of logical statements. The format of these logical statements is based on a simple structure of "If, Then, Else" conditional and executable statements. These features should be utilized whenever a special case of operation is needed.

System Operation Modes

Once a set of operation strategies has been designed and finalized, the engineer needs to determine the mode of operation of the traffic signal. The two major mode of operations in the NEMA controllers are time-of-day (TOD) and Traffic Responsive Plan Selection (TRPS).

Time-of-Day Operation

In the TOD mode of operation, timing plans are activated based solely on clock time. This mode of operation is obviously an open loop control system that does not depend on feedback received from the field (via system detectors). The mode, as such, is simplistic, and is not responsive to changes in traffic demand. However, most engineers tend to use this mode of operation due to its simplicity of configuration. The major decision the engineer has to make to use this mode, is how many instances of traffic states are needed to design the timing plans for. Once the timing plans are designed, their activation becomes a straightforward process.

Traffic Responsive Control

The TRPS mode of operation has had recent advances and further development in coordinated system operation. Initial research in this area presented in literature from Diakaki, Papageorgiou, and Aboudolas[69] formulated a control problem based on the store-and-forward modeling approach and the linear-quadratic regulator theory. Further research based the control on artificial intelligence algorithms.[70] Additional development from the same authors led to broadening the research to include neural networks and field evaluation across Europe.[71,72] In parallel with this research, several other reports and publications by the author of this chapter also addressed the issue of TRPS, [73–76] but from a more practical perspective. Research presented in these publications used clustering of traffic states, detector threshold mechanisms, and available options in traffic signal controllers, to provide guidelines and perform field implementations for both threshold-based and pattern matching-based TRPS modes.

Adaptive Control

The subject of adaptive control is probably one of the most nonstandardized, and perhaps misleading, in the field of traffic control. The source of confusion stems from the fact that every system vendor reflects a different meaning for the term "adaptive control." Some systems that are in essence different implementations of TRPS can be found in the market under the name of adaptive control systems. In addition, there is no guarantee that a so-called adaptive system would perform better than a TOD or TRPS system due to the lack of detailed documentation of what each system actually does.

Table 5.1 Summary of existing adaptive signal control systems

System	*Architecture*	*Controller*	*Detection*
SCOOT	Centralized	NEMA (EPAC)	Downstream loops
SCATS	Hierarchical (plan based)	2070	Stop bar loops
OPAC	Decentralized	NEMA	Downstream loops
RHODES	Decentralized	2070	Fully actuated design
ATCS	Centralized	2070 with LADOT firmware	Fully actuated with system detectors
TUC	Central System detectors	European	
UTOPIA	Distributed Fully actuated	European	

Perhaps the most agreed definition of adaptive control systems is that they are systems that "adapt" to changes in traffic demand by calculating/adjusting their control parameters in real time. The literature contains a long list of adaptive control systems.[77–86] An overview of the adaptive control strategies is presented in Table 5.1.

Traffic Operation And Safety

High-speed, isolated intersections are generally found on rural arterials. These intersections are usually fully actuated, and are therefore controlled primarily to address safety problems related to right-angle and rear-end crashes. The following section presents a quick summary of different safety measures and issues that need to be taken into account when designing optimal control for these intersections.

Dilemma Zone Issues

At a high-speed intersection, drivers travel fast with an expectation of proceeding through the intersection without stopping. A driver approaching a high speed intersection who is presented with the onset of yellow indication has to decide whether to stop or go. A wrong decision to go can possibly lead to running a red light and potentially being involved in a right-angle crash. A wrong decision to stop suddenly can possibly lead to a severe rear-end crash with the following vehicle. The zone where the risk of making an erroneous decision is high is termed the "dilemma zone".[87] In practice, the total number of vehicles in the dilemma zone has been used as a surrogate measure for safety at rural high-speed intersections. The dilemma zone was initially defined as the area where the driver can neither stop comfortably nor clear safely on the onset of yellow. This approach[88,89] uses deterministic design values such as perception reaction time, comfortable deceleration rate, and length of yellow interval to determine the location of the dilemma zone. The main limitation of this approach is that it assumes a driver has perfect knowledge of all variables. In reality, drivers have a perception of various variables, like distance from the stop bar, yellow duration, etc., but lack perfect knowledge.[90–92] The other limitation of this approach is that there is no conversion metric to convert the presence of the vehicle in its dilemma zone to the magnitude of the risk of crash.

Surrogates for Safety at High-Speed Intersections

Researchers characterized the surrogate measure as the decision dilemma zone to take into account the variability in human perception.[93] The decision dilemma zone is defined as the approach area within which the probability of deciding to stop on the display of yellow is within the range of 10–90 percent. This zone is considered to have a higher risk for rear-end collisions and red light violations. There have been several attempts to ascertain the dilemma zone boundaries.[93–99] Binary discrete choice models were also proposed to determine the probability of stopping for a given distance and speed to better understand the underlying human decision models and explain the variation in the observed dilemma zone boundaries.[93,95,100]

Some recent studies[101–104] also advocate the use of traffic conflict as a surrogate measure for traffic safety in microsimulation packages. Gettman and Head[105] provided a detailed use-case analysis for using traffic conflict as a surrogate measure for safety in a simulation package. The author of this chapter conducted several research projects on dilemma zone definition, quantification, and design as well.[106–110]

Intergreen Durations

The duration of intergreen times are determined using the dilemma zone boundaries. In the deterministic approach, intergreen times are designed so that a "reasonable" driver within the speed limit will have no dilemma. This approach tries to avoid any liability claims involving a red light collision by using a simple kinematic formula and is currently widely practiced because of its simplicity. This approach also is the approach suggested by the ITE *Traffic Engineering Handbook*.[90] The stochastic approach uses the probability of a stopping curve, at the onset of yellow, to limit the number of red light runners below a predetermined threshold. Typically, the threshold is set to 10 percent, and the yellow and all-red intervals are designed so that 90 percent of all vehicles will be able to proceed through the intersection without requiring any acceleration.

Green Durations at a High-speed Intersection

There are two main real-time strategies for determining safe green durations at high-speed intersections: green extension and green termination.

Green extension strategies use an advance detector at the start of the dilemma zone for the high-speed approaches. Minimum green duration is assigned to each signal phase. For high-speed approaches, the green is extended by an interval long enough for the vehicle to cross the dilemma zone if a vehicle is detected on the advance detector. The phase is terminated if no vehicle is detected during the vehicle extension. Engineering judgment is used to determine the value of the maximum green time. High-speed vehicles are provided with complete protection against dilemma zone incursions before the maximum green is reached. However, if the maximum green is reached, no protection is provided.

The green termination strategies, on the other hand, are relatively new and have been deployed at only a few intersections for evaluation purposes. These systems attempt to identify an appropriate time to end the green phase by predicting the value of a performance function for the near future. The objective is to minimize the number of vehicles in their dilemma zone as well as the total delay. Some examples of these strategies are Texas Transportation Institute's Detection-Control System[111] and Sweden's self-optimizing signal control.[112]

Final Remarks

The description of road traffic control concepts presented in this chapter are brief, and are intended to provide an overall understanding of important concepts, techniques, and algorithms currently used in practice. There are some concepts that are not discussed in this chapter, and are left to be discussed in other chapters. The field of Transportation Engineering is very well connected and is rapidly evolving. Some of the concepts discussed here are anticipated to remain for a long time in the future. Other concepts might be deprecated with the widespread use of automated vehicles and large computation powers in the near future.

References

1. *Traffic Signal Operation Self Assessment*. 2007, National Transportation Operations Coalition.
2. *National Traffic Signal Report Card: Executive Summary*. 2007, National Transportation Operations Coalition.
3. Bullock, D. and Urbanik, T., *Addressing Diverse Technologies and Complex User Needs*. 2000, Transportation in the New Millennium.
4. Koonce, P. et al., *Traffic Signal Timing Manual*. 2008, Federal Highway Administration: Washington, DC.
5. *Traffic Control Systems Handbook*. 2005, Dunn Engineering Associates.
6. *Improving Traffic Signal Operations—Primer*. 1995, Federal Highway Administration.
7. *Cross-jurisdictional Signal Coordination Case Studies*. 2002, Federal Highway Administration.
8. Rodegerdts, L. et al., *Signalized Intersections: Informational Guide*. 2004, Federal Highway Administration.
9. *Making Intersections Safer: A Toolbox of Engineering Countermeasures to Reduce Red-Light Running*. 2003, Federal Highway Administration, Institute of Transportation Engineers.
10. Bonneson, J., Pratt, M., and Zimmerman, K., *Development of Traffic Signal Operations Handbook*. 2009, Texas Transportation Institute.
11. Gordon, R., *NCHRP Synthesis 307: Systems Engineering Processes for Developing Traffic Signal Systems*. 2003, Transportation Research Board.
12. Brehmer, C. et al., *NCHRP Report 493: Evaluation of Traffic Signal Displays for Protected/Permissive Left-Turn Control*. 2003, Transportation Research Board.
13. Yin, Y., Robust optimal traffic signal timing. *Transportation Research Part B: Methodological*, 2008. 42(10): 911–924.
14. Stathopoulos, A., A multivariate state space approach for urban traffic flow modeling and prediction. *Transportation Research Part C: Emerging Technologies*, 2003. 11(2): 121–135.
15. Vanlint, J., Hoogendoorn, S., and Vanzuylen, H., Accurate freeway travel time prediction with state-space neural networks under missing data. *Transportation Research Part C: Emerging Technologies*, 2005. 13(5–6): 347–369.
16. van Lint, J. and van der Zijpp, N., Improving a travel-time estimation algorithm by using dual loop detectors, *Transportation Research Record*, 2003. 1855: 41–48.
17. Chumchoke Nanthawichit, T.N. and Suzuki, Hironori, Application of probe-vehicle data for real-time traffic-state estimation and short-term travel-time prediction on a freeway. *Transportation Research Record*, 2003. 1855: 49–60.
18. Wang, Y. and Papageorgiou, M., Real-time freeway traffic state estimation based on extended Kalman filter: a general approach. *Transportation Research Part B: Methodological*, 2005. 39(2): 141–167.
19. Wang, Y., et al., An adaptive freeway traffic state estimator. *Automatica*, 2009. 45(1): 10–24.
20. van Lint, J.W.C., Hooqendoorn, S.P., and van Zuylen, H.J., Freeway travel time prediction with state-space neural networks modeling state-space dynamics with recurrent neural networks. *Transportation Research Record*, 2002. 1811: 30–39.
21. van Lint, J.W.C., Reliable real-time framework for short-term freeway travel time prediction, *Journal of Transportation Engineering* © ASCE, 2006.
22. Ruimin Li, G.R. and Sarvi, Majid, Evaluation of speed-based travel time estimation models, *Journal of Transportation Engineering* © ASCE, 2006.

23. Ni, D. and Wang, H., Trajectory reconstruction for travel time estimation. *Journal of Intelligent Transportation Systems*, 2008. 12(3): 113–125.
24. Yeon, J., Elefteriadou, L., and Lawphongpanich, S., Travel time estimation on a freeway using discrete time Markov chains. *Transportation Research Part B: Methodological*, 2008. 42(4): 325–338.
25. van Lint, J.W.C., Empirical evaluation of new robust travel time estimation algorithms. *Transportation Research Board*, 2010. 2160: 50–59.
26. Mazzamatti, M. et al., Benefits gained by responsive and traffic adaptive systems in Sao Paulo. In *Road Transport Information and Control*, 1998. 9th International Conference on (Conference Publication No. 454).
27. Chaudhary, N., Kovvali, V., and Alam, M., *Guidelines for Selecting Signal Timing Software*. 2002, Texas Transportation Institute.
28. Stevanovic, A. and Martin, P., Assessment of the suitability of microsimulation as a tool for the evaluation of macroscopically optimized traffic signal timings. *ASCE Journal of Transportation Engineering*, 2008. 134(2) 2.
29. Stevanovic, A., Stevanovic, J., Zhang, K., and Batterman,S., Optimizing traffic control to reduce fuel consumption and vehicular emissions. *Transportation Research Record*, 2009. 2128: 105–113.
30. Kühlwein, J. and Friedrich, R., Traffic measurements and high-performance modelling of motorway emission rates. *Atmospheric Environment*, 2005. 39(31): 5722–5736.
31. Pandian, S., Gokhale, S., and Ghoshal, A.K., Evaluating effects of traffic and vehicle characteristics on vehicular emissions near traffic intersections. *Transportation Research Part D: Transport and Environment*, 2009. 14(3): 180–196.
32. Jayaratne, E.R. et al., Increase in particle number emissions from motor vehicles due to interruption of steady traffic flow. *Transportation Research Part D: Transport and Environment*, 2009. 14(7): 521–526.
33. Coelho, M.C., Farias, T.L., and Rouphail, N.M., Impact of speed control traffic signals on pollutant emissions. *Transportation Research Part D: Transport and Environment*, 2005. 10(4): 323–340.
34. Umbreras, J. et al., Assessment of vehicle emissions projections in Madrid (Spain) from 2004 to 2012 considering several control strategies. *Transportation Research Part A: Policy and Practice*, 2008. 42(4): 646–658.
35. Krawack, S., Traffic management and emissions. *The Science of the Total Environment*, 1993. 134(1–3): 305–314.
36. Höglund, P.G., Alternative intersection design – a possible way of reducing air pollutant emissions from road and street traffic? *Science of The Total Environment*, 1994. 146–147: 35–44.
37. Gram, F., Time variations in traffic and traffic emissions. *Science of The Total Environment*, 1996. 189–190: 115–118.
38. Boulter, P.G., Latham, S., and Ainge, M., Driving cycles for measuring passenger car emissions on roads with traffic calming measures. *The Science of the Total Environment*, 1999. 235(1–3): 77–89.
39. Zhu, Z., Li, T., and Li, W., Optimal total vehicle pollutants emission quantity based on link traffic capacity constraints. *Journal of Transportation Systems Engineering and Information Technology*, 2008. 8(2): 80–84.
40. Chen, K. and Yu, L., Microscopic traffic-emission simulation and case study for evaluation of traffic control strategies. *Journal of Transportation Systems Engineering and Information Technology*, 2007. 7(1): 93–99.
41. Oda, T., Signal control for reducing vehicle NOx and CO_2 emissions based on prediction of arrival traffic flows at intersections. *IEEJ Transactions on Industry Applications*, 2008. 128: 1418–1425.
42. Smit, R., Brown, A.L., and Chan, Y.C., Do air pollution emissions and fuel consumption models for roadways include the effects of congestion in the roadway traffic flow? *Environmental Modelling & Software*, 23(10–11): 1262–1270.
43. Gordon, R.L., Technique for control of traffic at critical intersections. *Transportation Science*, 1969. 3(4): 279–288.
44. Prosser, N. and Dunne, M., A procedure for estimating movement capacities at signalized paired intersections. In *2nd International Symposium on Highway Capacity*, 1994. Sydney, Australia.
45. Koonce, P., Urbanik II, T., and Bullock, D., Evaluation of diamond interchange signal controller settings by using hardware-in-the-loop simulation. *Transportation Research Record*, 1999. 1683: 59–66.
46. Tian, Z., Urbanik II, T., and Gibby, R., Application of diamond interchange control strategies at closely spaced intersections. *Transportation Research Record*, 2007. 2035: 32–39.
47. Kovvali, V.G. et al., Program for optimizing diamond interchanges in oversaturated conditions. *Transportation Research Record*, 2002. 181: 166–178.

48. Hanbali, R.M. and Fornal, C.J., Emergency vehicle preemption signal timing: constraints and solutions. *ITE Journal* (Institute of Transportation Engineers), 2004. 74(7): 32–40.
49. Obenberger, J. and Collura, J., Transition strategies to exit preemption control: state-of-the-practice assessment. *Transportation Research Record: Journal of the Transportation Research Board,* 2001. 1748: 72–79.
50. Obenberger, J. and Collura, J., Methodology to assess traffic signal transition strategies for exit preemption control. *Transportation Research Record*, 2007. 2035: 158–168.
51. Nelson, E. and Bullock, D., Impact of emergency vehicle preemption on signalized corridor operation: an evaluation. *Transportation Research Record*, 2000. 1727: 1–11.
52. Brennan Jr, T.M. et al., Performance measures for railroad-preempted intersections. *Transportation Research Record*, 2009. 2128: 20–34.
53. Marshall, P.S. and Berg, W.D., Design guidelines for railroad preemption at signalized intersections. *ITE Journal* (Institute of Transportation Engineers), 1997. 67 (2): 20–25.
54. Yohe, J.R. and Urbanik II, T., Advance preempt with gate-down confirmation solution for preempt trap. *Transportation Research Record*, 2007. 2035: 40–49.
55. Korve, H., Synthesis of Highway Practice 271—traffic signal operations near highway-rail grade crossings. 1999, *Transportation Research Board.*
56. Meng, L. et al., Field operation tests of adaptive transit signal priority systems, in *PATH Research Report #1055–1425*, 2010.
57. Martin, P. and Zlatkovic, M., Evaluation of transit signal priority strategies for bus rapid transit on 5600 West Street in Salt Lake County, Utah, in *Mountain Plains Consortium Report #09–213A*, 2010.
58. Hedden, C.K., *Transit Signal Priority: System Applications and Technology Investigation.* 2009. New Jersey Department of Transportation.
59. Koonce, P., Urbanik, T., and Mishra, A., Application of NTCIP Standard 1211 framework for upgrading downtown Baltimore light rail transit signal priority, in Experience, Economics, and Evolution–From Starter Lines to Growing Systems. 9th National Light Rail Transit Conference, *Transportation Research E-Circular*, 2003: Portland, OR.
60. Yue Li, Peter Koonce, Meng Li, Kun Zhou, Yuwei Li, Scott Beaird, Wei-Bin Zhang, Larry Hegen, Kang Hu, Alex Skabardonis, and Z. Sonja Sun, *Transit Signal Priority Research Tools.* May 2008, US Department of Transportation, Washington, DC, 20590.
61. Smith, H.R., Hemily, B., and Ivanovic, M., *Transit Signal Priority (TSP): A Planning and Implementation Handbook.* 2005, ITS America.
62. Chada, S. and Robert, N., *Effectiveness of Bus Signal Priority Final Report*, January 2002.
63. Skabardonis, A., Control strategies for transit priority. *Transportation Research Record*, 2000. 1727: 20–26.
64. Chang, J. et al., Evaluation of service reliability impacts of traffic signal priority strategies for bus transit. *Transportation Research Record*, 2003. 1841: 23–31.
65. Balke, K.N., Dudek, C.L., and Urbanik II, T., Development and evaluation of intelligent bus priority concept. *Transportation Research Record*, 2000. 1727: 12–19.
66. Oliveira-Neto, F.M., Loureiro, C.F.G., and Han, L.D., Active and passive bus priority strategies in mixed traffic arterials controlled by SCOOT adaptive signal system: assessment of performance in Fortaleza, Brazil. *Transportation Research Record*, 2009. 2128: 58–65.
67. Ekeila, W., Sayed, T., and El Esawey, M., Development of dynamic transit signal priority strategy. *Transportation Research Record*, 2009. 2111: 1–9.
68. Altun, S.Z. and Furth, P., Scheduling buses to take advantage of transit signal priority. *Transportation Research Record*, 2009. 2111: 50–59.
69. Diakaki, C., Papageorgiou, M., and Aboudolas, K., A multivariable regulator approach to traffic-responsive network-wide signal control. *Control Engineering Practice*, 2002. 10(2): 183–195.
70. Spall, J.C. and Chin, D.C., Traffic-responsive signal timing for system-wide traffic control. *Transportation Research Part C: Emerging Technologies*. 5(3–4): 153–163.
71. Azzam ul, A. et al., Traffic responsive signal timing plan generation based on neural network. In *Automation Science and Engineering, 2008.* CASE 2008. IEEE International Conference.
72. Kosmatopoulos, E. et al., International comparative field evaluation of a traffic-responsive signal control strategy in three cities. *Transportation Research Part A: Policy and Practice*, 2006. 40(5): 399–413.
73. Abdelaziz, S. and Abbas, M.M., Evaluation of pattern-matching traffic responsive control mode in a large arterial network. In *Transportation Research Board*, 88th Annual Meeting 2009. Washington, DC.
74. Abbas, M.M., Chaudhary, N., Pesti, G., and Sharma, A., Configuration methodology for traffic-responsive plan selection: a global perspective. *Transportation Research Record*, 2005. 1925: 195–204.

75. Abbas, M.M. and Sharma, A., Multiobjective plan selection optimization for traffic responsive control. *ASCE Journal of Transportation Engineering*, 2006. 132(5): 376–384.
76. Abbas, M.M. et al., Illustrative field configuration and evaluation of traffic-responsive control. *Journal of Transportation Engineering*, 2009. 135(9): 591–599.
77. Gaon, B.N. Operational control requirements of future ATC systems. In *Proceedings of the 1972 IEEE Conference on Decision Control and 11th Symposium on Adaptive Processes*, December 13–15. New York: IEEE.
78. Head, K.L., Mirchandani, P.B., and Shepherd, D., A hierarchical framework for real-time traffic control. *Transportation Research Record*, 1992. 1360: 82–88.
79. Head, L., An event-based short-term traffic flow prediction model. *Transportation Research Record*, 1995. 1510: 45–52.
80. Hu, K., LADOT's adaptive traffic control system. In *Workshop on Adaptive Signal Control Systems*. 2000.
81. Khoudour, L., Lesort, J.B., and Farges, J.L., *PRODYN—Three years of trials in the ZELT Experimental Zone*. Recherche—Transports—Securite, Special Traffic Management, 1991. 6 (English issue): 89–98.
82. Bretherton, R.D. and Bowen, G.T., Recent enhancements to SCOOT. SCOOT Version 2.4. In *Third International Conference on Road Traffic Control*, May 1, 1990–May 3, 1990. London: IEE.
83. Lowrie, P.R., SCATS: The Sydney Co-ordinated Adaptive Traffic System, in *IEE International Conference on Road Traffic Signaling*, 1982: London.
84. Gartner, N.H. and OPAC, A demand responsive strategy for traffic signal control. *Transportation Research Record*, 1983. 906: 75–81.
85. Powell, R.J., SCOOT in Southampton. Paper presented at the *PTRC Annual Summer Meeting* at Seminar M, 1985. p. 269, pp. 97–110.
86. Adaptive traffic control systems: domestic and foreign state of practice. *NCHRP Synthesis 403*. 2010, United States. p. 114.
87. Parsonson, P., Signalization of high speed, isolated intersections. *Transportation Research Record*, 1978. 681: 34–42.
88. Konecni, V., Konecni, D., and Ebbesen, E., Decision processes and risk taking in traffic: driver response to the onset of yellow light. *Journal of Applied Psychology*, 1976. 61: 359–367.
89. May, A.D., Clearance interval at flashing systems. *Highway Research Record*, 1968. 221: 41–71.
90. Engineers, I.o.T., *Traffic Engineering Handbook*, 5th ed., 1999. Institute of Transportation Engineers: Washington, DC.
91. Shanteau, R., Considerations in the length of the yellow interval. In *69th Annual Road School*, 1983. Purdue University, West Lafayette, IN.
92. Wortman, R. and Matthias, J., Evaluation of driver behavior at signalized intersections. *Transportation Research Record*, 1983. 904: 10–20.
93. Sheffi, Y. and Mahmassani, M., A model of driver behavior at high speed signalized intersections. *Transportation Science*, 1981. 15(1): 51–61.
94. Bonneson, J., McCoy, P., and Moen, B., Traffic detector design and evaluation guidelines, in *Report No. TRP-02-31-93*, 1994, Nebraska Department of Roads: Lincoln, Nebraska.
95. Chang, M., Messer, C., and Santiago, A., Timing traffic signal change intervals based on driver behavior. *Transportation Research Record*, 1985. 1027: 20–30.
96. ITE technical committee, 18, P. P., Chairman, Small-area detection at intersection approaches. *Traffic Engineering*, 1974: 8–17.
97. Olson, P. and Rothery, R., Driver response to amber phase of traffic signals. *Highway Research Board*, 1962. *Bulletin*, 330: 40–51.
98. Webster, F. and Ellson, P., Traffic signals for high speed roads. *Road Research Laboratory*, 1965.
99. Zegeer, C., Effectiveness of green-extension systems at high-speed intersections. In *Research Report*, 1977. Bureau of Highways, Kentucky Department of Transportation: Lexington, KT.
100. Gates, T., Noyce, D., and Laracuente, L., Analysis of dilemma zone driver behavior at signalized intersections. *Transportation Research Record*, 2007. 2030: 29–39.
101. Chin, H., Quek, S., and Cheu, R., Quantitative examination of traffic conflicts. *Transportation Research Record*, 1992. 1376: 86–91.
102. Chin, H. and Quek, S., Measurement of traffic conflicts. *Safety Science*, 1997. 26(3): 169–185.
103. Fazio, J., Holden, J., and Rouphail, N., Use of freeway conflict rates as an alternative to crash rates in weaving section safety analysis. *Transportation Research Record*, 1993. 1401: 61–69.

104. Horst, V., *A time-based analysis of road user behavior at intersections.* In *Third Workshop of International Cooperation on Theories and Concepts in Traffic Safety*, 1990. Cracow, Poland.
105. Gettman, D. and Head, L., Surrogate safety measures from traffic simulation models. In *FHWA-RD-03-050*, 2003. FHWA, US Department of Transportation: Washington, DC.
106. Machiani, S.G. and Abbas, M., Assessment of driver stopping prediction models before and after the onset of yellow using two driving simulator datasets. *Accident Analysis & Prevention*, 2015.
107. Adam, Z., Abbas, M., and Li, P., Evaluating green-extension policies with reinforcement learning and Markovian traffic state estimation. *Transportation Research Record*, 2009. 2128: 217-225.
108. Li, P. and Abbas, M., Stochastic dilemma hazard model at high-speed signalized intersections. *Journal of Transportation Engineering*, 2009. 136: 448–456.
109. Machiani, S.G. and Abbas, M., Safety surrogate histograms (SSH): A novel real-time safety assessment of dilemma zone related conflicts at signalized intersections. *Accident Analysis & Prevention*, 2015.
110. Li, P., Abbas, M., and Pasupathy, R., A stochastic dilemma zone protection algorithm based on the vehicles' trajectories. *Journal of Intelligent Transportation Systems*, 2015. 19:181–191.
111. Bonneson, J., Middleton, D., Zimmerman, K., Charara, H., and Abbas, M., Intelligent detection-control system for rural signalized intersections, in *Report No. FHWA/TX-03/4022-2*, 2002. Texas Department of Transportation: Austin, TX.
112. Kronborg, P., Davidsson, F., and Edholm, J., SOS—Self Optimising Signal control development and field trials of the SOS algorithm for self optimising signal control at isolated intersections, in *TFK Report 1997:2E*, 1997. TFK–Transport Research Institute: Stockholm, Sweden.

6

EVOLUTION OF INTELLIGENT TRANSPORTATION SYSTEMS

Praveen Edara[1] *and Jalil Kianfar*[2]

(1) UNIVERSISTY OF MISSOURI, (2) SAINT LOUIS UNIVERSITY

Intelligent Transportation Systems (ITS) are an important aspect of today's transportation system. From traveler information to advanced traffic-control systems, ITS plays a role in alleviating congestion, mitigating crashes, and reducing pollution. This chapter provides an overview of the different ITS technologies that have made an impact on transportation. Commencing with the formative years of ITS in the 1970s, the chapter discusses the development of the National ITS architecture in the US, Connected Vehicle technology, and other state-of-the-art ITS technologies such as crowd-sourced route guidance applications and autonomous vehicles. The strides made by ITS in the last four decades offer enough evidence that the ITS will continue to evolve and assist transportation agencies and users in realizing a congestion-free and accident-free traffic environment.

Technology in Transportation

Technology has long played an important role in transportation. Different transportation modes, be it the automobile, the locomotive, or others have used many technological innovations in the past. Construction technology has helped build transportation infrastructure such as roads, bridges, and intermodal facilities. The management of traffic operations including monitoring, control, and traveler information provision has witnessed increased use of technology since the 1970s. The advent of intelligent transportation systems (ITS) has laid a solid foundation for the future applications of technology to manage transportation facilities. In this chapter, the evolution of ITS since its inception in the 1970s to the current day is examined. The impact of ITS on travel has been significant. From traveler information to advanced traffic control systems, ITS has transformed the way we communicate with travelers, manage traffic congestion, improve traffic safety, and reduce the impact of travel on our environment. ITS technology has permeated into several areas of transportation. These areas include traffic operations, freight and commercial vehicle operations, public transit operations, traveler information, traffic safety applications, tolling and congestion pricing, driver assistance systems, connected vehicles, and autonomous vehicles. Due to the space limitations, it is impossible to discuss all ITS application areas in this chapter. Instead, we focus on providing coverage of three areas of ITS: 1) the formative years of ITS, 2) the establishment of a standard ITS architecture, and 3) the potentially transformative ITS applications of the modern day and the near future. Other topics, though important, are out of scope of this chapter and the reader is referred to other sources that cover them (Chowdhury and Sadek, 2003; Sussman, 2005; Gordon, 2009; US DOT, 2014).

The Initial Years

Existing literature on ITS reports the early research and development from the 1970s in the US, Japan, and Europe. Japan was one of the earliest in the world to deploy ITS technologies. Whelan (1995) reported that an in-vehicle dynamic route guidance system was tested in Tokyo in 1977. It was not until the late 1980s when a slew of ITS activities happened in these countries. Whelan

Table 6.1 Early ITS developments in Japan

ITS initiative	*Year*	*Brief description*
Comprehensive Automobile Traffic Control System (CACS)	1973	Develop and test an in-vehicle dynamic route guidance
Road/Automobile Communication System (RACS)	1980s	Research, development, and operations for near-term deployment of ITS
Advanced Mobile Traffic Information and Communication System (AMTICS)	1980s	Research, development, and operations for near-term deployment of ITS
Advanced Road Transportation System (ARTS)	1989	Apply RACS technology to develop next generation traffic systems
Vehicle Information and Communication System (VICS)	1990	Achieve intelligent coordination between cars and road infrastructure
Super Smart Vehicle System (SSVS)	1990	Explore concepts and develop technology for fully automatic driving 20 to 30 years into the future
Vehicle, Road, and Traffic Intelligence Society (VERTIS)	1994	Government supported public-private partnership organization was formed in the ITS area

Table 6.2 Early ITS developments in Europe

ITS initiative	*Year*	*Brief description*
Prototype dynamic route-guidance system	1980s	Demonstration of route-guidance system in Germany and Britain
Program for a European Traffic with Highest Efficiency and Unprecedented Safety (PROMETHEUS)	1986	Achieve intelligent vehicle-highway systems through an eight-year program with emphasis on vehicle systems
Dedicated Road Infrastructure for Vehicle Safety in Europe (DRIVE)	1980s	Advanced transport telematics (ATT) through road infrastructure projects (smart highways)
System of Cellular Radio for Traffic Efficiency and Safety (SOCRATES)	1989	Delivering ATT services via cellular radio communications
European Road Transport Telematics Implementation Coordination Organization (ERTICO)	1990s	Public-private organization to coordinate ongoing ATT projects and support implementation in Europe

Table 6.3 Early ITS developments in the USA

ITS initiative	*Year*	*Brief description*
Smart Corridor Project	1980s	Demonstration of computer and communication technologies on a freeway corridor
Pathfinder Experiment	1980s	Measure driver response to real-time in-vehicle traveler information system
Mobility 2000 Group	1987	Promote the field of Intelligent Vehicle Highway System (IVHS)
IVHS America	1990	Federal advisory committee to US DOT for the emerging national IVHS program
IVHS Act	1991	Authorized federal funding to support IVHS programs

(1995) provides a fascinating history of the ITS in Japan, Europe, and the US through these years, which is summarized in Tables 6.1, 6.2, and 6.3. In the tables, the first column lists the title of the ITS initiative, the second column lists the year, and the third column provides a brief description of the initiative undertaken. Traveler information provision and route guidance through the use of in-vehicle technology were the initial ITS initiatives that were deployed. The public–private partnership organizations for advancing ITS in each region were established in the early 1990s. These organizations are: VERTIS in Japan, ERTICO in Europe, and IVHS America in the US.

Development of ITS Architecture

A significant initiative of the US DOT, ITS America, and other public and private stakeholders was the development of a national ITS architecture that would serve as a framework for the design of ITS in the US. Such a common framework will allow for consistency in the ITS designed by different state transportation agencies across the country. The architecture presents a structure for the functions performed by different subsystems, the physical location where the functions exist, the interactions between subsystems, and the communication requirements for these interactions. The development of ITS standards for interaction between subsystems and the communication requirements is also facilitated by the national ITS architecture. The European Union, Canada, and South Korea have also established ITS architectures for their regions.

Figure 6.1 shows the latest national ITS architecture used by the US Department of Transportation (DOT). The national ITS architecture consists of three layers: the institutional layer, the transportation layer, and the communications layer. The institutional layer consists of the stakeholder institutions, policies, and funding options necessary for the deployment and maintenance of ITS. The transportation layer defines the various transportation services in the form of subsystems and their functions, whereas the communications layer defines the communication services necessary to support these transportation services.

Figure 6.2 presents the 22 transportation subsystems, shown as white rectangles, grouped into four classes—Travelers, Centers, Vehicles, and Field. For example, the traffic-management center subsystem functions include traffic monitoring, traffic control such as ramp metering, dynamic message signs, incident management, etc. The equipment needed to perform these functions

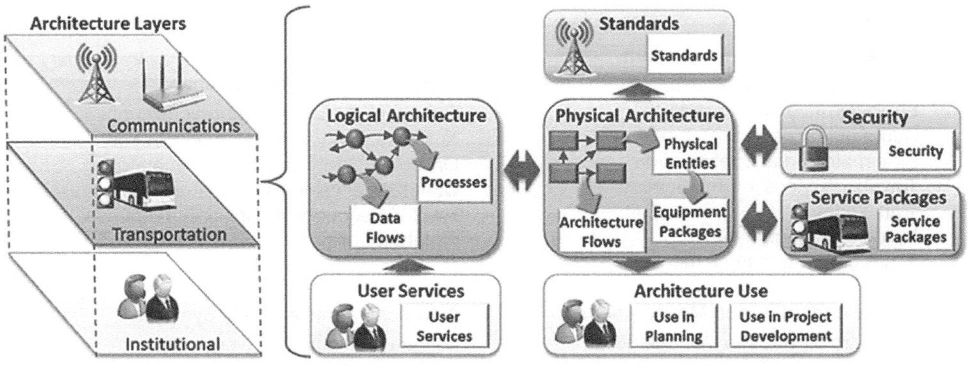

Figure 6.1 National ITS architecture (from National ITS Architecture 7.0, www.iteris.com/itsarch/html/menu/hypertext.htm)

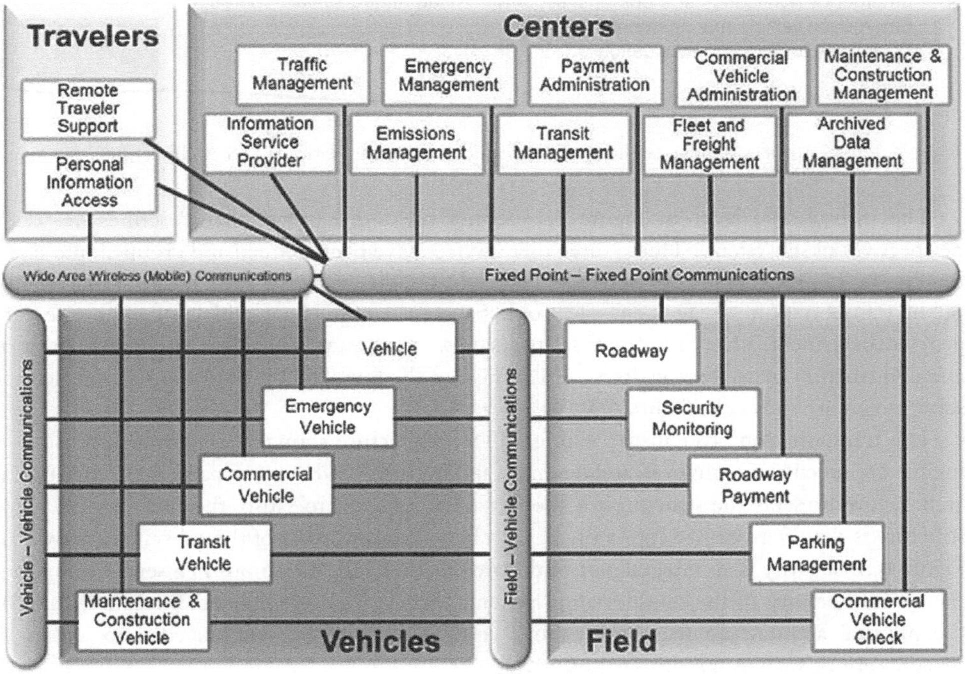

Figure 6.2 Subsystems and communications in the National ITS architecture (from National ITS Architecture 7.0, www.iteris.com/itsarch/html/entity/paents.htm)

such as computer hardware, video displays, and traffic-control equipment are located inside a traffic-management center. The roadway subsystem, on the other hand, would use in-field equipment such as traffic signal controllers, loop detectors, video surveillance, etc. The communication between the traffic-management "centers" subsystem and the roadway "field" subsystem is made via fixed-point communications such as fiber optics cables. Three other types of communication are shown in Figure 6.2—vehicle–vehicle communications, wireless communications, and field–vehicle communications.

Travel and Traffic Management
Pre-trip traveler information
En-route driver information
Route guidance
Ride matching and reservation
Traveler services information
Traffic control
Incident management
Travel demand management

Public Transportation Management
Public transportation management
En-route transit information
Personalized public transport
Public travel security

Emergency Management
Emergency notification and personal security
Emergency vehicle management
Disaster response and evacuation

Commercial Vehicle Operations
Commercial vehicle electronic clearance
Automated roadside safety inspection
On-board safety and security monitoring
Commercial vehicle administrative processes
HAZMAT security and incident response
Freight mobility

Advanced Vehicle Safety Systems
Longitudinal collision avoidance
Lateral collision avoidance
Intersection collision avoidance
Vision enhancement for crash avoidance
Safety readiness
Pre-crash restraint deployment
Automated vehicle operation

Figure 6.3 User services bundles with more than one service from National ITS Architecture 7.0

The institutional layer in Figure 6.1 defines the user services that describe the user's perspectives of subsystems. Thirty-three user services have been defined and grouped into eight bundles by US DOT and ITS America. The eight bundles are: travel and traffic management, public transportation management, electronic payment, commercial vehicle operations, emergency management, advanced vehicle safety systems, information management, and maintenance and construction management (Iteris, 2013). Five of the bundles consist of more than one user services and are shown in Figure 6.3.

The transportation layer in the national ITS architecture shown in Figure 6.1 consists of logical architecture, a physical architecture, and security and service packages. The logical architecture presents the structure for the functions required to satisfy the user services. This structure is then transformed into a physical architecture consisting of the subsystems shown in Figure 6.2. Security is an integral part of the architecture and the various ITS services improve the overall security of the transportation system. Service packages collect all relevant parts of the physical architecture (e.g., subsystems, equipment packages, etc.) needed to address a specific service such as transit signal priority.

Connected Vehicles

The safety, convenience, and efficiency of vehicles of the twenty-first century are incomparable to the first automobile invented by Karl Benz in 1886. Despite the drastic improvements in vehicle performance, the human element in driving still remains mostly the same. Humans are still in charge of controlling the automobile. While researchers have pursued the idea of self-driving cars and intelligent highways since the 1950s (*Science Digest*, 1958), the idea of making them available for public use in the not too distant future was not explored until the last decade. Connected vehicle technology is aimed at improving safety and efficiency, and reducing the environmental impacts of a transportation system.

Connected vehicle technology provides a driving environment where each vehicle is fully connected to the other vehicles and to the infrastructure in its surrounding area. The technology is aimed at improving safety and efficiency of a transportation system, and provides applications that can improve the environment. The development of this technology is being pursued collaboratively by public agencies, automobile manufacturers, communication industry, and application developers.

Connected vehicle efforts in the US are occurring in parallel to similar efforts by Europe's Cooperative Intelligent Transport Systems (C-ITS) (Burke and Williams, 2012) and Japan's SmartWay (Makino, 2006). The various applications provided by the connected vehicle technology are fashioned using a series of devices, systems, and technologies that collect, exchange, and process data. These include (Kianfar and Edara, 2013):

1. *On-Board Equipment (OBE)*: The OBE collects and transmits vehicle status data to other vehicles and nearby infrastructure. The data is collected according to established standards and protocols (e.g. SAE J2735—Chen et al., 2014) and includes data from vehicle sensors such as vehicle speed, location, brake status, and airbag activation. OBE also receives information from other vehicles and infrastructure. Vehicles could be equipped with OBE by automakers or through after-market devices.
2. *Roadside Equipment (RSE)*: This equipment is installed at roadway intersections, highway interchanges and other infrastructure locations to support data exchange between vehicles' OBE and infrastructure. RSE provides connection between vehicles and roadside devices such as signal controllers, and between vehicles and remote applications such as traveler information systems. When a vehicle enters a RSE coverage area, the information stored in its OBE is transmitted to the RSE, and the vehicle may also receive data from the RSE for any supported applications.
3. *Communication technologies*: Communication technology enables the necessary interaction between vehicles and between infrastructure and vehicles. Cellular communications, wireless communications, and dedicated short-range communications (DSRC) are examples of such technologies. DSRC is a protocol explicitly developed for short-range to medium-range communications between vehicles and infrastructure. The US Federal Communications Commission (FCC) allocated a 75 MHz communication spectrum for the development of DSRC-based secure connected vehicle safety applications (RITA, 2014a). DSRC is not affected by adverse weather conditions, has a very low latency of 200 microseconds (Guo, 2006), and can protect privacy and security of data exchange while providing fast communications between vehicles and infrastructure. In addition to SAE J2735 standard, "*IEEE 1609 Family of Standards for Wireless Access in Vehicular Environments (WAVE)*" (IEEE, 2014a) and "*IEEE 802.11p Standard for Information Technology*" (IEEE, 2014b) provide requirements for secure V2I and V2V communications and specifications for data exchange in wireless vehicular networks.

Connected vehicle applications are typically grouped into two categories: vehicle-to-vehicle applications (V2V) and vehicle-to-infrastructure applications (V2I). In V2V applications, each vehicle exchanges data with other vehicles in its vicinity (e.g. trucks, buses, passenger cars, and eventually motorcycles and bicycles). The V2I applications involve data exchange between vehicles' OBE and RSEs installed on the infrastructure. While V2I applications require significant investment in installing new equipment and developing communication backbone for the transportation system, V2V applications are independent of infrastructure and thus likely to be implemented in a shorter period of time. For this reason, V2V safety applications are currently the main focus of transportation agencies (NHTSA, 2014). Examples of V2V safety

Integrated Dynamic Transit Operations (IDTO):

- Connection Protection (T-CONNECT)
- Dynamic Transit Operations (T-DISP)
- Dynamic Ridesharing (D-RIDE)

Applications for the Environment, Real-Time Information Synthesis (AERIS):

- Eco-Signal Operations,
- Dynamic Eco-Lanes,
- Dynamic Low Emissions Zones

Dynamic Mobility Applications:

- Enable Advanced Traveler Information System (EnableATIS),
- Freight Advanced, Traveler Information Systems (FRATIS),
- Integrated Dynamic Transit Operations (IDTO),
- Intelligent Network Flow Optimization (INFLO),
- Multi-Modal Intelligent Traffic Signal Systems (MMITSS),
- Response, Emergency Staging and Communications, Uniform Management, and Evacuation (R.E.S.C.U.M.E.)

Figure 6.4 Applications provided by V21 communications

applications include emergency brake-light warning, forward collision warning, intersection movement assist, blind spot and lane change warning, do not pass warning, and control loss warning (RITA, 2014b). These applications are based on a connected vehicle's "Here I am" message that communicates an individual vehicle's data such as its location and speed every tenth of a second (Zeng et al., 2012). On the other hand, the V2I communication technology can serve a unique set of applications distinct from the V2V applications. The V2I enabled applications include the integrated dynamic transit operations, real-time environment-friendly applications, and dynamic mobility applications as enumerated in Figure 6.4.

Testing and demonstration of connected vehicle technology has been conducted primarily through test-beds and pilot projects. There are seven connected vehicle test-beds in the US with specific testing capabilities focusing on commercial vehicles, traffic and mobility, and other connected vehicle applications. These test-beds are located at Manhattan, NY; Maricopa County, AZ; Mclean, VA; Oak Ridge, TN; Orlando, FL; Palo Alto, CA; and Novi, MI. The Michigan test-bed provides researchers and developers with a facility to test and evaluate connected vehicle equipment and applications. In-vehicle and after-market safety devices, DSRC standards and security certificate credential management programs, and applications that use geometric intersection description (GID) and signal phase and timing (SPaT) data are a few examples of products and programs that can be evaluated at the Michigan test-bed (US DOT, 2014). Real-world, large-scale deployments of connected vehicle technology have not occurred yet. Even though several connected vehicles applications are undergoing the final stages of field evaluation, there are policy decisions and technical issues that should be addressed prior to a large-scale implementation of connected vehicles (PIARC, 2012; Hill, 2014). Some of these issues are:

- *Legal*: Ownership of data collected by connected vehicles, liability of drivers, application developers, automakers and transportation agencies.
- *Security*: Ensuring security of data exchange between vehicles and roadside equipment.
- *Privacy*: Protecting the privacy of users given the unprecedented amounts of data collected.
- *Standards*: Developing guidelines and standards to ensure the interoperability of systems, and vehicles manufactured in the US, Europe, Japan, and other regions.
- *Implementation of policies and strategies*: Developing financial and investment strategies, identifying high priority connected vehicles' applications, developing standard and testing guidelines for third-party equipment, considering impacts of market penetration on performance of the system.

Transportation agencies are making significant progress in addressing these issues towards real-world implementation of the connected vehicle technology. In February 2014, the US National Highway Traffic Safety Administration (NHTSA) made a decision to move towards enabling V2V applications in light vehicles (NHTSA, 2014). In Europe, the two standards organizations, ETSI and CEN, have developed and completed their first set of connected vehicle standards for the European Union (ETSI, 2014). Furthermore, the European Commission (EC), Japan and US have signed three bilateral agreements to collaborate on developing connected vehicles (RITA, 2014c). Such developments provide confidence that the promise of connected vehicle technology will become a reality in the not too distant future.

Autonomous Vehicles

One of the most ambitious ITS innovations in the past decade has arguably been the self-driving, autonomous, or automated vehicle. In the US, the DARPA Grand Challenge, a competition

Table 6.4 NHSTA levels of automation (NHTSA, 2014)

Level	*Automation Level*	*Description*
Level 0	No Automation	• Driver is in full control of the vehicle • All the control functions require driver's input • Safety features available to warn driver about road hazards but will not take any control action (e.g. blind spot monitoring system)
Level 1	Function-specific Automation	• Driver is responsible for safe operation and has overall control of the vehicle • One or more control functions could be automated e.g. adaptive cruise control, electronic stability control, or dynamic brake support in crash eminent situations • Driver is constantly engaged in physically controlling the vehicle using steering wheel and pedals
Level 2	Combined Function Automation	• Two or more primary control functions are automated • Driver is responsible for monitoring roadway and is in charge of safe operation of the vehicle • Driver is expected to be available to take control all the time and on short notice. • Unlike level 1, driver could be disengaged from physically controlling the vehicle using steering wheel and pedals during specific operating conditions
Level 3	Limited Self-Driving Automation	• Under certain traffic and environmental conditions driver can cede control of safety-critical functions to the vehicle • Driver is expected to be available for occasional control; however, the transition occurs at a comfortable transition time • Unlike level 2, driver is not responsible for constantly monitoring the roadway conditions
Level 4	Full Self-Driving Automation	• Vehicle performs all safety-critical driving functions and monitors the road conditions during the entire trip • Driver only has to provide destination or route preference information • Driver is not expected to be engaged in any control task during the trip

held by the United States Defense Advanced Research Projects Agency, kick-started the demonstration of fully autonomous ground vehicles on long-distance tracks. The first Grand Challenge was held in 2004 where vehicles had to successfully navigate around 150 miles in the Mojave Desert region in California (an off-road track). None of the participants completed the course. The second edition of the Grand Challenge was conducted in 2005, with five vehicles completing the course in the allotted time. The third edition of the Grand Challenge was different from the first two challenges in that it was in an urban environment with real-life conditions such as other vehicles on the road and the participants were required to follow the regular driving laws. Six teams successfully completed the challenge with the winner, Carnegie Mellon University's BOSS, completing the 60-mile course in 4 hours and 10 minutes (Schwarz et al., 2013; Luettel et al., 2012).

Several automobile manufacturers currently have an active autonomous vehicle program. Google's self-driving car has received much attention lately due to its extensive testing in real-world traffic conditions. It has accumulated over 700,000 accident-free miles of combined highway and city street mileage (IEEE Spectrum, 2014) as of the writing of this article. Such developments have led some researchers to anticipate that the autonomous vehicles may become mainstream in 10 to 15 years (Luettel et al., 2012). The automation of vehicles is expected to occur in several phases or levels. NHTSA has defined five levels of vehicle automation ranging from Level 0 to Level 4 (NHTSA, 2014). Table 4 provides a summary of automation levels suggested by NHTSA.

In Level 0, the driver is in full control of the vehicle and all the control functions require driver input. Safety features such as blind-spot monitoring, lane-departure warning and connected vehicle V2V applications that only warn drivers about the possible hazards belong to Level 0 automation. In Level 1, the driver is constantly engaged in physically controlling the vehicle using the steering wheel and pedals, but one or more control functions could be automated. In Level 2, two or more primary control functions are automated, but the driver is expected to be available to take control of the vehicle on short notice. In Level 3, the driver can cede control of safety-critical functions to the vehicle under certain traffic and environmental conditions, but is expected to take control of the vehicle over a comfortable transition period. In Level 4, the vehicle performs all the safety-critical driving functions and the driver only inputs the destination and route preferences.

To achieve Level 4 automation, a diverse set of challenges should be overcome. Schwarz et al. (2013) present a detailed discussion of these challenges including technical, human factors, legal, security, privacy, and long-term impacts of automated vehicles. Figure 6.5 provides a list of these challenges. Technical challenges of automated vehicles include identifying the location of the vehicle and detecting the road (i.e. sensing, localization and mapping), identifying the static and moving objects on the road (i.e. sensing and object detection), and traveling from origin to destination (i.e. path planning, decision-making).

Technical:	**Human Factors:**	**Societal and Economic:**
Sensors	Out of the Loop Performance Loss	Legal and Liability
Localization and Mapping	Driver Vehicle Interface	Security
Object Detection	Trust in Automation	Privacy
Path Planning		Long-Term Impacts
Decision Making		

Figure 6.5 Technical, societal, economic and human factors: related challenges of automated vehicles
Source: Schwarz et al., 2013

The human factors aspects of automated vehicles represent the second set of challenges. When drivers are not actively involved in the control of a vehicle, they will be subject to out of the loop (OOTL) performance loss (Endsley and Esin, 1995; Sheridan and Oarasuraman, 2005), which might negatively affect the driver performance when the control is transitioned back to the driver. Degradation of the driver's skill is another concern that might occur if drivers rely on the automated system for a long period of time (Schwarz et al., 2013). Designing a driver–vehicle interface (DVI) that is effective in warning the driver about the road hazards and clear in conveying the automated driving features is another challenge (Lee et al., 2001, 2002). The promising benefits of automated vehicles can only be realized if the drivers trust and use the automated driving features. However, a combination of trust and attention is required in the early levels of automation until a fully automated vehicle is developed.

Transportation professionals and policy makers are also facing questions related to societal and economic impacts of automated vehicles. Regulations for licensing and operating automated vehicles, liability of owners, manufacturers and developers in case of possible future crashes, developing insurance plans, protecting the automated vehicles from cyber-attacks and protecting the privacy of the automated vehicles users are some of the issues being raised. Schwarz et al. (2013) discuss the possible long-term impacts of automated vehicles. The automated vehicles might change the concept of car ownership and with the elimination of driver role in the vehicle, the future generation might prefer to use a car-sharing system rather than owning a personal vehicle or may be more inclined to participate in ridesharing. This could impact the urban land-use by reducing the need for parking spaces and parking lots. In a fully automated vehicular network, one can expect the roadway capacities to increase due to the shorter following distances between vehicles. The shorter following distances are expected due to the elimination of driver reaction time. However, the question of induced demand resulting from the higher capacities can only be answered after the full adoption of automated vehicles.

Other ITS Technologies

Advances in Traveler Information

Historically, in the US, traveler information was predominantly offered by public transportation agencies such as state departments of transportation. In the past few years, advances in traveler information systems have been driven by innovation in the private sector. Online maps have allowed travelers to learn about the possible routes to their destination both pre-trip and en-route. Companies such as MapQuest, Yahoo, Google, have offered free online map services to users. Currently, these Web services provide traffic maps indicating the current traffic speed and level of traffic on the routes. The accuracy and ease of use of the online maps has also driven developments in the route guidance systems such as those offered by WAZE in the US (and in some countries in Europe, South America, and Asia). These route guidance systems provide real-time information on hazards on the travel route. Such systems show a fundamental shift in the focus in information provision from public agencies to a crowd-sourced system that relies on feedback from other drivers in the road network. The proliferation of smartphones that offer the capability to be connected to the Internet and satellites is also playing a big role in the traveler information arena. Smartphone application programs have made it easier for travelers to access travel information such as preferred routes, travel times, and traffic updates. Social media such as Twitter allows users to post snippets of information, called tweets, about travel conditions including incidents. Transportation agencies and individual users can make other users in the social network aware of traffic conditions by posting tweets. In fact, most state

DOTs in the US have dedicated Twitter sites to provide real-time updates on traffic conditions in urban areas.

Finally, advances in Cloud computing technology is transforming the way the data needs to be stored and processed on board a vehicle. In Cloud computing, data and applications using the data are stored and executed on a remote server, thus obviating the need for any significant on board storage. For example, a lane departure warning system relies on location information obtained from satellites that is processed on a remote server and the desirable action is then passed on to the vehicle—either information provided to the driver or the vehicle takes a corrective action (US DOT, 2014).

Conclusion

Tremendous advances have emerged in infrastructure, in-vehicle, and traveler information technologies since the inception of ITS in the early 1970s. Although some ad hoc ITS deployments and research occurred in the 1970s, concerted efforts started to occur in the 1980s and continued into the 1990s. The establishment of entities such as IVHS America and its equivalents ERTICO in Europe and VERTICO in Japan launched ITS projects like never before. Several route guidance, traveler information, and traffic control applications were explored in the 1990s. To have consistency across the different ITS deployments in the US, a national ITS architecture was developed. The current version of the architecture consists of 22 subsystems grouped into four classes, and 33 user services grouped into eight bundles by the US DOT and ITS America. The 2000s have witnessed the concept of connected vehicles where they communicate data with other vehicles and the infrastructure. The most significant benefit of the connected vehicle technology is anticipated from its vehicle safety applications intended to prevent crashes. Mobility benefits offered by connected vehicle technology are also unprecedented.

Several new technologies have made their way into transportation this past decade. Online mapping services, crowd-sourced route guidance systems provide traveler information like never before. Robotics and artificial intelligence technologies were demonstrated in the development of autonomous vehicles. At this rate, the field of ITS will continue to evolve. The vision of accident-free and congestion-free driving conditions looks more promising than ever before. A common theme emerged from the review of different evolutionary stages of ITS. All major ITS advances have occurred through concerted collaborations of automobile companies, state and federal transportation agencies, users, and other stakeholders. Such collaborative relationships between different stakeholders will continue to play an important role in shaping the future of ITS.

References

Burke, M. and Williams, J. 2012. Cooperative ITS Regulatory Policy Issues: Discussion Paper. National Transport Commission, Austria.

Chen, C., Kianfar, J., and Edara, P. 2014. New Snapshot Generation Protocol in a Connected Vehicle Environment For Travel Time Estimation. *Transportation Research Record: Journal of the Transportation Research Board* No. 2424. pp. 1–10.

Chowdhury, M. and Sadek, A. 2003. Fundamentals of Intelligent Transportation Systems Planning. Artech House.

Endsley, M. R. and Esin, O.K. 1995. The out-of-the-loop performance problem and level of control in automation. *Human Factors: The Journal of the Human Factors and Ergonomics Society*, Vol. 37, No. 2, pp. 381–394.

European Telecommunications Standards Institute (ESTI). 2014. CEN and ETSI deliver first set of standards for Cooperative Intelligent Transport Systems (C-ITS). Available at: www.Etsi.Org/News-Events/News/753–2014–02-Joint-News-Cen-And-Etsi-Deliver-First-Set-Of-Standards-For-Cooperative-Intelligent-Transport-Systems-C-Its. Accessed April 30, 2014.

Gordon, R. 2009. *Intelligent Freeway Transportation Systems Functional Design*. Springer, New York.
Guo, J. 2006. Vehicle Safety Communications in DSRC. Presented at 2006 US Army VI Winter Workshop.
Hill, C. 2014. ITS ePrimer-Module 13: Connected Vehicles. Research and Innovative Technology Administration (RITA).
IEEE Spectrum, 2014. Available at: http://spectrum.ieee.org/cars-that-think/transportation/self-driving/google-autonomous-cars-are-smarter-than-ever. Accessed February 15, 2014.
IEEE Standard Association. 2014a. 1609 WG - Dedicated Short Range Communication Working Group. Available at: http://standards.ieee.org/develop/wg/1609_WG.html, Accessed July 30, 2014.
IEEE Standard Association. 2014b. 802.11p-2010 - IEEE Standard for Information technology. Available at: http://standards.ieee.org/findstds/standard/802.11p-2010.html 1/. Accessed July 30, 2014.
ITERIS. 2013. Key Concepts of the National ITS Architecture 7.0. Available at: www.iteris.com/itsarch/documents/keyconcepts/keyconcepts.pdf. Accessed April 19, 2014.
Kianfar, J. and Edara, P. 2013. Placement of Roadside Equipment in Connected Vehicle Environment for Travel Time Estimation. *Transportation Research Record: Journal of the Transportation Research Board* No. 2381, pp. 20–27.
Lee, J. D., Caven, B., Haake, S., and Brown, T.L. 2001. Speech-based interaction with in-vehicle computers: The effect of speech-based e-mail on drivers' attention to the roadway. *Human Factors: The Journal of the Human Factors and Ergonomics Society*, Vol. 43, No. 4, pp. 631–640.
Lee, J. D., McGehee, D.V., Brown, T.L. and Reyes, M. 2002. Collision warning timing, driver distraction, and driver response to imminent rear-end collisions in a high-fidelity driving simulator. *Human Factors: The Journal of the Human Factors and Ergonomics Society*, Vol. 44, No. 2, pp. 314–334.
Luettel, T., Himmelsbach, M., and Wuensche, H-J. 2012. Autonomous Ground Vehicles—Concepts and a Path to the Future. Proceedings of the IEEE. Vol. 100, May 13, 2012.
Makino, H. 2006. SmartWay Project Cooperative Vehicle Highway Systems. Presented at Vehicle Highway Automation Committee meeting at Transportation Research Board 85th Annual Meeting, Washington, DC, January 22–26.
National Highway Traffic Safety Administration (NHTSA). *Preliminary Statement of Policy Concerning Automated Vehicles*. Available at: www.nhtsa.gov/staticfiles/rulemaking/pdf/Automated_Vehicles_Policy.pdf. Accessed July 30, 2014. *Vehicles*. www.nhtsa.gov/staticfiles/rulemaking/pdf/Automated_Vehicles_Policy.pdf Accessed July 30, 2014.
PIARC (World Road Association). 2012. *The Connected Vehicle*. Available at: www.piarc.org/en/order-library/13731-en-/The%20connected%20vehicle.htm
Research and Innovative Technology Administration (RITA), 2014a. Dedicated Short Range Communications. Available at: www.its.dot.gov/DSRC/dsrc_faq.htm. Accessed May 2, 2014.
Research and Innovative Technology Administration (RITA), 2014b. Vehicle-to-Vehicle (V2V) Communications for Safety. Available at: www.its.dot.gov/research/v2v.htm. Accessed April 29, 2014.
Research and Innovative Technology Administration (RITA) 2014c. International Research. Available at: www.its.dot.gov/connected_vehicle/international_research.htm. Accessed April 30, 2014.
Schwarz, C., Thomas, G., Nelson, K., McCrary, M., Schlarmann, N., Powell, M. 2013. *Towards Autonomous Vehicles*. Mid-America Transportation Center. Report # MATC-UI: 25–1121–0003–117.
Science Digest (April, 1958). Electronic Highway of the Future. Available at: http://blog.modernmechanix.com/electronic-highway-of-the-future/. Accessed April 30, 2014.
Sheridan, T. B. and Parasuraman, R. 2005. *Human–Automation Interaction. Reviews of Human Factors and Ergonomics*, Vol. 1, No. 1, pp. 89–129.
Sussman, J. 2005. *Perspectives on Intelligent Transportation Systems*. Springer, New York.
US DOT, 2014. Connected Vehicle TEST BED: Testing Connected Vehicle Technologies in a Real-World Environment. Available at: www.its.dot.gov/connected_vehicle/pdf/DOT_CVBrochure.pdf. Accessed July 30, 2014.
Whelan, R. 1995. Smart Highways, Smart Cars. Artech House Publishers, Boston, MA.
Zeng, X., Kevin Balke, K., and Songchitruksa, P. 2012. Potential Connected Vehicle Applications to Enhance Mobility, Safety, and Environmental Security. Southwest Region University Transportation Center, Report 161103–1.

7

AUTONOMOUS INTERSECTION MANAGEMENT FOR SEMI-AUTONOMOUS VEHICLES

Tsz-Chiu Au,[1] *Shun Zhang,*[2] *and Peter Stone*[2]

(1) ULSAN NATIONAL INSTITUTE OF SCIENCE AND TECHNOLOGY,
(2) UNIVERSITY OF TEXAS AT AUSTIN

1. Introduction

Recent robotic car competitions and demonstrations have convincingly shown that autonomous vehicles are feasible with the current generation of hardware.[1] Looking ahead to the time when autonomous cars will be common, Dresner and Stone proposed a new intersection control protocol called Autonomous Intersection Management (AIM) and showed that by leveraging the control and network capabilities of autonomous vehicles it is possible to design an intersection control protocol that is much more efficient than traffic signals.[2] By removing human factors from control loops, autonomous vehicles, with the help of advanced sensing devices, can be safer and more reliable than human drivers. The AIM protocol exploits the fine control of autonomous vehicles to allow more vehicles simultaneously to cross an intersection, thus effectively reducing the delay of vehicles by orders of magnitude compared to traffic signals.[3]

AIM is designed for the time when all vehicles are fully autonomous. We anticipate, however, that there will be a long transition period during which most vehicles have some but not all capabilities of fully autonomous vehicles. In fact, this transition period has already begun. Since the late 1990s, adaptive cruise control systems and lane departure warning systems have become widely available as optional equipment on luxury production vehicles. The National Highway Traffic Safety Administration acknowledges that fully autonomous vehicles represent just the top level in five levels of vehicle automation.[4] Indeed, they define a level below this top level with vehicles that have limited self-driving automation. The main motivation of this chapter is to propose a new intersection control system called Semi-Autonomous Intersection Management (SemiAIM) that can accommodate both fully autonomous vehicles and semi-autonomous vehicles with limited self-driving automation. There is a high likelihood that human-driven vehicles, semi-autonomous vehicles, and fully autonomous vehicles will coexist on the road in the future. SemiAIM takes advantage of this trend and allows autonomous intersections to handle a traffic mixture with different types of vehicles.

The main contributions of this chapter are: 1) the introduction of the concept of SemiAIM; 2) a specification of constraint-based reservation requests for semi-autonomous vehicles; and

3) detailed empirical results demonstrating the effectiveness of this protocol, especially in moderate traffic levels with a mix of human-driven, semi-autonomous, and fully autonomous vehicles. The remainder of the chapter is organized as follows. Section 2 outlines the architecture of AIM which forms the basis of SemiAIM. Sections 3 and 4 categorize semi-autonomous vehicles and discuss the interaction between human drivers and semi-autonomous vehicles in SemiAIM. Section 5 describes the constraint-based reservation system in SemiAIM. Section 6 presents the results of the simulation experiments we used to evaluate SemiAIM. The related work and the conclusion are given in Sections 7 and 8, respectively.

2. Autonomous Intersection Management

As laid out in detail in Dresner and Stone (2008),[2] an ideal autonomous intersection management protocol must satisfy seven desiderata. It must:

1. allow for fully distributed and autonomous control by the vehicles;
2. have low communication complexity;
3. assume non-expensive vehicle sensors found in production;
4. use a standardized communication protocol;
5. be incrementally deployable;
6. be safe; and
7. be efficient.

Modern-day traffic signals and stop signs completely satisfy all but the last one of these properties. Traffic signals and stop signs are very inefficient—not only do vehicles traversing intersections equipped with these mechanisms experience large delays, but the intersections themselves can only manage a limited traffic capacity—much less than that of the roads that feed into them.

The Reservation Idea

With these seven desiderata in mind, we have introduced a novel approach to efficient intersection management that is a radical departure from existing traffic signal optimization schemes. The solution is based on a reservation paradigm, in which vehicles "call ahead" to reserve space-time in the intersection.[3] In the approach, we assume that computer programs called "driver agents" control the vehicles, while an arbiter agent called an "intersection manager" is placed at each intersection. The driver agents attempt to reserve a block of space-time in the intersection. The intersection manager decides whether to grant or reject requested reservations according to an intersection control policy. In brief, the paradigm proceeds as follows.

- An approaching vehicle announces its impending arrival to the intersection manager. The vehicle indicates its size, predicted arrival time, velocity, acceleration, and arrival and departure lanes.
- The intersection manager only considers the request if the vehicle in front, if any, has been granted a reservation.
- The intersection manager simulates the vehicle's path through the intersection, checking for conflicts with the paths of any previously processed vehicles.
- If there are no conflicts, the intersection manager issues a reservation. It becomes the vehicle's responsibility to arrive at, and travel through, the intersection as specified (within a range of error tolerance).

- The car may only enter the intersection once it has successfully obtained a reservation.
- Upon leaving the intersection, the car informs the intersection manager that its passage through the intersection was successful.

The best way to get a sense of the paradigm is to watch it in action (the video is available online at: www.cs.utexas.edu/~aim/video/fcfs-insanity.mov). Note that this video includes only autonomous cars. We have also extended the system to work when some of the cars are driven by people.[4] As one would expect, the improvements over current systems are smaller; but they are still significant.

Figure 7.1 diagrams the interaction between driver agents and an intersection manager. A key feature of this paradigm is that it relies only on vehicle-to-infrastructure (V2I) communication. In particular, the vehicles need not know anything about each other beyond what is needed for local autonomous control (e.g., to avoid running into the car in front). The paradigm is also completely robust to communication disruptions: if a message is dropped, either by the intersection manager or by the vehicle, delays may increase, but safety is not compromised. Safety can also be guaranteed in mixed mode scenarios when both autonomous and manual vehicles operate at intersections. The intersection efficiency will increase with the ratio of autonomous vehicles to manual vehicles in such scenarios.

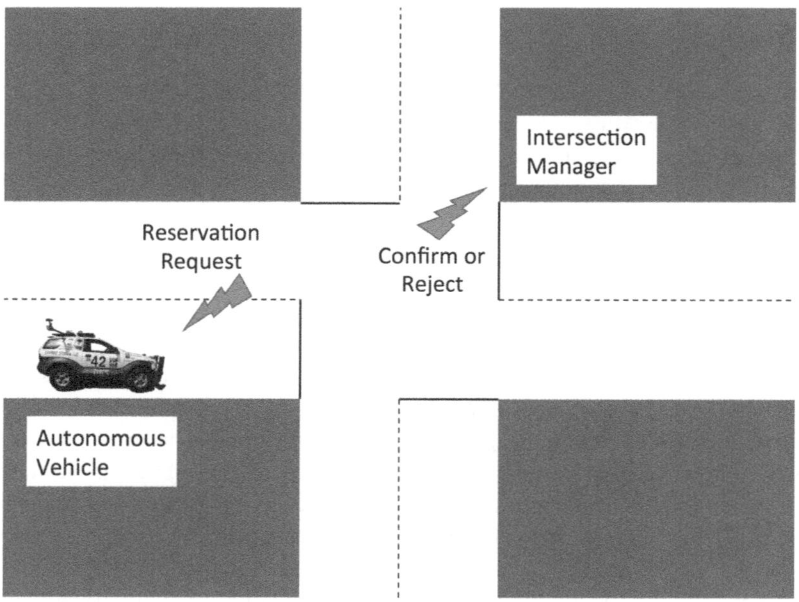

Figure 7.1 Diagram of the intersection system

Intersection Control Policy

Our prototype intersection control policy divides the intersection into a grid of reservation tiles, as shown in Figure 7.2 (this notation can be generalized for rectangular and irregularly shaped intersections). When a vehicle approaches the intersection, the intersection manager uses the data in the reservation request regarding the time and velocity of arrival, vehicle size, etc. to

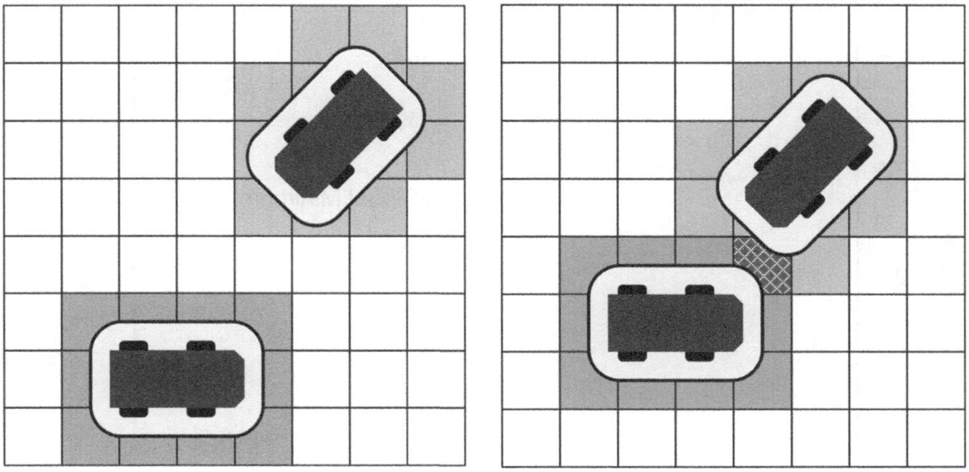

(a) (b)

Figure 7.2 (a) The vehicle's space–time request has no conflicts at time t (b) The black vehicle's request is rejected because at time t of its simulated trajectory, the vehicle requires a tile already reserved by another vehicle

simulate the intended journey across the intersection. At each simulated time step, the policy determines which reservation tiles will be occupied by the vehicle.

If at any time during the trajectory simulation the requesting vehicle occupies a reservation tile that is already reserved by another vehicle, or could be occupied by a human-driven vehicle, the policy rejects the driver's reservation request, and the intersection manager communicates this to the driver agent. Otherwise, the policy accepts the reservation and reserves the appropriate tiles. The intersection manager then sends a confirmation to the driver. If the reservation is denied, it is the vehicle's responsibility to maintain a speed such that it can stop before the intersection. Meanwhile, it can request a different reservation.

Simulation Results

Empirical results in simulation demonstrate that our proposed reservation system can dramatically improve the intersection efficiency when compared to traditional intersection control mechanisms. To quantify efficiency, we introduce delay, defined as the amount of added travel time incurred by the vehicle as the result of passing through the intersection.

In experiments, our fully implemented custom micro-simulator simulates three lanes in each of the four cardinal directions. The speed limit in all lanes is per hour. Different management systems (reservation system, stop sign, traffic signal) are run for at least 100,000 steps, which corresponds to approximately half an hour of simulated time. A video demonstration at the URL referred to in section 2 shows cars driving from all directions and barely slowing down as they traverse the intersection with no collisions.

The results for the experiments are shown in Figure 7.3. The reservation system (labeled FCFS) performs very well, nearly matching the performance of the optimal policy which represents a lower bound on delay should there be no other cars on the road. At higher levels of traffic, the average delay for a vehicle using the reservation system gets as high as 0.35 seconds, but it is never more than one second above optimal. Under none of the tested conditions does

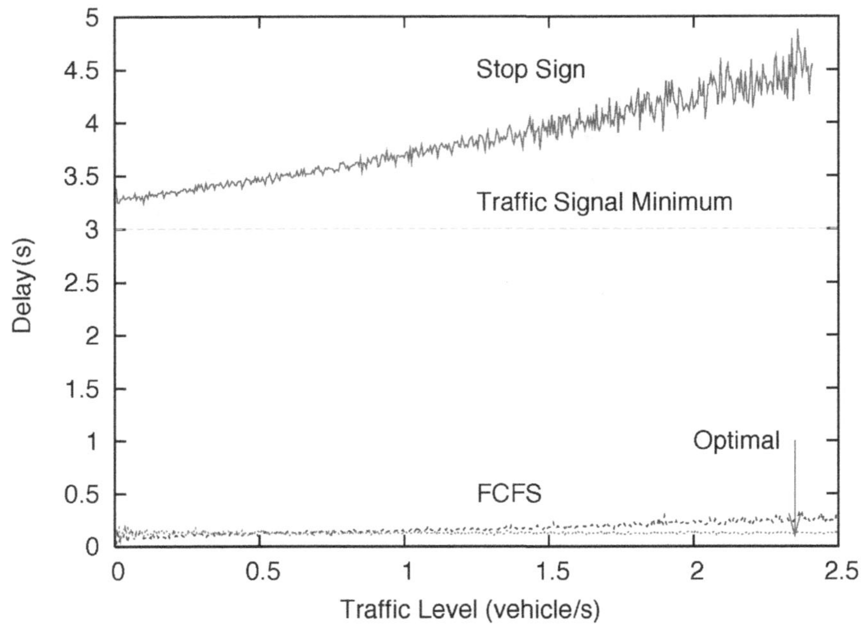

Figure 7.3 Traffic delay for the reservation system, stop sign, optimized traffic signal, and optimal lower bound when collisions are ignored

the reservation system approach the trip times of the traffic signal system—the best empirical performance over a range of tested cycle times is shown. Likewise, using stop signs increases delays over the reservation system, especially as the traffic level increases. Overall, by allowing for a much finer-grained coordination, the simulation-based reservation system can dramatically reduce per-car delay by two orders of magnitude in comparison to traffic signals and stop signs.

Worst-Case Analysis and V2V interactions

Additionally, we have studied the impacts on travel times for emergency vehicles; the potential benefits across networks of intersections; and the worst-case effects of vehicle mechanical failure in intersections, showing that despite the apparent potential for "pile-ups" when there is an accident, our protocol prevents such disastrous situations even under conservative assumptions.[5] We have also shown that the benefits of our system can be replicated at low-traffic intersections without any infrastructure in place via vehicle-to-vehicle (V2V) communication.[6]

3. Semi-Autonomous Vehicles

A drawback of FCFS is that it does not work with any vehicles that require human operations. Dresner and Stone proposed FCFS-signal (a.k.a. FCFS-Light) to make AIM compatible with human drivers.[7] According to their study, however, when more than 5–10 percent of traffic is comprised of human-driven vehicles, the average delay time of all vehicles increases significantly under FCFS-signal. The reason for this phenomenon is twofold. First, human-driven vehicles cannot enter the intersection during red signal phases, because they are *not* allowed to make any

Table 7.1 Features of semi-autonomous vehicles

Vehicle type	*Communication device*	*Cruise control*	*Adaptive cruise control*
SA-ACC	X	X	X
SA-CC	X	X	
SA-Com	X		

reservation. Second, as a consequence of the first, autonomous vehicles often fail to obtain reservations when they are behind the human-driven vehicles stalled at red signals. These issues stem from the fact that FCFS-signal prohibits human-driven vehicles from entering intersections during red signal phases. SemiAIM sets out to overcome these issues by allowing semi-autonomous vehicles—vehicles with limited autonomous driving and wireless communication capabilities—to use the AIM reservation system to enter an intersection during red signal phases. While these vehicles are not fully autonomous, they are able to follow a limited number of predictable trajectories at intersections more precisely than human drivers. This ability allows them to utilize our constraint-based reservation system to make reservations in the same manner as fully autonomous vehicles and enter an intersection at red signals.

Our proposed reservation system is general enough to accept reservation requests from any semi-autonomous vehicles that are capable of following some trajectories and communicating with the IM. To facilitate our discussion, we will focus on semi-autonomous vehicles that use the following set of equipment that is readily available today.

- *Communication device (Com)*: A component in a vehicle's on-board electronic system that enables the vehicle to wirelessly communicate with the transportation infrastructure including the IM. The communication is bidirectional.
- *Simple Cruise Control (CC)*: An optional speed control subsystem in vehicles' drivetrain that automatically controls the vehicle speed by taking over the throttle of the vehicles.
- *Adaptive Cruise Control (ACC)*: An advanced cruise control system that automatically adjusts the speed of a vehicle in order to maintain a certain distance from vehicles ahead.

All this equipment gives semi-autonomous vehicles some of the functionality of autonomous vehicles, although human drivers still retain some control of the vehicles. We can equip a semi-autonomous vehicle with more than one of these devices. In this chapter we introduce three types of semi-autonomous vehicles that we envision utilizing this equipment: Type SA-ACC Vehicles, Type SA-CC Vehicles and Type SA-Com Vehicles. Table 7.1 summarizes the equipment used on these different types of semi-autonomous vehicles. Figure 7.4 shows a screenshot of the SemiAIM simulator developed to simulate semi-autonomous vehicles at intersections. In the figure, there are five types of vehicles: fully autonomous vehicles, Type SA-ACC vehicles, Type SA-CC vehicles, Type SA-Com vehicles, and human-driven vehicles.

4. Interaction Model

Safety is a main concern when involving human drivers in the control loop of semi-autonomous vehicles. For the semi-autonomous vehicles defined above to be able to go through an intersection safely, we need to define a simple and clean interface for negotiating with the Interaction Model (IM) and passing control between the human driver and the driver agent. In

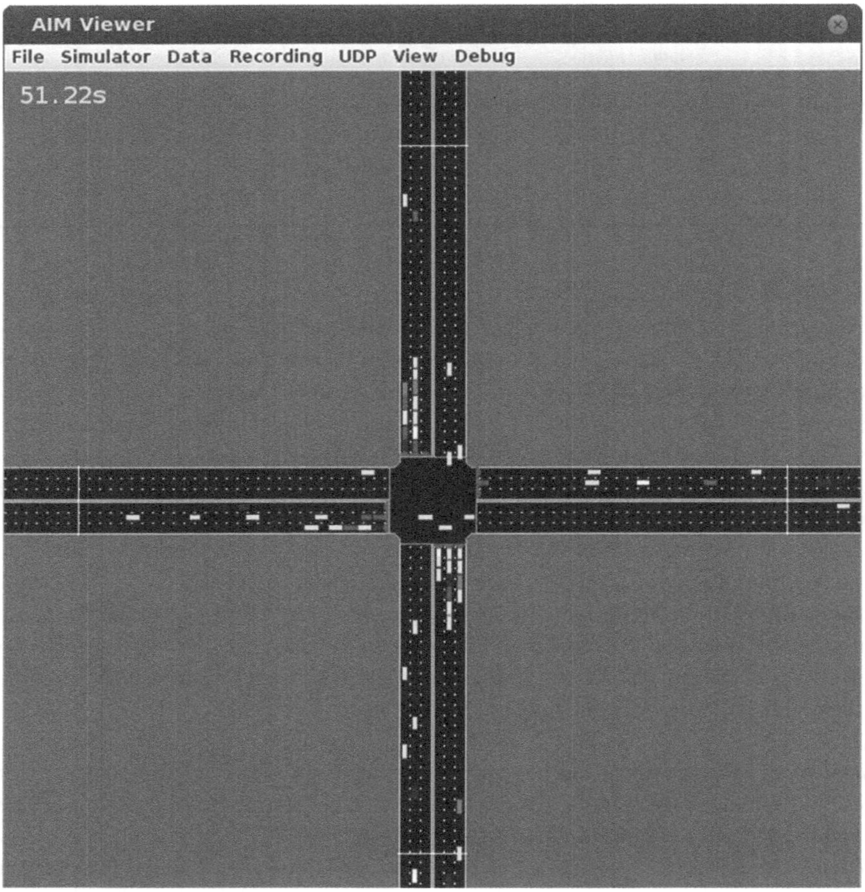

Figure 7.4 Screenshot of the simulator developed for the experiments on Semi-AIM. This simulator is modified from the open-source AIM4 simulator

this section, we describe how our proposed protocol can be realized safely by having an interaction model between human drivers and driver agents that simplifies the task of the human drivers.

Figure 7.4 summarizes the interaction model between human drivers, driver agents, and the IM. We require the inclusion in the vehicle of a single button that signals the driver agent to ask for a reservation. After pressing the button, the driver agent will automatically send a request message to the IM on behalf of the human driver. It is also important that there is a clear "okay" indicator (such as a green light) installed in the car which indicates when the request has been confirmed. After seeing the okay signal, the driver would have to actively pass control to the driver agent, again by pressing a single button. This way the driver will not be surprised by any sudden autonomous actions of the vehicle.

This interaction model only requires the human driver to perform relatively simple driving maneuvers such as holding the steering wheel at a certain angle (Types SA-ACC and SA-CC vehicles) or driving as if under a traffic signal (Type SA-Com vehicles). These tasks are much simpler than other maneuvers such as lane changing and passing other vehicles, and thus should not be taxing to experienced human drivers.

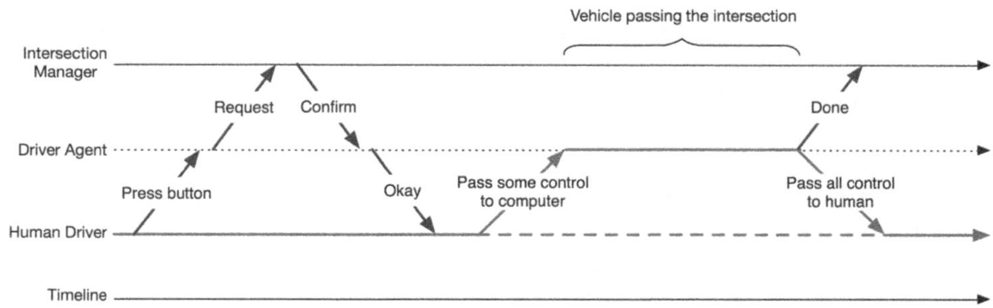

Figure 7.5 Interaction between human drivers, driver agents, and the IM. The solid lines are message passing, and the double stroked lines are transfer of control. Note that human drivers retain some control of the vehicle inside the intersection (the dashed line)

5. Constraint-Based Reservation Systems

SemiAIM allows human-driven vehicles and semi-autonomous vehicles to make reservations in the same way as fully autonomous vehicles. The key idea of SemiAIM is to turn AIM into a *constraint-based reservation system*, which allows vehicles to make reservations in terms of constraints over 1) their driving profiles such as their arrival time and arrival velocity, and 2) the relationships with other vehicles.

In AIM, a reservation request is a five-tuple, (l_1, l_2, t_0, v_0, p) where l_1 is the entry lane, l_2 is the exit lane, t_0 is the arrival time, v_0 is the arrival velocity, and p is the physical characteristics of the vehicle. This information allows the IM to compute the exact trajectory of the vehicle and reserve tiles on the trajectory. However, this computation assumes that the vehicle can be controlled precisely in the intersection so that it can meet the reservation constraints exactly. Human drivers cannot control their vehicles as precisely, and semi-autonomous vehicles may only be able to control certain aspects of their trajectories. Therefore, we need a new kind of reservation requests that does not rely on this assumption.

Maneuverability Profiles

Semi-autonomous vehicles only have limited self-driving automation. For example, adaptive cruise control can only provide autonomous longitudinal control, but still depends on the human driver to use the steering wheel to maintain the lateral control. Hence, the longitudinal control of Type SA-ACC vehicles can be very precise, but the lateral control cannot be as precise. Furthermore, the human driver cannot make a sharp turn while the adaptive cruise control is active, because the human must slow down the vehicle a lot in order to be able to make a sharp turn precisely. In our system, we use a maneuverability profile to encode the constraints over the control of human drivers when automation devices are activated. For simple cruise control, the maneuverability profile is written in Lisp syntax as follows:

```
(cc-profile (v v_error angle)
   (is-auto-speed-control)
   (not is-auto-steering)
   (< velocity (+ v v_error))
   (> velocity (- v v_error))
   (< steer-angle angle) (> steer-angle -angle))
```

where v is the target velocity, v_{error} is the maximum error of the target velocity, and angle is the maximum feasible steering angle for the human driver when the cruise control is turned on.

The maneuverability profile for adaptive cruise control is:

```
(acc-profile (vin d d_error angle)
    (is-auto-speed-control)
    (not is-auto-steering)
    (< (dist-from vin) (+ d d_error))
    (> (dist-from vin) (- d d_error))
    (< steer-angle angle) (> steer-angle -angle))
```

where vin is the vehicle id of the vehicle ahead, d is the target distance from the vehicle ahead, d_{error} is the maximum error for acc to maintain the target distance from vin, and angle is the maximum feasible steering angle for the human driver.

Other types of semi-autonomous vehicles have their own maneuverability profiles. When a driver agent makes a reservation request, it is his or her responsibility to ensure that the trajectories encoded by the reservation request satisfy all constraints in the maneuverability profile.

Constraint-Based Reservation Requests

Different maneuverability profiles can have different sets of constraints. To facilitate communication with all kinds of semi-autonomous vehicles, SemiAIM uses a unified language for vehicles to express their constraints in the same format in their requests. We define constraint-based reservation requests as follows. A request message consists of four components:

- An intention statement states the direction in which the vehicle wants to exit from an intersection. An intention statement is formally a disjunction of lane and road identifiers $(l_1, \vee l_2 \vee \ldots \vee r_1 \vee r_2)$,where l_i is an exit lane and r_i is an exit road. For every lane l_i, there exists only one r_j such that $l_i \in r_j$. Examples of legal intentions are $(l_1 \vee l_3)$ or (r_1). This feasibility facilitates different path planning strategies the vehicle might use.
- A vehicle type indicates the type of the vehicle. The IM maintains a database of all car types in existence, with information such as sizes, shapes, and kinematics of different car types. This information allows the IM to determine how the vehicle will move inside an intersection.
- An entry statement describes how a vehicle enters an intersection. An entry statement is a three-tuple $<(l_1 \vee l_2 \vee \ldots \vee l_n), [t_1, t_1], [v_1, v_1]>$, where l_i is a possible lane from which the vehicle enters the intersection, $[t_1, t_1]$ is the time interval during which the vehicle will arrive at the intersection, and $[v_1, v_1]$ is the interval of the arrival velocity of the vehicle.
- An acceleration profile is a schedule of acceleration a vehicle can use to accelerate through the intersection on a trajectory. An acceleration profile is a list of pairs $<(t_1, a_1), (t_2, a_2), \ldots, (t_n, a_n)>$, where a_i is the acceleration the vehicle will use from time t_i to time t_{i+1}. Note that the vehicle may or may not provide the list of acceleration profiles of all possible trajectories in a request message. If the list of acceleration profiles is missing, the IM will generate an acceleration profile based on a simulation of the vehicle movement given the vehicle type, the entry statement, and the intention statement.

A vehicle must generate constraint-based requests that satisfy its maneuverability profile only. In this chapter, we do not fully explore the space of how vehicles could generate such requests. However, we define the following two requests that will be used by Type SA-CC and Type SA-Com vehicles:

- A constant-velocity request is (*Intent, Type, Entry, AP*) where $Intent = l_1 \vee l_2 \vee \ldots \vee l_n)$ in which l_i is a possible lane from which the vehicle exits the intersection; *Type* is the vehicle type; $Entry = <(l_1' \vee l_2' \vee \ldots \vee l_n'), [t_1, t_1], [v_1, v_1]>$ is the entry statement; and $AP = (<(t_1, 0>)$ is the acceleration profile list. Since the acceleration in *AP* is always zero, the vehicle will move at a constant velocity.
- A whole-row request is <*Intent, Type, Entry, AP*>, where $Intent = (l_1 \vee l_2 \vee \ldots \vee l_n)$ in which l_i is a possible lane from which the vehicle exits the intersection; *Type* is the vehicle type; $Entry = <(l_1' \vee l_2' \vee \ldots \vee l_n'), [t_1, t_1], [v_1, v_1]>$ is the entry statement; and *AP* is the acceleration profile list. In order to reserve the entire row in an intersection, the difference between t_1 and t_2 as well as the difference between v_1 and v_2 must be large enough to cover the entire row, regardless of the acceleration profiles.

Unlike AIM's reservation requests which always correspond to a particular trajectory, a constraint-based request is an incomplete description of the trajectory. The IM interprets a constraint-based request as a description of a set of possible trajectories that the vehicle may follow, and reserves all the tiles on the possible trajectories. Once the IM receives a request, it conducts an internal simulation of the vehicle movement, as described in Dresner and Stone,[2] to determine the set of tiles on the possible trajectories and their availability.

Our hypothesis is that more restrictive constraint-based requests will perform better because SemiAIM reserves fewer tiles for them, meaning that some types of semi-autonomous vehicles will work better with SemiAIM. We will test this hypothesis in Section 6.

Anchor Requests

Semi-autonomous vehicles with adaptive cruise control can use a special constraint-based request called "anchor requests" to make reservations. An anchor request is (*Type, vin, d*), where *Type* is the vehicle type, *vin* is the vehicle id of the vehicle ahead, and *d* is the following distance from the vehicle of vin. Currently, anchor requests are designed specifically for semi-autonomous vehicles with adaptive cruise control. However, this formulation is general enough to enable some sophisticated cooperative maneuvers such as platooning.[8]

6. Experimental Evaluation

To demonstrate the feasibility of SemiAIM as well as evaluate the hypothesis that SemiAIM can offer substantial improvements over traffic signals and FCFS-Signal, we conducted several experiments using a modified AIM4 simulator that simulates the behavior of vehicles in our constraint-based reservation system. The implementations of fully autonomous vehicles (Type A) and human-driven vehicles (Type H), as described in Dresner and Stone, are provided in the simulator. The concrete implementations of the semi-autonomous vehicles in Section 3 are as follows.

- *Type SA-Com vehicles*: Send a whole-row request to the IM. If the entire lane is available, the request is confirmed; otherwise, the vehicle resends the request after a few seconds and decelerates enough to be able to stop before the intersection. If it keeps failing to obtain a reservation, it stops at the intersection and returns control to the human driver, who must then obey the traffic signal.

- *Type SA-CC vehicles*: Send a constant-velocity request to the IM when it is close enough to the intersection. If it can enter the intersections by keeping the current velocity, the request is confirmed; otherwise, the vehicle follows rules for Type SA-Com vehicles.
- *Type SA-ACC vehicles*: If there is a vehicle ahead which is either autonomous or semi-autonomous and is going in the same direction, it sends an anchor request to the IM. If the request is confirmed, the vehicle follows the vehicle ahead and enters the intersection. If the request is denied or there is no such vehicle ahead, the vehicle follows rules for Type SA-CC vehicles.

Our experiments were conducted in a 3 × 3 intersection (three lanes in each direction), as shown in Figure 7.6. Unless otherwise specified, the simulator spawns vehicles in each lane according to a Poisson distribution with an expectation of 360 vehicles per hour (i.e., traffic level = 360 vehicles/hour/lane). We chose this traffic level as being heavy enough to cause significant delays at signals, but light enough to allow for benefits even if cars are not precisely controlled.

The performance of an intersection is measured by the average delay, where delay is computed as the increase in travel time compared to traversing the intersection without slowing down at all, as if no other vehicles were on the road. Thus, lower delays correspond to better performance. This measure is the same one used in Dresner and Stone.[2] For all the vehicles, both the static buffer and the edge time buffer are 0.25 meters (see Dresner and Stone[2] for details) when auto-controlled.

Traffic signals are used as a fallback strategy when vehicles cannot get a reservation. We used an optimized six-phase signal schedule that includes relatively long green times (roughly

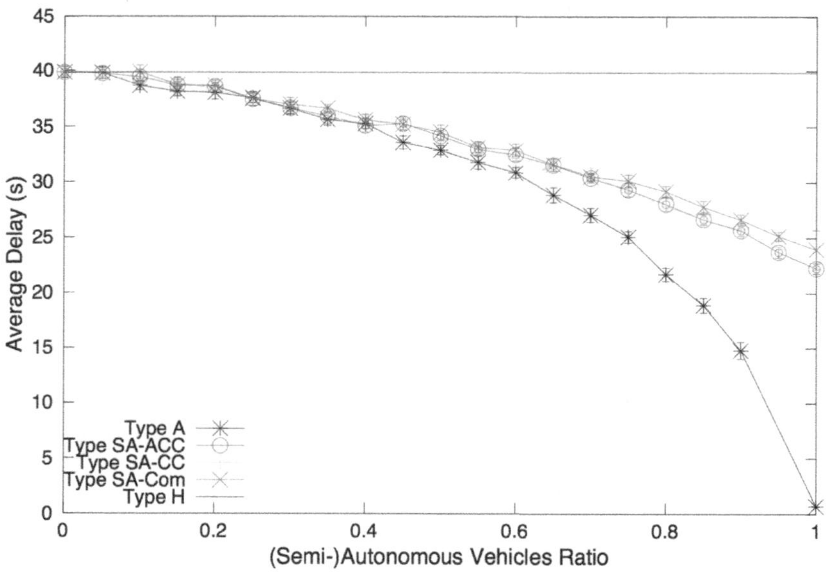

Figure 7.6 (Semi-)autonomous vehicles vs. human-driven vehicles. Traffic level = 360 vehicles/lane/hour

35 seconds) for East–West and North–South phases, and shorter phases for each of the four single-direction phases (roughly 9 seconds). It was optimized using SYNCHRO, a commercial traffic optimization package.

Technology Penetration

Our first experiment studies the effect of an increasing penetration rate of vehicular automation technology. Suppose human-driven vehicles are gradually replaced by a particular type of semi-autonomous vehicle or fully autonomous vehicle until all vehicles become that type. We examine how much benefit SemiAIM provides during the transition period.

In this experiment, the traffic consisted of one of the three types of vehicles we defined in Section 3 as well as Type A and Type H vehicles. We measured the traffic delay as we gradually increased the ratio of (semi-)autonomous vehicles to human-driven vehicles (Type H), while keeping the traffic level at 360 vehicles/hour/lane. As an example, consider the ratio of Type SA-ACC vehicles. At the beginning, there are 0 percent Type SA-ACC vehicles and 100 percent human-driven vehicles, and the ratio is 0. As the number of Type SA-ACC vehicles increases, the ratio increases and eventually becomes 1, which means there are 100 percent Type SA-ACC vehicles and 0 percent human-driven vehicles. We repeated the simulation 30 times for 1800s during each time. For each run, we measured the average delay of all vehicles. The average delays are shown in Figure 7.7. Each data point in the figure is an average of 30 values, and the error bar is the 95 percent confident interval of the average delay.

According to Figure 7.7, the performance of semi-autonomous vehicles is very similar to fully autonomous vehicles when the ratio to human-driven vehicles is below 40 percent. However, when the ratio increases beyond 40 percent, fully autonomous vehicles increasingly outperform semi-autonomous vehicles. Previous studies showed that FCFS-Signal needs at least 90 percent of fully autonomous vehicles in the traffic in order to be fully effective.[5] We successfully replicate the result, observing that the average delay drops rapidly when the traffic has more than 90 percent fully autonomous vehicles, and approaches zero when all vehicles are fully autonomous. Semi-autonomous vehicles cannot achieve the same dramatic decrease in traffic delay, but they, with the help of SemiAIM, manage to reduce the delay by 46 percent (from 39.9s to 22.4s) compared to human-driven vehicles. As expected, in the presence of semi-autonomous vehicles, SemiAIM provides significant advantages even when there are no fully autonomous vehicles on the road.

Another observation is that both Type SA-ACC and Type SA-CC vehicles have a significantly lower average delay than Type SA-Com vehicles. The difference is consistent with our hypothesis that the use of more constrained requests can increase the performance of intersections since the footprints of the vehicles are smaller and more vehicles can enter the intersection at the same time. Nonetheless, the difference is small because the simulated human drivers in the simulator can control their vehicles quite well.

Table 7.2 Deployment schedule in Figure 7.6

Schedule	*1*	*2*	*3*	. . .	*17*
Type H	10%	10%	10%	. . .	10%
Type SA-{ACC, CC, Com}	85%	80%	75%	. . .	5%
Type A	5%	10%	15%	. . .	85%

Incremental Deployment

Our second experiment considered realistic scenarios of incremental deployment of autonomous vehicle technology. We believe that the adoption of semi-autonomous vehicles will be much faster than fully autonomous vehicles since the cost of ownership of semi-autonomous vehicles are lower. Thus we foresee there will be a period of time during which semi-autonomous vehicles dominate the roads. However, this domination will be short-lived—eventually, fully autonomous vehicles will displace semi-autonomous vehicles due to their greater convenience. This section studies the effects of different levels of incremental deployment of (semi-) autonomous vehicles over time.

We consider two plausible deployment schedules, one for the period of time before the domination of semi-autonomous vehicles, and the other for the subsequent period. In the first deployment schedule (Table 7.2), Type H vehicles on the road are gradually displaced by (semi-) autonomous vehicles. Assume that two out of three car buyers who abandoned their vehicles for newer technology will buy semi-autonomous vehicles rather than a fully autonomous vehicle. Then eventually about two-thirds of vehicles on the road will be semi-autonomous, and one-third will be fully autonomous. Table 7.3 shows the second deployment schedule under which semi-autonomous vehicles are gradually displaced by fully autonomous vehicles. We assume that even when most vehicles are fully autonomous, a small number of drivers still prefer driving themselves.

Table 7.3 Deployment schedule in Figure 7.7

Schedule	*1*	*2*	*3*	...	*31*
Type H	90%	87%	84%	...	0%
Type SA-{ACC, CC, Com}	9%	11%	13%	...	69%

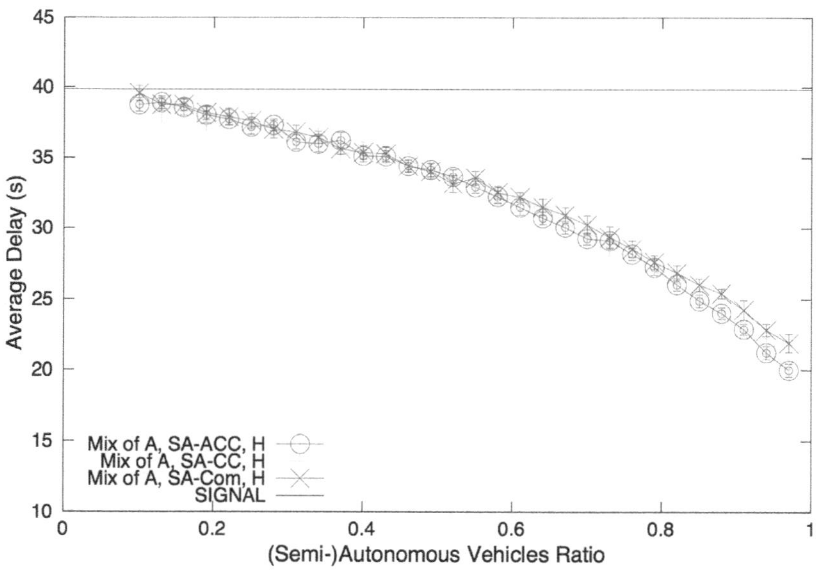

Figure 7.7 Average delay according to the deployment schedule in Table 7.2.
Traffic level = 360 vehicles/lane/hour

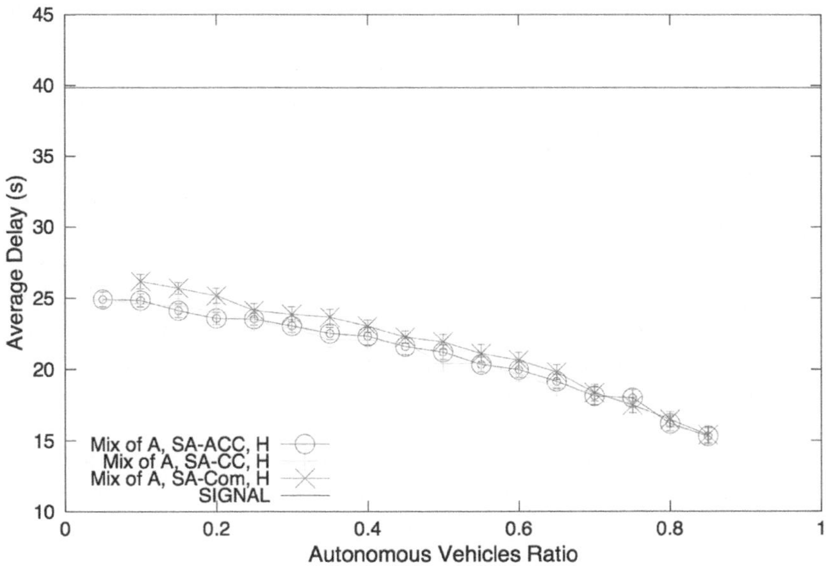

Figure 7.8 Average delay according to the deployment schedule in Table 7.3.
Traffic level = 360 vehicles/lane/hour

Figures 7.6 and 7.7 show the results of the simulations according to these deployment schedules. The simulation setup is exactly the same as in the first experiment. For each vehicle distribution, we ran the simulation 30 times for 1800s each run. Thus, each data point in the figures is an average of 30 delay times. The 95 percent CIs of the average are shown as the error bars in the figures. According to these figures, the average delay of the vehicles keeps decreasing as (semi-)autonomous vehicles are deployed, and the choice of the semi-autonomous vehicle technology does not matter much. Ultimately, the average delay depends on the number of drivers who insist on driving themselves.

7. Related Work

The main contribution of our work is an extension to the FCFS-Signal policy proposed by Dresner and Stone.[2] The FCFS-Signal policy allows autonomous vehicles to enter the intersection if there is no potential collision with trajectories reserved for human-driven vehicles. Their experimental results indicated that a mixture of human-driven vehicles and autonomous vehicles is possible, and leads to better performance than having all human-driven vehicles, which is the current status quo. However, their experiments indicated that the impact of autonomous vehicles is expected to be relatively small until almost all (90–95 percent) of the vehicles on the road are autonomous.[5] Our extension to embrace semi-autonomous vehicles shows significant performance benefits all along the technology penetration curve.

Our work is similar to the analysis of adaptive cruise control performance by Jerath and Brennan, who showed that by introducing adaptive cruise control vehicles into traffic, the vehicles would have a more condensed formation, thus increasing the efficiency of traffic.[9] They concretely explained the impact of adaptive cruise control (ACC) on congestion due to the formation of self-organizing traffic jams (or "phantom" jams). While the analysis of Jerath and Brennan is focused on highways, we are focusing on the intersection—a place more critical both from the points of view of congestion and safety. Selvakumar and Muthukumaran also make use of adaptive cruise control technology, but with a different focus on the context of driving.[10] It assumes a GPS equipped ACC system that (apart from performing normal ACC functions) slows down the vehicle intelligently when it enters speed restricted zones such as schools and colleges. It is also capable of detecting the speed breakers ahead and controls the vehicle dynamically according to the speed limit set for that part of the road.

There is limited literature on understanding changes to road infrastructure that can facilitate vehicular autonomy. One such project on jointly optimizing autonomous vehicles and road infrastructure is the PATH program, which relies on magnetic markers in the roadway for measuring steering angle and vehicle movements.[11] Kolodko and Vlacic used golf-cart-like Imara vehicles in evaluating an autonomous intersection.[12] Our work differs from these previous research by incorporating semi-autonomous vehicles into the mix, along with autonomous and human-driven vehicles. We do not utilize the potential change of the infrastructure. However, this can be done by dedicating certain lanes to certain types of vehicles.

Our work assumes that the vehicles run towards their destination lanes with smooth trajectories. They are mostly straight lines or smooth curves. Wuthishuwong et al. redesigned the trajectories for safety concerns.[13] They also used Vehicle to Infrastructure communication (V2I) approach. By dividing the space in the intersection into discrete grids, they used a dynamic programming method to determine the optimal trajectory. On the other hand, Au et al. did similar work on designing trajectories when a collision is about to happen.[14] In our work, we mark the cases when the vehicles lose control as emergencies and cause the vehicles to break. However, it is possible to devise paths for the colliding vehicles to evade. The plan devised by Au et al. guarantees that once an evasion plan is executed, there will be no collision among vehicles that have not yet collided. This can be incorporated into our work which involves semi-autonomous vehicles.

The complexity of communication between the intersection and autonomous vehicles is discussed in Yu et al.[15] If a request is rejected by the intersection manager, the autonomous vehicle has to resend the request, which has much redundancy. Yu et al. proposed a ksync policy, which is inspired by the rsync algorithm that solves such redundancy. Only necessary information is transmitted at each time step. They have shown empirical results that the average compression rate of data transferred by each autonomous vehicle can be improved by over 80 percent based on the experimental data on the platform of the open-source AIM simulator. Such a policy can also be applied to the semi-autonomous vehicles defined in this chapter.

Cooperation of multiple intersections are also discussed in the literature. As a continuum of Dresner and Stone,[2] Hausknecht et al.[16] apply the AIM policy on multiple intersections. Hausknecht et al. devised a navigation policy that determines the route the vehicle takes from the source to the destination. For example, one planning approach is time-based A⋆. As considering a network of intersections, the authors considered the effects of network topology and lane reversal. The concept of Infrastructure-to-Infrastructure (I2I) communication is proposed for an intersection to actively collaborate with its neighbor intersections.[17] An intersection can gather information from its neighbors, so a system that controls all the intersections may not be required.

Another way to manage autonomous vehicles at intersections is based on Vehicle-to-Vehicle (V2V) communication, such that vehicles coordinate in a peer-to-peer fashion when crossing the intersection.[6] The anchor requests in Section 5 can be implemented using V2V communications. Naumann et al. investigated a distributed policy that uses virtual "tokens" that a vehicle must possess to cross certain contested areas of the intersection[18] and formally evaluated it using petri-nets. The information of the locations, velocities and accelerations are exchanged among the vehicles for the decision of entering the intersection. VanMiddlesworth et al. developed an AIM-inspired protocol that enables vehicles to "call ahead" to reserve space-time in the intersection.[11] They assumed an agent placed at the intersection that functions like a stop sign, which does not do simulation as the intersection manager does in this chapter. Such a policy is only efficient for small traffic level. However, only considering small traffic level may yield more efficient solutions. Neuendorf et al. designed a vehicle platoon controller in decentralized autonomous intersection management, which is platoon-stable and has enlarged functionality in comparison to a conventional one.[19] Some other works use Vehicle-to-Vehicle approach with different focus. For example, we can require that vehicles cross the intersections without stopping.[20] The authors also use a simulation approach, but using a stronger mathematical model.

8. Conclusions and Future Work

This chapter introduces SemiAIM, a new constraint-based autonomous intersection management system that enables human-driven vehicles and semi-autonomous vehicles, in addition to fully autonomous vehicles, to make reservations and enter an intersection in an AIM-like style. To the best of our knowledge, SemiAIM is the first intersection control protocol to leverage limited autonomy of robotic cars to enable smooth interactions between human-driven, fully autonomous, and semi-autonomous vehicles at intersections. Our experimental results showed that our system can greatly decrease traffic delay when most vehicles are semi-autonomous, even when few (if any) are fully autonomous. Our incremental deployment study shows that traffic delay keeps decreasing as more vehicles employ features of autonomy. In the future, we intend to devise better constraint-based reservation requests using more accurate profiling of the vehicles' physical behavior.

Acknowledgements

This work has taken place in 1) the Learning Agents Research Group (LARG) at the Artificial Intelligence Laboratory, The University of Texas at Austin, and 2) the Agents and Robotic Transportation Lab (ART) at Ulsan National Institute of Science and Technology (UNIST). LARG research is supported in part by grants from the National Science Foundation (CNS1330072, CNS1305287), ONR (21C18401), AFOSR (FA87501410070, FA9550141 0087), and the Federal Highway Administration (DTFH6107H00030).

References

1. M. Buehler, K. Iagnemma, and S. Singh, "The DARPA Urban Challenge: Autonomous Vehicles in City Traffic." Springer, Berlin, 2009.
2. K. Dresner and P. Stone, "A multiagent approach to autonomous intersection management." In *Journal of Artificial Intelligence Research (JAIR)*, March 2008.
3. D. Fajardo, T.-C. Au, S. T. Waller, P. Stone, and D. Yang, "Automated intersection control: Performance of a future innovation versus current traffic signal control." In *Transportation Research Record: Journal of the Transportation Research Board*, no. 2259, pp. 223–232, 2012.
4. NHTSA, "Preliminary statement of policy concerning automated vehicles." National Highway Traffic Safety Administration, Technical Report, 2013.
5. K. Dresner and P. Stone. "Mitigating catastrophic failure at intersections of autonomous vehicles." In *AAMAS Workshop on Agents in Traffic and Transportation*, pp. 78–85, Estoril, Portugal, May 2008.
6. M. VanMiddlesworth, K. Dresner, and P. Stone, "Replacing the stop sign: Unmanaged intersection control for autonomous vehicles." In *AAMAS Workshop on Agents in Traffic and Transportation*, Estoril, Portugal, pp. 94–101, May 2008.

7. K. Dresner and P. Stone, "Sharing the road: Autonomous vehicles meet human drivers." In *Proceedings of the International Joint Conference on Artificial Intelligence (IJCAI)*, 2007.
8. S. Sheikholeslam and C. A. Desoer, "Longitudinal control of a platoon of vehicles." In *American Control Conference*, 1990, pp. 291–296.
9. K. Jerath and S. N. Brennan, "Adaptive cruise control: Towards higher traffic flows, at the cost of increased susceptibility to congestion." In *Proceedings of the 10th International Symposium on Advanced Vehicle Control (AVEC'10)*, Loughborough, UK, August 22–26, 2010.
10. M. Selvakumar, and D. Muthukumaran. "Real-time location based adaptive cruise control system." In *International Journal of Engineering Sciences and Research Technology*, pp. 1359–1362, vol. 4, no. 4, 2014.
11. S. Shladover, C. Desoer, J. Hedrick, M. Tomizuka, J. Walrand, W.-B. Zhang, D. McMahon, H. Peng, S. Sheikholeslam, and N. McKeown. "Automated vehicle control developments in the path program." In *IEEE Transactions on Vehicular Technology*, vol. 40, no. 1, pp. 114–130, 1991.
12. J. Kolodko and L. Vlacic, "Cooperative autonomous driving at the intelligent control systems laboratory." In *IEEE Intelligent Systems*, vol. 18, no. 4, pp. 8—11, July–August 2003.
13. C. Wuthishuwong and A. Traechtler. "Vehicle to infrastructure based safe trajectory planning for Autonomous Intersection Management." In *Proceedings of IEEE International Conference on ITS Telecommunications (ITST)*, pp. 175–180, 2013.
14. T.-C. Au, C.-L. Fok, S. Vishwanath, C. Julien, and P. Stone. "Evasion planning for autonomous vehicles at intersections." In *IEEE/RSJ International Conference on Intelligent Robots and Systems (IROS)*, pp. 1541–1546, 2012.
15. C. Yu, G. Tan, and Y. Yu. "Make driver agent more reserved: An AIM-based incremental data synchronization policy." In *Proceedings of IEEE International Conference on Mobile Ad-hoc and Sensor Networks*, pp. 198–205, 2013.
16. M. Hausknecht, T.-C. Au, and P. Stone. "Autonomous intersection management: Multi-intersection optimization." In *Proceedings of IEEE/RSJ International Conference on Intelligent Robots and Systems (IROS)*, pp. 4581–4586, 2011.
17. C. Wuthishuwong and A. Traechtler. "Coordination of multiple autonomous intersections by using local neighborhood information." In *Proceedings of IEEE International Conference on Connected Vehicles and Expo (ICCVE)*, pp. 48–53, 2013.
18. R. Naumann and R. Rasche, "Intersection collision avoidance by means of decentralized security and communication management of autonomous vehicles." In *Proceedings of the 30th International Symposium on Automotive Technology & Automation* (ISATA) —ATT/IST, 1997.
19. N. Neuendorf, and T. Bruns. "The vehicle platoon controller in the decentralised, autonomous intersection management of vehicles." In *Proceedings of the IEEE International Conference on Mechatronics (ICM)*, pp. 375–380, 2004.
20. M. Tlig, O. Buffet, and O. Simonin. "Decentralized traffic management: A synchronization-based intersection control—extended version", Research Report No. 8500, INRIA (2014).

8

BEHAVIORAL APPROACHES FOR COMBATING AGGRESSIVE DRIVING

David L. Wiesenthal,[1] *James E. W. Roseborough,*[1] *and Christine M. Wickens*[2]

(1) YORK UNIVERSITY, TORONTO, (2) CENTRE FOR ADDICTION AND MENTAL HEALTH, TORONTO

Aggressive Driving: Combating the Problem

Aggressive driving is a serious public health concern because of the stress and frustration it can cause, and the dangerous and potentially fatal consequences that may arise from these unsafe behaviors. Currently, in psychology and in the general public, there is increasing concern with the problem of aggressive driving. There have been several recent reviews of the problem appearing in the psychological literature (Wickens et al. 2014; Wickens, Mann, & Wiesenthal, 2013; Wiesenthal, Lustman, & Roseborough, in press; Wiesenthal & Singhal, 2012). The US National Highway Traffic Safety Administration (NHTSA, 2008) has estimated that approximately two-thirds of roadway fatalities can be attributed to behaviors associated with aggressive driving. Concern over driver aggression is an international phenomenon (Björklund, 2008; Chliaoutakis et al., 2002; Elliot, 1999; Félix, Kluppels, Meulemans, Vandenreijt, & Wissur, 2000; Golias & Karlaftis, 2002; Lupton, 2002; Parker, Lajunen, & Summala, 2002), having received attention from a variety of media (Lupton, 2001), motoring organizations (see the American Automobile Association report by Rathbone & Huckabee, 1999), safety organizations (e.g., Canada Safety Council), and governmental bodies (Goehring, 2000; Ministry of Transportation, Ontario, 2003; National Highway Traffic Safety Administration, 1998).

In 1995, the Automobile Association in Great Britain conducted a study of 526 drivers in an effort to quantify the extent of aggressive driving (Joint, 1997). The results indicated that 88 percent of the respondents had experienced aggressive behaviors by other drivers in the last 12 months. Specific aggressive driving behaviors experienced included: aggressive tailgating (62 percent), high beaming (59 percent), receiving obscene or rude gesturing (48 percent), deliberately being blocked by another vehicle (21 percent) and receiving verbal abuse (16 percent). Around 1 percent of drivers said that they had been physically assaulted by another driver. Men (52 percent) were more likely than women (42 percent) to have received aggressive or rude gestures and were also more likely to have received verbal abuse (19 percent and 10 percent respectively). Some 12 percent of the respondents indicated that they had not experienced any aggressive driving behaviors. Respondents were also asked about the aggressive behaviors they may have displayed toward other drivers. A total of 40 percent of respondents indicated that they had never behaved aggressively toward other motorists. However, 60 percent admitted to aggressive behaviors such as those listed above. Men were more likely than women to admit to having engaged in any of these behaviors (64 percent and 54 percent respectively). The majority of respondents also felt that the behavior of drivers had changed for the worse in recent years. A national Canadian survey found that the public viewed aggressive driving as an increasingly dangerous activity, and indicated a need for greater police enforcement (Vanlaar, Simpson, Mayhew, & Robertson, 2007).

This chapter will not present an overview of the various factors of stress, roadway situations, attributional issues and personality, and the role they have played in the instigation of aggressive driving. Interested readers are directed to the reviews previously cited for an evaluative summary of these variables. Instead, this chapter will highlight the role that psychology can play in understanding and coping with aggressive roadway behaviors, and will conclude with an appeal for evaluation research as a necessary component of any initiative to reduce aggression. We argue that psychological strategies should be employed along with the tactics developed by highway engineering, police enforcement, education, and vehicle design. Psychological initiatives will be presented along with their limitations and recommendations for conducting evaluation research.

Defining Aggressive Driving

Defining aggressive driving has been a controversial issue for the literature, which is replete with vague terms and categories (e.g., "road rage"). Comparison of findings across research studies can be problematic, as much of the writing and research on the topic varies in terms of how broad or narrow a definition of aggressive driving has been adopted. We propose defining *aggressive driving* as representing behaviors that are prohibited by traffic legislation (e.g., following too closely, failure to dim high beams, speeding, weaving between lanes, etc.) and rude actions that cross the boundary of polite behavior (e.g., swearing, making obscene gestures, etc.). *Driver violence* consists of actions that violate the criminal code such as assaults, shootings, and the pointing of firearms at other vehicles, to name a few. The problem of defining aggressive driving is mirrored in the literature of aggression, with two classes of definition: 1) response form and 2) behavioral intent (Goranson, 1970). Response form definitions operationalize definitions in terms of specific actions (e.g., hitting a vehicle, kicking), while behavioral intent definitions posit that there has to be an *intent* to harm another regardless of the nature of the observable action. We are proposing a response form definition of aggressive driving because it is difficult, if not impossible, to establish intent on the part of other drivers in many roadway situations.

Elements in Aggressive Driving

Different disciplines have approached aggressive driving from the framework of their practice. Highway engineers have incorporated design elements in roads aimed at controlling drivers, while vehicle designers have examined vehicle design and developed safety devices to reduce the likelihood of collisions. Traffic psychology has examined psychological constructs such as stress, personality variables, behavior modification, attributional processes, perceptual variables, etc. (Glendon, 2011; Hennessy & Wiesenthal, 2005).

The Road Environment

Traffic calming techniques are design elements in roadways with the primary purpose of reducing vehicle speed or eliminating the problem of drivers not stopping at signed intersections. Traffic calming techniques consist of roundabouts replacing intersections with stop signs, rumble strips placed either at right angles to the pavement to indicate to drivers to reduce vehicle speed, or parallel to the road shoulders to signal that vehicles are departing from the lane, speed bumps, pinch points, and chicanes (forced turns in the road) to slow vehicles. Traffic calming measures have been around for more than half a century (Kjemtrup & Herrstedt, 1992; see Bunn, Collier, Frost, Roberts, & Wentz, 2003; Elvik, 2001 for meta-analyses on their effectiveness), but until recently, traffic calming has not been extensively employed in North America.

Vehicle Design

A variety of devices have been developed to control aspects of vehicle operation such as speed governors, GPS-based speed recorders, blind-spot warnings, lane departure indicators and adaptive cruise control. These devices are now found on trucks and automobiles, and may serve to control aggressive driving and reduce the provocative events that stimulate hostility in other road users. Smart, Cannon, Howard, Frise, and Mann (2005) proposed that vehicle design may play a role in reducing aggressive roadway behaviors. They suggested that to reduce repeated horn blowing and headlight flashing, switches controlling these functions could be altered to provide a waiting period before they could be activated again. Smart et al. have also argued that vehicles could be equipped with an electronic sign that drivers could activate to communicate an apology to offended motorists. Voice-synthesized feedback could signal the driver that he/she is following too closely or is speeding. Active cruise control would also serve to prevent tailgating. Voice stress analyzers or heart rate sensors in the steering wheel could activate the music player to soothe stressed drivers. Research has established that as heat rises, so do aggressive behaviors (Anderson, 1989; Baron & Bell, 1975), thus air-conditioned vehicles may serve to control the tempers of frustrated drivers. Noise is another stressor related to aggression (Glass & Singer, 1972), so the use of air conditioners would also serve to reduce vehicle noise from open windows as well as reduce exposure to exhaust and other noxious odors.

Enforcement

Photo radar installations along roads and at intersections automatically record speeders and vehicles running red signals or entering intersections during the amber phase of traffic signals. It might be interesting to have a changeable message sign at these intersections or sections of roadways indicating to passing motorists the number of speeding drivers or red-light runners detected by photo radar installations (Roseborough & Wiesenthal, 2014). These devices may be augmented by either portable or fixed radar-activated speed warning signs (Van Wagner, Van Houten, & Betts, 2011). Police may use both marked and unmarked patrol cars equipped with radar guns to detect speeding drivers, as well as vehicles equipped with illegal radar detectors.

Another tactic used by police involves the use of public service messages urging drivers to use their cellular telephones (once they have pulled off the roadway) to contact police after observing problem drivers, reporting the vehicle, its license plate number, along with the vehicle's location and the nature of the offense. Drivers so identified receive a warning letter cautioning them not to engage in the specific risky actions (Wickens, Wiesenthal, & Rippey, 2005). The encouragement of reporting aggressive drivers to highway police is aimed at both deterrence (the aggressive driver is aware that his/her driving is liable to receive police attention) and censuring the offending driver (by means of a warning letter and the opening of a police file to monitor subsequent complaints). It has also been suggested that the act of reporting may also serve to reduce tendencies to retaliate against the offending motorist (Brian Jonah, Transport Canada, personal communication, 2011).

Psychological Strategies

Psychological approaches to the problem of aggressive driving have been drawn from a variety of subdisciplines including social, clinical, cognitive, applied psychology and learning theory. Applied psychology's techniques have found widespread adoption in educational and organizational settings, as well as in childrearing and clinical institutions because of their low cost, ease of implementation, and efficacy. The following discussion will provide a brief overview of how psychological theory and practice have already been used to control objectionable driver behavior citing evidence-based reports.

Stress Reduction

Research has demonstrated that drivers encountering traffic congestion experience stress, and that stress is intensified for those drivers whose baseline, or resting level, of stress is already high (Hennessy & Wiesenthal, 1999). As stressed drivers are more prone to drive aggressively (Hennessy & Wiesenthal, 1999), reducing stress should lower the probability of aggressive responding. Listening to music has been demonstrated to be an effective stress reducer, especially when the driver is allowed to select his/her favorite music as opposed to when the driver's choice may be limited by what is broadcast over automobile radios (Wiesenthal, Hennessy, & Totten, 2000). Deep breathing, a core element in yoga, has also been identified as an effective stress reducer in a variety of contexts, including the reduction of measured stress levels in different traffic conditions from freely moving to congested rush hour traffic (Wiesenthal, Hennessy, & Lubertacci, 2003). Urging drivers experiencing rush-hour delays to play their favorite music over their vehicle's sound system and teaching breathing exercises when stressed is likely to lower the risk of aggressive actions. For drivers unable or unwilling to adopt these tactics, public service announcements should urge the use of mass transportation, or car/van pooling with other drivers.

Modifying Attributional Processes

Driving research has examined how a motorist's attributions regarding another motorist's driving behavior contribute to subsequent thoughts, feelings, and behavior (Britt & Garrity, 2006; Roseborough, Wiesenthal, Flett, & Cribbie, 2011). Wickens, Wiesenthal, Flora, and Flett (2011) examined the influence of attributions of controllability, intentionality, and locus of causality, on judgments of responsibility, driving anger and driving aggression. Participants responding to offensive behavior described in a written vignette attributed more responsibility to a driver's actions if: 1) the driver was believed to be in control of behavior; 2) the behavior was due to an internal factor of the driver, and 3) if the behavior was perceived as intentional. Additionally, attributions of intentionality and judgments of responsibility were related to increased driving anger. Participants who completed a diary of anger-provoking driving behaviors also attributed more responsibility to actions that were believed to be controlled, intentional, and due to an offending driver's personality patterns. Furthermore, more anger was experienced when participants judged another driver as responsible for his/her behavior.

It is conceivable that motorists witnessing dangerous or illegal behaviors and who attribute responsibility to the offending motorist may desire punishment. Furthermore, if punishment is not applied via a collision or legal action, witnesses may try to apply justice or teach the offender a lesson, via retaliatory aggressive driving (Lennon & Watson, 2011; Roseborough & Wiesenthal, 2014; Wiesenthal, Hennessy, & Gibson, 2000). In line with Wickens et al. (2011), several studies have found support for the importance of attributions of intentionality in the development of roadway retaliation. Lustman, Wiesenthal, and Flett (2010) and Roseborough (2014) used vignettes to examine the influence of attributions about offensive driving behavior on retaliatory behavior. Lennon and Watson (2011) examined qualitative interviews to identify cognitions and motivations related to retaliatory aggressive driving. All three studies found that participants who attributed intentionality to another driver's behavior were more likely to retaliate. Lennon and Watson (2011) also identified that drivers who attributed intention felt retaliation was justified.

Over the past 50 years, the driving literature has investigated the influence of attributions on aggressive driving. Driving research exploring the role of attributions in aggressive driving is providing a better understanding of the mechanisms contributing to aggressive driving and the results may aid in the design of driver education programs. Training drivers to attribute offensive behaviors to situational causes rather than dispositional causes, or to recognize that their own roadway behaviors may be misperceived as offensive by others, may reduce the likelihood of subsequent driver aggression (Wickens et al., 2014).

Behavior Modification/Applied Behavior Analysis

Behavior modification, or applied behavior analysis, is the application of learning principles to cause behavioral changes in a variety of spheres (Kazdin, 1994). Prompts are events that help to initiate, facilitate, or encourage a desired response and by doing so, discourage undesirable behaviors (Kazdin, 1994; Wiesenthal & Hennessy, 2005). Prompts can be instructions, gestures, signs, or other cues that generate a specific behavior. Examples of prompting strategies are the anti-littering messages on product wrappers, or the health warnings on both cigarette packets and bottles of wine or liquors intended to discourage or limit the consumption of those products. The effectiveness of prompts depends on their clear specification of desired behaviors, the proximity of the prompt to where the behavior will be performed, the convenience of the requested behavior, the message's polite phrasing and its noticeability (Kazdin, 1989). Prompts have been found to be highly effective tools to influence a range of roadway behaviors in a variety of settings (Arnold & Van Houten, 2011; Crowley-Koch, Van Houten, & Lim, 2011; Hopper & McCarl-Nielsen, 1991; Okinaka & Shimazaki, 2011). In terms of road safety concerns, prompts have increased seat-belt use by up to 30 percent (Austin, Alvero, & Olson, 1998; Cox, Cox, & Cox, 2000; Geller, Bruff, & Nimmer, 1985; Gras, Cunhill, Planes, Sullman, & Oliveras, 2003; Thyer, Geller, Williams, & Purcell, 1987). Simple prompt signs have encouraged the use of designated drivers in a college pub (Brigham, Meier, & Goodner, 1995). Van Houten and Nau (1983) demonstrated that speeding on city streets could be reduced by distributing written prompts containing positive feedback regarding safe driving speeds to individuals prior to driving. Prompt signs have increased the number of vehicles stopping at pedestrian crosswalks, especially when used in combination with clearly marked stop lines on the road (Van Houten, 1988; Van Houten, Malenfant, & Rolider, 1985; Van Houten & Retting, 2001). Van Houten and Malenfant (1992) reported the maintenance of these effects one year following the removal of the prompt sign, although the stop lines may have served as a continued prompt to drivers.

Wiesenthal and Ford (1990; see Wiesenthal & Hennessy, 2005) successfully prompted drivers to make complete stops at four-way stop signs located in a variety of residential and commercial streets in Toronto. The behavior change was negatively related to how long the prompt sign was present. Complete stops peaked during the initial introduction of the prompt, but decreased steadily over the following six weeks, although even after the passage of the six weeks, the effectiveness of the prompt was still greater than during the baseline control period. Wiesenthal and Ford (1992) have suggested that behavior change may not persist when prompts are used for prolonged periods, thus bringing the issue of behavior sustainability into question. Prompts produce behavior change, but most previous research has employed the operant conditioning design in which prompts are introduced and removed in cycles so as to continually maintain a level of novelty regarding the prompt. These ABAB designs (where A is the time period assessing the baseline and B is the period where a reinforcement or stimulus is introduced) are useful for demonstrating that behavior is under the control of a reward or a specific stimulus. This experimental design reduces the likelihood of respondents habituating to the changed stimulus or reward, as the introduction of these changes is usually for a brief duration with A and B frequently alternated. If one is going to base public policy decisions on research derived from these designs, it is vital that the research provide assurance that the behavioral changes are enduring rather than transitory. Over time, the novelty of the prompt is likely to decline and its noticeability diminishes with the prompt becoming just another unseen traffic sign (Naatanen & Summala, 1976). Frequently rotating a series of prompts seems to enhance their noticeability (Wiesenthal & Hennessy, 2005). By using different prompts for stopping at stop signs and turn signaling, both messages had demonstrable effects on drivers. Habituation to the message was delayed by using the changeable messages with the effects as strong at the end of the sixth week as it was at the beginning (Wiesenthal & Hennessy, 2005). With the increased usage of changeable message signs on major highways that are currently used to display information about traffic delays, construction sites or lane and exit closures, there is the possibility of using these electronic message boards to promote safer driving by presenting variable safety prompts that offer resistance to habituation. By presenting a series of simply worded safety messages (e.g., "Observe speed limits," "Signal lane changes," "Always wear seat belts," "Don't text and drive") that change every few minutes, the driver who frequently travels over the same roadway section is less likely to habituate to the sign by reading the same unchanging message.

Passengers may also provide safety-related prompts to drivers. In one study, passengers were able to promote safer driving by providing feedback to drivers regarding errors observed during a 9 km ride over varied roads and highways (Hutton, Sibley, Harper, & Hunt, 2002). The major errors were inappropriate speed choice, tailgating, failure to check mirrors, and noticing hazards. Mirror checking increased for both drivers, along with choosing appropriate speeds. Unfortunately, the issue of sustainability of the safer behaviors was not addressed.

Incentive Programs

A major focus of applied behavior analysis has been the study of the role of reinforcement or reward in motivating desirable behavior. As part of the Fun Theory competition sponsored by Volkswagen, a novel project was implemented in Stockholm (Sweden) which involved the use of speed cameras and monetary prizes. Vehicles obeying the speed limit were photographed by a speed camera and their drivers automatically entered in a lottery that provided monetary rewards for obeying the speed limit. Project funding came from the fines collected from speeding drivers. A total of 24,857 vehicles passed the speed camera. Prior to the program's implementation, the average speed was 32 km/hr, but after the start of this initiative, the average speed decreased to 25 km/hr, a drop of 22 percent. One motorist was reported to receive a SEK 20,000 reward (US $3,000—see www.thefuntheory.com).

Battista, Burns, and Taylor (2010) described a successful replication of Dutch research which involved providing rewards for drivers of instrumented vehicles who refrained from either speeding or tailgating. The automobiles had sensing equipment and a display panel showing reward points (redeemable for a variety of gift cards). During the 12-week long reward phase of the project, both speed and headway compliance increased. Rewards ranged from $120 to $935, averaging $307. Not surprisingly, the improvement in safer driving was not sustained once the rewards were withdrawn. Aside of the excessive cost of the driver rewards, there is the further issue of generalizing the findings because it was not conducted over a longer time frame which might have indicated a diminution of effects. Recently, insurance companies have begun incorporating an incentive program in order to improve their customers' roadway behavior, offering rate reductions to customers who demonstrate continued safe driving practices (see www.allstate.com/drive-wise.aspx). What remains to be seen is whether monitoring of customer behavior by insurance companies will eventually result in increased premiums, as companies discover that their customers are a greater risk or liability than the company originally believed (Bell, 2011).

The Media's Role in Reducing Aggressive Driving

Cognitive psychology has found that being able to recall an incident seems to heighten the subjective probability that the event will occur (Tversky & Kahneman, 1973). This means that news media publicizing police actions in apprehending aggressive drivers might lower the prevalence of risky driving. Interviews with convicted aggressive drivers receiving widespread coverage may lead the public to believe they are likely to be detected should they engage in aggressive driving. It would also be useful for the police and the media to regularly publicize pro-social actions of drivers who have assisted other road users or have demonstrated notable acts of courtesy.

An award-winning multimedia advertising campaign commissioned by the Roads and Traffic Authority of New South Wales (Australia) employed a shaming strategy to target young male drivers. Focusing on the slogan "Speeding, no one thinks big of you," public service advertisements and commercials featured both young and older women pedestrians responding to a young aggressive male driver (who swerves and speeds on local streets) by wiggling their pinky finger at the driver. The wiggling finger is supposed to convey a contemptuous sexual insult, suggesting that the young man's macho display of reckless speed is overcompensation for other presumed deficits. The Roads and Traffic Authority found that 50 percent of males (aged 17–25) and 46 percent of the general population responded that, following viewing of the commercial, they would criticize another's driving. A total of 59 percent of the public and 54 percent of young males recognized the message's meaning that speeding is stupid and fails to impress the opposite sex (Roads and Traffic Authority, 2011). While previous campaigns featuring fear arousing messages and depicting wrecked vehicles and injured motorists were ineffective, this campaign resulted in a significant drop in speeding tickets (207,732 down from 226,239 the previous year). A similar drop occurred for "high risk" speeding tickets (a drop from 2,747 to 2,618). Crashes also decreased following the campaign (Anonymous, 2009). Unfortunately, there was no control condition studied, so there is no assurance that policing practices had not changed over the time period, or that some event (e.g., a celebrity killed during an episode of aggressive driving) had not occurred during the period under study that might serve to inhibit drivers.

Another form of media suited perfectly for the driving environment is the radio. In Germany, usage of radio ghosts has been employed in an attempt to reduce drink driving (Macleod, 2012). Radio ghosts are small radio transponders placed within roadside memorials. When motorists drive by listening to the radio, the transponder interrupts their program and substitutes the narrated story of what caused the roadside death of the victim at that location. Similar transponders could be placed at intersections equipped with red-light cameras or locations equipped with speed cameras to provide messages regarding the equipment's efficacy or near dangerous stretches of road to provide safety messages. The messages could be varied to alert drivers to delays caused by road construction, collisions, and lane blockages. The implementation of such devices should coincide with program evaluation research to assess their efficacy.

Entertainment Media Portrayals of Aggressive Driving

Following the release of the movie *Bullitt* (D'Antoni & Yates, 1968) featuring an exciting police chase over the streets of San Francisco, dangerous driving has become a common scenario in both motion pictures and television. Greenberg and Atkin (1983) performed a content analysis of 223 prime time broadcasts seen in 1975–1980 television seasons in the United States, assessing dangerous driving practices broadcast on popular television programming. Irregular driving acts (e.g., quick braking, screeching brakes, excessive speeding, etc.) occurred at the rate of 7.5 per hour. Additional research was conducted during the 1988 season. Atkin (1989) concluded that this type of programming was particularly problematic because: 1) novel actions were portrayed that would be unlikely to have been viewed outside of these programs; 2) the public may become disinhibited if these events were seen as both normative and justifiable, and; 3) the negative consequences (e.g., social disapproval, physical harm, punishment, etc.) of these risky driving behaviors were rarely depicted.

Shin, Hallett, Chipman, Tator, and Granton (2005) performed a content analysis of automobile and truck commercials airing in the United States and Canada between 1998 and 2002. Out of a total of 250 commercials, 45 percent depicted an episode of dangerous driving. Aggressive driving accounted for 85 percent of the dangerous driving sequences, with speed violations totaling 56 percent of the unsafe driving scenes. Currently, in Canada, there is an agreement (frequently violated) not to depict risky driving in television advertising. Given these realities, it might be reasonable to suggest a rating system for depictions of aggressive driving in a manner analogous to the showing of aggressive or sexual content in other viewing contexts.

Video games provide an additional genre of aggressive driving scenarios (Fischer, Kastenmüller & Greitmeyer, 2010; Fischer, Vingilis, Greitmeyer, & Vogrincic, 2011; Vingilis et al., 2013) where gamers may achieve high scores by displaying extremely risky simulated driving. In the context of video gaming, it is the active gamer, rather than those observing players, who displays the heightened tendency to later engage in risky driving in real life contexts. Given the research linking the playing of video games eliciting aggressive driving to the gamers' behavior on the road, it may also be worthwhile to consider restricting the distribution of such games to young drivers. Video games might also be rated for the level of risk-taking and aggression that play evokes.

Combating Aggressive Driving by Reducing Anonymity

Anonymity occurs when an individual cannot be identified by others and therefore cannot be judged, evaluated, criticized and, perhaps most importantly, punished, increasing the likelihood of engaging in aggressive behavior (Novaco, 1998; Zimbardo, 1969). Automobiles act as a physical and psychological shield from other motorists. By providing protection and insulation from the outside world, those in vehicles may be experiencing detachment and anonymity. The anonymity provided by this insulation weakens restraints and inhibitions that are usually exercised in public interactions (Ellison-Potter, Bell, & Deffenbacher, 2001; Wiesenthal & Janovjak, 1992).

Ellison, Govern, Petri, and Figler (1995) found that drivers with greater anonymity (e.g., drivers of convertibles with the top up, roof closed) compared to drivers in an identifiable condition (e.g., drivers of convertibles with the top down, roof open), honked sooner and for a longer duration when they were delayed at an intersection by a driver who failed to advance when the light turned green. While driving on urban roads there is little probability of repeated interactions with other motorists and such conditions lead to feelings of anonymity, power, and control (Hennessy, 1999; Hennessy & Wiesenthal, 2001a, 2001b; Lightdale & Prentice, 1994; Wiesenthal & Janovjak, 1992). The ease with which one can drive away, and the limited probability that one will be recognized by other motorists, provides optimal conditions for the emergence of discourteous and unsafe driving behaviors. One suggestion to inhibit or curb nasty behavior occurring because of anonymity would be to have the driver's name, telephone number, and email address on the back of a vehicle so that if the driver was behaving aggressively, he or she might be inhibited by the ease at which he or she could be reported. Local police would have contact information as well.

Intervention Strategies and Evaluation Research

Whatever technique or series of initiatives are implemented, program evaluation should be an essential component of any attempt to reduce or control aggressive driving. A budget for program implementation should have an automatic, mandated proportion of the total expenditure earmarked for conducting evaluation research. The program evaluation should be assessing actual behavioral change in terms of crashes, documented offenses, speed reductions, lane discipline, etc. rather than merely surveying motorists about their perception of the problem. While public opinion surveys and focus groups are useful for gauging reactions of road users, they are not substitutes for data on the effectiveness and costs of implementing and maintaining various programs. Evaluation research needs to be conducted over a reasonably long period to establish whether effects are merely short-term or indicative of a lasting, more enduring behavior change that is truly cost-effective. It is no longer sufficient to simply demonstrate an effect or establish a relationship between several variables. Longer periods of evaluation research lend increased confidence to the durability or sustainability of public policies and safety initiatives (Linden et al., 2010).

The safety initiative should involve planning well before the actual start of the program to collect data both prior to, during, and over a sufficient time period. A cost/benefit analysis of the intervention is needed because the cost of the intervention may be grossly out of line with the cost and extent of the problem specific to the community. Engaging in a systematic analysis of the extent of the problem, providing a clear description of the nature/methodology of the program, and then statistically documenting the initiative's outcome over a reasonable time period would also benefit other jurisdictions tackling a similar problem. The results of the evaluation research, regardless of whether the program is a success or a failure, need to be reported in transportation research journals and conferences so that workers in the field can benefit from the initiative or modify the program to strengthen its effects or adapt it to local realities and needs. By demonstrating the outcome of the intervention(s), public policy measures can be considered and the less effective procedures currently in place can be dropped. Publicizing the program using a variety of media (including open access Internet publications) serves to prevent other jurisdictions from making mistakes or going in ineffective directions.

Finally, the training of students is vitally important to provide the next generation of researchers. Governmental agencies need to open themselves to outside researchers from universities and host students as interns. These departments need to make the data they have already collected (which belongs to the public and should be in the public domain) available to graduate student researchers for their training. Universities need to teach graduate students in engineering, human factors, social science, and policy studies, the requisite methodologies and statistical analyses for performing program evaluations (Drenth, 2008).

Conclusions: Dealing with Aggressive Drivers

Kazdin (2009) described important characteristics of "wicked problems" which are problems that require novel ways of thinking, evaluation, and intervention strategies. As such, this approach may prove useful in dealing with the problem of aggressive driving. Aggressive driving may be conceptualized as resulting from a series of intersecting trends. Increased automobile ownership, limited highway facilities, inadequate mass transportation have all contributed to highway congestion. Congestion has increased the number of drivers competing for quick commutes who may behave aggressively due to the delays they encounter, along with other stressors like heat, noise, odors, etc. Drivers are experiencing longer commuting times as they live at increasing distances from workplaces. Stress and budgetary crises have resulted in declining maintenance of existing highways and restrictions in building additional roadways, further increasing the stresses experienced by motorists. Other "wicked problems" stem from political and economic challenges in building efficient mass transit systems and in resistance to flexible working times. Given the abundance of environmental contributors to roadway stresses and the variety of motives for engaging in aggressive driving, it seems reasonable to suggest a multifaceted approach in dealing with roadway aggression that draws upon engineering, policing, psychology, and vehicle design. Such an approach would have a greater likelihood of success because each of the techniques that have been implemented targets different motivational roots and offers diverse strategies. No single solution will likely be effective in reducing aggressive behavior. Thus, it is important to consider public education initiatives, enhanced enforcement techniques, incentive programs, teaching stress reduction techniques, and other potential initiatives. Another matter to be considered is how society will be able to judge when aggressive driving is no longer a problem. Like other "wicked problems," it is likely always to be present to some extent and all that might be hoped for would be a reduction in the frequency and severity of aggressive driving encounters.

Kazdin (2009) points out that multiple stakeholders will formulate their own idiosyncratic suggestions to deal with "wicked problems." Police, automobile clubs, insurers, media, educators, and government officials are likely to have very different ideas and solutions for aggressive driving. Their values and organizational cultures are likely to hold differing criteria for judging the worthiness, as well as how to measure the outcome, of various reduction strategies. Concern with reducing or eliminating media depictions of aggressive driving may evoke concern over censorship. Different countries, as well as regions within a given country, are likely to have their own history and tactical set for dealing with roadway issues. What constitutes aggressive driving in one region, may not be identified as aggressive in another; traffic safety authorities are well aware of the concept of "driving culture" which defines normative behavior on roads in a given community (Zaidel, 1992).

As we are only now examining the role of personality and individual difference variables in the attribution process where motorists' behavior is perceived as hostile and potentially threatening, there may be new and innovative recommendations for specific therapeutic tactics for managing anger and reducing stress while driving. New developments in other relevant subdisciplines are also possible. Yet, information is likely to be incomplete and it may very well be the case that policy makers will have to arrive at decisions without complete knowledge of all the variables affecting aggressive driving. Complexities may stem from rural–urban differences, dealing with young drivers, insurance providers, automobile manufacturers, and the criminal justice system, along with political issues arising from the needs of these constituencies. While the overall problem of aggressive driving may seem comparable cross-nationally, closer study may reveal a host of differences requiring different solutions. Certainly, the previous strategies of public education, along with media campaigns will be one of the first ideas that policy makers will develop, but it may very well be the case that these previously tried solutions may not always be helpful if the motoring public does not identify itself as the root of the aggressive driving problem.

References

Anderson, C. A. (1989). Temperature and aggression: Ubiquitous effects of heat on occurrence of human violence. *Psychological Bulletin, 106*, 74–96.

Anonymous. (2009). Roads and traffic authority. Speeding. No one thinks big of you. Sydney, Australia: Clemenger BBDO.

Arnold, M. L. & Van Houten, R. (2011). Increased following headway with prompts, goal setting, and feedback in a driving simulator. *Journal of Applied Behavior Analysis, 44*, 245–254.

Atkin, C. K. (1989). Television, socialization and risky driving by teenagers. *Alcohol, Drugs, and Driving, 5*, 1–11.

Austin, J., Alvero, A. M., & Olson, R. (1998). Prompting patron safety belt use at a restaurant. *Journal of Applied Behavioral Analysis, 31*, 655–657.

Baron, R. A. & Bell, P. A. (1975). Aggression and heat: Mediating effects of prior provocation and exposure. *Journal of Personality and Social Psychology, 31*, 825–832.

Battista, V., Burns, P., & Taylor, G. (2010). Using rewards to influence driving behaviour: A field operational trial. Presented at the Canadian Multidisciplinary Road Safety Conference XX, Niagara Falls, ON.

Bell, C. (2011, August 13). Car insurer to monitor driving. Retrieved March 12, 2014 from: www.bankrate.com/financing/cars/car-insurer-to-monitor-driving/

Björklund, G. M. (2008). Driver irritation and aggressive behaviour. *Accident Analysis and Behavior, 40*, 1069–1077.

Brigham, T. A., Meier, S. M., & Goodner, V. (1995). Increasing designated driving with a program of prompts and incentives. *Journal of Applied Behavior Analysis, 28*, 83–84.

Britt, T. W. & Garrity, M. J. (2006). Attributions and personality as predictors of the road rage response. *British Journal of Social Psychology, 45*, 127–147.

Bunn, F., Collier, T., Frost, C., Roberts, I., & Wentz, R. (2003). Traffic calming for the prevention of road traffic injuries: Systemic review and meta-analysis. *Injury Prevention, 9*, 200–204.

Chliaoutakis, J. El., Demakakos, P., Tzamalouka, G., Bakou, V., Koumaki, M., & Darvin, C. (2002). Aggressive behaviour while driving as predictor of self-reported car crashes. *Journal of Safety Research, 33*, 431–443.

Cox, B. S., Cox, A. B., & Cox, D. J. (2000). Motivating signage prompts safety belt use among drivers exiting senior communities. *Journal of Applied Behavior Analysis, 33*, 635–638.

Crowley-Koch, B. J., Van Houten, R., & Lim, E. (2011). Effects of pedestrian prompts on motorist yielding at crosswalks. *Journal of Applied Behavior Analysis, 44*, 121–126.

D'Antoni, P. (Producer) & Yates, P. (Director). (1968). *Bullitt* (motion picture). United States, Warner Brothers/Seven Arts.

Drenth, P. J. D. (2008). Psychology: Is it applied enough? *Applied Psychology: An International Review, 57*, 524–540.

Elliot, B. J. (1999). Road rage—media hype or serious road safety issue? Third National Conference on Injury Prevention and Control, May 9–12, 1999, Brisbane, Australia. Retrieved from: www.Drivers.com/cgi-bin/go.cgi?type=ART&id=000000165&static=1

Ellison, P. A., Govern, J. M., Petri, H. L., & Figler, M. H. (1995). Anonymity and aggressive driving behavior: A field study. *Journal of Social Behavior and Personality, 10*, 265–272.

Ellison-Potter, P., Bell, P., & Deffenbacher, J. (2001). The effect of trait driving anger, anonymity, and aggressive driving behavior. *Journal of Applied Social Psychology, 31*, 431–443.

Elvik, R. (2001). Area-wide traffic calming schemes: A meta-analysis of safety effects. *Accident Analysis & Prevention, 33*, 327–336.

Félix, B., Kluppels, L., Meulemans, C., Vandenreijt, B., & Wissur, A. (2000). An educational programme for aggressive drivers. Aggressive Driving Issues Conference, October 16 to November 30, 2000, Ontario Ministry of Transportation, Ontario, Canada. Downloaded December 20, 2010 from: www.aggressive.drivers.com/main.html

Fischer, P., Kastenmüller, A., & Greitmeyer, T. (2010). Media violence and the self: The impact of personalized gaming characters in aggressive video games on aggressive behavior. *Journal of Experimental Social Psychology, 46*, 192–195.

Fischer, P., Vingilis, E., Greitmeyer, T., & Vogrincic, C. (2011). Risk-taking and the media. *Risk-Analysis: An International Journal, 31*, 699–705.

Geller, E. S., Bruff, C. D., & Nimmer, J. G. (1985). "Flash for life": Community-based prompting for safety belt promotion. *Journal of Applied Behavior Analysis, 18*, 309–314.

Glass, D. C. & Singer, J. E. (1972). *Urban Stress: Experiments on Noise and Social Stressors.* New York: Academic Press.

Glendon, A. I. (2011). Traffic psychology. In P. R. Martin, F. M. Cheung, M. C. Knowles, M. Kyrios, J. B. Overmier, & J. M. Prieto (Eds.), *IAAP Handbook of Applied Psychology*, pp. 543–558. Malden, MA: Wiley-Blackwell.

Goehring, J. B. (2000). Aggressive driving: Background and overview report. Washington, DC: National Conference of State Legislatures. Retrieved from: www.ncsl.org/programs/esnr/aggrdriv.htm

Golias, I. & Karlaftis, M. G. (2002). An international comparative study of self-reported driver behavior. *Transportation Research Part F, 4*, 243–256.

Goranson, R. E. (1970). Media violence and aggressive behavior: A review of experimental research. In L. Berkowitz (Ed.), *Advances in Experimental Social Psychology* (Vol. 5), 1–31. New York: Academic Press.

Gras, M. E., Cunhill, M., Planes, M., Sullman, M. J. M., & Oliveras, C. (2003). Increasing safety belt use in Spanish drivers: A field test of personal prompts. *Journal of Applied Behavior Analysis, 36*, 249–251.

Greenberg, B. S. & Atkin, C. K. (1983). The portrayal of driving on television, 1975–1980. *Journal of Communication, 33*, 44–55.

Hennessy, D. A. (1999). The influence of driving vengeance on aggression and violence. In D. Faulkner (Ed.), *Proceedings of the Canadian Multidisciplinary Road Safety Conference XI.* Halifax, Canada: Daltech Vehicle Safety Institute, Dalhousie University.

Hennessy, D.A. & Wiesenthal, D.L. (1999). Traffic congestion, driver stress, and driver aggression. *Aggressive Behavior, 25*, 409–423.

Hennessy, D.A. & Wiesenthal, D.L. (2001a). Gender, driver aggression, and driver violence: An applied evaluation. *Sex Roles, 44*, 661–676.

Hennessy, D.A. & Wiesenthal, D.L. (2001b). Further validation of the Driving Vengeance Questionnaire. *Violence and Victims, 16*, 565–573.

Hennessy, D.A. & Wiesenthal, D.L. (2005) Driving vengeance and willful violations: Clustering of problem driving attitudes. *Journal of Applied Social Psychology, 35*, 61–79.

Hopper, J. R. & McCarl-Nielsen, J. (1991). Recyling as altruistic behaviour: Normative and behavioral strategies to expand participation in a community recycling program. *Environment & Behavior, 23*, 195–220.

Hutton, K. A., Sibley, C. G., Harper, D. N., & Hunt, M. (2002). Modifying driver behavior with passenger feedback. *Transportation Research Part F, 4*, 257–269.

Joint, M. (1997). Road rage. In AAA Foundation for Traffic Safety (Ed.), Aggressive driving: Three studies. Retrieved June 22, 2010 from: www.aaafts.org/Text/research/agdrtext.btm

Kazdin, A. E. (1989). *Behavior Modification in Applied Settings* (4th ed.). Pacific Grove, CA: Brooks/Cole.

Kazdin, A. E. (1994). *Behavior Modification in Applied Settings* (5th ed.). Pacific Grove, CA: Brooks/Cole.

Kazdin, A. E. (2009). Psychological science's contributions to a sustainable environment. *American Psychologist, 64*, 339–356.

Kjemtrup, K. & Herrstedt, L. (1992). Speed management and traffic calming in urban areas in Europe: A historical view. *Accident Analysis & Prevention, 24*, 57–65.

Lennon, A. & Watson, B. (2011). "Teaching them a lesson?" A qualitative exploration of underlying motivations for driver aggression. *Accident Analysis & Prevention, 43*, 2200–2208.

Lightdale, J. R. & Prentice, D. A. (1994). Rethinking sex differences in aggression: Aggressive behavior in the absence of social roles. *Personality and Social Psychology Bulletin, 20*, 34–44.

Linden, R., Mann, R. E., Smart, R. G., Vingilis, E., Solomon, R., Chamberlain, E., Ashbridge, M., Rehm, J., Fischer, B., Stoduto, G., Wilk, P., Roerecke, M., Trayling, C., & Wiesenthal, D. L. (2010). Research, policy development and progress: Antisocial behaviour and the automobile. *Canadian Public Policy, 26*, S81–S93.

Lupton, D. (2001). Constructing road rage as news: An analysis of two Sydney newspapers. *Australian Journal of Communication, 28*, 23–36.

Lupton, D. (2002). Road rage: Drivers' understandings and experiences. *Journal of Sociology, 38*, 275–290.

Lustman, M., Wiesenthal, D. L., & Flett, G. L. (2010). Narcissism and aggressive driving: Is an inflated view of the self a road hazard? *Journal of Applied Social Psychology, 40*, 1423–1449.

MacLeod, D. (2012). Retrieved November 13, 2012 from: http://theinspirationroom.com/daily/2012/radio-ghosts/

Ministry of Transportation, Ontario (2003). "Aggressive driving and road rage". Retrieved December 2003 from: www.mto.gov.on.ca/english/safety/quicknotes/aggressive.htm

Naatanen, R. & Summala, H. (1976). *Road User Behavior and Traffic Accidents*. Amsterdam: North Holland Publishing.

National Highway Traffic Safety Administration (NHTSA) (1998). Final report: Capital beltway update: Beltway user focus groups. Washington, DC: US Department of Transportation. Retrieved from: www.nhtsa.dot.gov/people/injury/research/aggressive/final.rpt.html

National Highway Traffic Safety Administration (NHTSA). (2008). *Traffic Safety Facts 2008*. Washington, DC: National Center for Statistics & Analysis.

Novaco, R. W. (1998) *Roadway Aggression*, Institute of Transportation Studies Review, University of California, Vol. 21, No. 4, pp. 1–4.

Okinaka, T. & Shimazaki, T. (2011). The effects of prompting and reinforcement on safe behavior of bicycle and motorcycle riders. *Journal of Applied Behavior Analysis, 44*, 671–674.

Parker, D., Lajunen, T., & Summala, H. (2002). Anger and aggression among drivers in three European countries. *Accident Analysis & Prevention, 34*, 229–235.

Rathbone, D. B. & Huckabee, J. C. (1999). *Controlling Road Rage: A Literature Review and Pilot Study*. Washington, DC: The AAA Foundation for Traffic Safety.

Roads and Traffic Authority (2011). Speeding! No one thinks big of you campaign fact sheet. Retrieved March 12, 2014 from: www.rms.nsw.gov.au/newsevents/downloads/070803_pinkyfactsheet_dl1.html

Roseborough, J. E. W. (2014). Retaliatory aggressive driving: A justice perspective. (Unpublished doctoral dissertation). York University, Toronto, ON.

Roseborough, J. E. W. & Wiesenthal, D. L. (2014). Roadway justice—Making angry drivers, happy drivers. *Transportation Research Part F, 24*, 1–7.

Roseborough, J., Wiesenthal, D., Flett, G., & Cribbie, R. (2011). The belief in an unjust world and narcissism-the influence of personality on perceptions of injustice, driving anger and aggressive driving. In: Proceedings of the XXI Canadian Multidisciplinary Road Safety Conference. Transport Canada and Canadian Association of Road Safety Professionals, Canada.

Shin, P. C., Hallett, D., Chipman, M. L., Tator, C., & Granton, J. T. (2005). Unsafe driving in North American automobile commercials. *Journal of Public Health, 27*, 318–325.

Smart, R. G., Cannon, E., Howard, A., Frise, P., & Mann, R. E. (2005). Can we design cars to prevent road rage? *International Journal of Vehicle Information and Communication Systems, 1*, 44–55.

Thyer, B. A., Geller, E. S., Williams, M., & Purcell, E. (1987). Community-based "flashing" to increase seat belt use. *Journal of Experimental Education, 55*, 155–159.

Tversky, A. & Kahneman, D. (1973). Availability: A heuristic for judging frequency and probability. *Cognitive Psychology, 5*, 207–232.

Van Houten, R. (1988). The effect of advanced stop lines and sign prompts on pedestrian safety in a crosswalk on a multilane highway. *Journal of Applied Behavior Analysis, 21*, 245–251.

Van Houten, R. & Malefant, L. (1992). The influence of signs prompting motorists to yield before marked crosswalks on motor vehicle-pedestrian conflicts at crosswalks with flashing amber. *Journal of Applied Behavioral Analysis, 24*, 217–225.

Van Houten, R. & Nau, P. (1983). Feedback interventions and driving speed: A parametric and comparative analysis. *Journal of Applied Behavior Analysis, 12*, 253–281.

Van Houten, R. & Retting, R. A. (2001). Increasing motorist compliance and caution at stop signs. *Journal of Applied Behavior Analysis, 34*, 185–193.

Van Houten, R., Malefant, L., & Rolider, A. (1985). Increasing driver yielding and pedestrian signaling with prompting, feedback, and enforcement. *Journal of Applied Behavior Analysis, 18*, 103–110.

Vanlaar, W., Simpson, H., Mayhew, D., & Robertson, R. (2007). *The Road Safety Monitor 2006: Aggressive Driving*. Ottawa, ON: Traffic Injury Research Foundation.

Van Wagner, M., Van Houten, R., & Betts, B. (2011). The effects of a rectangular rapid flashing beacon on vehicle speed. *Journal of Applied Behavior Analysis, 44*, 629–633.

Vingilis, E., Seeley, J., Wiesenthal, D. L., Wickens, C. M., Fischer, P., & Mann, R. E. (2013). Street racing video games and risk-taking driving: An Internet survey of automobile enthusiasts. *Accident Analysis & Prevention, 50*, 1–7.

Wickens, C. M., Mann, R. E., Roseborough, J. E. W., Wiesenthal, D. L., & Smart, R. G. (2014). The straight 'A's of road rage: Attribution, anger, and aggressive behaviour. In M. G. Penrod & S. N. Paulk (Eds.), *Psychology of Anger: New Research*, (pp. 49–69). Hauppauge, NY: Nova Science Publishers.

Wickens, C. M., Mann, R. E., & Wiesenthal, D. L. (2013). Addressing driver aggression: Contributions from psychological science. *Current Directions in Psychological Science, 22*, 386–391.

Wickens, C. M., Wiesenthal, D. L. & Rippey, K. (2005). Motorists' perceptions of aggressive driving: A comparative analysis of Ontario and California drivers. In D. A. Hennessy & D. L. Wiesenthal (Eds.), *Contemporary Issues in Road User Behavior and Traffic Safety*, pp. 23–36. Hauppauge, NY: Nova Science Publishers.

Wickens, C. M., Wiesenthal, D. L., Flora, D. B., & Flett, G. L. (2011). Understanding driver anger and aggression: Attributional theory in the driving environment. *Journal of Experimental Psychology: Applied, 17*, 354–370.

Wiesenthal, D. L. & Ford, D. (1990). Driver behaviour and prompts. Highway Safety Conference, Ministry of Transportation, Toronto, ON.

Wiesenthal, D. L. & Ford, D. (1992). The effects of prompts on driver behaviour at signs. Third Annual Conference on Highway Safety, Ministry of Transportation, Toronto, ON.

Wiesenthal, D. L. & Hennessy, D. A. (2005). The effects of multiple variable prompt messages on stopping and signaling behaviors in motorists. In D. A. Hennessy & D. L. Wiesenthal (Eds.), *Contemporary Issues in Road User Behavior and Traffic Safety*, pp.185–198. Hauppauge, NY: Nova Science Publishers.

Wiesenthal, D. L. & Janovjak, D. P. (1992). Deindividuation and automobile driving behaviour. *LaMarsh Research Programme Report Series*, No. 46, 2, LaMarsh Research Programme on Violence and Conflict Resolution, York University, North York, Ontario, Canada.

Wiesenthal, D. L. & Singhal, D. M. (2012). Evolutionary psychology, demography and driver safety research: A theoretical synthesis. In S. Craig Roberts (Ed.), *Applied Evolutionary Psychology*, pp. 399–413. Oxford: Oxford University Press.

Wiesenthal, D. L., Hennessy, D.A., & Gibson, P.M. (2000). The Driving Vengeance Questionnaire (DVQ): The development of a scale to measure deviant driver attitudes. *Violence and Victims, 15*, 115–136.

Wiesenthal, D.L., Hennessy, D.A., & Lubertacci, F.S. (2003, June). The effectiveness of music and relaxation training in reducing driver stress. Presented at the Canadian Multidisciplinary Road Safety Conference XIII, Banff, Alberta.

Wiesenthal, D.L., Hennessy, D.A., & Totten, B. (2000). The influence of music on driver stress. *Journal of Applied Social Psychology, 30*, 1709–1719.

Wiesenthal, D. L., Lustman, M., & Roseborough, J. (in press). Aggressive driving. In A. Smiley (Ed.), *Human Factors in Traffic Safety* (3rd ed.). Tucson, AZ: Lawyers & Judges Publishing Company.

Zaidel, D. M. (1992). A modeling perspective on the culture of driving. *Accident Analysis & Prevention, 24*, 585–597.

Zimbardo, P.G. (1969). The human choice: Individuation, reason and order versus deindividuation, impulse and chaos. In W.J. Arnold, and D. Levine (Eds.), *Nebraska Symposium on Motivation* (Vol. 17), pp. 237–307. Lincoln: University of Nebraska Press.

9

PEDESTRIAN TRANSPORTATION

Christopher M. Monsere,[1] *Nick Foster,*[1] *Todd Borkowitz,*[1] *Sirisha Kothuri,*[1] *Matthew Downey, and Robert L. Bertini*[2]

(1) PORTLAND STATE UNIVERSITY, (2) CALIFORNIA POLYTECHNIC STATE UNIVERSITY

Introduction

In many locations, streets have been designed and operated primarily for use by motor vehicles. In many urban areas, however, it is becoming more common for streets to be viewed as significant public places. Bicycling, walking and other human-powered forms of travel—commonly referred to as active transportation—are gaining increased design attention for their associated economic, environmental and health benefits. New designs are being applied that modify streets to better meet the mobility needs of all users, and to be safer, more sustainable, resilient, multimodal and economically beneficial. Active transportation facilities that focus on moving people have a key role to play in this transformation.

This chapter provides an overview of the design assumptions and considerations for pedestrian facilities. Following a brief review of safety trends and discussion of general design philosophy, the chapter presents design inputs (dimensions and performance) and an overview of typical facilities for pedestrians. The design review includes operations, primarily at signalized intersections. The chapter primarily references US-based design standards and documents.

Safety Trends

The National Highway Traffic Safety Administration (NHTSA) reports that 4,432 pedestrians were killed in crashes in the US in 2011. Figure 9.1 shows an 18-year trend. In parallel with overall fatality trends, there has been a general decline in the numbers of fatalities as shown in the figure. However, over this period a total of 87,617 pedestrians were killed—an unacceptable number. The overall temporal and spatial trends of non-motorized fatalities have some similarities, as shown in Table 9.1. Around 70 percent of the fatalities occur in urban areas (where people are walking and bicycling). Most pedestrian (70 percent) fatalities occur away from intersections. An important distinction is that about 70 percent of pedestrian fatalities occur at night.

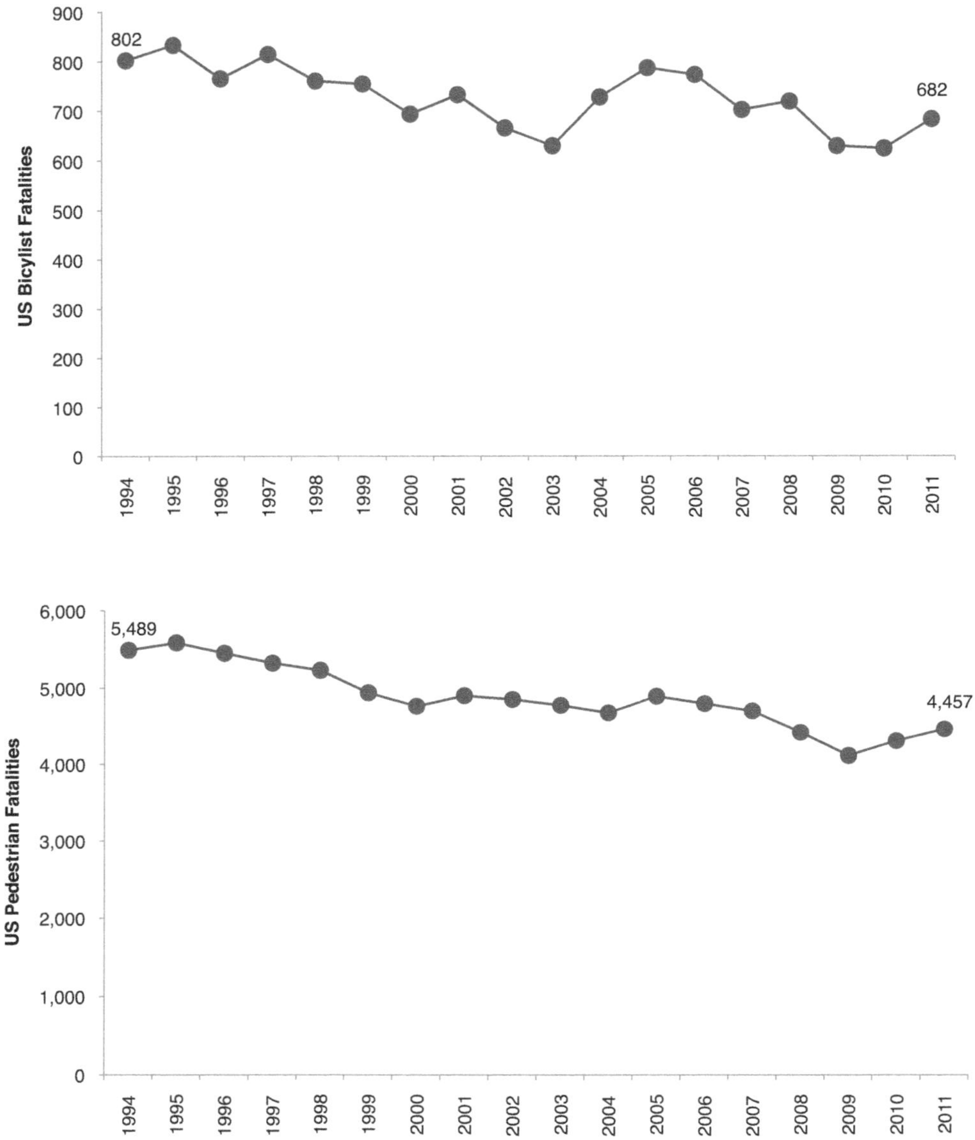

Figure 9.1 US Pedestrian fatality trends, 1994–2011

Crash Analysis Tools and Estimates of Safety Effectiveness

In the 1970s, methods for typing motor vehicle crashes with pedestrians were developed by NHTSA to better define the sequence of events and precipitating actions leading to pedestrian–motor vehicle crashes. These methods were applied in subsequent work by the University of North Carolina Highway Safety Research Center to develop the PEDSAFE toolkit (including the PBCAT tool, which does the actual crash typing. These tools help diagnose crash problems based on type and suggest mitigations. The tools for PEDSAFE can be found at the following website: www.walkinginfo.org/pedsafe/.

Table 9.1 Locations and context of US pedestrian fatalities (NHTSA)

Context	*Percentage of pedestrians killed*	
	2010	*2011*
Land use		
Rural	27	27
Urban	73	73
Non-motorist location		
Intersection	21	19
Non-intersection	68	70
Other	10	10
Weather		
Clear/cloudy	88	88
Rain	9	9
Snow	1	1
Fog	1	1
*Time of day**		
Daytime	32	30
Nighttime	68	70

Note: Percentage of unknown values are not displayed.
* Daytime: 6 a.m.–5.59 p.m. Nighttime: 6 p.m.–5.59 a.m.

In addition to these tools, the *Highway Safety Manual* and the Countermeasure Clearinghouse can be used to search for crash modification factors that can be used to estimate the safety effectiveness of treatments or design modifications. They can be found at the following: *Highway Safety Manual* (AASHTO, 2010): safety.fhwa.dot.gov/hsm; FHWA Countermeasure Clearinghouse: www.cmfclearinghouse.org/index.cfm.

Principles of Design

Effective network planning and facility design are critical for making walking safe, convenient and attractive. In the following sections, design minimums are presented. While employing minimum standards may be unavoidable in some locations, it is often preferable, even recommended to exceed them. As use increases, minimum designs are more likely to be inadequate.

The Sustainable Safety vision for road safety developed in the Netherlands (SWOV, 2012) is a helpful framework that can lead to a traffic system where serious crashes are prevented; facilities that recognize human vulnerability and cognitive capabilities and limitations; people, vehicles and the road are treated in an integrated manner; and safety gaps are bridged in a proactive way. With this in mind, a Sustainable Safety approach considers the following five principles:

- *Road functionality*: Roads are developed for single functionality (arterial vs. collector vs. local) in a hierarchically structured network.
- *Homogeneity of mass and/or speed and direction*: Speed, direction and mass should be equalized at moderate and high speeds.

- *Predictability of road course and road-user behavior by a recognizable road design*: The user's expectations should be supported through consistency and continuity of road design.
- *Forgiveness of the environment and of road users*: Injuries will be limited through the provision of a forgiving road environment and anticipation of road user behavior.
- *State awareness by the road user*: Users should be able to assess their capacity to handle the driving task.

The walking and bicycling modes are commonly connected in many transportation analysis, design, and planning activities. Indeed, there are many similarities including (ODOT, 2011):

- *Location*: While often physically separated from each other, both are typically located between motor vehicle lanes and right-of-way lines, often competing for space with transit, streetscape amenities and utilities.
- *Exposure*: Both are very vulnerable in crashes with motor vehicles and are exposed to the environment (both pleasant and unpleasant).
- *Wide range of users*: Bicyclists and pedestrians may be any age and there are also no licensing requirements. Pedestrian and bicyclist performance and behaviors vary, and change with age and other factors.
- *User preferences*: Both prefer side-by-side travel, separation from automobiles and each other, and well-maintained, safe and well-illuminated facilities.

However, it is clear that bicycling and walking also have key differences, and in most analysis, design, and planning it is advisable to treat these modes separately. Some key differences between the modes are (ODOT, 2011):

- *Accessibility*: Pedestrian facilities are expected to meet American with Disabilities Act (ADA) requirements given their more frequent use by persons with disabilities. Bicycle facilities (except shared-use paths) are not.
- *Speed and travel distance*: Bicyclists typically travel between 10 and 15 miles per hour, while pedestrians typically travel between 2 and 3.5 miles per hour. Bicyclists will typically travel up to 5 miles in urban areas, while pedestrians will typically travel up to 1 mile—each trip generally taking about 20 to 25 minutes.
- *Tolerance of out of direction travel*: Bicyclists generally tolerate some indirect travel while pedestrians may demand the most direct route.
- *Legal use of road space*: Bicyclists are generally allowed to operate with motor vehicle traffic. In many locations, a bicyclist can operate as either a vehicle or pedestrian under certain conditions. Pedestrians are generally only allowed in roadways when accessing or leaving a vehicle, in marked or unmarked crosswalks and on roadway shoulders.
- *Parking*: Bicyclists require safe, secure and convenient destination parking.

Analysis of Pedestrian Facility Performance

Facility performance is assessed in two ways: by examining how many people are using it or, in the case of a planned facility, the forecasted use; and/or by assessing the quality of service provided by the facility to its users, via some measure of performance. Both are important to consider.

Measuring Pedestrian Activity (Counting)

A pedestrian-traffic counting program should consist of long-term, permanent counters and short-term counts at key locations. Permanent counters should be sited so as to provide data useful for creating factors for adjusting short-term counts. Relying only on short-term counts can lead to large errors (Nordback et al., 2013), as pedestrian flows can vary widely based on the day of the week, season, time-of-day, and weather (Figliozzi et al., 2014; El Esaway, 2013). Automated technology for permanent non-motorized traffic counters is still evolving and some technology is better for bicyclists while some is better for pedestrians (Schneider et al., 2008). For more information, see Chapter 4 of FHWA's *Traffic Monitoring Guide* (2013a).

Predicting Activity (Demand Forecasting)

Accurately predicting future volumes can help transportation professionals make educated decisions about facility design and prioritization. Forecasting pedestrian volumes is a rapidly evolving practice. Some agencies include pedestrians in their travel demand models (Singleton and Clifton, 2013); however, they typically do not assign mode-specific trips to the network, as is done with motor vehicle traffic. Demand analysis is also done at the individual facility or site level for specific projects or developments. For more information on the state of the practice in non-motorized demand forecasting, see NCHRP Report 770—*Estimating Bicycling and Walking for Planning and Project Development* (Kuzmyak et al., 2014).

Assessing Quality of Service

Quality of service for pedestrians typically involves assessing the perceived quality of service from the user's perspective (i.e. comfort/satisfaction, speed or travel time). There are several methods in practice that quantify user perceptions, allowing for objective evaluation of projects and design trade-offs and communication of system performance. Most measures are primarily based on roadway geometry and motor vehicle volumes and speeds.

Highway Capacity Manual (HCM) 2010 Multimodal Level-of-Service

The 2010 edition of the *HCM* introduced new multimodal level-of-service analysis procedures, which include pedestrians, based on surveys of user perceptions. The models provide a Level of Service (LOS) grade on an "A"–"F" scale, similar to what is used for motor vehicles, for links (sections of road between signalized intersections), signalized intersections, and segments (intersections and links combined). The *HCM* also includes capacity-based measures for dense urban environments or other high-activity areas where available space is likely the overarching concern for pedestrians.

Danish Road Directorate Level-of-Service

Denmark has LOS models for pedestrians. Similar to the *HCM*, they are mathematical models that produce "A"–"F" letter grades and are based on user perceptions of satisfaction. The Danish models include some factors not considered by the *HCM*.

Bicycle/Pedestrian Environmental Quality Index

The San Francisco Department of Public Health used surveys of national experts to develop the Bicycle and Pedestrian Environmental Quality Indices. These methods award points to a facility for certain characteristics, with a final score on a 0–100 scale (SFDH, 2009).

Design Characteristics of Pedestrians

Nearly all roadway users, at one time in their travels, are pedestrians. Bicyclists must get to and from their bicycles, transit users to and from their stops, and drivers to and from their vehicles. At different times in their travels, pedestrians demand different amounts of space. Pedestrians carrying grocery bags, pushing their wheelchair, pushing a stroller, pulling a piece of wheeled

Table 9.2 Design considerations for mobility impairments (AASHTO, 2004)

Mobility impairment	*Design considerations*
Wheelchair users	• Firm, stable and non-slippery surfaces are easier to maneuver wheelchairs over. • Cross slopes and grades should be kept to a minimum; when over 2 percent, they create difficulty for wheelchair users to travel a straight route. • A wheelchair user, particularly when motorized, takes up more space than a standing person and demands additional radii for turning.
Walking-aid users (including crutches, canes, and walkers)	• Cracks and grates can catch walking aids. • Slippery and uneven surfaces can be hazardous. • Walking aid users often demand additional width for passing and turning. • Walking aid users often demand longer signal timing.
Prosthesis users	• Need longer signal timing. • Cross slopes and grades should be kept to a minimum as prosthesis users often have difficulty maintaining balance.
Visual impairments	• Benefit from detectable warnings at crossings that can be sensed by feet or other mobility devices. • Benefit from physical barriers located at construction areas, ending to crosswalks and sidewalks, or other potentially hazardous situations. • Benefit from directional guidance at detours, non-standard or complex intersections, pedestrian-actuated signals and beacons, and other locations. • Benefit when provided with information communicated via multiple senses. • Benefit from consistent sound, color and/or texture.
Hearing impairments	• Challenged in areas like complex intersections where hearing potential dangers like a moving vehicle may be the first sense that one is approaching or that there are multiple potential dangers. • Benefit when provided with information communicated via multiple senses. • Benefit from consistent sound, color and/or texture.
Cognitive impairments	• People with cognitive impairments can benefit from consistent and universal design of signage that utilizes symbols, colors, and/or pictures with minimal or no text (e.g. a person walking, instead of the word "WALK"); this is also helpful for children and non-English speakers.

luggage, running, riding a bicycle, waiting in line or in conversation with a friend may each have their own unique spatial needs. People of different ages and abilities may also travel at different speeds, demanding that infrastructure-like signals account for slower speeds and adequate space for passing. In addition, sidewalks that are contiguous, have safe crossings, maintain horizontal and vertical clearances free of obstructions and have amenities like benches and curb ramps, accommodate pedestrians with impairments to provide access and travel for all pedestrians.

The percentage of trips made by walking is distinct from that of other modes. Walking comprises 15.2 percent of all trips for people between the ages of 5 and 15, and 8.9 percent of people over 65 (Pucher and Renne, 2003). Nearly 40 percent of walking trips in the US are for errands, much higher than the 14 percent for bicycling trips. The Pedestrian and Bicycle Information Center reports that over 75 percent of pedestrian trips are, on average, 2 miles or less—about a 30-minute walk (PBIC, 2014). Overall, people walk for 10 percent of their trips, and also 5 percent walk for recreation and exercise (PBIC, 2014). Personal safety (both perceived and actual) is an important issue.

Pedestrians include three critical groups that require additional attention: 1) children, 2) seniors, and 3) people with mobility impairments. The following list summarizes some of the issues to be aware of in these three groups:

- Children are vulnerable due to their small size, which makes them more difficult for others to see. They can also be less predictable than adults and have less experience in assessing potential roadway dangers and developed skills with which to avoid them. Their cognitive abilities are less developed and gap judgment for street crossing can be particularly challenging.
- Seniors often experience less ability to detect, evaluate and react to roadway dangers. Also, their fragility may make them less likely to easily recover from crashes once struck.
- Many people have mobility impairments, whether ambulatory, visual, auditory, or mental disabilities, and whether temporary or permanent. While minimum standards required by the ADA help reduce barriers to publicly accessible facilities, an understanding of the specific needs of common impairments can be helpful to engineers and planners when designing facilities. Some key considerations, gleaned from the AASHTO guidance, are listed below in Table 9.2. Additional guidance can be found in the *AASHTO Guide for the Planning, Design, and Operation of Pedestrian Facilities.*

Physical Dimensions

Spatial needs for the average pedestrian are considered when designing pedestrian facilities or adjacent features to provide adequate comfort, safety and visibility. General unobstructed visual buffers for different pedestrian activities include:

- Public event: 6 feet.
- Shopping: 9–12 feet.
- Normal walk: 15–18 feet.
- Pleasure walk: 35 feet or more.

In addition, about 5 feet of obstruction-free surface width is needed to accommodate two people walking side-by-side or passing one another. Additional guidance on pedestrian facility design can be found in the *AASHTO Guide for the Planning, Design, and Operation of Pedestrian Facilities* (2004).

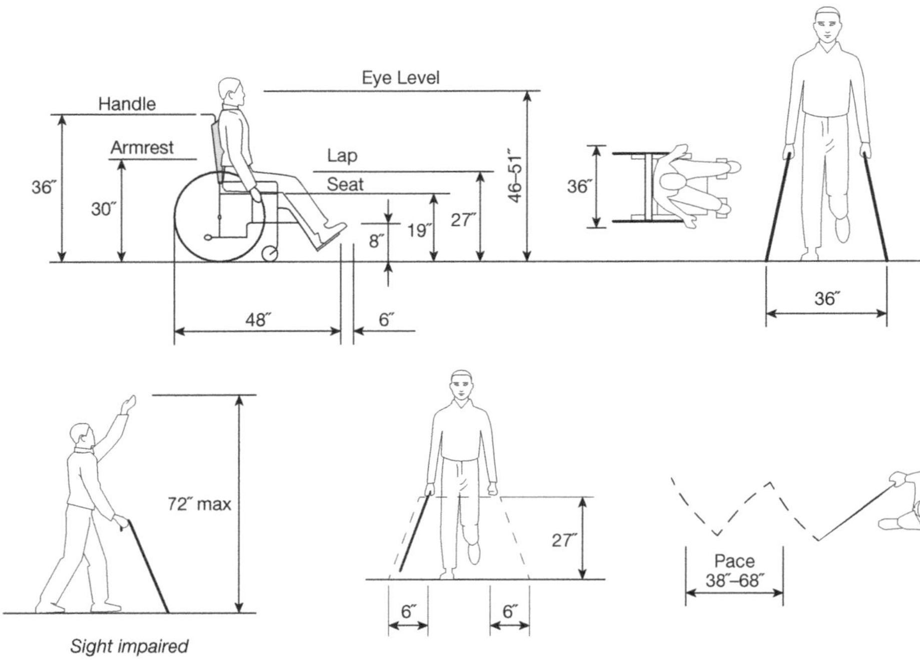

Figure 9.2 Spatial dimensions for pedestrians with disabilities

Source: Guide for Planning, Design, and Operation of Pedestrian Facilities. Copyright 2004 by the American Association of State Highway and Transportation Officials, Washington DC. Used by permission.

Performance

A person's physical capabilities and environmental factors like topography can have significant impacts on the performance values selected for design. The primary assumption for performance relates to walking speed.

Walking Speed

The MUTCD recommends a walking speed of 3.5 feet per second for signal timing. AASHTO recommends a walking speed of 3 feet per second in areas with large senior populations. Additional guidance on walking speed-responsive facility design can be found in the *AASHTO Guide for the Planning, Design, and Operation of Pedestrian Facilities*. In addition to personal characteristics, environmental factors may also impact walking speed. These may include: air temperature and precipitation; time of day; walk purpose (i.e. work/exercise, commuting, pleasure/shopping); slope; and congestion.

Pedestrian Facility Design and Operation

In urban areas, pedestrians should be given preference wherever possible. In designing pedestrian facilities, there are some general considerations that should be taken into account:

- *Accessibility*: Designers must meet a wide range of accessibility requirements to accommodate all users.

- *Social space:* Many pedestrian spaces are also social areas. Designers should seek to create desirable places.
- *Shortest path*: Designers should recognize that the crossing paths should be direct with minimal out-of-direction travel. Unnecessary diversions will be ignored and long delays will result in non-compliant behavior.

This section briefly reviews accessibility requirements, sidewalks and shared use paths. A discussion on intersection considerations follows. The majority of the section is devoted to midblock crossings. Finally, a discussion of timing traffic signals for pedestrians is presented. Further information regarding the facilities discussed in this chapter can be found in:

- AASHTO *Guide for the Planning, Design, and Operation of Pedestrian Facilities*: Design guidance and planning considerations, including rural applications.
- *Manual on Uniform Traffic Control Devices* (MUTCD): Guidance on markings, signs, and signals/beacons.
- *Evaluation of Pedestrian-Related Roadway Measures: A Summary of Available Research* by J. Mead, C. Zeeger, and M. Bushnell is a comprehensive summary of the related literature (Mead et al., 2013).

Accessibility Requirements

Among the most important provisions of the ADA is that all federal, state, and local government agencies must provide pedestrian access for people with disabilities to streets and sidewalks wherever pedestrian facilities exist (Kirschbaum et al., 2001; Markesino and Barlow, 2007). For public right-of-way, this is covered in the Proposed Guidelines for Pedestrian Facilities in the Public Right-of-Way (PROWAG), 2011. Accessibility provision is regulated through the imposition of standards covering features such as ramps, accessible signals, and detectable warnings. Private individuals and businesses with responsibility for a public right-of-way are responsible for providing disability access on that right-of-way. The projects covered by the ADA include all projects for construction or alteration of a facility that provides access to pedestrians, and the law mandates that accessibility improvements for an alteration project are made at the same time as the rest of the alteration. As evidenced by the 2011 case of *Frame* v. *City of Arlington*, in which a disabled man sued Arlington, Texas, for failing to provide accessible sidewalks and won in a decision that was upheld by the Supreme Court, public agencies are subject to litigation if they fail to comply with ADA standards. In addition to making all new facilities accessible, the ADA also required public agencies to develop transition plans to make existing pedestrian facilities accessible to disabled citizens.

The key provisions of the applying to pedestrian facilities include the following:

- For accessibility purposes, a sidewalk must maintain a continuous width of 4 feet. There cannot be any obstructions within this 4 feet minimum space.
- If the continuous width is less than 5 feet, then 5 feet wide by 5 feet long passing spaces need to be provided at least every 200 feet.
- The grade of a sidewalk cannot exceed that of the adjacent road surface.
- If the sidewalk is not next to a road and is in its own *row*, then the maximum grade cannot exceed 5 percent.
- The cross slope at driveways and other areas where the sidewalk may be sloped cannot exceed 2 percent.

Segments

The following provides a brief overview of common pedestrian facilities found along road segments. In urban areas, all road segments should have provisions for persons walking. This is commonly done with sidewalks, which have a hard, smooth surface and are separated with a curb and/or planting strip. Sidewalks are sometimes used by bicyclists (especially children) but some cities may ban bicycle riding on sidewalks. In general, dual use can be expected but pedestrian accommodation should drive design.

Paths can also be designed for pedestrians' use. These off-street facilities will be used by multiple types of users (pedestrians, cyclists, skaters and joggers) and are often called shared-use paths. It is not realistic to plan and design a path for exclusive pedestrian use, as others will be attracted to the facility. In rural areas, low pedestrian volumes may not justify the installation of sidewalks and road shoulders can be used for pedestrians (though not ideal). Typical shoulder widths for vehicle operations are usually adequate for this type of pedestrian use. In rural recreational areas where higher (seasonal) pedestrian activity is expected, improved and accessible facilities should be provided.

Sidewalks

Sidewalk spaces generally have four zones: 1) curb, 2) furniture (also called a buffer space), 3) pedestrian (the walkable space), and 4) frontage zone. These zones are shown in Figure 9.3.

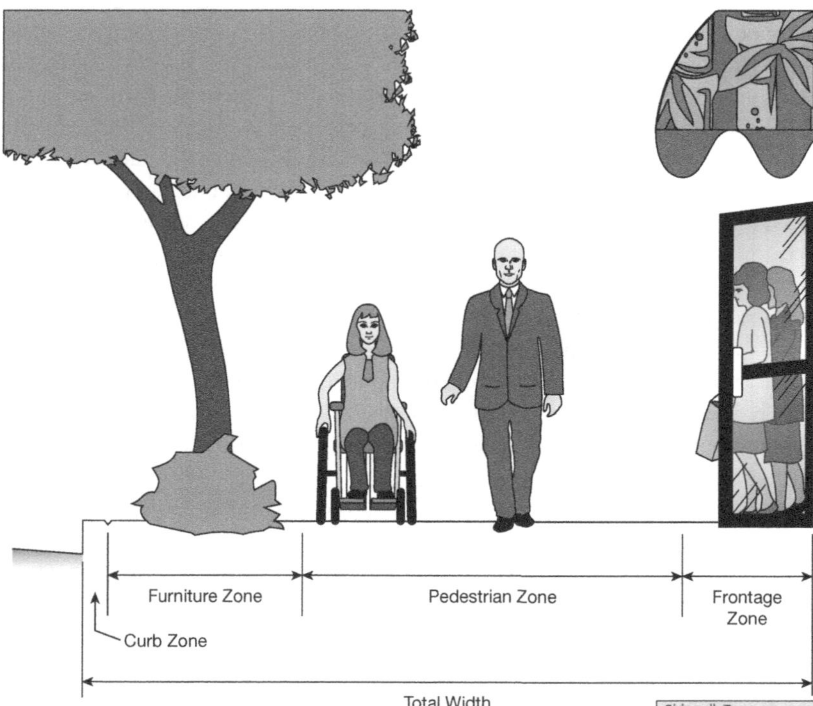

Sidewalk Zone	Minimum Width
Curb Zone	6 in.
Planter/Furniture/Utility Zone	24 in. (48 in. if planting trees)
Pedestrian Zone	60 in.
Frontage Zone	30 in.
Total Sidewalk Corridor	10 ft.

Figure 9.3 Pedestrian sidewalk zones

Source: Designing Sidewalks and Trails for Access: Part II of II: Best Practices Design Guide

The furniture area (buffer space) can be paved (e.g. decorative pavers, an area for planters/furniture) or landscaped (e.g. trees/shrubs). The benefits of providing a buffer area include further separation from traffic, increasing comfort, safety and air quality. Research also shows that a physical buffer provides greater comfort than just open space (Dowling et al., 2008). A buffer space can add to the visual element of the roadway and make for a more pleasant experience for all roadway users. In addition, a buffer space can make it easier to comply with ADA requirements as it allows the sidewalk to remain level across driveways and sidewalks, and curb cuts will more properly align at intersections. Recommended buffer widths vary depending on the type of buffer, classification of the street, surrounding land-uses, and the presence of a bicycle lane or on-street parking (which also serve as buffers).

The walking zone must be kept obstruction-free, and should be smooth and even to satisfy the needs of users. If a paved buffer is used, it should be distinguished from the walking area—this will help prevent obstacles from being placed in the walking area. It is desirable for the walking area to be wide enough for two wheelchairs to pass side-by-side, about 6 feet. In certain situations, 5 feet may be acceptable (e.g. local residential areas). An absolute minimum width of 4 feet allocated for walking space should only occur at pinch points where extra width is infeasible. AASHTO recommends 6–8 feet width on arterials with a planted buffer, or 8–10 feet if there is no buffer. A 10 feet or greater width is recommended for Central Business Districts or other areas with active pedestrian use.

The frontage zone is only necessary where buildings or other vertical barriers (e.g. fences, retaining walls) are adjacent to the sidewalk. The frontage zone allows for placement of street furniture, gives people space to window shop and socialize, and provides space for entering and exiting buildings. The frontage zone should be at least 2 feet wide with 1 foot as the absolute minimum (ODOT, 2011). A minimum of 4 feet or wider is recommended for CBDs and other areas where sidewalk cafés and merchandise may be found.

RAMPS

PROWAG (Chapter 4) contains an extensive section about requirements and standards for sidewalk ramps. Ramp slopes are limited to a range of between 5 percent and 8.3 percent, with a maximum allowable rise for a single ramp of 2.5 feet. For locations that require more than a 2.5-foot change in elevation, landing areas at least 5 feet long and with a maximum slope of 2 percent must be provided between each 2.5 foot rise.

DRIVEWAY CROSSINGS

When sidewalks cross driveways, care must be taken to provide sufficient width for a level walkway and ADA-compliant cross-slopes must be maintained. Sidewalks with buffer areas make this easier to accommodate.

Shared-Use Paths

Shared-use paths provide safe and direct access for non-motorized users of all ages and abilities including pedestrians, runners, bicyclists of all types (upright, recumbent, tandem and with trailer), roller skaters and skateboarders. These off-street paths provide safe and relatively stress-free cycling routes and are especially important for children, the elderly and women, and for anyone with special needs due to any sort of disability (Pucher and Buehler, 2008). Shared-use paths should function and be viewed as part of the transportation system for pedestrians and bicyclists. In addition, shared-use paths should not be used to preclude on-road bicycle facilities, but rather to supplement a network of on-road bike lanes, shared roadways, bicycle boulevards, and paved shoulders (AASHTO, 2012).

As with sidewalks, provisions of the PROWAG requirements apply to shared-use paths. Designs should provide adequate width for the various users to mix. An 11-foot wide pathway (minimum) is needed to enable a bicyclist to pass another path user traveling in the same direction, at the same time a path user is approaching from the opposite direction (AASHTO 2012). If the volumes of pedestrians and bicyclists are significant, then separation of the modes may be necessary but it must be shown in a clear and straightforward way. Pedestrians are typically provided with a bi-directional walking lane on one side of the pathway, while bicyclists are provided with directional lanes of travel (AASHTO 2012). Extensive design guidance on minimum horizontal curve radius, fencings, intersection crossings, entrance to pathways and other design features are provided in AASHTO.

Intersections

Intersections should be designed to accommodate safe pedestrian crossings. This includes provisions for crosswalks, pedestrian signals, curb ramps, easy pathfinding, good visibility and other amenities as warranted.

Curb Ramps

Street crossings can be particularly difficult for wheelchair users, given the common necessity to turn around and to change grades, so a considerable portion of PROWAG's sidewalk standards are devoted strictly towards ramps and crossings. Chapter 3 of PROWAG defines three types of accessible curb ramps, which are perpendicular, parallel, and combination ramps. Perpendicular ramps, which have slopes cutting through the curb at right angles, can only be used when sidewalks are 12 feet wide, and parallel ramps, which have slopes in line with the direction of the sidewalk, require only 4 feet of width. For both ramp types, the minimum acceptable running slope is 5 percent and the maximum is 8.3 percent. At all street crossings, it is important to provide detectable warning surfaces, particularly so that visually impaired pedestrians can determine where it is appropriate to cross. They are required at all curb ramps, refuge islands, and at grade rail crossings. PROWAG standardizes these surfaces by mandating that they be marked with truncated domes and visually contrasting colors. These domes should be 0.2 inches tall and have center-to-center spacing of between 1.6 and 2.4 inches. Figure 9.4 demonstrates the appropriate size and placement of detectable warning surfaces for the various types of curb ramps.

Curb Extensions

Curb extensions, also known as bulb-outs, are found on streets with on-street parking. These extensions increase the visibility of the crossing and pedestrians waiting to cross, and improve the sight distance for pedestrians by allowing them to get out past parked cars. Extensions also narrow the road, thereby potentially slowing traffic, and reduce the length of the crossing. Extensions should be the full width of the parking lane but should not extend into the bike lane, if one is present.

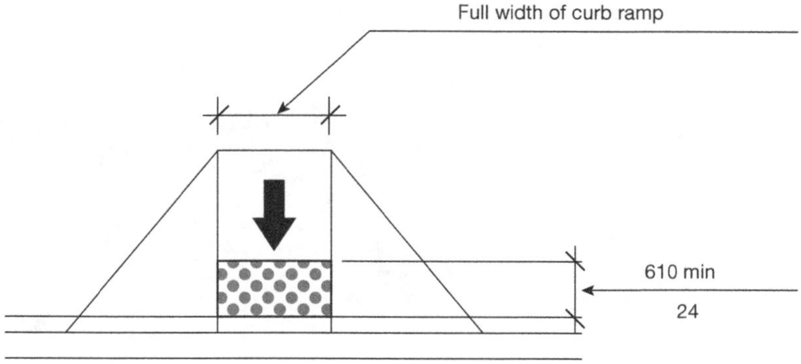

(a) perpendicular

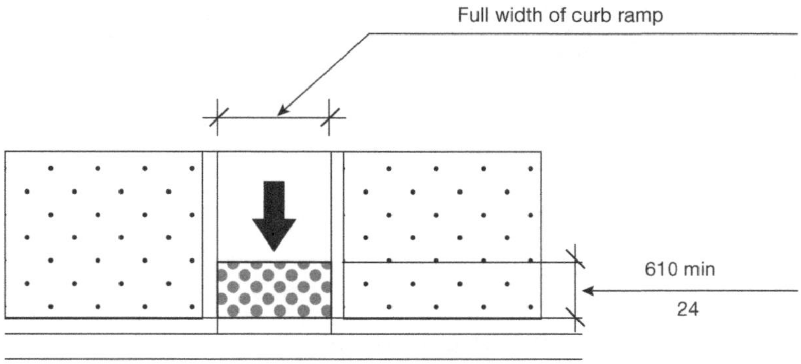

(b) returned curb

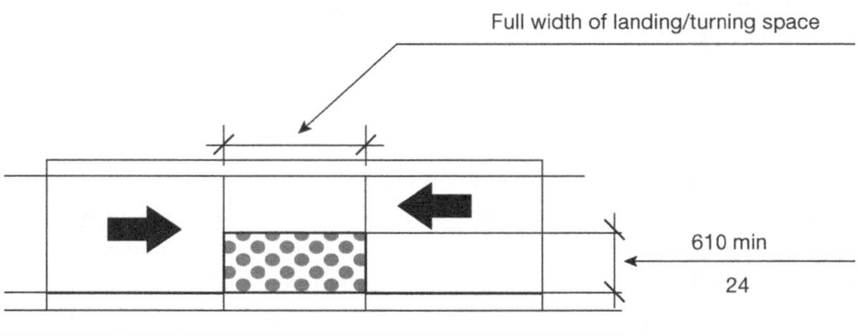

(c) parallel

Figure 9.4 Detectable warning surface placement
Source: PROWAG, Chapter 3, 2011

Figure 9.5 Curb extension, Portland, OR
Source: C. Monsere

Mid-Block Pedestrian Crossings

On many roadways, the distance between controlled intersections is too far to expect pedestrians to divert to those crossings. Assuming sufficient demand, a midblock crossing may be warranted. Selection of the appropriate crossing treatment is an important and challenging process. There are a number of factors that should be considered, including speed and volume of motor vehicle traffic, the distance to the nearest crossing, and the demand (potential or actual). Note that due to high speed, volumes and widths at difficult crossing locations, counting existing use may not adequately capture the potential demand.

Mid-block crossings should have adequate sight distance so that pedestrians can adequately judge oncoming traffic and drivers can see pedestrians waiting to cross the road. The addition of advance stop lines is recommended to improve yielding distances and to mitigate the multiple threat crash where one vehicle stops but the other does not, and the pedestrian does not notice approaching vehicles due to an obstructed view (Van Houten et al., 2001). Additional guidance and descriptions (including selection worksheets) are provided in TCRP Report 112/NCHRP Report 562, Fitzpatrick et al., 2007). There are five basic types of midblock crossings which are described in more detail in the following subsections:

- Marked crosswalks that have only standard crosswalk markings and signs.
- Enhanced crossings where the visibility of the crosswalk and waiting pedestrians is increased through the use of signs, markings, and/or passively flashing (on all the time) beacons.
- Active crossings where the warning device is only active when a pedestrian is present (e.g. a pushbutton beacon).

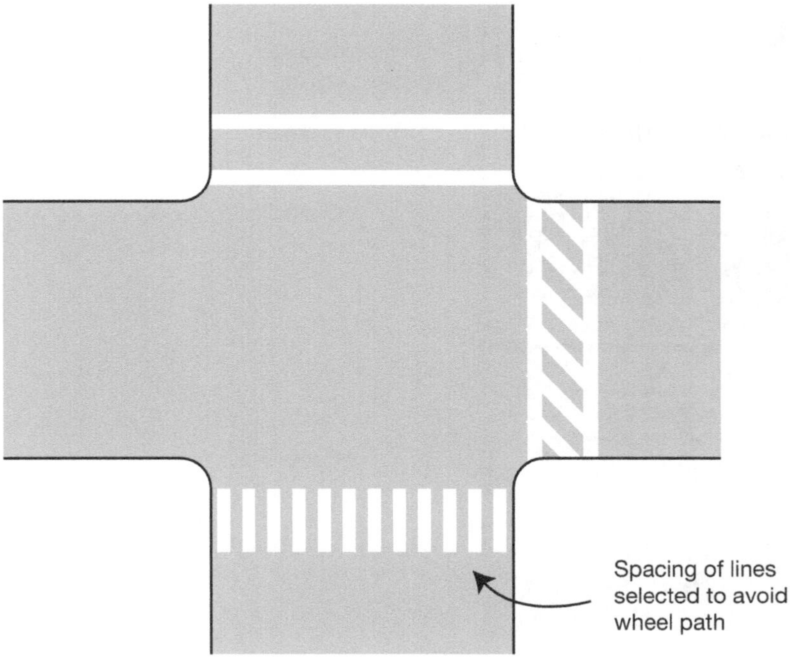

Figure 9.6 Examples of crosswalk markings
Source: MUTCD 2009, Figure 3B–19

- Crossings that are controlled with a red signal or beacon (e.g. a Pedestrian Hybrid Beacon).
- Crossings that are controlled with a standard traffic signal.

Marked Crosswalks

Marked crosswalks are the most basic type of treatment. Markings can be either standard, continental or diagonal as shown in Figure 9.6 from the MUTCD (2009). Continental markings can be placed out of the wheel path, limiting maintenance costs. A landmark study by Zeeger et al. (2002) for the FHWA "Safety Effects of Marked Versus Unmarked Crosswalks at Uncontrolled Locations" found that just marking a crosswalk alone may increase the risk of a crash involving a pedestrian if the ADT is greater than 15,000, the speeds are higher than 40 mph and/or the crossing is multilane. In these situations, designers should provide additional enhancements to crossing such as median islands, in street signs, beacons, or rapid flash devices. If the ADT is less than 12,000 and the speeds are lower than 35 mph and only two vehicle lanes are being crossed, pavement markings only can be sufficient.

Median Refuge Islands

Median refuge islands shorten crossings by allowing pedestrians to cross only one direction of traffic at a time and simplify the crossing as the pedestrian only needs to watch one direction of traffic at a time. The islands can serve a traffic calming purpose, as they narrow the visual field for drivers, which could cause them to slow down. The median is also an ideal space to add additional enhancements such as signs. Studies have shown that drivers yield anywhere from

Figure 9.7 Mid-block crosswalk with median refuge
Photo: N. Foster

7 to 75 percent of the time at sites with median islands, with the average between 29 and 34 percent (Turner et al., 2006). To allow pedestrian access, a cut through (at grade) is preferred because there is typically not sufficient width for two ramps and a landing. The island itself should be at least 6 feet wide, though wider is preferable, especially if bikes are expected to use the facility where 8 feet wide is suggested.

In-Street Signs

In-street signs improve the visibility of the crosswalk to drivers and can also remind drivers of the state law (either yield to or stop for crossing pedestrians). Studies have found that drivers yield between 44 and 97 percent of the time when these signs are present, with an average of between 77 and 90 percent. This is only for roads with speed limits of 25 to 30 mph (Turner et al., 2006). They are to be placed in the roadway at the crosswalk on the centerline, the lane line, and/or on a median island. They should be designed so that they can be hit by a car and bounce back up. A recent evaluation found significantly increased yielding with this "gateway" type of approach (Bennett et al., 2014). Maintenance can be an issue as they can be ripped out of the ground, especially by large trucks and snowplows.

In-Roadway Warning Lights

In-roadway warning lights are flashing lights mounted in the roadway surface designed to alert drivers to the presence of a pedestrian. They are pedestrian-activated by pushbutton or passive detection. Their effectiveness in daylight is likely limited. For this reason, they are often not a recommended treatment in many jurisdictions.

Figure 9.8 Rectangular rapid flash beacon assembly
Photo: N. Foster

Overhead Beacons

Overhead flashing beacons may be passive (continuously flashing) or active (activated by pushbutton or pedestrian detection). They are more effective if they are pushbutton activated. They improve the visibility of the crosswalk. If they are actuated only, then they also warn drivers that someone is crossing, or is about to cross, the road. Driver yield rates at passive devices ranged from 25 to 73 percent (31 to 67 percent average.) At active devices, driver yield rates ranged from 29 to 92 percent with a 47 to 52 percent average (Turner et al., 2006).

Rectangular Rapid Flash Beacon

Rectangular Rapid Flash Beacons (RRFBs) improve the visibility of crosswalks by adding rapid-staccato flash LED devices beneath the crossing warning signs (Van Houten et al., 2008). They warn drivers that a pedestrian is crossing or about to cross as they are pedestrian-activated and flash as soon as they are pushed. They flash in an irregular manner, which further helps attract drivers' attention. Studies have shown them to be quite effective in increasing driver yielding about 75–88 percent on average (Foster et al., 2013). They are significantly less expensive to install than PHBs and traffic signals. They are not currently in the MUTCD, but under FHWA Interim Approval (FHWA, 2008).

Pedestrian Hybrid Beacon

Pedestrian Hybrid Beacons (PHBs) (Figure 9.9) are typically found on long crossings where motor vehicle traffic volumes and/or speeds are high. The devices must be pedestrian-activated.

Figure 9.9 Pedestrian hybrid beacon
Source: Photo by N. Foster

It is desirable to serve pedestrians as soon as possible once the button is pushed (though the MUTCD suggests that if installed in a system it should be coordinated). When activated, the display sequence starts with a flashing yellow indication followed by the top two beacons displaying steady red, with the pedestrian signal indicating WALK. Beacons alternate flashing red and the pedestrian signal indicates FLASHING DON'T WALK. Finally, the beacons go dark and pedestrian signal indication goes to steady DON'T WALK. Drivers stay stopped on the steady red but proceed after stopping on the alternating red phase, as long as the crosswalk is clear. This operation reduces driver delay (when compared to a full signal). This desired behavior from motorists is not necessarily intuitive and single installations in a jurisdiction will likely require public education campaigns.

Driver yield rates are between 94 percent and 100 percent, with an average of 97–99 percent (Turner et al., 2006). The 2009 MUTCD includes a warrant based on crossing volumes, speeds and motor vehicle volumes. There needs to be a minimum of 20 crossings/hour. The MUTCD suggests that they are not placed within 100 feet from side streets or driveways that are stop or yield controlled.

Pedestrian Traffic Signal

The 2009 MUTCD contains warrants for installing traffic signals based on pedestrian demand. The lowest warrant requires 75 crossings/hour for four or more hours to even consider a signal, while the peak hour warrant requires 133 crossings but cannot be applied if the distance to the nearest traffic signal or stop-controlled intersections is less than 300 feet (MUTCD, 2009). Full traffic signals for midblock crossings are typically found in areas with high crossing demand across multi-lane high-speed/volume crossings. Like PHBs, they are pedestrian-activated and it is desirable that the activation is "hot" response to improve pedestrian service. If the signal is within a coordinated corridor, it should be coordinated with the other signals. Driver yield rates are between 91 percent and 100 percent, with an average of 95–99 percent (Turner et al., 2006).

Grade Separated Crossings

A final option is to separate pedestrian crossing with a structure—a bridge or tunnel. These should be used in only the most difficult crossings due to the expense and the desire of most pedestrians to take the shortest path. Each option has distinct advantages and disadvantages. For tunnels, one clear disadvantage is perceptions of personal safety issues.

Traffic Operations for Pedestrians

Pedestrian Signal Heads

Pedestrian signal heads are used to provide indications for pedestrians exclusively. The MUTCD recommends that engineering judgment should guide the need for a separate pedestrian signal head. A pedestrian signal head typically consists of three signal indications: 1) a steady walking person (WALK); 2) a flashing upraised hand (FLASHING DON'T WALK or FDW); and 3) a steady upraised hand (DON'T WALK). The MUTCD provides clear guidance on the pedestrian signal head indications and their meaning. The WALK is displayed in white and

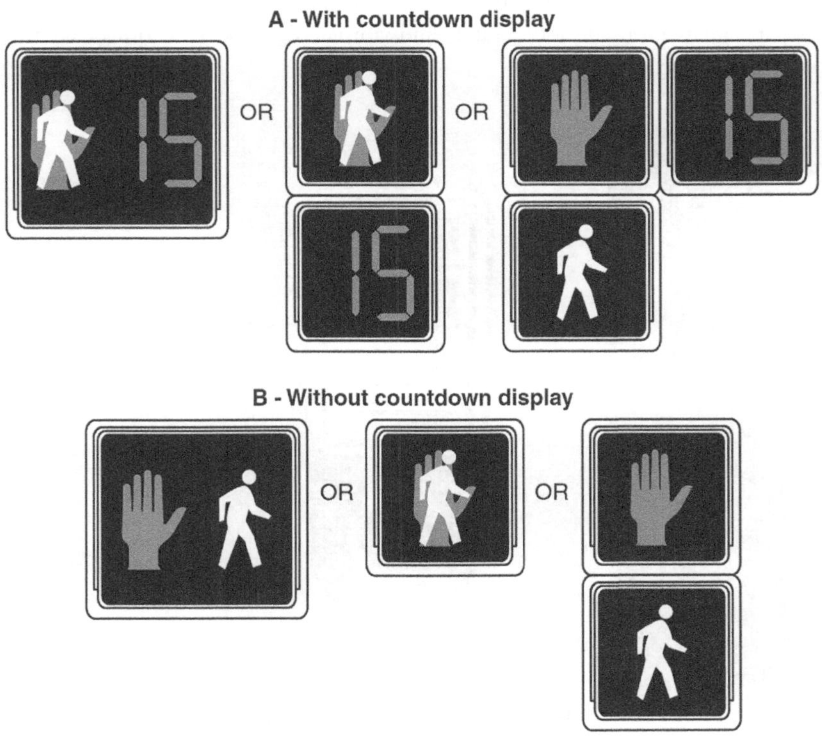

Figure 9.10 Pedestrian signal head with and without countdown timers
Source: MUTCD, 2009

indicates to the pedestrian that they are permitted to start crossing the roadway. The FDW is displayed in Portland orange and indicates that the pedestrian should not start crossing the roadway and those that have already started to cross should complete their crossing. The DON'T WALK is also displayed in Portland orange and indicates that the pedestrian should not enter the roadway in the direction of the signal indication. Countdown timers provide a more intuitive and clear indication of the timing of the pedestrian change interval in more than 7 seconds (MUTCD, 2009).

Pedestrian signal heads should be clearly visible and mounted on the bottom of the signal-housing bracket. Physical separation from vehicle signal heads is required, if they are mounted on the same support. The mounting height should not be less than 7 feet or greater than 10 feet above the sidewalk level (MUTCD, 2009). The signal indication symbols should be at least 6 inches in height; when crossing distances are greater than 100 feet, they are required to be at least 9 inches in height.

Detection and Pushbuttons

Pedestrian detection at intersections is achieved automatically using passive detection devices or manually using pushbuttons. When pushbuttons are used, they should be placed adjacent to the sidewalk they are controlling and no more than 10 feet from the edge of the curb. Accessibility is key and the mounting height should not be more than 4 feet above the sidewalk. Clear signage should be provided to indicate the crosswalk that is being controlled by the pushbutton. Pilot lights that provide feedback to the pedestrian that their call for service has been registered may be used. If they are used, the light is required to stay on until the walk has been served. If accessible pedestrian signals are used, both audible and vibrotactile walk indications have to be provided.

(a) (b)

Figure 9.11 (a) Typical push button. (b) Accessible pushbutton
Source: Photos by C. Monsere

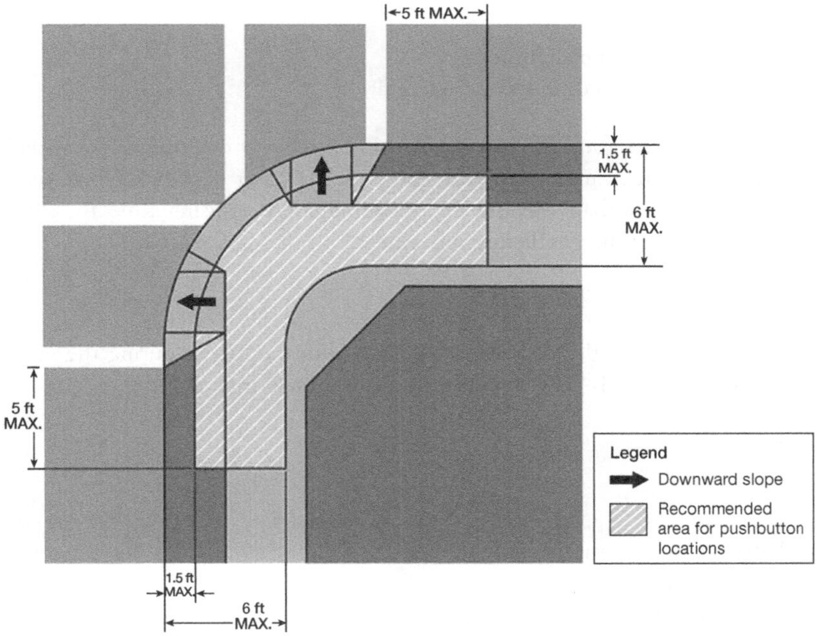

Notes:

1. Where there are constraints that make it impractical to place the pedestrian pushbutton between 1.5 feet and 6 feet from the edge of the curb, shoulder, or pavement, it should not be further than 10 feet from the edge of the curb, shoulder or pavement.
2. Two pedestrian pushbuttons on a corner should be separated by 10 feet.
3. This figure is not drawn to scale.
4. Figure 4E-4 shows typical pushbutton locations.

Figure 9.12 Pushbutton locations
Source: MUTCD, 2009

Timing of Traffic Signals for Pedestrians

Pedestrian timing consists of three intervals as outlined earlier: WALK, FDW, and DON'T WALK. The length of the walk interval is dictated by agency policy (Koonce et al., 2008). Factors to be considered while establishing a walk time include pedestrian demand, school zones and the presence of elderly pedestrians. MUTCD recommends a walk time between 4 and 7 seconds depending on the pedestrian demand. Further guidance on walk times is provided in the Traffic Signal Timing Manual (Koonce et al., 2008), MUTCD (FHWA, 2009) and Traffic Control Devices Handbook (Pline, 2001).

The MUTCD currently stipulates that the pedestrian clearance interval must be calculated assuming the distance from the curb to the far side of the opposing travel way, or to a median of sufficient width for pedestrians to wait. The clearance time is computed as

$$PCT = \frac{D_c}{V_p}$$

where

D_c = pedestrian crossing distance (in feet)
V_p = pedestrian walking speed (in feet per second)

Pedestrian clearance parameter setting (PC) in the signal controller is determined by agencies in one of two ways. Pedestrian clearance time can end with the onset of the vehicular yellow interval. This approach is pedestrian friendly by providing more crossing time for slower pedestrians prior to the onset of the conflicting movement.

$$PC = PCT$$

In the second approach, a part of the pedestrian clearance time can occur during the yellow interval, or yellow and red intervals. This approach is more vehicle friendly.

$$PC = PCT - (Y + R)$$

where

PC = pedestrian clearance interval duration (s)
PCT = pedestrian clearance time (s)
Y = yellow change interval (s)
R = red clearance interval (s)

The MUTCD stipulates that if countdown timers are used, pedestrian clearance interval must finish timing before the onset of the yellow clearance interval.

Pedestrian friendly signal timing strategies such as pedestrian progression, recall phasing (where pedestrians are served every cycle irrespective of demand), short cycle lengths, protected left turns, leading pedestrian intervals (pedestrians are allowed to start crossing a few seconds before the parallel vehicle green indication), and exclusive pedestrian phase (pedestrians are permitted exclusive use of the intersection with lateral and diagonal crossings) may be used where the situation warrants, to promote walking and improve conditions for pedestrians.

Conclusion

Designers, planners and operators of transportation systems have the responsibility to consider all users. Not every design can, or should, be expected to optimize all performance measures for each mode. Better public spaces result when practitioners consciously consider the elements and principles of design highlighted in this chapter.

References

American Association for State Highway and Transportation Officials (AASHTO) (2004), *Guide for the Planning, Design, and Operation of Pedestrian Facilities*, 1st ed. AASHTO, Washington, DC.

American Association for State and Highway Transportation Officials (AASHTO) (2010). *Highway Safety Manual*, Washington, DC.

American Association for State Highway and Transportation Officials (AASHTO) (2012). *Guide for the Development of Bicycle Facilities*. AASHTO, Washington, DC.

Bennett, M. K., Manal, H., and Van Houten, R. G. (2014). A Comparison of Gateway In-Street Sign Treatment to Other Driver Prompts to Increase Yielding to Pedestrians at Crosswalks. In Transportation Research Board 93rd Annual Meeting (Paper No. 14–0222).

Dowling, R.G., Reinke, D. B., Flannery, A., Ryus, P., Vandehey, M., Petritsch, T. A., Landis, B. W., Rouphail, N. M., and J. A. Bonneson (2008). Multimodal Level of Service Analysis for Urban Streets. NCHRP Report 616. Transportation Research Board of the National Academies, Washington, DC.

El Esaway, M., Lim, C., Sayed, T., and Mosa, A. I. (2013). Development of Daily Adjustment Factors for Bicycle Traffic. *Journal of Transportation Engineering*, 139(8), 859–871.

FHWA (2008). Interim Approval for Optional Use of Rectangular Rapid Flashing Beacons (IA-11). Available at: http://mutcd.fhwa.dot.gov/resources/interim_approval/ia11/fhwamemo.htm (accessed July 16, 2008).

FHWA (2009). *Manual on Uniform Traffic Control Devices* (MUTCD), US Department of Transportation, Washington, DC.

FHWA (2013). *Traffic Monitoring Guide*. US Department of Transportation, Washington, DC.

Figliozzi, M., Johnson, P., Monsere, C., and Nordback, K. (2014). A Methodology to Characterize Ideal Short-term Counting Conditions and Improve AADT Estimation Accuracy Using a Regression-based Correcting Function. *Journal of Transportation Engineering*, American Society of Civil Engineers.

Fitzpatrick, K., Turner, S., Brewer, M., Carlson, P., Ullman, B., Trout, N., Park, E. S., Whitacre, J., Lalani, N. and Lord, D. (2007). Improving Pedestrian Safety at Unsignalized Intersections. TCRP Report 112/NCHRP Report 562: Transportation Research Board of the National Academies, Washington, DC.

Foster, N., Monsere, C., and Carlos, K. (2013). Evaluating Driver and Pedestrian Behaviors at Enhanced Multilane Midblock Pedestrian Crossings: Case Study in Portland, Oregon. Transportation Research Record: *Journal of the Transportation Research Board*, Transportation Research Board of the National Academies, Washington, DC.

Highway Capacity Manual (2010). Transportation Research Board, National Research Council, Washington, DC.

Kirschbaum, J. B., Axelson, P. W., Longmuir, P. E., Mispagel, K. M., Stein, J. A., and Yamada, D. A. (2001). Designing Sidewalks and Trails for Access, Part 2, Best Practices Design Guide. Report No FHWA-EP-01–027. Washington, DC.

Koonce, P., Rodegerdts, L., Lee, K., Quayle, S., Beaird, S., Braud, C., Bonneson, J., Tarnoff, P., and Urbanik, T. (2008). *Traffic Signal Timing Manual*. FHWA-HOP-08–024. US Department of Transportation, Washington, DC.

Kuzmyak, J. R., Walters, J., Bradley, M., Kockelman, K. (2014). *Estimating Bicycling and Walking for Planning and Project Development: A Guidebook*. NCHRP Report, 770, 161pp.

Markesino, J. and Barlow, J. (2007). *Special Report: Accessible Public Rights-of-Way Planning and Design for Alterations*. Rep. United States Access Board, August 2007. Available at: www.access-board.gov/guidelines-and-standards/streets-sidewalks/public-rights-of-way/guidance-and-research/accessible-public-rights-of-way-planning-and-design-for-alterations (accessed March 12, 2014).

Mead, J., Zegeer, C., and Bushell, M. (2013). Evaluation of Pedestrian-Related Roadway Measures: A Summary of Available Research. Pedestrian and Bicycle Information Clearinghouse (PBIC), Chapel Hill, NC, for Federal Highway Administration, DTFH61-11-H-00024, April, pp. 126.

Nordback, K., Marshall, W. E., Janson, B. N., and Stolz, E. (2013). Estimating Annual Average Daily Bicyclists: Error and Accuracy. Transportation Research Record: *Journal of the Transportation Research Board*, No. 2339, Transportation Research Board of the National Academies, Washington, DC, pp. 90–97.

Oregon Department of Transportation (ODOT) (2011). *Bicycle and Pedestrian Design Guide*. Salem, OR.

PBIC (2014). Who's Walking and Bicycling. Available at: www.pedbikeinfo.org/data/factsheet_general.cfm (accessed June 10, 2014).

Pline, J. L. (Ed.). (2001). *Traffic Control Devices Handbook*. Institute of Transportation Engineers.

Proposed Guidelines for in the Public Right-of-Way (PROWAG). *Federal Register* (2011): United States Access Board, July 26.

Pucher, J. and Renne, J. L. (2003). Socioeconomics of Urban Travel: Evidence from the 2001 NHTS. *Transportation Quarterly*, 57(3), 49–77.

Pucher, J. and Buehler, R. (2008). Making Cycling Irresistible: Lessons from The Netherlands, Denmark and Germany, *Transport Reviews*, 28:4, 495–528.

Schneider, R., Arnold, L., and Ragland, D. (2008). A Methodology for Counting Pedestrians at Intersections: Using Automated Counters to Extrapolate Weekly Volumes from Short Manual Counts. Transportation Research Record: *Journal of the Transportation Research Board*, 2140(1), 1–12.

Singleton, P. A. and Clifton, K. J. (2013). Pedestrians in Regional Travel Demand Forecasting Models: State-of-the-Practice. Presented at the 92nd Annual Meeting of the Transportation Research Board, Washington, DC.

SWOV (2012) Background of the five Sustainable Safety Principles. Available at: www.swov.nl/rapport/Factsheets/UK/FS_Sustainable_Safety_background.pdf (accessed June 9, 2014).

Turner, S., Fitzpatrick, K., Brewer, M., and Park, E. S. (2006). Motorist Yielding to Pedestrians at Unsignalized Intersections: Findings from a National Study on Improving Pedestrian Safety. Transportation Research Record: *Journal of the Transportation Research Board*, 1982(1), 1–12.

Van Houten, R. V., Malenfant, J. E. L., and McCusker, D. (2001). Advance Yield Markings: Reducing Motor Vehicle—Pedestrian Conflicts at Multilane Crosswalks with Uncontrolled Approach. Transportation Research Record: *Journal of the Transportation Research Board*, 1773(1), 69–74.

Van Houten, R., Ellis, R., and Marmolejo, E. (2008). Stutter-flash light-emitting-diode Beacons to Increase Yielding to Pedestrians at Crosswalks. Transportation Research Record: *Journal of the Transportation Research Board*, 2073(1), 69–78.

Zegeer, C. V., Stewart, R., Huang, H., and Lagerwey, P. (2002). Safety Effects of Marked Versus Unmarked Crosswalks at Uncontrolled Locations: Executive Summary and Recommended Guidelines. FHWA-RD-01–075, FHWA, US Department of Transportation, Washington DC.

10

BICYCLE TRANSPORTATION

Christopher M. Monsere,[1] *Nick Foster,*[1] *Todd Borkowitz,*[1]
Sirisha Kothuri,[1] *and Robert L. Bertini*[2]

(1) PORTLAND STATE UNIVERSITY, (2) CALIFORNIA POLYTECHNIC STATE UNIVERSITY

Introduction

In many locations, streets have been designed and operated primarily to accommodate motor vehicles. In many urban areas, however, it is becoming more common for streets to be viewed as significant public places. Bicycling, walking, and other human-powered forms of travel—commonly referred to as active transportation—are gaining increased design attention for their associated economic, environmental, and health benefits. New designs are being applied that modify streets to better meet the mobility needs of all users, and to be safer, more sustainable, resilient, multimodal and economically beneficial. Active transportation facilities that focus on moving people have a key role to play in this transformation.

This chapter provides an overview of the design assumptions and considerations for bicycle facilities. Following a brief review of safety trends and discussion of general design philosophy, the chapter presents design inputs (dimensions and performance) and an overview of typical facilities for bicycles. The design review includes operations, primarily at signalized intersections. The chapter primarily references US-based design standards and documents.

Safety Trends

The National Highway Traffic Safety Administration (NHTSA) reports that 682 bicyclists were killed in crashes in the US in 2011. Figure 10.1 shows an 18-year trend. In parallel with overall fatality trends, there has been a general decline in the number of fatalities as shown in Figure 10.1. However, over this period a total of 13,084 bicyclists were killed—an unacceptable number. The overall temporal and spatial trends of non-motorized fatalities have some similarities as shown in Table 10.1. Around 70 percent of the fatalities occur in urban areas (where people are walking and bicycling). Most bicyclist (60 percent) fatalities occur away from intersections.

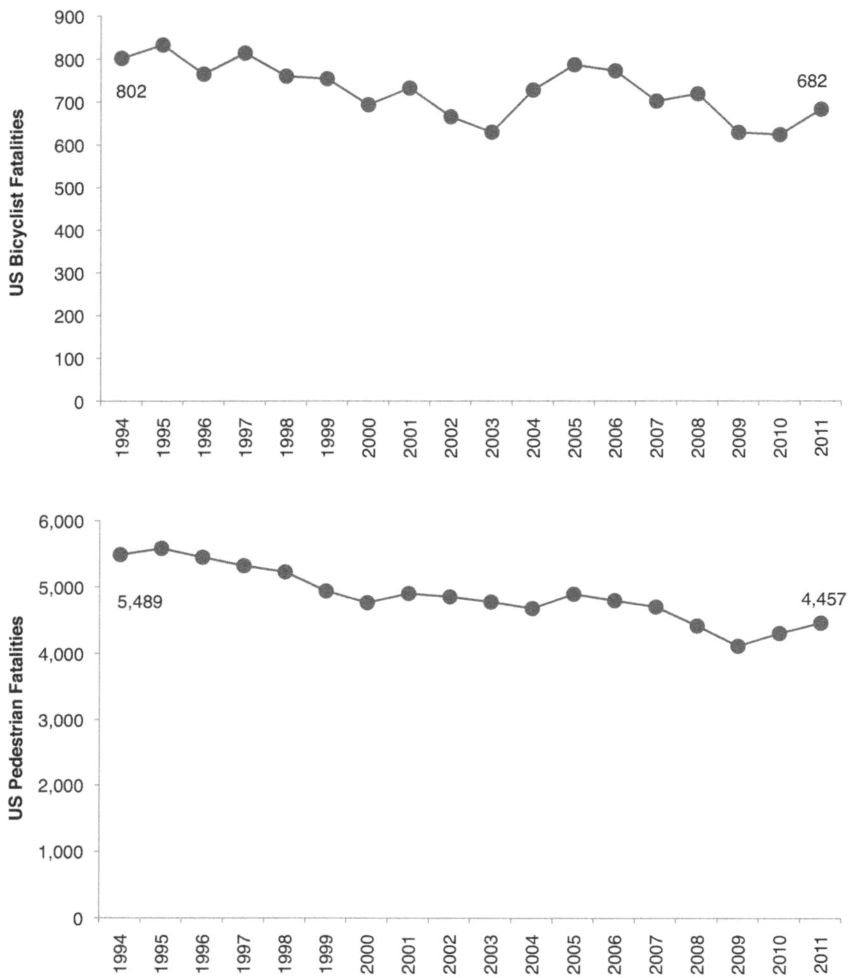

Figure 10.1 US bicycle fatality trends, 1994–2011

Crash Analysis Tools and Estimates of Safety Effectiveness

In the 1970s, methods for typing bicycle crashes with motor vehicles were developed by NHTSA to better define the sequence of events and precipitating actions leading to bicycle-motor vehicle crashes. These methods were applied in subsequent work by the University of North Carolina Highway Safety Research Center to develop the BIKESAFE toolkit (including the PBCAT tool which does the actual crash typing). These tools help diagnose crash problems based on type and suggest mitigations. The tools can be found at: BIKESAFE: www.bicyclinginfo.org/bikesafe/.

In addition to this tool, the Highway Safety Manual and the Countermeasure Clearinghouse can be used to search for crash modification factors that can be used to estimate the safety effectiveness of treatments or design modifications. They can be found at: *Highway Safety Manual* (AASHTO, 2010): safety.fhwa.dot.gov/hsm/; FHWA Countermeasure Clearinghouse: www.cmfclearinghouse.org/index.cfm.

Table 10.1 Locations and context of US bicycle fatalities

Pedalcyclists killed	*Percentage of pedalcyclists killed*	
	2010	*2011*
Land use		
Rural	28	31
Urban	72	69
Non-motorist location		
Intersection	33	31
Non-intersection	57	59
Other	8	10
Time of day		
Midnight–3:59 a.m.	7	8
4 a.m.–7:59 a.m.	11	9
8 a.m.–11:59 a.m.	13	14
Noon–3:59 p.m.	17	18
4 p.m.–7:59 p.m.	28	30
8 p.m.–11:59 p.m.	25	21

Source: NHTSA

Note: Percentage of unknown values are not displayed.

Principles of Design

Effective network planning and facility design are critical for making bicycling safe, convenient and attractive. In the following sections, design minimums are presented. While employing minimum standards may be unavoidable in some locations, it is often preferable, even recommended to exceed them. As use increases, minimum designs are more likely to be inadequate.

The Sustainable Safety vision for road safety developed in the Netherlands (SWOV, 2012) is a helpful framework that can lead to a traffic system where serious crashes are prevented; facilities that recognize human vulnerability and cognitive capabilities and limitations; people, vehicles and the road are treated in an integrated manner; and safety gaps are bridged in a proactive way. With this in mind, a Sustainable Safety approach considers the following five principles:

- *Road functionality*: Roads are developed for single functionality (arterial vs. collector vs. local) in a hierarchically structured network.
- *Homogeneity of mass and/or speed and direction*: Speed, direction and mass should be equalized at moderate and high speeds.
- *Predictability of road course and road-user behavior by a recognizable road design*: The user's expectations should be supported through consistency and continuity of road design.
- *Forgiveness of the environment and of road users*: Injuries will be limited through the provision of a forgiving road environment and anticipation of road-user behavior.
- *State awareness by the road user*: Users should be able to assess their capacity to handle the driving task.

Bicycling and walking modes are commonly connected in many transportation analysis, design, and planning activities. Indeed, there are many similarities including (ODOT, 2011):

- *Location*: While often physically separated from each other, both are typically located between motor vehicle lanes and right-of-way lines, often competing for space with transit, streetscape amenities and utilities.
- *Exposure*: Both are very vulnerable in crashes with motor vehicles and are exposed to the environment (both pleasant and unpleasant).
- *Wide range of users*: Bicyclists and pedestrians may be any age and there are also no licensing requirements. Pedestrian and bicyclist performance and behaviors vary and change with age and other factors.
- *User preferences*: Both prefer side-by-side travel, separation from automobiles and each other, and prefer well-maintained, safe and well-illuminated facilities.

However, it is clear that bicycling and walking also have key differences and in most analysis, design, and planning it is advisable to treat these modes separately. Some key differences between the modes are (ODOT, 2011):

- *Accessibility*: Pedestrian facilities are expected to meet Americans with Disabilities Act (ADA) requirements given their more frequent use by persons with disabilities. Bicycle facilities (except shared-use paths) are not.
- *Speed and travel distance*: Bicyclists typically travel between 10 and 15 miles per hour, while pedestrians typically travel between 2 and 3.5 miles per hour. Bicyclists will typically travel up to 5 miles in urban areas, while pedestrians will typically travel up to 1 mile—each trip generally taking about 20–25 minutes.
- *Tolerance of out of direction travel*: Bicyclists generally tolerate some indirect travel while pedestrians may demand the most direct route.
- *Legal use of road space*: Bicyclists are generally allowed to operate with motor vehicle traffic. In many locations, a bicyclist can operate as either a vehicle or pedestrian under certain conditions. Pedestrians are generally only allowed in roadways when accessing or leaving a vehicle, in marked or unmarked crosswalks and on roadway shoulders.
- *Parking*: Bicyclists require safe, secure and convenient destination parking.

Analysis of Facility Performance

Facility performance is assessed in two ways: By examining how many people are using it or, in the case of a planned facility, the forecasted use; and/or by assessing the quality of service provided by the facility to its users, via some measure of performance. Both are important to consider.

Measuring Activity (Counting)

A non-motorized traffic counting program should consist of long-term, permanent counters and short-term counts at key locations. Permanent counters should be sited so as to provide data useful for creating factors for adjusting short-term counts. Relying on short-term counts can lead to large errors (Nordback et al., 2013), as non-motorized traffic can vary widely based on the day of the week, season, time-of-day, and weather (Figliozzi et al., 2014; El Esaway, 2013). Figure 10.2 shows the daily count in bicycle traffic in both directions on a key commute route in Portland, Oregon, for about one month. The day-to-day variations are evident. Automated technology for permanent non-motorized traffic counters is still evolving and some technology is better for bicyclists, while some is better for pedestrians (Peiris, 2014). For more information, see Chapter 4 of FHWA's *Traffic Monitoring Guide* (2013).

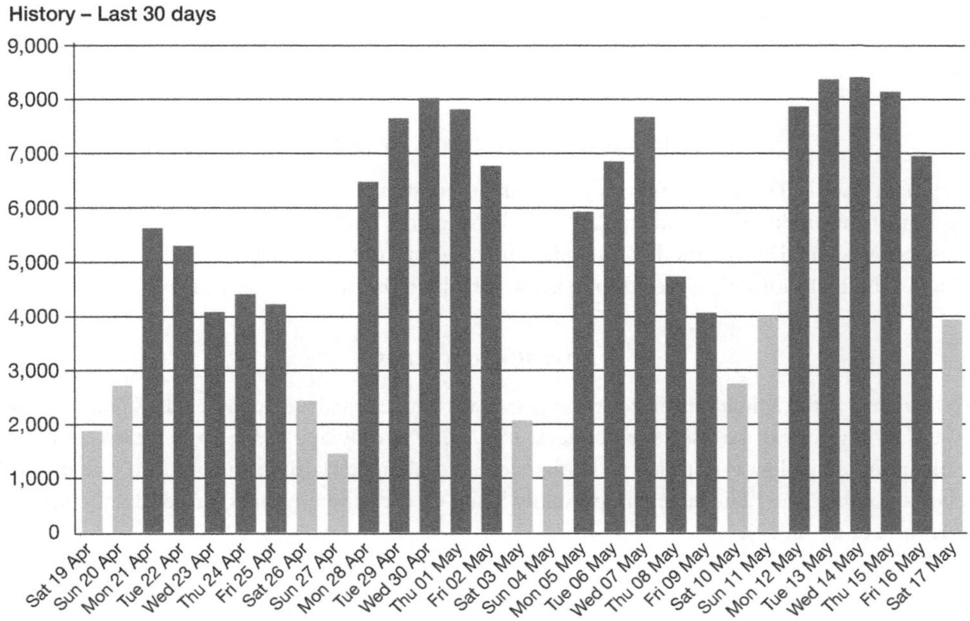

Figure 10.2 Daily bicycle counts across Hawthorne Bridge, Portland, OR (April 19, 2014–May 17, 2014)

Predicting Activity (Demand Forecasting)

Accurately predicting future volumes can help transportation professionals make educated decisions about facility design and prioritization. Forecasting bicyclist traffic is a rapidly evolving practice. Some agencies include bicycles in their travel demand models; however, they typically do not assign mode-specific trips to the network, as is done with motor vehicle traffic. Demand analysis is also done at the individual facility or site level for specific projects or developments. For more information on the state of the practice in non-motorized demand forecasting, see NCHRP Report 770—*Estimating Bicycling and Walking for Planning and Project Development* (Kuzmyak et al., 2014).

Assessing Quality of Service

Quality of service for bicyclists typically involves assessing the perceived quality of service from the user's perspective (i.e. comfort/satisfaction, speed or travel time). There are several methods in practice that quantify user perceptions, allowing for objective evaluation of projects and design trade-offs and communication of system performance. Most measures are primarily based on roadway geometry and motor vehicle volumes and speeds.

Highway Capacity Manual (HCM) 2010 Multimodal Level-of-Service

The 2010 edition of the *HCM* introduced new multimodal level-of-service analysis procedures, which include bicyclists, that are based on surveys of user perceptions. The models provide a Level of Service (LOS) grade on an "A"–"F" scale, similar to what is used for motor vehicles, for links (sections of road between signalized intersections), signalized intersections, and segments (intersections and links combined). The *HCM* also includes capacity-based measures for dense urban environments or other high-activity areas where available space is likely the overarching concern for bicyclists.

Danish Road Directorate Level-of-Service

Denmark has LOS models for bicyclists and pedestrians. Similar to the *HCM*, they are mathematical models that produce "A"–"F" letter grades and are based on user perceptions of satisfaction. The Danish models include some factors not considered by the *HCM*, most notably the inclusion of protected bike lanes, which are more common in Denmark.

Level of Traffic Stress

Researchers at the Mineta Institute at San Jose State University developed Level of Traffic Stress as a planning tool to be used for network analysis. It is only applicable to bicyclists and uses look-up index tables, as opposed to mathematical models, to classify a facility into one of four levels of traffic stress. The tables are primarily based on the report authors' professional opinions (Mekuria et al., 2012).

Bicycle/Pedestrian Environmental Quality Index

The San Francisco Department of Public Health used surveys of national experts to develop the Bicycle and Pedestrian Environmental Quality Indices. These methods award points to a facility for certain characteristics, with a final score on a 0–100 scale (BEQI, 2009).

Design Characteristics of Bicyclists

The design of bicycling facilities can be greatly informed by answering one basic but key question: Who should the facility be designed for? Characteristics that define a person on a bicycle for design may include whether a bicyclist is riding for transportation or recreation, as well as his or her age, skill, experience, physical ability, route familiarity, weight and bulk of cargo being transported, and the style of bike being ridden. The AASHTO *Guide for the Development of Bicycle Facilities* (2012) simplifies these types into two categories: 1) experienced/confident riders and 2) casual/less confident riders. Table 10.2 summarizes the contrasting user profiles between these populations.

It is also worth highlighting the key differences between the bicycle and the motor vehicle as in design. Some key points to remember based on the CROW Manual from the Netherlands include (*Design Manual for Bicycle Traffic*, CROW, 2007):

1. *The bicycle is powered by muscle power (usually).* This impacts design since minimizing energy loss for a rider can lead to bicycle-friendly design. A range of factors lead to energy loss, including: Friction losses in the bearings and chain; rolling resistance from friction between the tires and road; air resistance; vibration losses in the frame, saddle, and tires; braking and acceleration; and gravity when riding uphill.
2. *The bicycle is unstable.* Stability is affected by wind (due to weather and passing vehicles), imperfections in riding surface (rough surfaces, surface transitions, potholes, cracks), and involuntarily low speeds.
3. *The bicycle has no crumple zone.* Cyclists are very likely to be injured or killed in conflicts with vehicles.

Table 10.2 Summary of types of cyclists

Experienced/confident riders	*Casual/less confident riders*
Most are comfortable riding with vehicles on streets and are able to navigate streets like a motor vehicle, including using the full width of a narrow travel lane when appropriate and using left-turn lanes.	Prefer shared use paths, bicycle boulevards, or bike lanes along low-volume, low-speed streets.
While comfortable on most streets, some prefer on-street bike lanes, paved shoulders, or shared use paths when available.	May have difficulty gauging traffic and may be unfamiliar with rules of the road as they pertain to bicyclists; may walk bike across intersections.
Prefer a more direct route.	May use less direct route to avoid arterials with heavy traffic volumes.
Avoid riding on sidewalks. Ride with the flow of traffic on streets.	If no on-street facility is available, may ride on sidewalks.
May ride at speeds up to 25 mph on level grades, up to 45 mph on steep descents.	May ride at speeds around 8 to 12 mph.
May cycle longer distances.	Cycle shorter distances: 1–5 miles is a typical trip distance.

Source: AASHTO, 2012.

4. *The bicycle has hardly any suspension.* Smooth road surfaces are essential for bicycle friendliness.
5. *The cyclist rides in the open air.* Advantages of the open air should be retained in design; wind and rain impact the attractiveness and safety of cycling.
6. *Cycling is a social activity.* Ideally, two cyclists should be able to ride side-by-side, especially if recreational cycling is common or for parents to escort children.
7. *People have limitations.* Designs should attend to limitations in experience, physical capabilities, and information processing.

Physical Dimensions

The typical width of the bicycle with a rider is about 30 inches (0.75 m). This width is insufficient for bicycle facility design since some additional space is needed to avoid obstacles and provide for the natural in-lane variation of most cyclists. Design guidance suggests that the minimum width of a bicycle facility should be 4 feet, with 5 feet being preferred. The length of a typical adult bicycle is 70 inches. Recumbent, bicycles with trailers (for children or cargo) and tandem bicycles are longer. These types of bicycles are more likely on recreational routes and shared-use paths. Typical design dimensions are shown in Figure 10.3 (AASHTO 2012).

Performance

A bicyclist's physical capabilities, type of bicycle, and environmental factors such as topography can have significant impact on the performance values selected for design. Three primary user performance inputs needed for design are: 1) speed (cruising or rolling); 2) acceleration and deceleration capabilities; and 3) reaction times. A summary of typical performance variables is presented in Table 10.3 and discussed below.

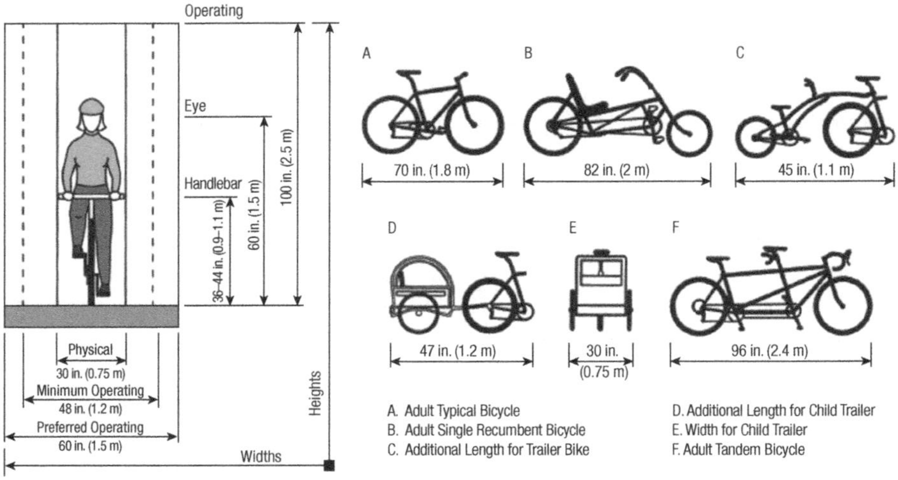

Figure 10.3 Physical dimensions of bicyclists

Source: Guide for the Development of Bicycle Facilities. Copyright 2012 by the American Association of State Highway and Transportation Officials, Washington DC. Used by permission.

Table 10.3 Summary performance data

Bicyclist type	*Feature*	*Value*	
		US Customary	*SI*
Typical upright adult bicyclist	Speed, pave level terrain	8–15 mph	13–24 km/h
	Speed, downhill	20–30 plus mph	32–50 plus km/h
	Speed, uphill	5–12 mph	8–19 mph
	Perception reaction time	1.0–2.5s	1.0–2.5s
	Acceleration rate	1.5–5.0 ft/s^2	0.5–1.5 m/s^2
	Coefficient of friction for braking, dry level pavement	0.32	0.32
	Deceleration rate (dry level pavement)	16.0 ft/s^2	4.8 m/s^2
	Deceleration rate for wet conditions (50–80% reduction in efficiency)	8.0–10.0 ft/s^2	2.4–3.0 m/s^2
Recumbent bicyclist	Speed, level terrain	11–18 mph	18–29 km/h
	Acceleration rate	3.0–6.0 ft/s^2	1.0–1.8 m/s^2
	Deceleration rate	10.0–13.0 ft/s^2	3.0–4.0 m/s^2

Source: AASHTO, 2012

Speed

Bicycling speed is a design input for signal timing, stopping sight distance, and horizontal curve radius for shared-use paths. In general, the upright bicyclist on level terrain will have a speed between 8 and 15 miles per hour, with 10 mph being a common assumed value. A bicyclist traveling downhill may be able to maintain a speed comparable to motor vehicles (30 mph) and not require as much maneuverability space as a bicyclist traveling uphill and at a slower speed. Particularly in scenarios with limited right-of-way, an engineer may consider installing wider and more protected facilities for uphill bicyclists than for ones traveling downhill. Bicyclists traveling at higher speeds also require greater stopping distance. Electric bicycles (as defined by the Consumer Product Safety Commission) are gaining popularity and allow for higher speeds (up to 20 mph) but limited data exist on in-street performance to date.

Acceleration and Deceleration

Acceleration values determine how quickly a cyclist can obtain cruising speed from a standing start. AASHTO suggests a range of acceleration rates of 1.5 to 5 ft/s^2. Deceleration values, coupled with surface friction assumptions, determine stopping distance. AASHTO recommends a deceleration rate of 16 ft/s^2 on dry level pavement and 8–10 ft/s^2 on wet level pavement.

Reaction Time

Reaction time is a function of decision-making and complexity, and tends to increase as the amount of information to be processed increases. Reaction time is less for expected events than for unexpected events. AASHTO recommends 1.0–2.5 seconds as appropriate assumptions.

Bicycle Facility Design and Operation

In designing bicycle facilities, there are a number of considerations that generally apply, including (ODOT, 2011):

- Drainage inlets, grates and manholes should be oriented so that bike tires will not fall in them and utility covers should be flush with the pavement.
- Smooth pavement is always preferable.
- For on-street facilities, the design should reinforce that it is an exclusive space for bicyclists. If it looks like a space for motor vehicles and is wide enough to accommodate them, it may be used by motor vehicles.
- On one-way streets, it is preferable to place the bicycle lane on the right-hand side, unless there are fewer conflicts on the left-hand side.
- The design should allow for sweeping and debris and snow removal. This is particularly important for protected bike lanes, where specialized equipment may be required due to the widths and barriers.

This section briefly reviews bicycle facilities on segments and at intersections. Operations of traffic signals, maintenance, parking and bike sharing are also discussed. The key references and design guides for engineers in the field include:

- *Guide for the Development of Bicycle Facilities* (AASHTO, 2012).
- *Urban Bikeway Design Guide*, 2nd ed. (NACTO, 2014).
- *Urban Streets Design Guide* (NACTO, 2013).
- *Manual on Uniform Traffic Control Devices* (MUTCD) (FHWA, 2009).
- *Highway Capacity Manual* (TRB, 2010).
- *Bicycle Parking Guidelines* (Association of Pedestrian and Bicycle Professionals, 2010).
- *Design Manual for Bicycle Traffic* (CROW, 2007).
- *Bicycle and Pedestrian Design Guide, Appendix L* (Oregon DOT, 2011).

Segments

The following provides a brief overview of common bicycle facilities found along road segments. Selecting the proper facility for a road depends on the context, including surrounding land-use, posted speed, traffic volume, and roadway width. Figure 10.4 presents an example selection matrix used by the Oregon Department of Transportation (ODOT). Note that at high vehicle speeds and even low volumes, physical separation is preferred.

Shared Roadways

Shared roadways are any public road on which bicycles are allowed. Ideally, these are found on local and minor collector roadways with speeds less than 30 mph and average daily traffic (ADT) volumes less than 4,000 vehicles per day. Treatments that can be applied to shared roadways include shared lane markings, bicycle route signs, wide curb lanes, and bike boulevards.

Bicycle Boulevards

Bicyclists are given priority over motor vehicles on bicycle boulevards (or neighborhood greenways), which are best suited for continuous local streets with low motor vehicle volumes (see Figure 10.5). Common features of bicycle boulevards are traffic calming, wayfinding, and stop signs turned to favor continuous bicycle traffic, while traffic diverters are used to prevent through motor vehicle movements. These provide a high level of comfort, are relatively inexpensive, and provide benefits to pedestrians and residents in terms of traffic calming.

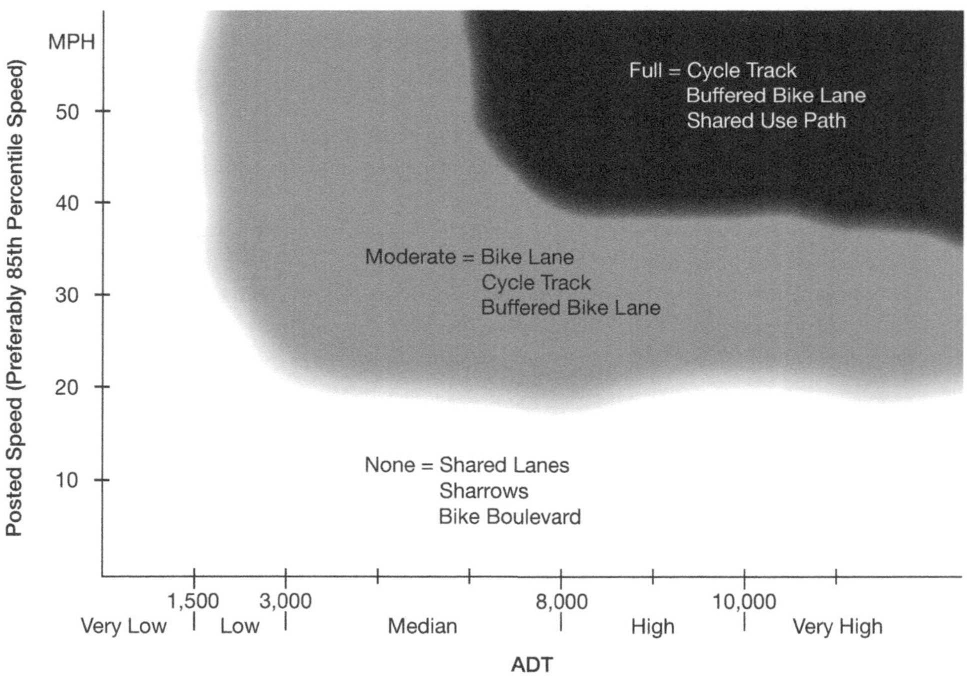

Figure 10.4 Oregon DOT bike facility selection matrix (ODOT, 2011)

Figure 10.5 Speed hump and sharrow on a bicycle boulevard, SE Spokane Street, Portland, OR
Source: Photo, TREC

Paved Shoulders

Paved shoulders are similar to bicycle lanes, but they are not designated for exclusive bicycle use. They are typically used in rural settings, where traffic volumes are lower than in urban areas. A consideration with rural settings is the use of rumble strips, which can be problematic for bicyclists. ODOT recommends placing rumble strips with 4 feet of shoulder to the right of the strip and providing 12-foot breaks in the strips every 40 to 60 feet (ODOT, 2011).

Bicycle Lanes

Bicycle lanes are typically used for moderate to high volume/speed roads in urban/suburban areas. They provide an exclusive space for bicyclists separate from motor vehicle traffic and serve to inform drivers to expect bicyclists. Bicycle lanes may be adjacent to the curb or between a parking lane and travel lane (Figure 10.6). Design guidance from AASHTO and NACTO is generally similar, recommending a width of 5 or 6 feet, with a minimum width of 3 or 4 feet where space is constricted. If adjacent to parked cars, additional width should be provided so that the cyclists can avoid car doors.

Buffered Bicycle Lanes

Buffered bicycle lanes are similar to bicycle lanes, but include a painted buffer between the bike, motor vehicle, and if applicable, parking lanes (Figure 10.7). They are typically found on urban arterials and should be considered for higher speed/volume roadways or areas with a high proportion of heavy vehicles or on-street parking with frequent turnover. The buffer provides increased separation, comfort, and visibility for bicyclists. The NACTO guide recommends the

Figure 10.6 Bike lane on busy street, Boise, ID
Source: Photo by N. Foster

Figure 10.7 Buffered bike lane, Milwaukee Avenue, Chicago, IL
Source: Photo by N. McNeil

Figure 10.8 Contra-flow bike lane, 8th Street, Boise, ID
Source: Photo by N. Foster

same minimum width as a standard bike lane; with the combined buffer and bike lane counting as the total width (e.g. a 2-foot buffer and 4-foot lane meets the desired 6-foot bike lane width).

Contra-flow Bicycle Lanes

Contra-flow bicycle lanes allow for bicycle travel in the opposite direction of motor vehicle travel on a one-way street, typically with low volumes and speeds (Figure 10.8). They improve bicycle connectivity and reduce wrong-way and sidewalk riding. Conventional bicycle lane width standards apply for contra-flow lanes. Intersections and driveways require special attention as motorists may not expect two-way bicycle travel. Bicyclists should have their own signals in the contra-flow direction. Alternatively, they can be instructed to follow the pedestrian signal.

Protected Bicycle Lanes

Protected bicycle lanes, also referred to as cycle tracks or separated bicycle lanes, are lanes that are buffered from motor vehicle traffic by a physical barrier (e.g. plastic flexposts, parked cars, curb, raised surface, or planters). They are typically found on urban collectors and arterials. Protected bike lanes offer physical separation from motor vehicle traffic and the greatest level of comfort of on-street bike infrastructure (Winters and Teschke, 2010). They may offer one (Figure 10.9) or two-way (Figure 10.10) travel.

NACTO recommends a minimum width of 5–7 feet for the lane for one-way facilities and 8–12 feet for the lanes in a two-way facility. The guide also recommends 3 feet for the buffer space (NACTO, 2014). Protected bicycle lane designs require special attention at intersections, driveways, and transit stops where there are likely to be high numbers of pedestrians and/or

Figure 10.9 One-way protected bike lane, NE Multnomah Street, Portland, OR
Source: Photo by N. McNeil

Figure 10.10 Two-way protected bike lane, Dearborn Street, Chicago, IL
Source: Photo by N. McNeil

motor vehicles crossing the path of bicyclists in the lane. Two-way operation requires special attention to the contra-flow direction, as motor vehicle drivers turning off the facility will not generally be expecting two-way traffic. Due to their increased width, a common issue is vehicles parking in the lane while loading/unloading.

Intersections

The following provides a brief overview of common bicycle intersection treatments.

Bicycle Boxes

Bicycle boxes provide a space for bicyclists to queue in front of motor vehicles at signalized intersections. In addition to reducing delay for bicyclists, they increase their visibility and reduce the risk of a turning vehicle crashing into a bicyclist at the start of the green phase. They are currently considered experimental in the *Manual of Uniform Traffic Control Devices* (MUTCD). Initial research shows positive results (Dill et al., 2012) related to driver comprehension, bicycle use, and conflicts. Note the design does not "solve" right-hook crashes that occur later in the green phase.

Two-Stage Turn Boxes

Two-stage turn boxes, also called a "Copenhagen Left," allow bicyclists to make a left-turn in two stages, to avoid having to merge into motor vehicle traffic lanes. They are most common on multi-lane roads or streets with protected bike lanes where merging is most difficult. To use the boxes, bicyclists travel through the intersection to the box, located in front of the stopped motor vehicle traffic, where they wait until the green indication for the through movement in the direction they want to travel.

Combined Bicycle Lane/Turn Lane

In situations where there is not sufficient width for a dedicated bicycle lane and a right-turn lane, combining the two together may be appropriate. The combined lane includes standard right-turn lane markings, along with a suggested bike lane that is delineated by either bike lane stencils and a dashed line or shared lane markings (NACTO, 2014).

Protected Bicycle Lane Intersection Approach

The buffer between a protected bicycle lane and motor vehicle traffic necessarily drops through intersections. The protected bicycle lane may be maintained all the way to the signalized intersection. In this case, the bicycle lane may be separated in time from right-turning traffic using a bike-specific signal, or if right-turning volumes are low, may be designed as in Figure 10.11. In areas where it is desirable to have a right-turn lane and there is not sufficient space for a dedicated turn lane and a protected bicycle lane, bicycle and motor vehicle traffic may share a mixing zone (see Figure 10.11) or the bicycle lane may be carried forward as a conventional bicycle lane to the left of the right-turn lane in a turning zone design as shown in Figure 10.12 which requires right-turning motor vehicles to merge across the lane.

Figure 10.11 Mixing zone, NE Multnomah Street, Portland, OR
Source: Photo by N. McNeil

Figure 10.12 Turning zone with restricted entry, L Street, Washington, DC
Source: Photo by T. Goddard

Traffic Operations for Bicycles

The following section provides a brief overview of traffic operations to accommodate bicycles.

Detection

The primary purpose of detection is to inform the signal controller a user desires service (Koonce et al., 2008). Both manual (pushbuttons) and automated technologies (inductive loops, video imaging, microwave and infrared) exist for detecting bicycles at signalized intersections. Inductive loops are the most commonly used technology and work by detecting relative change in inductance when a bicycle is present over the loop. Efficient detection requires clear communication and guidance to the user on how to actuate detection. A summary of the advantages and disadvantages of each detection technology is shown in Table 10.4. Further design guidance on bicycle detection can be found in the NACTO *Urban Bikeway Design Guide*. Ideally, feedback to the user should be provided to indicate when they are detected (as shown in Figures 10.13a and b).

Table 10.4 Summary of bicycle detection technologies

Type	*Advantages*	*Disadvantages*
Inductive loop	• Low cost	• Marking location is important, ineffective if location is unknown • Regular maintenance and calibration is necessary to maintain accuracy
Video	• Detection zones can be easily changed • Pavement condition is immaterial	• Less reliable in adverse weather • Cost
Microwave	• Can be used with any surface	• Maintenance is complex
Infrared	• Lane closures not necessary during installation • Can be used on any surface	• Difficult to detect smaller masses
Pushbutton	• Can be used for bicycle and pedestrian detection	• Bicyclists may have to dismount to activate the button, if not properly located

Source: Peris, Sam City of Palo Alto

Bicycle-Specific Signals

Bicycle-specific signals are being increasingly adopted in communities to enable safe bicycle crossings through intersections. While the 2009 MUTCD allowed circular indications for bicycle movements, interim approval for using bicycle signal displays was granted by the FHWA in 2013 (FHWA, 2013b).

Timing of Signals for Bicycles

Traffic signals are often coordinated along corridors to allow motor vehicles to progress with limited stops. It is advantageous to coordinate traffic signals for bicycle speeds to reduce travel

Figure 10.13a Use of green ball and sign for bicycle signal, Austin, TX

Source: Photo by C. Monsere

Figure 10.13b Use of bicycle signal in Lens, Portland, OR

Source: Photo by P. Singleton

times and stops. While challenging, it may also be possible to achieve multimodal progression along a corridor for both motor vehicles and bicycles at different speeds by ensuring simultaneous progression bands (Taylor, 2000).

Bicycles typically need longer minimum green times and clearance intervals than vehicles due to their slower operating speeds. These parameters for bicycle specific signals should be set considering bicycle accelerations and speeds, which can be estimated using existing guidance (NACTO, 2014; AASHTO, 2012). The AASHTO guide contains equations to determine minimum green time for cyclists starting from stop and clearance intervals based on crossing time for rolling cyclists. AASHTO provides a formula to estimate minimum green time for bicycles from a standing position:

$$BMG + Y + R_{\text{clear}} = PRT + \frac{V}{2a} + \frac{(W + L)}{\text{V}}$$

where:

BMG	=	Bicycle minimum green interval (sec)
PRT	=	Perception and reaction time, 1 (sec)
Y	=	Length of yellow interval (sec)
R_{clear}	=	Length of red interval (sec)
W	=	Intersection width (feet)
L	=	Typical bicycle length = 6 (feet)
a	=	Bicycle acceleration = 1.5 (feet/sec^2)
V	=	Bicycle crossing speed = 14.7 (feet/sec)

For rolling cyclists, AASHTO presents an equation for determining the rolling crossing time that can be used to check the length of the change and clearance interval. A cyclist who enters the intersection just at the end of green should have sufficient time to clear the intersection during the yellow change and all-red clearance intervals. Rolling time is presented as the sum of the braking distance, intersection width, and length of bicycle divided by the assumed rolling speed (suggested as 10 mph or 14.7 feet/second):

$$BCT_{\text{rolling}} = PRT + \frac{V}{2a} + \frac{(BD + W + L)}{\text{V}}$$

$$BD = PRT + V + \frac{V^2}{2a}$$

where:

BCT	=	Bicycle crossing time (sec)
PRT	=	Perception and reaction time, 1.0 (sec)
BD	=	Braking distance (feet)
W	=	Intersection width (feet)
L	=	Typical bicycle length = 6 (feet)
a	=	Bicycle deceleration rate for wet pavement = 5 (feet/sec^2)
V	=	Bicycle crossing speed = 14.7 (feet/sec) or 10 miles per hour (mph)

Application of the equation can generate unreasonably long clearance intervals for wide intersections. However, the guide makes it clear that “the yellow interval is based on the approach speeds of automobiles, and therefore, should not be adjusted to accommodate bicycles” (AASHTO, 2012, pp. 4–46). The guide suggests modifying the all-red time or, if that is insufficient, providing for extension time using dedicated bicycle-detector and bicycle-controller settings to add sufficient time to clear the intersection.

Similarly, NACTO requires that an “adequate clearance interval (i.e., the movement’s combined time for the yellow and all-red phases) should be provided to ensure that bicyclists entering the intersection during the green phase have sufficient time to safely clear the intersection before conflicting movements receive a green indication” (NACTO, 2014). An equation is provided to calculate the total clearance interval using the intersection width (W) and cyclist velocity (V):

$$C_i = 3 + \frac{W}{V}$$

In determining this minimum interval, field investigation of bicyclists’ speed is recommended. The guide suggests intervals sufficient for 15th percentile speeds should be used. Absent field data, NACTO suggests that “14 feet per second (9.5 miles per hour) may be used as a default speed” (NACTO, 2014).

Maintenance

Maintenance of bicycle lanes requires regular sweeping to remove debris. In areas with snow, many cyclists commute year round. Design of facilities should accommodate snow removal. Accommodation of bicycle travel should also be done in temporary work zones.

*Figure 10.14*a On-street bicycle coral parking outside restaurant, Coronada, CA

Source: Photo by C. Monsere

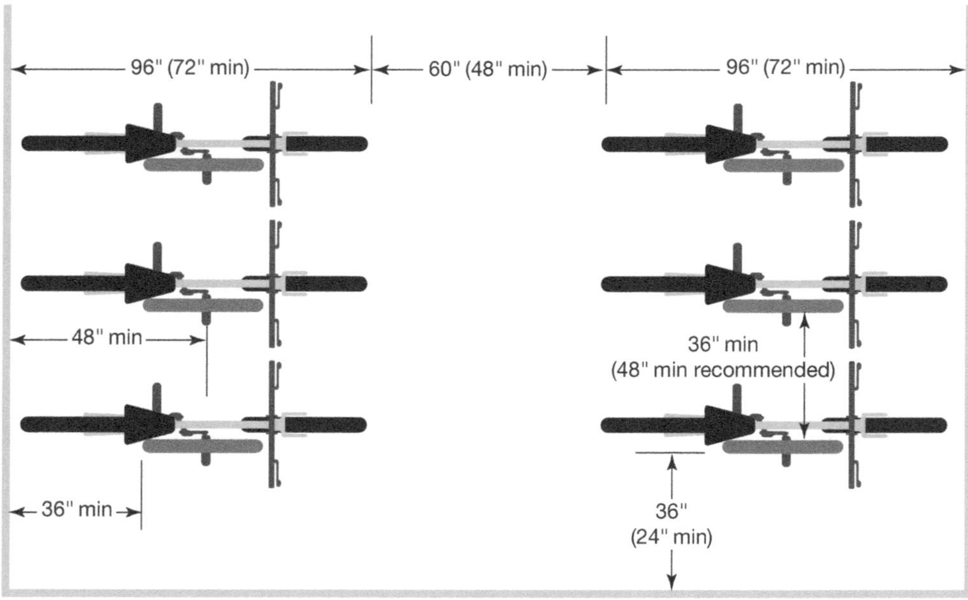

Figure 10.14b Minimum bicycle rack arrangement using inverted U staple

Source: *APBP Bicycle Parking Guidelines*, 2nd ed., pp. 2–36), www.apbp.org

Bicycle Parking

Bicycle parking falls under one of two categories: short-term or long-term. It is important to provide well-located, secure bicycle parking at popular bicycling destinations. For short-term parking, racks are typically provided. Figure 10.14 shows an inverted U staple that is deployed as an on-street bicycle coral parking in front of a restaurant. Guidance on spacing and other racks can be found in *Bicycle Parking Guidelines: A Set of Recommendations from the Association of Pedestrian and Bicycle Professionals* by the Association of Pedestrian and Bicycle Professionals. Long-term parking facilities offer superior security by preventing non-users from accessing bicycles and preventing theft and damage. These can be integrated with transit or in facilities to accommodate by the transit agency or private companies, such as BikeStation.

Bicycle Sharing

Modern bicycle-sharing systems involve the card-based readers to rent bikes on a short-term basis. Bicycles are equipped with electronically locking racks, telecommunication systems, smartcards, mobile phone access, on-board computers, and GPS tracking systems. Users can check out a bicycle at one station and return to another station. Charges are usually minimal for initial use and increase thereafter. A substantial effort is usually needed to rebalance the station bikes. The largest US fleet is located in New York, as shown in Figure 10.15.

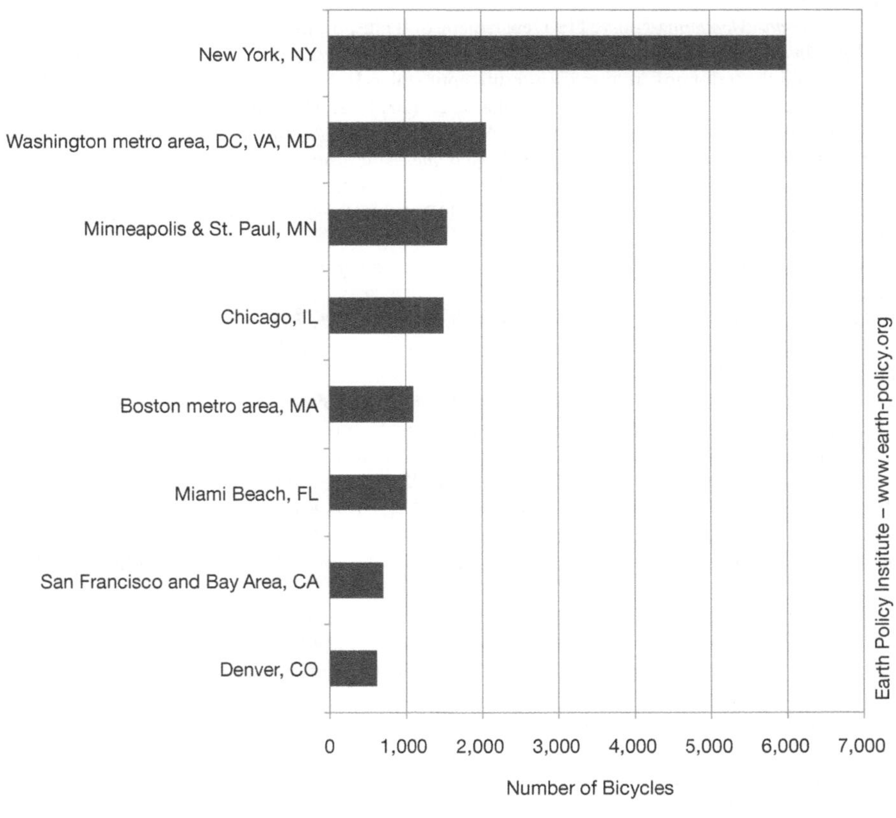

Figure 10.15 US bike-share

Source: Earth Policy Institute Data Highlights (Larson, 2014)

Conclusion

Designers, planners and operators of transportation systems have the responsibility to consider all users. Not every design can, or should, be expected to optimize all performance measures for each mode. Better public spaces result when practitioners consciously consider elements and principles of design highlighted in this chapter.

References

American Association for State and Highway Transportation Officials (AASHTO) (2010). *Highway Safety Manual*, Washington, DC.

American Association for State Highway and Transportation Officials (AASHTO) (2012). *Guide for the Development of Bicycle Facilities.*

Association of Pedestrian and Bicycle Professionals (2010). *Bicycle Parking Guidelines: A Set of Recommendations from the Association of Pedestrian and Bicycle Professionals.* Cedarburg, WI.

Bicycle Environmental Quality Index (BEQI), Draft Report (2009). San Francisco Department of Public Health.

Dill, J., Monsere, C., and N. McNeil (2012). Evaluation of Bike Boxes at Signalized Intersections. *Accident Analysis and Prevention*, 44(1), 126–134.

Dutch Design Manual for Bicycle Traffic (CROW, 2007).

El Esawey, M., Lim, C., Sayed, T., and Mosa, A. I. (2013). Development of Daily Adjustment Factors for Bicycle Traffic. *Journal of Transportation Engineering*, 139(8), 859–871.

FHWA (2009). *Manual on Uniform Traffic Control Devices* (MUTCD), US Department of Transportation, Washington, DC.

FHWA (2013a). *Traffic Monitoring Guide.* US Department of Transportation. Washington, DC, 468pp.

FHWA (2013b). *Interim Approval for the Optional Use of Bicycle Signal Faces* (IA–16). December 24. Available at: http://mutcd.fhwa.dot.gov/resources/interim_approval/ia16/index.htm (accessed June 15, 2014).

Figliozzi, M., Johnson, P., Monsere, C., and Nordback, K. (2014). A Methodology to Characterize Ideal Short-term Counting Conditions and Improve AADT Estimation Accuracy Using a Regression-based Correcting Function. *Journal of Transportation Engineering*, American Society of Civil Engineers.

Koonce, P., Rodegerdts, L. Lee, K. Quayle, S., Beaird, S., Braud, C., Bonneson, J., Tarnoff, P., and Urbanik, T. (2008) *Traffic Signal Timing Manual.* FHWA-HOP-08–024, US Department of Transportation, Washington, DC.

Kuzmyak, R. et al. (2014). *Estimating Bicycling and Walking for Planning and Project Development: A Guidebook.* NCHRP Report, 770, 161pp.

Larson, J. (2014) U.S. Bike-Sharing Fleet More than Doubles in 2013. *Highlights.* Earth Policy Institute, August 28, 2013; Web, February 15, 2014. Available at: www.earth-policy.org/data_highlights/2013/highlights40 (accessed June 15, 2014)

Mekuria, M. C., P. G. Furth, P. G., and Nixon, H. (2012) Low-Stress Bicycling and Network Connectivity. MTI Report 11–19. Mineta Transportation Institute, May.

NACTO (2013). *Urban Streets Design Guide*, Island Press, Washington, DC.

NACTO (2014). *Urban Bikeway Design Guide*, Island Press, Washington, DC.

Nordback, K., Marshall, W.E., Janson, B.N., and Stolz, E. (2013). Estimating Annual Average Daily Bicyclists: Error and Accuracy. Transportation Research Record: *Journal of the Transportation Research Board*, No. 2339, Transportation Research Board of the National Academies, Washington, DC, pp. 90–97.

Oregon Department of Transportation (ODOT) (2011). Bicycle and Pedestrian Design Guide. Salem, OR.

Peiris, S. (2014). Bicycle Detection at Signalized Intersections. Available at: www.mtc.ca.gov/services/arterial_operations/downloads/3-1-10/Bicycle_Detection-City_of_PaloAlto.pdf (accessed June 5, 2014).

SWOV (2012). Background of the five Sustainable Safety Principles. Available at: www.swov.nl/rapport/Factsheets/UK/FS_Sustainable_Safety_background.pdf (accessed June 9, 2014).

Taylor, D.B. and Mahmassani, H. S. (2000). Coordinating Traffic Signals for Bicycle Progression. Transportation Research Record 1705, pp. 85–92.

Winters, M. and Teschke, K. (2010). Route Preferences Among Adults in the Near Market for Bicycling: Findings of the Cycling in Cities Study. *American Journal of Health Promotion*, 25, 40–47.

11

TRAFFIC CRASH ANALYSIS AND PREVENTION

Gudmundur F. Ulfarsson

UNIVERSITY OF ICELAND

Introduction

Transportation, in particular vehicular road traffic, is a vital component of the global economy and a mobility enabler for billions of individuals and businesses. Yet, road traffic crashes remain one of the leading causes of death in the world. In a global review on road traffic crashes, there were about 1.2 million fatalities due to road traffic crashes in 2002, making traffic crashes the 11th leading cause of death in the world (WHO, 2004). The problem is especially high among the young, with road traffic injuries being the second leading cause of death for 5–29-year-olds, and the third leading cause of death for 30–44-year-olds in 2002 (WHO, 2004). These results were worse in the 2009 World Health Organization (WHO) report, when road traffic injuries had become the tenth leading cause of death (WHO, 2009). World governments at the United Nations (UN) General Assembly declared 2011–2020 as the Decade of Action for Road Safety on March 2, 2010, and called for UN member states to implement road safety activities (UN, 2010). The most recent WHO (2013) report shows that global fatalities due to traffic crashes remain at 1.2 million, with 20–50 million people injured each year. The report states that 88 countries saw a reduction in their number of fatalities between 2007 and 2010 but 87 countries saw an increase in fatalities (WHO, 2013).

On a global scale, five primary factors are linked to traffic fatalities: Speed, drink-driving, lack of helmet usage, lack of seat-belt usage, lack of child restraints. Yet only 28 countries have laws that the WHO considers adequate in terms of addressing all these factors (WHO, 2013). The middle-income (by global standards) nations suffer the largest losses due to traffic crashes (WHO, 2013). These nations are increasingly using motorized transport but do not have the resources to apply the safety improvement strategies of the high-income nations.

Global data show that 27 percent of fatalities due to traffic crashes are among pedestrians and bicyclists; in low- and middle-income nations the share is much higher (WHO, 2013). The protection of vulnerable road users is of key importance and this is an area where engineering solutions can offer help, since the design of infrastructure can lead to a safer transport environment for pedestrians and cyclists. Furthermore, various countermeasures are available, or can be designed, to reduce crash frequency and severity.

Taken together, the global evidence shows that traffic crashes take a serious toll on societies as the victims are most frequently people in their formative or most productive years. Traffic crashes cause a vast amount of global economic loss in addition to the incalculable personal loss due to lives lost and injuries. It is clear that engineering solutions—in addition to regulation, education and enforcement—can provide a positive impact on global public health through reductions in traffic crash frequencies and severities. Traffic safety is an engineering area where there is great room for individual engineers to come up with new ideas or implementations that can have a global impact and save millions of people from injury or death. For engineers wanting to make their mark on the world, there are few areas which provide more direct opportunities to save lives.

This chapter is intended to provide a transportation engineering overview of the field of traffic crash analysis and prevention. Engineering is of key importance in reducing the toll of traffic crashes. Engineering analysis and solutions work directly through improved roadway, roadside, and vehicle design and work indirectly by providing information to education and enforcement efforts, and can help guide legislative action and policy making.

Engineering Analysis of Traffic Crashes

The fundamental factor underlying the number of traffic crashes is the exposure to the risk of a crash and furthermore the risk of injury in a crash. This leads to three logical ways to reduce traffic crashes by 1) reducing exposure, 2) reducing the probability of a crash occurring, and 3) reducing the probability of a crash leading to injuries (TØI, 2012). The process that follows is then to 1) select locations for safety treatment, 2) diagnose the problems at the locations, 3) select or design countermeasures which reduce the probability of crashes occurring and/or reduce the severity of crashes, 4) perform cost-benefit analyses of each treatment and location, 5) develop a prioritized list of projects, 6) implement the projects, and 7) evaluate the safety impacts due to each implemented treatment. The *Highway Safety Manual* (AASHTO, 2010) is intended to provide a practical system for this safety treatment process.

The high-income nations of the world have available to them the resources to collect vast amounts of data on the operation of the transport system which engineers can use with sophisticated methods to determine safety problems and prioritize projects. However, the low- and middle-income nations, which suffer a large share of the fatalities and injuries (WHO, 2013), do not have the resources to collect and maintain the necessary databases for sophisticated analyses. Further, national capacity weaknesses in terms of a lack of national strategy for safety improvement and the lack of a strong national agency are global problems that affect numerous nations and that need to be alleviated (Bliss and Breen, 2009). Also, for all nations, investigating road crash data is a reactive response. The crashes have occurred and analysis is used to react to these events in order to reduce such occurrences in the future. The reactive approach must not be the only approach and proactive solutions need to be applied. Low- and middle-income nations can in many ways build on the experience of the high-income nations and use international research results with their local analysis to develop countermeasures, many of which are low cost. The five key traffic safety factors globally—speed, drink-driving, lack of helmet usage, lack of seat-belt usage, lack of child restraints—can, to a certain degree, be improved with legislation but laws may remain ineffective without public education and awareness of the laws, the reasons for the laws, and ultimately also law enforcement. Engineering solutions can, first and foremost, contribute by making the road infrastructure and the vehicles safer, whereas education and enforcement of laws contribute to making the people safer. All three are important. The high number of global fatalities among vulnerable road users also shows that a focus needs to be placed on the safety of pedestrians and cyclists early in the design process.

Proactive Safety

Proactive traffic safety is currently growing in international applications. The road assessment programs (RAP) are such ventures. The RAPs are non-government organizations but are now active in 70 nations and working towards the goals of safer drivers, safer vehicles, and safer roads (EuroRAP, 2014). The RAPs started with EuroRAP in 1999 (EuroRAP, 2014) but have gone international with IRAP (IRAP, 2012). Their focus is on the roads and in showing the great value and relative low cost by using engineering solutions proactively to improve road safety (EuroRap, 2014). The RAPs' standard practice is to risk map roads based on crash and traffic data, track safety performance of road networks, star rate roads in terms of expected safety using drive-through inspections, and provide road investment plans based on the roads' ratings. The RAPs' road safety toolkit provides free information on ways to improve the safety of roads, vehicles, and road users (IRAP, 2010).

Governments have employed a different proactive approach with the road safety audits (HA, 2003; FHWA, 2006; RTA, 2011). Road safety audits are formal reviews that are performed by a team of experts, independent of the design team, in order to identify possible road safety concerns or risk inherent in the design, taking into account the perspective of all road users (HA, 2003; FHWA, 2006; RTA, 2011). This is a proactive process, as it is not a crash history that calls for the audit. The road safety audit in the UK is performed at four stages (HA, 2003):

1. At the completion of preliminary design before publication. This is an important and cost-effective stage to find safety issues with an intended project because at this stage everything can still be readily changed, even land requirements.
2. At the completion of detailed design. At this stage the audit can consider particular design decisions such as position of signs and exact intersection design. This is the last chance for improvement in design, before construction, and is therefore still a cost effective time to identify and correct designs that might increase crash risk or severity.
3. At the completion of construction but before the road is opened for use. Now is the last chance to identify potential risk factors before actual crashes can occur. It is costly to make changes after construction but still better than to experience crashes.
4. The new construction is monitored during the first year and then audited with 12 months of crash data and again with 36 months of crash data in order to find whether the new construction is operating within the expected norms. There is variation between nations in terms of regulations of these audits, when they are called for, and if they are performed at all the four stages (HA, 2003; FHWA, 2006; RTA, 2011). The road safety audits can also be performed for roads already in use in a proactive way as a part of a review of a roadway network, or as a reaction to crashes.

A slightly less formal procedure, the road safety check, is also in use and it serves a similar purpose but it is performed by a road safety professional as opposed to a road safety audit team. The road safety check combines principles from the road safety audit and crash investigation. The road safety check, like the audit, should be performed independently and from the perspective of all road users. It can be performed before, during, and after construction (RTA, 2011). It is a less costly process than the full road safety audit and is therefore more readily used in lower-income nations, but it has the same objective, of identifying possible risks and providing suggestions for countermeasures or corrective actions.

Countermeasures

The objective of all road safety reviews, whether they are proactive or reactive, audits, checks, or other research, is to provide useful countermeasures and corrective actions that reduce risk, reduce severity, and improve the safety of the road infrastructure, the vehicles, and/or the road users themselves. In the early days of driving, road safety was viewed primarily as a driver safety issue. If drivers were more careful, crashes would not occur. It is still possible to maintain that nearly all crashes are due to human factors, behavior or mistakes. However, a road system must be designed given the realization that the users are human and humans are not perfect, they will make mistakes. The transport system as a whole must therefore be designed to reduce the chances of humans making mistakes, yet be as forgiving as possible when inevitable mistakes occur. Forgiving roads is one of the objectives of the RAPs (IRAP, 2012) and they provide a website of information about countermeasures with the road safety toolkit (IRAP, 2010). The *Highway Safety Manual* provides crash modification factors for various treatments which allow an evaluation of the safety impacts of projects in advance of their construction (AASHTO, 2010). The *Handbook of Road Safety Measures* (Elvik et al., 2009; TØI, 2012) provides a wealth of information on countermeasures and road safety based on extensive meta-analyses of published research results.

When considering vulnerable road users, pedestrians and bicyclists, vehicle speed is a key issue. Areas with much pedestrian traffic are increasingly speed limited at 30 km/h or less and research shows that at and beyond that speed the probability of fatality for vulnerable users hit by a vehicle rises sharply (Kim et al., 2007; IRAP, 2012). On a global scale, the situation can be simplified down to these major points (IRAP, 2012): Bicyclists and pedestrians are typically killed when walking or biking along the road, but also when crossing and at intersections. Drivers, vehicle passengers, and motorcyclists are most seriously injured or killed in run-off-road or head-on crashes, along with high-speed intersection crashes, e.g. due to red-light running or another kind of right-of-way violation. Each of these has engineering solutions that can reduce the injuries and fatalities. Sidewalks for pedestrians, bicycle paths, and even motorcycle lanes increase the separation of these road users from each other and automobiles, and can greatly reduce the crash risk between an automobile and pedestrian, bicyclist or motorcycle.

Median barriers are among the most cost-effective countermeasures against the risk of a head-on crash. Now there are lower cost alternatives available than previously with wire barriers, which also take less space than the heavier barriers most commonly in use before. The placement and choice of barriers depends on the location and traffic volume. Their installation is not worthwhile in all locations as they themselves can contribute to an increase in other kinds of same-direction crashes. Run-off-road crashes can be mitigated with barriers but also through an improvement of the roadside environment, e.g. by removal of nearby rocks, trees, or reduction in the size of ditches, or their complete enclosure underground with a drainage pipe and drainage rock on top (IRA, 2012). For some intersections, roundabouts can virtually eliminate fatal crashes. There are extensive available lists of available treatments for various safety issues, some of which are quite cost-effective (Elvik et al., 2009; AASHTO, 2010; IRAP, 2010; TØI, 2012).

The fundamental safety problems related to drunk and drugged driving, seat-belt usage, child-restraint usage, distracted and drowsy driving require more of a legislative, education, and law-enforcement approach (Goodwin et al., 2013). Speeding, however, can be handled both with such measures and engineering approaches (Goodwin et al., 2013). Speed cameras can be viewed as an engineering solution to provide automated enforcement, but speed can also be handled with the roadway and roadside design. Traffic-calming design can be used to give drivers a sense of discomfort as speed increases and such measures are an important part of the countermeasure toolkit because traffic-calming can lead to a reduction in crash risk and severity (Elvik, 2001; TØI, 2012). TØI (2012) and Elvik et al. (2009) provide a listing of expected traffic safety impacts due to various traffic-calming measures.

A popular countermeasure for red light running is the red light camera. A comprehensive recent meta-analysis indicates that such cameras are effective in reducing right-angle intersection collisions but that there is a statistically significant increase in rear-end crashes (Høye, 2013). This shows the potential duality of countermeasures. They can reduce one type of crash but increase another. This is a common conclusion for countermeasures and red light cameras are noted here as only one example. Median barriers, for example, can also increase the number of same-direction crashes as they reduce the frequency of head-on crashes. Countermeasures to the countermeasures may therefore be necessary. In the case of red light cameras, an example could be signposting an area monitored with red light cameras (Høye, 2013), with the aim of both increasing the deterrent of the cameras, as their presence becomes more known, and alerting drivers to expect more sudden stops at red lights. The meta-analysis found some evidence that this type of signposting is better done for a larger area, rather than only intersection by intersection (Høye, 2013).

Young drivers are also known to have a much greater crash risk than the middle-aged and this issue must primarily be handled through behavioral controls based on license restrictions, education, and enforcement (Goodwin et al., 2013). Older drivers are consistently found to have higher injury severities and the rate increases with age above 70 (NHTSA, 2013; TØI, 2012). Research, however, shows that in terms of behavior, older drivers have a safe profile; for example, they are most likely to obey speed limits and are the least likely group to drive while intoxicated (Lyman et al., 2002; Langford and Koppel, 2006). However, their increased fragility contributes to their high fatality and injury rate (Ridella et al., 2012) and they as a group are more likely to have health conditions that can increase crash risk and impair driving ability (Carr, 2000). Various dementing illnesses have been identified as likely contributors to increased crashes (Carr and Ott, 2010) and as a major contributor to driver's license revocation among drivers reported as potentially medically unfit to drive (Meuser et al., 2010).

Disease, rather than age, is therefore the key word for the safety of the older driver. The high injury severity and fragility of the older person, whether pedestrian (Kim et al., 2008), bicyclist (Kim et al., 2007) or vehicle occupant (TØI, 2012), suggests the need for even more forgiving vehicles and road infrastructure as the older driver population continues to increase. Also, there is a need for better ways to measure expected crash risk in relation to medical impairments so that the medically impaired can be identified more readily, without relying only on simplistic age-based screens. There is also a need to specify the acceptable expected crash risk of drivers in order to compare the expected crash risk of medically impaired drivers to other groups, such as young drivers, in order to evaluate if the crash risk is truly higher. It is an open question, how much higher the expected crash risk of a medically impaired person can be before that risk warrants driver's license restriction or revocation.

Statistical Evaluation of Countermeasures, Crash Frequency, and Severity

Many nations collect large amounts of data on crashes, mostly through police records but also through medical records. This information can be used reactively to investigate the traffic crash experience of the road network. The *Highway Safety Manual* (AASHTO, 2010) provides a system for just such an evaluation in a practical framework. Using statistical methods, it is possible to screen a network for high crash locations which can then be diagnosed using road safety audit or check methods; countermeasures can then be suggested based on their cost and impact on crashes. Once a road section has been improved, it is monitored and the effectiveness of the improvement quantified.

In evaluating the impact of countermeasures, the effect of regression to the mean must be controlled for (Hauer, 1997; Elvik, 1997). This is because safety countermeasures are typically installed at locations where there have recently been a high number of crashes. However, basing decisions simplistically on recent crash frequencies leads to biased suboptimal decisions because there is no control for the statistical variation in crashes. Given a longer time frame of observations, the recent high-frequency crash location will have its crash frequency regress to the mean. A simplistic before and after study, based only on crash frequency, is therefore likely to overestimate the impact of countermeasures, since the observed high frequency before may have been due to statistical variation and not systematic high risk. As crash frequency regresses to the mean, a reduction in crashes is seen, even though the countermeasure could be ineffective. Proper control for regression to the mean is therefore of key importance to evaluate safety projects correctly and to make optimal decisions (Hauer, 1997). The crash modification factors in the *Highway Safety Manual* are developed from studies controlling for regression to the mean effects (AASHTO, 2010).

The methods in the *Highway Safety Manual* (AASHTO, 2010) and those primarily used by safety agencies are crash rates, e.g. the number of crashes per one million vehicle-kilometers driven, and typically as a function of the traffic volume on the road, normally the annual average daily travel volume. Some limited additional variables may be used as well. There is now a large gap between safety analysis as done in practice and the state-of-the-art in safety research. Mannering and Bhat (2014) warn against this gap being used to excuse holding back advances at the methodological frontier with the reason that those advances may not be practical. This is a timely but continually important warning. What is practical today was at the frontier some years ago. The frontier of research investigates numerous analytic methods that over time lead to the increasing practicality of ever more sophisticated methods, although some remain impractical. Research identifies new methods and as their usefulness is shown or rejected, the useful methods are made practical, incorporated into software packages, and put into mass use. For example, recent work has attempted to draw up a possible typology of methods that could extend the *Highway Safety Manual* (Venkataraman et al., 2014b).

Research into traffic safety has primarily been split between traffic crash frequency analysis and traffic crash severity analysis. Occasionally, researchers handle both frequency and severity simultaneously but this is then typically done on the basis of severity-conditional frequency modeling (e.g. Venkataraman et al., 2013). Lord and Mannering (2010) provide an overview of the history and status of crash frequency analysis. Researchers started with a simple Poisson model and quickly moved to the negative binomial model, which in many ways has become the workhorse of traffic crash frequency analysis. The negative binomial model of the probability of y_{it} crashes occurring in location segment i during time interval t is written (Cameron and Trivedi, 2013):

$$P\left(y_{it} \mid x_{it}, \beta\right) = \frac{r\left((1/\alpha) + y_{it}\right)}{r(1/\alpha) y_{it}!} \left(\frac{(1/\alpha)}{(1/\alpha) + \lambda_{it}}\right)^{1/\alpha} \left(\frac{\lambda_{it}}{(1/\alpha) + \lambda_{it}}\right)^{y_{it}},$$

where x_{it} is a vector of observable variables, β is a vector of estimable parameters on the variables, α is a single estimable overdispersion parameter, and the expected number of crashes is

$\ln \lambda_{it} = \beta x_{it} + \varepsilon_{it}$, where ε_{it} is independently and identically log-gamma distributed unobserved error term. The negative binomial model is implemented in numerous software packages; it can be practically estimated, used readily, and it offers more flexibility than the Poisson model because it allows data to be overdispersed, i.e. the variance is larger than the mean. The model results are interpreted easily as a positive estimated parameter indicates that the variable increases the expected crash frequency.

If the size of the road segment varies, it is important to include a variable to control for this size. Of vital importance in traffic crash models is the inclusion of a measure of traffic crash risk exposure, such as traffic volume for automobile crashes. Without a measure of exposure the model is likely to lead to biased results, since other variables, correlated with exposure, will suffer from omitted variables bias. Typically, the time period remains fixed for all observations and does not need to be controlled for. If this time period is too short, it can lead to a problem due to a preponderance of zero counts, which violates the Poisson or negative binomial model distributions. There are various ways to handle this issue if it occurs, although some are debated (Lord et al., 2007).

Researchers have developed countless variants and more sophisticated models that handle various types of correlations and other features of safety data, such as time varying or random parameters on the explanatory variables (Lord and Mannering, 2010). Crash frequency data are inherently messy, real world data that offer up impressive challenges for statistical analysis.

The statistical analysis of injury severity has for several years been primarily based on what can be called discrete outcome models (Savolainen et al., 2011). This branch of statistical modeling was developed actively as discrete choice models within econometrics (McFadden, 1981) and as such became quite popular in transportation (Ben-Akiva and Lerman, 1985). This field has since led to numerous model variants and techniques. A review of injury severity research and its numerous model variants is provided by Savolainen et al. (2011).

The fundamental basis for the use of discrete choice models for injury severity hinges on the idea that there is some kind of propensity for each severity in an occurred crash, and that this propensity has observable and systematic components but also an unobservable random component. The notion of propensity replaces utility in the discrete choice framework. Then, conditional on the crash occurring, the propensity function is used to describe the probability of discrete injury severity outcomes. A straightforward and commonly used discrete injury severity model is the multinomial logit model where the probability of person *n* experiencing injury severity category *i* is written (McFadden, 1981; Ben-Akiva and Lerman, 1985; Shankar and Mannering, 1996):

$$P_{ni} = \frac{\exp \beta_i x_{ni}}{\sum_{\forall j} \exp \beta_j x_{nj}},$$

where $\beta_i x_{ni}$ is the systematic and observable part of the propensity function

$$S_{ni} = \beta_i x_{ni} + \varepsilon_{ni} \text{ ,}$$

β_i is a vector of estimable parameters, *xni* is a vector of observable variables, and ε_{ni} is an independently and identically Gumbel type 1 extreme value distributed unobserved error term. Typically, the only observable variable that depends on the injury severity outcome *i* is the alternative-specific constant, or intercept, in each propensity function. The other variables are generally specific to the crash itself—e.g. driver, vehicle, roadway, and environmental characteristics—and are identical across injury outcomes. This model is readily used in a number of available software packages (e.g. SPSS, STATA, Limdep, R) and it is practical to implement and estimate using typical crash data, such as police reports (Washington et al., 2011).

To start implementing this model, the first step is to select an injury outcome as the base case. Its propensity function becomes zero and the other results are interpreted as differences from the base case. A further complication occurs if the injury outcomes are three or more. Then the sign of the estimated parameter on a particular explanatory variable may not necessarily correspond to the actual direction of change in probability of that outcome as the variable increases. This mathematical feature of the model is proven by Greene (2012) and has led to the use of marginal effects, or more commonly elasticities, to investigate the change in probability as a variable changes (Washington et al., 2011). Since many variables in injury severity analysis are binary indicator variables—e.g. 1 if paved road, 0 otherwise—a pseudo-elasticity has been used to calculate the average change in probability as the variable changes from 0 to 1 (Ulfarsson and Mannering, 2004).

A more recent statistical approach is to allow the estimable parameters to vary using a mixed logit model (McFadden and Train, 2000; Train, 2009). This is now increasingly being done when investigating injury severity and is proving to be a fruitful tool for such analysis (Milton et al., 2008; Kim et al., 2010). Interpreting random parameter model results is somewhat more complicated than for fixed parameter models and the mixed logit is no exception. This renders the model less practical for prediction but, given the experience so far, the model is quite useful for analysis (Kim et al., 2013).

The mixed logit model requires a distributional assumption for the random parameters, but several such assumptions can be tested in terms of their results on goodness-of-fit. The normal distribution is a logical starting point in the absence of a priori information. With that distribution, each parameter can be described by a mean and a standard deviation. If the standard deviation for a random parameter is not found to be statistically significantly different from zero, it suggests the parameter should remain fixed and not random given that other distributional alternatives have been tested as well, such as the uniform distribution.

There is special complexity to interpreting random parameter models. Now the parameters are not known in advance for practical application, only their distribution. The results generally even find that the parameter distributions are on both sides of zero, meaning that each random parameter can become either positive or negative, depending on the draw. With a normally distributed random parameter the weight on either side of zero is simply found through the p-value of the t-statistic for testing the significant difference of zero from the parameter mean based on its estimated random parameter standard deviation $(0 - \beta)/\sigma$, as it will give the share of the distribution, centered at β, larger than zero. Is the variable then increasing or decreasing safety if it can have a parameter on either side of zero? The answer is that it depends. To fully interpret the model requires simulations. It is therefore likely to be some time before these models enter practical use at transportation safety agencies, but they do provide rich interpretations for safety researchers.

A powerful feature of the mixed logit model for severity and the random parameter negative binomial model for frequency is the ability to specify the mean of each random parameter as a function of observable characteristics. This enables the model to capture unobserved heterogeneity and does not constrain the random parameters on the same variable to all be drawn from a distribution with the same mean. This is especially important in the safety context because this allows engineers to investigate complex interactions between location-specific characteristics and the influence of an explanatory variable. For example, the effect of roadway lighting can have different effects depending on surroundings such as interchange type (Venkataraman et al., 2014a). This can also aid the interpretation of the random parameters as the heterogeneity shows how certain features push the mean towards the negative or the positive.

Savolainen et al. (2011) and Mannering and Bhat (2014) provide a warning against the use of simple parsimonious models due to their omitted variables bias, and they warn against overly simple analyses which ignore self-selection and the endogeneity of variables. For example, a simple correlation test between an indicator variable for a particular countermeasure and injury severity may show positive significance but this can readily occur due to the correlation of the tested variable with an omitted variable which provides the true causal effect, quite possibly due to self-selection (Mannering and Bhat, 2014). Simplistic analyses, based on limited data and few variables, are at risk of overestimating the importance of the included variables for determining safety. Sound engineering judgment is necessary to avoid falling into a correlation-without-causation trap and thereby recommending countermeasures that are not useful in practice.

Malyshkina and Mannering (2010) demonstrate how to evaluate the impact of roadway design exceptions on both crash frequency and severity using sophisticated statistical modeling based on random parameters. They first used the mixed logit model for severity and a random parameter negative binomial model for frequency. The parameters turned out to be fixed in the frequency model and they concluded by using a regular negative binomial model for the crash frequencies. This gives an example of the importance of beginning with a fully specified, flexible model structure and if it turns out to be non-significant, then constraining the structure as appropriate. This will lead to more robust analysis than a priori, assuming it is sufficient to use a simple model with few variables.

Conclusions

This chapter has focused on traffic safety from a transportation engineer's point of view. We have seen the enormous size of the problem and that it is in large part borne by low- and middle-income nations. Cost-effective solutions to identify locations for improvements have been discussed and citations provided to publications that provide information about countermeasures and evaluations of the safety benefits of each. This is important in order to prioritize projects and make more optimal decisions. We have noted that there is a gap between practice and research. The research frontier is moving forward into especially exciting times with new, high quality data sources in sight or beginning to appear (Mannering and Bhat, 2014), after perhaps a relatively quiet period in the past decade that has been characterized by researchers developing ever more complex models in an attempt to squeeze every bit of information out of somewhat limited data. More advanced data sources are set to open up the traffic safety research field widely (Mannering and Bhat, 2014).

Despite the gap between what is currently practical and what is currently being developed and tested by researchers, it is important for engineers to continue performing advanced traffic safety research. At times, the research advances may appear unrealistic in the short term, but let us remember that the advanced research of today is tomorrow's practical, cost-effective engineering solution.

There is also an exciting world on the automotive engineering side as vehicle technology is set to make vehicles safer. There are a number of assisted driving technologies already available in vehicles, e.g. adaptive cruise controls that can maintain a safe distance to a vehicle in front (Rajamani, 2012). However, perhaps most promising in the long term are autonomous vehicles such as the Google Car (Google, 2010). It may seem like science fiction, but this technology is quite possibly closer to practice than we expect. This type of technology has the potential to reduce traffic crashes, reduce the need for private vehicles as autonomous private rapid transit becomes possible, and could make better use of resources (Folsom, 2011). Even given such technological vehicle advancements, vehicle drivers, pedestrians, bicyclists, and other road users need road infrastructure that is designed to help reduce traffic crash risk and, in case of the eventual crash, is forgiving and reduces severity.

The size of the traffic crash fatality and injury toll on humanity is so large that traffic safety is an important area where transportation engineers can contribute directly to the betterment of societies by reducing injuries and saving lives.

References

AASHTO, 2010. *Highway safety manual.* American Association for State Highway and Transportation Officials (AASHTO), Washington, DC.

Ben-Akiva, M. and Lerman, S., 1985. *Discrete choice analysis: Theory and application to travel demand.* MIT Press, Cambridge, MA.

Bliss, T. and Breen, J., 2009. *Implementing the recommendations of the world report on road traffic injury prevention.* World Bank Global Road Safety Facility, Washington, DC.

Carr, D.B., 2000. The older adult driver. *American Family Physician*, 61 (1), 141–148.

Carr, D. B. and Ott, B. R., 2010. The older driver with cognitive impairment "It's a very frustrating life". *Journal of the American Medical Association*, 303 (16), 1632–1641.

Elvik, R., 1997. Evaluations of road accident blackspot treatment: A case of the iron law of evaluation studies? *Accident Analysis and Prevention*, 29, 191–199.

Elvik, R., 2001. Area-wide urban traffic calming schemes: A meta-analysis of safety effects, *Accident Analysis and Prevention*, 33, 327–336.

Elvik, R., Høye, A., Vaa, T., and Sørensen, M., 2009. *The handbook of road safety measures*, 2nd ed. Emerald Group Publishing, Bingley.

EuroRAP, 2014. *About EuroRAP.* EuroRAP, Basingstoke, UK. Available at: www.eurorap.org/about-eurorap/, www.eurorap.org/about-eurorap/history/(accessed April 22, 2014).

FHWA, 2006. *FHWA road safety audit guidelines.* US Department of Transportation, Federal Highway Administration (FHWA), pub. FHWA-SA-06–06, Washington, DC.

Folsom, T. C., 2011. Social ramifications of autonomous urban land vehicles. *Proceedings of the IEEE International Symposium on Technology and Society* (ISTAS), May 23–25, Chicago.

Goodwin, A., Kirley, B., Sandt, L., Hall, W., Thomas, L., O'Brien, N., and Summerlin, D., 2013. *Countermeasures that work: Highway safety countermeasures guide for state highway safety offices.* 7th ed., US Department of Transportation, National Highway Traffic Safety Administration (NHTSA), pub. DOT-HS-811–727, Washington, DC.

Google, 2010. *What we're driving at.* Google. Available at: http://googleblog.blogspot.com/2010/10/what-were-driving-at.html/(accessed April 22, 2014).

Greene, W. H., 2012. *Econometric analysis*, 7th ed. Prentice Hall, NJ.

HA, 2003. HD 19/03 Road safety audit. In: *Design Manual for Roads and Bridges*, Vol. 5, Sec., 2, Part 2. Department for Transport, Highways Agency, London.

Hauer, E., 1997. *Observational before-after studies in road safety: Estimating the effect of highway and traffic engineering measures on road safety.* Pergamon Press, Oxford.

Høye, A., 2013. Still red light for red light cameras? An update. *Accident Analysis and Prevention*, 55, 77–89.

IRA, 2012. Roadside improvements (in Icelandic: *Lagfæringar á umhverfi vega*). Icelandic Road Administration (IRA), Reykjavik, Iceland.

IRAP, 2010. *Road safety toolkit.* International Road Assessment Programme (IRAP), Basingstoke, UK. Accessed April 22, 2014, at: http://toolkit.irap.org/.

IRAP, 2012. *Vaccines for roads.* 2nd ed. International Road Assessment Programme (IRAP), Basingstoke, UK.

Kim, J.-K., Kim, S., Ulfarsson, G. F., and Porrello, L. A., 2007. Bicyclist injury severities in bicycle-motor vehicle crashes. *Accident Analysis and Prevention*, 39 (2), 238–251.

Kim, J.-K., Ulfarsson, G. F., Kim, S., and Shankar, V. N., 2013. Driver-injury severity in single-vehicle crashes in California: A mixed logit analysis of heterogeneity due to age and gender. *Accident Analysis and Prevention*, 50 (1), 1073–1081.

Kim, J.-K., Ulfarsson, G. F., Shankar, V. N., and Kim, S., 2008. Age and pedestrian injury severity in motor-vehicle crashes: A heteroskedastic logit analysis. *Accident Analysis and Prevention*, 40 (5), 1695–1702.

Kim, J.-K., Ulfarsson, G. F., Shankar, V., and Mannering, F., 2010. A note on modeling pedestrian injury severity in motor vehicle crashes with the mixed logit model. *Accident Analysis and Prevention*, 42 (6), 1751–1758.

Langford, J. and Koppel, S., 2006. Epidemiology of older driver crashes—Identifying older driver risk factors and exposure patterns. *Transportation Research Part F*, 9, 309–321.

Lord, D. and Mannering, F., 2010. The statistical analysis of crash-frequency data: A review and assessment of methodological alternatives. *Transportation Research Part A*, 44, 291–305.

Lord, D., Washington, S.P., and Ivan, J.N., 2007. Further notes on the application of zero inflated models in highway safety. *Accident Analysis and Prevention*, 39 (1), 53–57.

Lyman, S., Ferguson, S.A., Braver, E.R., and Williams, A.F., 2002. Older driver involvement in police reported crashes and fatal crashes: Trends and predictions. *Injury Prevention*, 8, 116–120.

Malyshkina, N. V. and Mannering, F. L., 2010. Empirical assessment of the impact of highway design exceptions on the frequency and severity of vehicle accidents. *Accident Analysis and Prevention*, 42, 131–139.

Mannering, F. L., and Bhat, C. R., 2014. Analytic methods in accident research: Methodological frontier and future directions. *Analytic Methods in Accident Research*, 1, 1–22.

McFadden, D. F., 1981. Econometric models of probabilistic choice. In: Manski, C. and McFadden, D. F. (Eds.), *Structural analysis of discrete data with econometric applications*. MIT Press, Cambridge, MA.

McFadden, D. and Train, K., 2000. Mixed MNL models for discrete response. *Journal of Applied Econometrics*, 15 (5), 447–470.

Meuser, T. M., Carr, D. B., and Ulfarsson, G. F., 2009. Motor-vehicle crash history and licensing outcomes in older drivers reported as medically impaired in Missouri. *Accident Analysis and Prevention*, 41, 246–252.

Milton, J., Shankar, V., and Mannering, F., 2008. Highway accident severities and the mixed logit model: an exploratory empirical analysis. *Accident Analysis and Prevention*, 40 (1), 260–266.

NHTSA, 2013. *Safety in numbers*, 1(8). US Department of Transportation, National Highway Traffic Safety Administration (NHTSA), Washington, DC.

Rajamani, R., 2012. Adaptive cruise control. In: *Vehicle dynamics and control*. Mechanical Engineering Series, Springer, New York, pp. 141–170.

Ridella, S. A., Rupp, J. D., and Poland, K., 2012. Age-related differences in AIS 3+ crash injury risk. Types, causation and mechanisms. *Proceedings of the IRCOBI Conference on the Biomechanics of Impact*, September 12–14, Dublin.

RTA, 2011. *Guidelines for road safety audit practices*. Roads and Traffic Authority of New South Wales (RTA), pub. 11.291, New South Wales.

Savolainen, P. T., Mannering, F. L., Lord, D., and Quddus, M. A., 2011. The statistical analysis of highway crash-injury severities: A review and assessment of methodological alternatives. *Accident Analysis and Prevention*, 43, 1666–1676.

Shankar, V. and Mannering, F., 1996. An exploratory multinomial logit analysis of single-vehicle motorcycle accident severity. *Journal of Safety Research*, 27 (3), 183–194.

Train, K., 2009. *Discrete choice methods with simulation*. Cambridge University Press, New York.

TØI, 2012. *Trafiksikkerhetshåndboken*, 4th ed. Høye, A., Elvik, R., Sørensen, M. W. J., and Vaa, T., (Eds.) *Transportøkonomisk institutt* (TØI), Oslo, Norway (in Norwegian).

Ulfarsson, G. F. and Mannering, F. L., 2004. Differences in male and female injury severities in sport-utility vehicle, minivan, pickup and passenger car accidents. *Accident Analysis and Prevention*, 36 (2), 135–147.

UN, 2010. *Improving global road safety*. Resolution 64/255, United Nations General Assembly, New York.

Venkataraman, N., Ulfarsson, G. F., and Shankar, V. N., 2013: Random parameter models of interstate crash frequencies by severity, number of vehicles involved, collision and location type. *Accident Analysis and Prevention*, 59 (1), 309–318.

Venkataraman, N. S., Shankar, V. N., Ulfarsson, G. F., and Deptuch, D., 2014a: A heterogeneity-in-means count model for evaluating the effects of interchange type on heterogeneous influences of interstate geometrics on crash frequencies. *Analytic Methods in Accident Research*, 2, 12–20.

Venkataraman, N. S., Ulfarsson, G. F., and Shankar, V., 2014b: Extending the highway safety manual (HSM) framework for traffic safety performance evaluation. *Safety Science*, 64, 146–154.

Washington, S., Karlaftis, M., and Mannering, F., 2011. *Statistical and econometric methods for transportation data analysis*, 2nd ed. Chapman & Hall/CRC, Boca Raton, FL.
WHO, 2004. *World report on road traffic injury prevention*. Executive Ed. Peden, M., World Health Organization, Geneva.
WHO, 2009. *Global status report on road safety: Time for action*. World Health Organization, Geneva.
WHO, 2013. *Global status report on road safety 2013: Supporting a decade of action*. World Health Organization, Geneva.

12

TRAVEL BEHAVIOR IN RESPONSE TO NETWORK DISRUPTIONS

Praveen Edara and Timothy C. Matisziw

UNIVERSITY OF MISSOURI

Introduction

According to the Merriam-Webster dictionary, to disrupt is to "cause (something) to be unable to continue in the normal way: to interrupt the normal progress or activity of (something)" (Merriam-Webster, 2014). In surface transportation, a disruption could take many forms ranging from minor road accidents delaying traffic to geographically extensive road-capacity reductions emanating from an extreme event such as a major earthquake. Not only are the traffic characteristics of a region impacted by network disruptions, but so are the economic, social, energy, and the many other systems that are interdependent with our transportation activities. The focus of this chapter is on the effects of network disruptions on individual travel behavior and resulting travel patterns in the network. This chapter will capture the current state of practice in the understanding of the relationship between traveler behavior, system performance, and network disruptions. An understanding of how people adapt to changes in a network and how traffic flows manifest over the course of a disruptive event(s) is essential to the effective planning, response, and policy making for future disruptions. The implications of the observed travel behavior during network disruptions on transportation planning and policy making for future disruptions is also discussed in this chapter.

The location(s) and magnitude of disruptions in a transportation system depends in part on how vulnerable a system is to a triggering event. Grubesic and Matisziw (2013) provide a taxonomy of the ways in which infrastructure and system performance can be vulnerable to disruption. In particular, they attribute vulnerabilities to factors such as capacity and level of congestion, the structural condition of the infrastructure, the types of threats to which the infrastructure may be exposed, the topology of the system, and the extent to which the infrastructure is interdependent upon the functioning of other systems. They also describe other, less tangible aspects of a system that may contribute to vulnerability to disruption such as, the types of planning safeguards that have been implemented to mitigate disruption as well as political issues and policy. Perhaps one of the more complex factors underlying vulnerability to disruption is the many types of threats that can trigger disruptions to traffic flow in transportation systems.

Some threats and their potential impact to traffic flow can be somewhat predictable. For instance, coordinated and publicized events such as infrastructure repair and modification are common sources of disruption to traffic flow. According to the United States (US) Federal Highway Administration (FHWA, 2009), there are typically more than 3,000 road-work projects during the peak construction season in the US, traversed by an estimated 12 billion vehicle miles each year. These active work zones often necessitate closure of travel lanes and/or shoulders, or reduction of lane widths, resulting in a reduction in roadway capacity and a disruptive influence on traffic flow. Peak travel demand periods (i.e. rush hour, leisure travel) and special events (i.e. sports, concerts) are expected to degrade traffic flow with a certain degree of regularity. While not as predictable in their location, scope and magnitude, the disruptiveness of events triggered by inclement weather (i.e. flooding, ice and snow, tornados, hurricanes, etc.) can be anticipated to some extent by travelers. Likewise, events such as minor traffic incidents (i.e. debris, accidents, law-enforcement activities) and signal outages whose location and duration may be unknown, have become so common that they are anticipated to be part of the normal daily traffic operations and plans are often in place to effectively mitigate their impact on traffic flow. At the other end of the spectrum are extreme events whose location and magnitude are relatively uncertain, potentially giving rise to much severe impacts to traffic flow.

Travelers respond to disruptions in many ways given the amount of knowledge they have about the state of the transportation system—by electing to maintain their current travel patterns, enduring additional travel costs, choosing alternative routes or modes, shifting departure times, selecting different destinations, cancelling trips, among others. As a result of these responses, the travel demand and the traffic flow in a network can be extremely dynamic. The impact of a disruption or a series of disruptions on the total travel demand traversing the affected portion of the system depends on the amount of redundant or unused road capacity available in the affected area as well as the knowledge of the traveler about the current state of the transportation network. If there is a sufficient number of alternative routes or modes that can absorb the disrupted trips from the affected routes, the total travel demand to the region may remain unchanged. In fact, some of the empirical research reviewed in this chapter suggests this to be the case (Hunt et al., 2002; Zhu et al., 2010). Clearly, the total travel demand will be impacted more in areas without sufficient redundant capacity. The question of travelers' preferences among shifting departure times, choosing alternative routes, and choosing alternative modes to complete their disrupted trips is important. Another pertinent question is whether any of the changes in travel behavior witnessed during the disruption will result in a habitual change—i.e., longer-term changes to an individual's travel behavior. The answers to these questions have policy implications in decisions regarding the design and management of alternative routes, and diversion of travelers to transit and other high-occupancy modes (ridesharing) that are more sustainable.

Analyzing models of flow in transportation systems is an important component of planning for perturbations in the system, whether they are planned or unexpected events. Knowledge of the location and level of flow in a system can assist transportation planners in tasks such as strategically dispatching emergency responders, notifying travelers of alternative routes, identification of detour routes and evacuation corridors, development of transportation management plans, traffic control strategies, rescheduling of transport services, adjusting network policies, and allocation of infrastructure improvement resources. Traffic flow though, is highly dynamic, depending on many factors such as infrastructure condition and characteristics, the location and magnitude of travel demand, the travel strategies and behaviors of users, routing alternatives, congestion, among others. Given that changes in any one or more of these factors can impact traffic flow, producing a good model of what flow may look like at any point in the future can be extremely challenging. This challenge is compounded even more since knowledge of the location and magnitude of events affecting traffic flow in the future is even less certain.

This chapter focuses primarily on long-term disruptions to network operation. Most of the existing research on how individual travel behavior is affected by network disruptions considers those events with a duration of six or more months (in many cases over a year). Short-term disruptions such as traffic incidents, snow/ice storms, and others that impact traffic only over the course of a few days are discussed to a lesser degree in this chapter since they do not typically generate the same type or magnitude of behavioral changes that long-term disruptions do. For example, a traffic accident may motivate motorists to choose alternative routes to avoid the affected roadway. However, it may not induce any significant mode shift or departure time shift. Even less likely is the possibility of motorists cancelling their trips in response to a short-term disruption. The effects of short-term disruptions are also more likely to be localized as compared to long-term disruptions that usually have a network-wide impact.

This chapter is organized as follows. Existing literature on travel behavior in response to network disruptions is reviewed. Research on travel behavior in response to long-term construction projects, bridge failures, and earthquakes is highlighted in this respect. A discussion of models and tools used to analyze traffic flows in a disrupted transportation network is also provided. The chapter concludes by discussing the implications of empirical evidence found in the literature on transportation planning for disruptions.

State-of-the-Practice in Understanding Travel Behavior During Network Disruptions

Coordinated Disruptions

Major infrastructure rehabilitation and construction projects typically induce long-term disruptions to network capacity. The long-term nature of such projects can therefore provide travelers with the information and experiences needed to induce changes to their travel behavior. If a project involves closure of an entire roadway, travelers must seek alternative options to complete their activities. Some travelers may choose to switch to other transportation modes, consider alternative destinations, or try carpooling for their trips. Traveler response to major roadwork projects has been documented in previous research. Typically, studies conducted a longitudinal analysis of travel behavior at different snapshots in time since the beginning of a project.

A seminal study by Goodwin et al. (1998) considers the effects of capacity-reduction scenarios at 100 locations across the world including the UK, Germany, Italy, Switzerland, Japan, Australia, Sweden, the US, and Canada. Among other things, the authors examine the overall reduction in traffic ("disappearing"), i.e., after excluding the rerouting (to alternative routes) traffic and retiming (or shifting departure times) traffic from the total traffic that used to use the affected route. They report that in over 50 percent of the scenarios, at least a 14 percent overall reduction in traffic due to the capacity reduction was realized. They further explain that the percentage reduction in traffic depends on the traveler choices in response to capacity reduction. Specifically, when an insufficient amount of unutilized capacity on alternative routes existed near the affected route, they find that a certain proportion of traffic would disappear from the affected area.

Hunt et al. (2002) examine the travel impacts of a major bridge closure in downtown Calgary, Canada. In 1999, the Centre Street Bridge across the Bow River connecting the downtown to the residential portion of the city was closed to traffic for 14 months for repairs. The average weekday traffic volume over the bridge prior to the closure was 34,000 vehicles/day. The authors report a 4.4 percent decrease in daily traffic volume during the closure on the North, East–West, and South corridors into downtown. However, they conclude that not all of this reduction in traffic may be attributed to the bridge closure; some of it may have been due to other factors. Traveler surveys indicate that 4.4 percent of automobile users switched to other modes with 2.7 percent of prior travelers deciding to no longer make the trip. Again, not all travelers switching modes attributed the bridge closure as the reason for their change in travel behavior. Only 0.2 percent of automobile users stated the bridge closure as the reason for cancelling their trips.

In the St. Louis, Missouri, metropolitan region, Interstate 64 is a major freeway carrying traffic to and from downtown St. Louis. Between January 2008 and January 2010 portions of Interstate 64 were closed for infrastructure improvements. Gentry et al. (2011) provide a comprehensive evaluation of the impacts of this closure on travel in the St. Louis region. Traveler surveys were conducted to obtain feedback from travelers in response to the road closures. While the majority of respondents appear to have been satisfied with the options to reroute their trips in St. Louis during the closure, a few reported making changes to their shopping and driving habits. The mobility impacts were also measured. The daily traffic volumes on designated alternative routes increased, by 10 percent to 23 percent for freeway routes and by 30 percent to 35 percent for arterial routes. Various best practices were used by the Missouri Department of Transportation (MODOT) to help advertise and improve alternative routes to accommodate the increased traffic volumes. The demand for a ridesharing program ("RideFinders") increased during the first year of closure but remained steady during the second year. The authors, however, caution that other factors such as the increase in gas prices and downturn in the economy may also have contributed to the increase in the ridesharing. Public transit, both bus and rail, also witnessed an increase in ridership during the closure.

Zhang et al. (2012) investigate the travel behavior and changes in traffic flow patterns in response to a full freeway closure in California. In 2008, Interstate 5, a major freeway connecting the downtown and suburbs of Sacramento, California, was closed for about two months for reconstructing the pavement and upgrading the drainage system, among other improvements. They report that the overall traffic demand near the closure area only dropped by about 3 percent to 7 percent over this period. They also found evidence that travelers switched to alternate routes including arterials and changed their departure times in response to the closure. Increase in carpooling was observed in the initial days of the closure, albeit a trend that did not last for long. The authors note that the road network redundancy, especially the presence of another major interstate, Interstate 80, and an efficient arterial network, helped absorb the additional traffic and mitigate the adverse impacts. All observed changes in travel behavior only lasted for the duration of the project. The authors reported that the traffic demand went back to pre-construction levels after the completion of the project, thus indicating no long-term habitual changes in travel behavior. Ye et al. (2010) analyze surveys of the residents in the region affected by the Interstate 5 closure. Their findings were similar to the observed travel behavior reported by Zhang et al. (2012). For example, Ye et al. (2010) report that most respondents cited avoiding rush hour and seeking alternative routes as their rationale for changing their travel patterns.

Unexpected Disruptions

The effects of unplanned network disruptions on travel behavior have been examined to a much lesser extent. Giuliano and Golob (1998) investigate the effects of the Los Angeles, California 1994 Northridge earthquake on travel in the region. In this event, portions of two main freeways, Interstate 5 and Interstate 10 were significantly damaged. The authors report that travelers primarily responded to this disruption by changing routes, and changing travel time schedules (departure times), and avoiding discretionary trips to the damaged areas. A short-term increase in transit ridership was observed at some locations but the overall transit use in the region appeared to remain relatively stable. Thus, in this case it seems that travelers were more reluctant to switch modes than to switch routes or departure times. Travel patterns in the region were found to return to pre-earthquake levels immediately after the infrastructure facilities were restored.

Zhu et al. (2010) evaluate the effects of the 2007 collapse of the I-35W bridge in Minneapolis, Minnesota. Prior to its collapse, the I-35W bridge carried approximately 140,000 vehicles per day to and from Minneapolis, the largest city in the state of Minnesota. Due to its heavy daily usage, the bridge collapse was feared to have substantial impacts on the travel patterns in the road network. The authors surveyed travelers likely affected by the bridge collapse and used the traffic flows measured across the network to infer the travel impacts of the bridge collapse. Based on this information, they find that: 1) total travel demand in the affected region did not vary significantly after the bridge collapse, 2) higher average travel times were reported after the collapse, 3) no significant increases in public transit ridership was observed, and 4) changing routes and trip departure times were believed to be the most common reactions of travelers in response to the bridge collapse. Woolley-Meza et al. (2013) attempt to characterize disruptions to global air travel given two events, the September 11, 2001 attacks in the US and the 2010 volcanic eruption of Mt. Eyjafjallajokull in Iceland. To describe how these events affected traffic flow, they approximate the extent to which passengers might have rerouted their trips to avoid closed airports.

Zhu and Levinson (2008) review different types of disruptions that have occurred in the past and studies that have attempted to quantify their impact to transportation activity. To this end, they provide many examples of the broad effects of disruptive events, such as increased costs, route changing, mode diversion, etc. However, they find that detailed studies on the actual location and magnitude of traffic flow change are lacking. To remedy this shortcoming, they suggest more fully tracking transportation flows via technologies such as global positioning systems (GPS). In an attempt to explore how travelers' behavior may be influenced by disruptions, Lu et al. (2011) evaluate how individuals route selection changes when presented with a simulated transportation system, disruptive events, and travel information.

Modeling Transportation Flows During Disruptions

Most models of transportation flow entail assumptions of perfect information about the network, uniform perceptions of travel cost, and stability of traffic flow over time (Ortuzar and Willumsen, 2011). However, given a disruption(s) to the system, this assumption would not be valid in most cases. One approach for evaluating the effect of an event on traveler behavior and traffic flow is to use the evidence obtained from previous studies of travel behavior to disruptions. He and Liu (2012) provide an approach for updating daily assignment of travelers to a network that accounts for how the travelers learn about and adjust for congestion after a disruption. Unnikrishnan et al. (2012) propose a multi-commodity network flow model for assigning origin/destination (OD) flows to network paths when travelers are faced with potential disruptions and are allowed to update their routing as they proceed along their trip. Kunimatsu and Hirai (2014) apply a multiple regression model to predict passenger flows given a disruption to a rail network. To accomplish this, they use detailed records of passenger flows and disruptions for a rail network over the previous year, so that for any time period of interest, they have a good representation of prior traffic flow. Kim and Mahmassani (2014) apply a gamma-gamma mixture model approach in an attempt to characterize traffic variability over time. In their application, traffic scenarios are simulated based upon actual weather events and traffic accidents occurring over the period of a year. Mahmassani et al. (2009) describe DYNASMART-X, a simulation software for modeling network characteristics, such as OD flows, based on a range of inputs such as the network, signalization, and real-time observations of traffic. In this simulation package, algorithms are provided for modeling how travelers react to certain stimuli, such as new information (e.g. from variable message signs) and changes in weather. However, the impacts of these events on flows are modeled globally on the network via predefined equations. Liao and Yen (2010) also leverage DYNASMART simulations to analyze traffic impact from earthquake scenarios. In Chang et al. (2009), the authors model transportation flows during an earthquake using the network loss analysis module of a comprehensive GIS-based visualization software, MAEViz. Edara et al. (2010) provide a summary of various simulation packages used in the literature for modeling hurricane evacuation scenarios with different road network treatments. Similarly, Henchey et al. (2014) examine prospects for simulating traffic flows as influenced by weather and traffic events to assist in planning for better responses by emergency medical services. Some studies have developed decision support system software specifically focusing on emergency evacuations. For example, Tufekci and Kisko (1991) developed the Regional Evacuation Modeling System (REMS) capable of evaluating different types of emergencies including natural disasters (e.g. hurricanes) and chemical or nuclear accidents using optimization techniques. In the context of commercial air travel, Berge et al. (2006) discuss simulation of disruptive events to air traffic and the use of this information to identify rescheduling alternatives for recovering the system.

A variety of approaches for modeling travel flow resulting from a disruption on a single link of a network have been proposed. For example, Baykal-Gursoy et al. (2009) discuss a queuing approach for modeling traffic flow on a road that experiences a single disruptive event, such as a traffic incident. Jiang and Adeli (2003) detail a model for estimating the traffic flow in a work zone based upon the characteristics of the work zone. The impacts of multiple, simultaneously present disruptions on traffic flow have been investigated to a lesser extent. Lee and Kim (2007) consider the economic impact of damage to multiple bridges resulting from an earthquake when assessing restoration strategies. In their modeling approach, origin-destination (OD) commodity flow data are used to represent the amount of interaction among analysis zones that could be disrupted given such an event and then model impact to flow. Aside from modeling flow disrupted by a few scenarios of damage, optimization approaches have also been developed to identify scenarios of disruption that could impact traffic flows the most in support of such tasks as searching for vulnerabilities or strategically targeting a network (Matisziw and Murray, 2009).

Given that the duration of many disruptions can vary quite widely, identifying/collecting data to adequately represent traffic patterns during a particular time frame can be a challenge. Since flows in transportation systems can exhibit tremendous variability throughout hours, days, weeks, months, etc., accounting for traffic flow during those periods is an important task (Ortuzar and Willumsen, 2011). For example, using high temporal resolution data on flows in a telecommunication system over a 24-hour period, Matisziw et al. (2012) model which sets of network arcs, if completely damaged, would impact OD flow the most. They find that even over short durations of time, small changes in network flows can dramatically influence the spatial distribution of vulnerabilities in a system.

Given the complexity of modern transportation systems, obtaining a direct observation of flow on all portions of a system at any point in time is typically not a feasible task. Hence, planners have to be reliant on using samples of observations to infer flow throughout the system.

To this end, several techniques for collecting data on transportation flows are commonly utilized. First, flow can be observed along selected segments of a system, such as measured by loop detectors in road networks and number of scheduled seats in air networks. Second, flow between nodes in a network can be collected by observing the origin and destination associated with the flow. For example, this information could be elicited via travel surveys in road networks or by evaluating weigh bills or automatic ticketing records in rail networks. Regardless of how flows are observed, some level of uncertainty always exists in their measurement. That is, flow may only be observed for sample of traffic, specific types of vehicles, at a certain temporal frequency, at a limited set of locations, and at some level of reliability. In cases where some link flows can be observed, flows on links that were not directly observed must be somehow inferred (Van Zuylen and Willumsen, 1980; Erlander et al., 1985). Likewise, in cases where origin/destination flows cannot be directly observed, the magnitude of interaction among OD pairs must be estimated (Ortuzar and Willumsen, 2011). Given the diversity of types of information collected about the functioning of transportation systems, transportation planners often have to build their representation of traffic flows by fusing data from a variety of sources (Tamin and Willumsen, 1989; List and Turnquist, 1994).

While existing protocols for collecting and inferring transportation flows have proven useful for representing broad flow characteristics in a system, their applicability in planning for short-term and/or more extreme changes to a system has not been extensively evaluated. For instance, in support of data-intensive procedures such as real-time incident detection, it is recommended that key traffic parameters at each sensor location be collected at 20–30 second epochs at minimum (FHWA, 2006). Measurement of seasonal fluctuation in traffic flow can be captured by averaging traffic over seven-day periods at sampling locations (Indiana LTAP Center 2005). In some instances, when deriving traffic parameters for portions of a network, one may wish to avoid collection during periods of inconsistent travel. For instance, travel on roads after noon on Fridays may have different characteristics than regular weekday travel after noon (Indiana LTAP Center 2005). For general engineering planning applications, less frequent measurement of traffic may be required. Generally speaking, it is recommended that all linkages within a planning region be evaluated at least once every six years (FWHA, 2013). The origin/destination flows obtained from surveys are spatially and temporally aggregate, reported for only a sample of individuals. For instance, the US Bureau of Transportation Statistics provides estimates of journey-to-work flows between zones in the US based upon census data collected every ten years. Many simulation programs such as CORSIM and VISSIM require observations of traffic parameters at relatively frequent intervals (e.g. 15 minutes). However, to what extent are these long-term estimates useful when it comes to applications such as planning for events that cause network disruptions? Also, observations and estimates of traffic flow are rarely if ever associated with the type of and location of disruptions influencing traffic at that time in the system. Therefore, it is important to consider the state of practice in the understanding of how traveler behavior and transportation flows are impacted by disruptive events and how these responses can be integrated into disaster planning. Consideration of traveler behavior in response to disruptions will ultimately allow network planners to better model transportation conditions when evaluating options for disaster preparedness and response.

Implications on Planning and Response to Disruptions

Some findings from the literature on traveler response to road network disruptions may be applicable to analysis of future disruptions. First, the total travel demand entering the region affected by a disruption can be expected to be lower than the demand prior to the disruption.

However, the magnitude of reduction in total demand can vary quite a bit from one location to another. With the exception of one study (Goodwin et al., 1998), most studies report demand reduction of less than 10 percent. In the case studies involving major urban areas, even a small reduction in travel demand has important implications on the congestion in the network. Second, some studies cite the existence of redundant capacity in the road network as an important factor driving traveler choices during disruptions. When sufficient unused capacity is available and accessible to traffic without significant increase in their trip times, travelers may be more motivated to switch to alternative routes than other options such as switching to another mode. This finding suggests that transportation agencies must consider providing sufficient redundancy while designing the network. For example, when planning for future network modifications, the provision of capacity redundancy may be included as a constraint to accommodate response to disruptions.

One other common finding across several studies was that most observed changes in traveler behavior during a disruption were short lived. Route changes, mode changes, departure time changes, and others occurred only during the disruption. As soon as the network returned to its normal pre-disruption condition, travelers reverted back to their old travel choices and habits, and the modal travel demand and route traffic flows returned to the pre-disruption levels. This finding should not come as a surprise. Route choice and mode choice decisions are fairly stable after travelers have experimented with them for a few times (i.e. attainment of user equilibrium). Once travelers are satisfied with their knowledge of a system, there is usually no reason for them, under normal conditions, to consider other alternatives since they perceive that they have already optimized their choices. Thus, unless the new travel experiences during the disruption are not better than their pre-disruption experiences, it is not surprising that they revert back to their previous choices. It is not clear in the reviewed case studies if the enhancements made to the alternative routes and/or modes remained after the disruptions ended. For example, signal timing changes on an alternative arterial street may have only been in place during a construction project and not continue beyond the project, thus not making it a competitive alternative for travelers post-disruption. The continuation (or discontinuation) of these alternative route and modal enhancements could be a factor in the travel decisions made after the disruption has been cleared.

The main behavioral responses to network disruptions were found to be route shift and departure time shift. Mode shift was less common or preferred by travelers. The modal share for transit and ridesharing modes increased in the early period of disruption but the increase was not sustained and their share reverted back to the pre-disruption levels. This finding indicates that transportation agencies may prioritize network enhancements such as improving the performance of alternative routes and providing the necessary traveler information over enhancements to transit or other modes. However, there may be other important reasons for enhancing transit operations such as equity concerns that were not specifically discussed in the disruption literature.

Finally, the role of transportation agencies in the overall management of transportation operations in the disruption-affected region was found to be quite important for alleviating the disruption impacts. For planned disruptions, an agency must plan designated alternate routes, modal enhancements, and disseminate that information to the travelers before the actual event takes place. En-route traveler information on routes is also typically provided using intelligent transportation systems. Traffic control technologies such as ramp metering, active traffic management, variable speed limits, and others may be used to improve the traffic operations. The role of a transportation agency during unexpected disruptions also involves information dissemination and traffic management. Thus, agencies planning or responding to disruptions in the future must be aware of their critical role in efficiently managing travel during disruptions.

Closing

This chapter presents an overview of the understanding of the impact of network disruptions on travel behavior. The literature on planned disruptions such as construction activities and unexpected disruptions such as earthquakes and bridge failures reported some common findings. While the magnitude of the reduction in total travel demand to the affected area varied across studies, most studies reported a reduction in the total travel demand due to a disruption. When redundant capacity was available in the affected area, switching to an alternate route appeared to be the most popular travel choice among individuals affected by the disruption. Regardless of the nature of the travel choices made by individuals in response to a disruption, i.e. route shift, mode shift, departure time shift, the changes only lasted the duration of the disruption. There is no significant evidence that the changes to travel behavior observed during a disruption induced long-term behavioral changes. Models that estimate traffic flows under network disruptions are an important component of planning for disruptions. While existing protocols for collecting and inferring transportation flows have proven useful for representing broad flow characteristics in a network, their applicability in planning for has not been extensively evaluated in the literature. Available approaches and data needs for modeling traffic flow during network disruptions were also reviewed in this chapter. While some work has begun to incorporate the effects of disruptions to traffic flows, many challenges remain. First, given the complex nature of flows in many transportation systems, improvements in the collection of observations of flows, network conditions, and response to disruption are needed. To address this need, advances in collecting network parameters consistently over large regions at a high spatio-temporal resolution are needed. Also, more work must be conducted in merging available data sources to better leverage existing data when creating a representation of a system and its characteristics. For instance, traffic count data obtained from sensors could be augmented with mobile phone trace data. Second, approaches for integrating a better representation of the state of a system in light of disruptions into planning models are needed. For instance, one might consider how to introduce broader network conditions into assessing the potential impact of a construction project or a series of projects.

References

Baykal-Gursoy, M., Xiao, W., and Ozbay, K. 2009. Modeling traffic flow interrupted by incidents. *European Journal of Operational Research*. 195, 127–138.

Berge, M.E., Carter, M.L., Haraldsdottir, A., Repetto, B., and Kang, L. 2006. Airline schedule recovery in flow management: an application for departure rerouting. 25th DASC Digital Avionics Systems Conference—Network-Center Environment: the impact on Avionics and Systems, Portland, OR. October.

Chang, L., Elnashai, A.S., and Spencer, B.F. 2009. Transportation system modeling and decision support for catastrophic event planning in the central United States: A case study of St. Louis area. Technical Council on Lifeline Earthquake Engineering Conference (TCLEE), Oakland, CA, pp. 1–8.

Edara, P., Sharma, S., and McGhee, C. 2010. Development of a large-scale traffic simulation model for hurricane evacuation: Methodology and lessons learned. *ASCE Natural Hazards Review Journal*, 11(4), 127–139.

Erlander, S., Jornsten, K.O., and Lundren, J.T. 1985. On the estimation of trip matrices in the case of missing and uncertain data. *Transportation Research Part B*. 19B, 123–141.

FWHA. 2006. *Traffic Detector Handbook*, 3rd ed. vol. 1. US Department of Transportation, Federal Highway Administration. FHWA-HRT-06–108.

FHWA. 2009. Work Zone Safety and Mobility Fact Sheet - 10th Annual National Work Zone Awareness Week. Available at: safety.fhwa.dot.gov/wz/wz_awareness/2009/factsht09.cfm (accessed May 25, 2014).

FWHA. 2013. *Federal Highway Administration Traffic Monitoring Guide*. US Department of Transportation, Federal Highway Administration, Office of Highway Policy Information. September.

FHWA Work Zone Safety and Mobility Fact Sheet—10th Annual National Work Zone Awareness Week. Available at: http://safety.fhwa.dot.gov/wz/wz_awareness/2009/factsht09.cfm (accessed May 25, 2014).

Gentry, L., Chilukuri, V., Hiles, T., Trueblood, M., Kinzel, C., Frazier, R., Bham, G., Baik, H., Lee, J., Hodge, D., Witkowski, M., and Ryan, T. *The New I-64 Economic and Regional Mobility Study*. Final Report CMR 12–010. Missouri DOT. November 2011.

Giuliano, G. and Golob, J. 1998. Impacts of the Northridge earthquake on transit and highway use. *Journal of Transportation and Statistics* 1, 1–20.

Goodwin, P., Hass-Klau, C. and Cairns, S. Evidence on the effects of road capacity reductions on traffic levels. Traffic Engineering and Control, June 1998, pp. 348–354.

Grubesic, T.H. and Matisziw, T.C.. 2013. A topological framework for categorizing infrastructure vulnerability. *GeoJournal*. 78, 287–301.

He, X. and Liu, H.X. 2012. Modeling the day-to-day traffic evolution process after an unexpected network disruption. *Transportation Research Part B*. 46, 50–71.

Henchey, M.J., Batta, R., Blatt, A., Flanigan, M., and Majka, K. 2014. A simulation approach to study emergency response. *Journal of Simulation*. 8(2), 115–128.

Hunt, J., Brownlee, A., and Stefan, K. 2002. Responses to Centre Street bridge closure—where the "disappearing" travelers went. Transportation Research Record: *Journal of the Transportation Research Board*, 1807, Transportation Research Board of the National Academies, Washington, DC, pp. 51–58.

Indiana LTAP Center. 2005. Guidelines for traffic counts on local roads and streets. Purdue University, SP-1–2005, December.

Jiang, X. and Adeli, H. 2003. Freeway work zone delay and cost optimization model. *Journal of Transportation Engineering*. May/June, 230–241.

Kim, J. and Mahmassani, H.S. 2014. A finite mixture model of vehicle-to-vehicle and day-to-day variability of traffic network travel times. *Transportation Research Part C*. 46, 83–97.

Kunimatsu, T. and Hirai, C. 2014. Methods for analyzing passenger flows during train traffic disruptions using accumulated passenger data. *QR of RTRI*. 55(2), 86–90.

Lee, J.S. and Kim,T.J. 2007. Implementation of spatialtemporal model for infrastructure reconstruction strategy under large-scale disaster. *Transportation Research Record*. 2022, 39–46.

Liao, D. and Yen. P.W. 2010. A linkage tool for analyzing earthquake traffic impact in micro level based on seismic risk assessment and traffic simulation. ACM International Conference Proceedings, 1st International Conference and Exhibition on Computing for Geospatial Research and Applications, COM. Geo 2010, Washington, DC. June.

List, G.F. and Turnquist, M.A. 1994. Estimating truck travel patterns in urban areas. *Transportation Research Record*. 1430, 1–9.

Lu, X., Gao, S. and Ben-Elia, E. 2011. Information impacts on route choice and learning behavior in a congested network. *Transportation Research Record*. 2243, 89–98.

Mahmassani, H.S., Dong, J., Kim, J., and Chen, R.B. 2009. Incorporating weather impacts in traffic estimation and prediction systems. US Department of Transportation, FHWA-JPO-09–065.

Matisziw, T.C., Grubesic, T.H., and Guo, J. 2012. Robustness elasticity in complex networks. *PLOS ONE*. 7(7), e39788.

Matisziw, T.C. and Murray, A.T. 2009. Modeling s-t path availability to support disaster vulnerability assessment of network infrastructure. *Computers & Operations Research*. 36, 16–26.

Merriam-Webster.com. 2014. "disrupt."Available at: www.merriam-webster.com (accessed May 20, 2014).

Ortuzar, J.D. and Willumsen, L.G. 2011. *Modelling Transport*, 4th ed.. John Wiley & Sons: Croydon.

Tamin, O.Z. and Willumsen, L.G. 1989. Transport demand model estimation from traffic counts. *Transportation*. 16, 3–26.

Tufekci, S. and Kisko, Y.M.. 1991. Regional evacuation modeling system (rems): a decision support system for emergency area evacuations. *Computers and Industrial Engineering*. 21(1–4), 89–93.

Unnikrishnan, A., Shahabi, M., and Figliozzi, M. 2012. Model for network assignment problem of capacitated freight with disruptions. *Transportation Research Record*. 2269, 11–19.

Van Zuylen, H.J. and Willumsen, L.G. 1980. The most likely trip matrix estimated from traffic counts. *Transportation Research Part B*.14B, 281–293.

Woolley-Meza, O., Grady, D., Thiemann, C., Bagrow, J.P., and D. Brockmann. 2013. Eyjafjallajokull and 9/11: The impact of large-scale disasters on worldwide mobility. *PLOS ONE*. 8(8), e69829.

Ye, L., Mokhtarian, P., Circella, G. Commuter impacts and behavior changes during a temporary freeway closure: The Fix I5 Project in Sacramento, California. TRB 89th annual meeting, January 10–14, 2010.

Zhang, H., Chen, Y-R., Li, R., and Qian, Z. 2012. What happens when a major freeway is closed for repair? The case of Fix I-5 in downtown Sacramento, California. Transportation Research Record: *Journal of the Transportation Research Board*, No. 2278, Transportation Research Board of the National Academies, Washington, DC., pp. 134–144.

Zhu, S. and Levinson, D. 2008. Planned and unplanned disruptions to transportation networks. *Transportation Research Synthesis*, Minnesota Department of Transportation, Office of Investment Management. August.

Zhu, S., Levinson, D., Liu, H., and Harder, K. 2010. The traffic and behavioral effects of the I-35W Mississippi River bridge collapse. *Transportation Research Part A: Policy and Practice*, 44(10), 771–784.

13

TRANSPORTATION PLANNING

Matthew Karlaftis, Konstantinos Kepaptsoglou, and Grigorios Fountas

NATIONAL TECHNICAL UNIVERSITY OF ATHENS

Introduction

Transportation systems are vital for socioeconomic prosperity and growth of modern communities. Human activities generate demand for transportation, while economic development, societal trends, and technological progress are major drivers affecting its growth. Transportation infrastructures are expected to provide adequate supply and efficient services, and meet demand and quality requirements. Often, transportation systems and infrastructures fail to accommodate growing demand; their capacity is exceeded and their services degrade, leading to phenomena such as congestion, air and noise pollution. In other cases, transportation systems suffer from overdesign: capacity exceeds potential demand, yielding an economically inefficient transportation system. It is evident that in both cases, proper planning is required for efficiently linking the demand and supply sides of a transportation system.

Transportation planning refers to those actions required for designing, improving, evaluating, and siting transportation systems and infrastructures. Typically, transportation planning follows a rational model approach (Meyer and Miller, 2001): the problem at hand is determined (design or improve a transportation system), goals and criteria are set, alternatives are formed and evaluated, and final plans are developed. For example, consider the case of upgrading a city's public transportation system. The associated problem could be that of introducing and designing a new light rail transit (LRT) system: LRT corridors and service characteristics should be determined. Relevant goals may include the increase of transit share among travelers, the improvement of air quality and the reduction of congestion; evaluation criteria such as transit ridership, modal shift, and volume to capacity ratio may also be defined. Different scenarios of LRT corridors are set up and evaluated; the most prominent among them is processed and final LRT plans are developed.

Models have a critical role in transportation planning; they are used for estimating current and future transportation system characteristics, and eventually for determining evaluation criteria and design variables of the system. For instance, in the aforementioned LRT example, ridership estimation and the determination of service frequencies would be elements derived by model application. In this context, the objectives of this chapter are twofold: a) to offer an overview of the typical transportation planning approach and framework, and b) to present most common models exploited in the context of the so-called "Four-step planning process." The remainder of the chapter is organized as follows: The next section presents a generic approach for analyzing transportation systems and links it to the transportation planning process. Next, the Four-Step Model (FSM) for planning transportation system and its elements are discussed. Finally, an example framework for applying the Four-Step Model is offered.

Transportation Analysis Framework

A transportation system consists of infrastructures and services; these form the supply side of the transportation planning process. From a practical perspective, the supply side refers to what the transportation system can efficiently and/or adequately handle with respect to transporting people and goods, based on its physical attributes and service characteristics. Depending on the type of infrastructures, different attributes may be used for estimating transportation system capabilities; these could include but are not limited to capacity, speed, physical structure, connectivity, vehicle occupancy, and so on. For instance, capabilities of a highway link can be determined by its capacity and speed, while a bus line may be characterized by its service frequencies and vehicle capacity.

A transportation system is also affected by its surrounding activity system. Land uses and socioeconomic activities form travel patterns and flows; these will interact with the transportation system and practically dictate its operational requirements and actual performance. For example, flows along a highway would determine its level of congestion, while ridership of a transit line is used for setting its operating frequencies. This interaction is clearly represented in the transportation system analysis framework of Manheim (1979) and Florian et al. (1988), as depicted by McNally (2008) in Figure 13.1.

The transportation system (T) comprises infrastructures and services, whose performance capabilities are estimated through (performance) procedures (P). Demand patterns are the outcome of the activity system (A), with appropriate procedures (D) used for that purpose. It is important to note that systems (T) and (A) refer to different spatial levels (McNally, 2008). Indeed, demand procedures (D) produce trips in an aggregated level (for example, between traffic analysis zones—TAZ), since, in most cases, the activity system (A) of an area can be only be represented in a similar, accumulated manner. On the contrary, performance procedures (P) correspond to actual transportation infrastructures and services (albeit relatively simplified for modeling purposes). A higher level of detail (in the mode and transportation link/node levels) is therefore possible, when calculating capabilities of the transportation system (T). Next, equilibration represents the interaction between the transportation and activity systems: demand (D) and performance (P) procedure outcomes are appropriately exploited in such a way so that trips are converted into network flows along transportation system (T) elements. This implies that different outputs from different spatial levels have to be combined: equilibration defines paths consisting of transportation system (T) elements (links and nodes) and assigns trips to these paths. These outputs can be subsequently used to enforce changes in the supply side (S) of the transportation system (T) or the location characteristics (L) of the activity system (A). The framework is clearly iterative: any change in the activity or transportation systems through procedures (S) and (L) would affect flows; these in turn could possibly require additional changes in the transportation and activity systems.

A possible question that may arise is how the Manheim/Florian framework can be linked to the rational model approach for planning a transportation system. Essentially, a transportation analysis framework is at the core of evaluating alternative transportation options. Once the

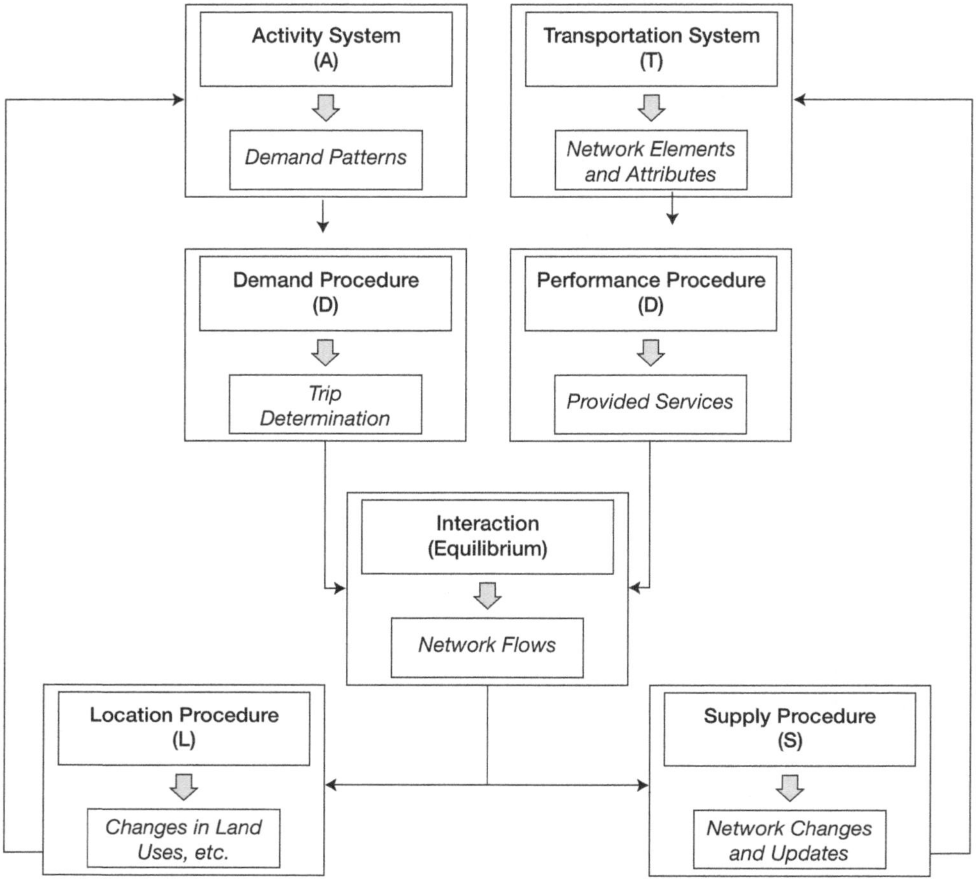

Figure 13.1 Presentation of the Manheim/Florian Transportation Systems Analysis Framework

problem is defined, specific objectives and evaluation criteria should be estimated. Each alternative *i* would correspond to a transportation system (T_i), an activity system (A_i) and associated performance and demand procedures (P_i) and (D_i). Equilibration for each alternative will yield system flows and allow the calculation of indicators corresponding to evaluation criteria. Changes and/or sensitivity analyses for an alternative would be depicted through its associated supply procedure (S_i). As such, the framework would be applied for deriving the impacts and performance of an alternative to a transportation system.

The Four-Step Model (FSM) for Transportation Planning

The Four-Step Model (FSM) has been the traditional approach for deriving and forecasting flows of a transportation system. It encompasses demand, performance and equilibration procedures in a sequential way; in accordance with the Manheim/Florian framework, outcomes, demand and performance procedures are combined and flows along system links are determined. Before proceeding to an overview of these steps, it is necessary to offer a brief view of attributes and terms required for proceeding into implementing the FSM model.

Figure 13.2 Sample network

Transportation System Representation

A transportation system is typically represented by a network, consisting of links and nodes. Depending on the number and types of transportation modes operating in the system, one or several networks may be developed for that purpose (such as a highway, a pedestrian, a transit, or a rail network). A network may be used by a single mode or multiple modes (for example, cars and transit may share the same highway system). Each link and node has associated attributes, related to their performance capabilities and type of mode considered: for example, highway system link attributes may include capacity, free-flow speed, travel times, and so on. For transit, operating frequencies, vehicle capacities could be such attributes. These attributes are used by the performance and equilibration related procedures and particularly for calculating performance and capabilities, and assigned flows for each mode of the system. Figure 13.2 presents a sample network used for a FSM.

Activity System Representation

Ideally, analysis and representation of trip-related decisions should be made at the disaggregated level of a single traveler or household; trip decisions of an individual differ according to his activities, tastes, past experiences, and so on. However, in conventional planning studies, it is usually difficult to implement such detail for a number of reasons, including computational complexity, data availability, etc. For that reason, the study area is divided into geographical units of relatively uniform socioeconomic characteristics (transportation analysis zones—zones—

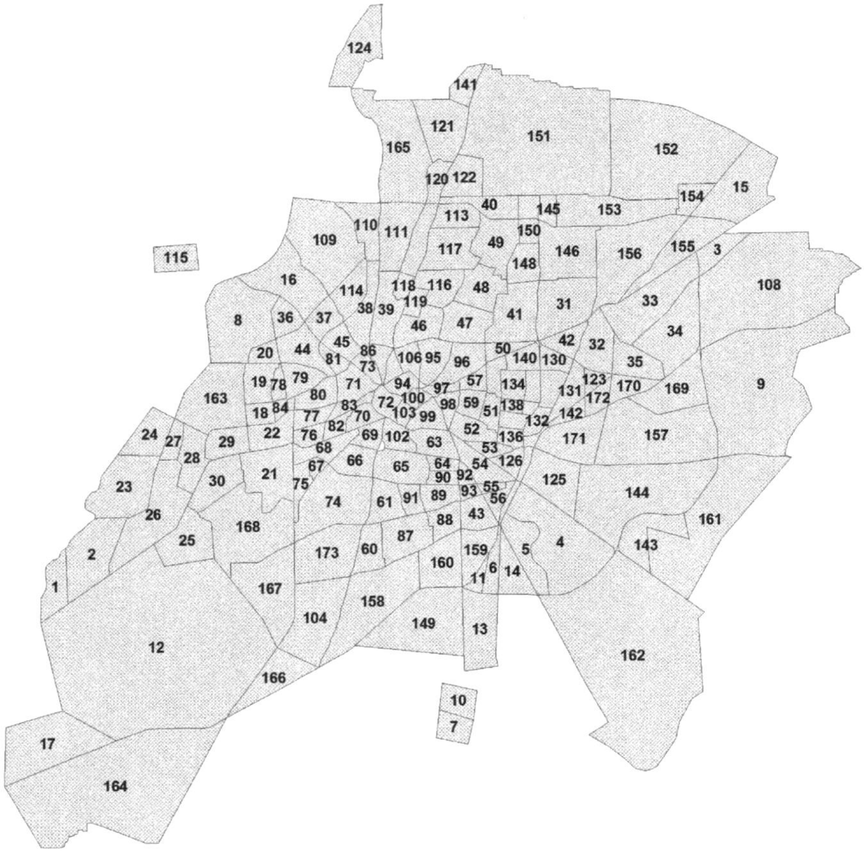

Figure 13.3 TAZ representation

TAZ), whose departing or originating travelers are assumed to have similar travel patterns. Zones allow for spatial aggregation of trip decisions and therefore simplify further analyses. A centroid node is assigned to each zone; all trips corresponding to a zone are assumed to depart or arrive to its centroid node. Centroids practically act as zone "entrances" to the transportation system (S): special links (called "connectors") are used for that purpose by connecting centroids to the transportation system networks. Figure 13.3 presents a typical zoning system and Figure 13.4 is a complete representation of a transportation and activity system, including centroids and connectors.

Trip Purpose

As trip decisions are directly related to the activity system, trip purposes are critical factors for representing demand patterns, since they dictate production and attraction between different land uses. For instance, residential areas tend to produce trips for both work and other purposes, while commercial areas attract shopping trips. Work-related trips are the most common ones; other purposes may include education (often considered a work-trip), shopping, leisure, business affairs etc. Most trips are home-based: a trip originates from home for some destination and its return

Figure 13.4 Complete transportation and activity system representation

Typical criteria for determining a TAZ

1. Socioeconomic characteristics should be homogeneous.
2. Intrazonal trips should be minimized.
3. Physical, historical, and political boundaries should be utilized where possible.
4. Zones should not be created within other zones (i.e., no "doughnuts").
5. The zone system should generate and attract approximately equal trips and/or contain approximately equal population, households, or area per zone.
6. Zones should be defined by census tract boundaries where possible.

(O'Neil, 1991)

counterpart has home as a destination. Other trips may be non-home-based, such as trips that go from a workplace to another, etc. A distinction by trip purpose is used for determining trips and mode choice: home-based trips are distinguished into those whose purpose is work (Home-Based Work—HBW) and those who have another purpose (Home-Based Other—HBO), while Non Home-Based trips (NHB) include all other trips whose origin is not home.

Temporal Attributes

Demand procedures and estimation are based on typically extensive household and roadside surveys, which extend on a daily basis. As such, demand estimates are produced for 24-hour periods. Obviously, demand fluctuates throughout the day, even for different trip purposes. On the contrary, transportation system operations are typically designed for hourly periods and particularly peak hours. It is therefore necessary to convert daily demand to peak-hour demand, using appropriate transformation factors (for instance, obtained from daily traffic counts), before equilibration.

The main steps of the FSM are the following (also shown in Figure 13.5).

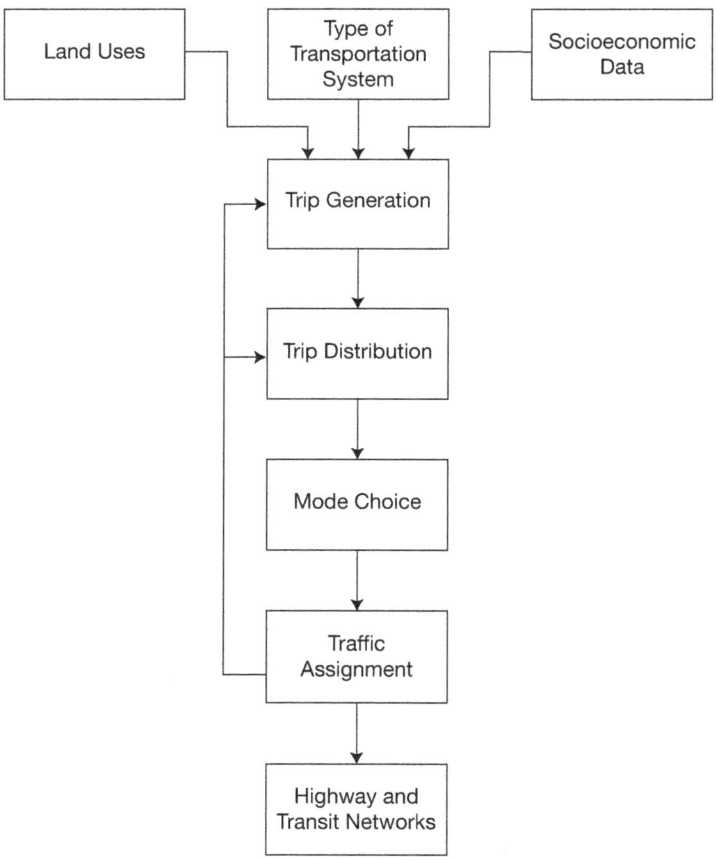

Figure 13.5 The Four-Step Model structure

1. *Trip generation:* In this first step, trips generated and attracted in zones of a study area are estimated. Total number of trips produced and attracted by each zone are estimated. Trip generation refers to the decision to travel.
2. *Trip distribution:* The second step matches origin and destination zones and allocates trips between them, according to their trip productions, attractions and some measure of spatial separation. Trip distribution is about the decision on trip destination. The outcomes of this step are the so-called "origin-destination" matrices (OD).
3. *Mode choice*: In the third step, trips are allocated to available travel modes, using mode characteristics and performance. The mode choice step refers to mode decision by travelers.
4. *Assignment*: The final, fourth step, is the equilibration step. Trips per mode are assigned to network links for each mode (traffic, transit etc) and total flows per link are determined. This final step is about path decision between zones and allocation of flows to path links.

Note: Sequential application of the FSM does not require internal consistency between its steps. This implies that travel time related inputs used for measuring separation in trip distribution, travel costs in mode choice, and traffic assignment outputs may differ. This is a shortcoming of the FSM, widely acknowledged in the literature (Miller, 2001). A common solution is to incorporate feedback loops to the FSM steps; traffic assignment results (such as transit accessibility measures, highway speeds and/or travel times) obtained from a first FSM iteration are fed again to trip generation, distribution and mode choice steps, and demand is re-estimated. However, there is no guarantee of convergence after a number of iterations, into a concise answer (Miller, 2001).

Trip Generation

The first step of the FSM focuses on predicting trips produced and attracted by transportation analysis zones (TAZ). A number of factors may affect productions and attractions for each zone, including characteristics of travelers (age, gender, income, etc.), household characteristics (household size, vehicle ownership, income), zone characteristics (land uses, population density, accessibility) and the transportation system level of service (McNally, 2008; Ortuzar and Willumsen, 2011; Caliper, 2008). For trip productions in particular, it is common to use a measure of population size and a measure of wealth (income or automobile ownership) as explanatory factors (Caliper, 2008). For attractions, these factors are related to land uses for each zone (McNally, 2008; Ortuzar and Willumsen, 2011). For instance, the number of employees in a zone is a good measure of work-trips attracted to that zone, while roofed space for retail, office, industry, etc. are possible explanatory factors for both work and non-work purpose trips (McNally, 2008). It is also widely acknowledged that usually productions can be better estimated given available explanatory factors, compared to attractions. Indeed, productions are estimated using demographic information (which are widely available and relatively easy to collect through household surveys). On the contrary, attractions are a function of land uses, for which representation and collection of related data is far more complex, particularly in areas with mixed land uses.

Typical modeling approaches used for trip generation include linear regression and category analysis (or cross-classification). These are discussed in detail below.

> *Note:* While accessibility may affect trip productions and attractions, transportation system-related factors are rarely incorporated in trip generation procedures. However, modern GIS-based software offer the ability of proceeding in relevant analysis and yield more realistic models (Caliper, 2008).

Linear Regression

In the case of linear regression use for trip generation, a statistical relationship is developed, which relates the number of zonal trips (produced or attracted) with explanatory factors, such as those discussed earlier. The linear regression model formulation is:

$$Y = \beta_0 + \beta_1 X_1 + \ldots + \beta_n X_n + \varepsilon \tag{1}$$

where Y = dependent variable (number of trips generated), $X_1, \ldots, X_n$ = independent variables (explanatory factors), $\beta_0, \beta_1, \ldots, \beta_n$ = model coefficients, and ε = an error term.

Regression may be zonal or disaggregated at the household level. Zonal regression uses average values obtained from data aggregated in the zone level. For example, in zonal regression, the dependent variable would be the average number of trips per household, with explanatory variables being average values of zone characteristics. Zonal regression suffers from some econometric issues raised in the literature (Ortuzar and Willumsen, 2011):

a. While dependent variables cannot obtain a value below zero, some regression models may produce negative values—a proper model transformation is then required prior to applying linear regression.
b. There may be zones without adequate data; these cannot be included in the modeling process.
c. Often, the problem of heteroscedasticity occurs, as aggregated variables reflect the size of zones.
d. Changes in the zoning system would require estimation of new models.

Disaggregate models use trips at the household level and relate them to individual or household-related parameters. These are preferable as they are practically zone independent. However, they require a considerable amount of observed data for calibration and validation. Overall, linear regression has proven to be an approach of good performance for estimating productions and attractions, provided that related modeling assumptions are met. Furthermore, the method is relatively easy and straightforward to apply, even with simple spreadsheet software.

Category Analysis

In the category analysis (or cross-classification) method, trip productions and attractions for each zone are straightforwardly estimated as a function of different household attributes, such as income level, automobile ownership, and household size. The method is disaggregate in the sense that it uses large samples for and detailed categories for calculating trip rates (Wright, 1996). The method assumes temporal stability of trip rates for households, within particular classes (Ortuzar and Willumsen, 2011). Households are allocated to homogeneous categories, according to some socioeconomic characteristics, and for each trip purpose and category, the number of trips (trip rate) is calculated. Detailed trip rates per category and trip purpose are used for estimating total trips; these are obtained either from local surveys in the study area or national studies—for example, NCHRP Report 365 (Martin and McGuckin, 1998) and NCHRP Report 716 (Cambridge Systematics et al., 2012). Trips for each category within a zone are calculated using the following equation:

$$T_{ipk} = H_i C_{ik} R_{ip} \tag{2}$$

where T_{ipk} = number of trips generated in zone i for purpose p, H_i = number of households in zone i, C_{ik} = coefficient which indicates the production of trips by households with characteristic k (category k), R_{ip} = trip rate for purpose p.

Category analysis is straightforward to apply and does not require any prior assumptions on the relationship between trips and socioeconomic characteristics; in fact, these may even defer from category to category. However, the method is expensive with respect to data requirements. In particular, it is often difficult to obtain the portion of households within a category for each zone. Furthermore, the constraint of not being able to forecast beyond predefined categories, along with an efficient way of selecting categories and their boundaries, still remain as method weaknesses (Willumsen and Ortuzar, 2011).

Balancing

Productions and attractions are estimated from independent models in the zone level. This implies that production and attraction totals may differ: a balancing process is then required for equalizing production and attraction totals. As noted earlier, trip production models are generally better predictors compared to trip attraction models, since the latter often lack adequate data for proper estimation of attractions. As such, balancing between productions and attractions is performed using productions. Therefore, lower accuracy of attraction models is typically acceptable in planning studies. Productions and attractions are balanced for each trip purpose, using total productions as constants, by multiplying attractions with the outcome of the following equation:

$$f = \frac{T}{\sum_j A_j} \tag{3}$$

where f = balancing factor, T = total number of trips produced by all zones, and A_j = attraction for each zone j.

> The final outcome of the first step is the total number of daily trips produced and attracted by a zone, ideally for different trip purposes, and/or other population characteristics.

External Trips

Apart from trips generated and attracted in a study area, there may by additional trips, whose origin or destination (or both) are outside the study area. Obviously, trip generation models cannot estimate these trips so additional traffic counts and roadside surveys in major corridors near the boundaries of the study area are required for capturing these trips. Usually, information from roadside surveys on the single edge (origin or destination) within a study area are collected and then expanded to the population using traffic counts.

Trip Distribution

The outcome of the first FSM step (trip generation) includes total trips originating from or arriving to a zone. A subsequent key question arises: "How are produced trips distributed to destinations?" In this context trip distribution allocates trips between zones, keeping in mind total zone productions and attractions. The outcome of the step is the so-called Origin-Destination (OD) matrix, whose elements refer to trips between zones. Obviously the sum of each matrix row/column is equal to the zone's produced/attracted trips. A typical OD matrix is shown in Figure 13.6 (where T_{ij} = trips between zones *i* and *j*, P_i = productions of zone *i*, P_j = attractions of zone *j*).

$$
\begin{array}{ccccccccc}
 & & & & \langle \text{Destination} \rangle & & & & \\
 & \text{Zone} & 1 & \ldots & j & \ldots & n & & P_i = \sum_j T_{ij} \\
 & 1 & T_{11} & \ldots & T_{1j} & \ldots & T_{1n} & & P_1 \\
 & \ldots & \ldots & \ldots & \ldots & \ldots & \ldots & & \ldots \\
\langle \text{Origin} \rangle & i & T_{i1} & \ldots & T_{ij} & \ldots & T_{in} & & P_i \\
 & \ldots & \ldots & \ldots & \ldots & \ldots & \ldots & & \ldots \\
 & n & T_{n1} & \ldots & T_{nj} & \ldots & T_{nn} & & P_n \\
 & A_j = \sum_i T_{ij} & A_1 & \ldots & A_j & \ldots & A_n & & \sum_i P_i = \sum_j A_j
\end{array}
$$

Figure 13.6 Typical OD matrix structure

As stated earlier, the OD matrix should fulfill the following conditions, implying that the total trip productions (attractions) of a zone should be equal to the sum of trips distributed to (from) other zones to that zone.

$$P_i = \sum_j T_{ij} \tag{4}$$

$$A_j = \sum_i T_{ij} \tag{5}$$

Most common methods for trip distribution include growth factor methods and gravity models. The former are simple, straightforward processes used for updating or forecasting existing OD matrices. On the contrary, gravity models (often called the "workhorse" of modeling flows) can be directly applied for estimating trips between pairs of zones, based on productions, attractions, and network and zone characteristics.

Gravity Model

Gravity models are inspired by the Newtonian law; they are often called synthetic as they attempt to determine trips between zones, without taking into account any previously observed trip patterns. The concept behind these models is that flows between zones are affected by conditions at trip ends (zones) and the degree of separation between these ends. A typical formulation of gravity models is the following:

$$T_{ij} = P_i \cdot A_j \cdot f(c_{ij}) \tag{6}$$

where T_{ij} = trips between zones i and j, P_i = produced trips by zone i, A_j = attracted trips to zone j, c_{ij} = trip cost between zones i and j, and $f(\,)$ = friction function between zones i and j.

Similarly to Newton's law of gravity, productions and attractions of zones act as attracting masses of the two trip ends. Friction (or deterrence) function $f(\,)$ represents the disincentive to travel between the two zones; c_{ij} may correspond to the distance, travel time or even generalized cost between the two zones. Table 13.1 presents common forms of friction function $f(\,)$ found in the literature and Table 13.2 suggested parameter values for the gamma type function (Martin et al., 1998).

Table 13.1 Popular friction functions

Type		*Form*	
Exponential		$f(c_{ij}) = \alpha \cdot \exp(-\gamma \cdot c_{ij})$	
Power		$f(c_{ij}) = \alpha \cdot c_{ij}^{-\beta}$	
Gamma (combined)		$f(c_{ij}) = \alpha \cdot c_{ij}^{-\beta} \cdot \exp(-\gamma \cdot c_{ij})$	
HBW:	$\alpha = 28507$,	$\beta = 0.020$,	$\gamma = 0.123$
HBO:	$\alpha = 139173$,	$\beta = 1.285$,	$\gamma = 0.094$
NHB:	$\alpha = 219113$,	$\beta = 1.332$,	$\gamma = 0.100$

Furthermore, the requirements posed by equations 4 and 5 led to the development of modified gravity models forms, the so-called singly and doubly constrained models. These models incorporate balancing factors for that purpose: singly constrained models attempt to fulfill either equation 4 or 5 (preserve productions and attractions only), while doubly constrained models require fulfillment of both equations 4 and 5 (preserve both productions and attractions). The common form of these constrained models is the following:

$$T_{ij} = A_i \cdot P_i \cdot B_j \cdot f(c_{ij}) \tag{7}$$

where T_{ij} = trips between zones i and j, P_i = produced trips by zone i, A_j = attracted trips to zone j, c_{ij} = trip cost between zones i and j, $f(\,)$ = friction function between zones i and j, A_i, B_j = balancing factors.

Factors A_i and B_j may be calculated using the following functions:

$$A_i = \frac{1}{\sum_j B_j D_j f(c_{ij})} \tag{8}$$

$$B_i = \frac{1}{\sum_i A_i O_i f(c_{ij})} \tag{9}$$

If either A_i or B_j is set equal to one, a singly (origin or destination) constrained gravity model is defined. Otherwise, as values of A_i and B_j are clearly interdependent, an iterative process is required for calculating balance factors: a value of 1 is set for B_js and A_is are calculated. These in turn are used for calculating new B_js. The process is continued up until convergence is met, with respect to total productions and attractions per zone.

Growth Factor Methods

Growth factor models are used for updating existing OD matrices and estimating their future values, based on overall or zone-specific information of demand growth. A major assumption of growth factor methods is that the transportation system will remain relatively unchanged in the future. The most common form is the uniform growth factor method: this is used when overall demand growth information is available for an area. Future trips are calculated as follows:

$$T'_{ij} = \tau \cdot T_{ij} \tag{10}$$

where T'_{ij}, T_{ij} = updated and base trips between zones i and j, and τ is the growth factor.

If specific growth rates exist either for productions or attractions, equation 10 is modified as follows:

$$T'_{ij} = \tau_i \cdot T_{ij} \tag{11}$$

$$T'_{ij} = \tau'_j \cdot T_{ij} \tag{12}$$

Obviously, individual growth factors τ_i, τ'_j are applied to corresponding rows of the origin—destination matrix. When growth rate information is available for both productions and attractions of a zone, a doubly constrained growth factor method is required for obtaining updated trip values. Such methods, often called by the names of their authors (Fratar and Furness) methods. The typical formulation used is:

$$T'_{ij} = A_i \cdot \tau_i \cdot B_j \cdot \tau'_j \cdot T_{ij} \tag{13}$$

where A_i, B_j are balancing factors.

If growth and balancing factors for productions and attractions are combined in single values a_i and b_j respectively, the formulation is transformed into:

$$T'_{ij} = a_i \cdot b_j \cdot T_{ij} \tag{14}$$

where a_i and b_j are final adjustment coefficients.

Coefficients a_i and b_j should be calculated by applying an iterative approach, so that equations 4 and 5 are fulfilled. The iterative approach for estimating adjustment coefficients is the following:

1. Set b_js equal to 1 and find a_is that meet constraints (4).
2. For a_is find b_js that meet constraints (5).
3. For new b_js, find updated a_is that meet constraints (4) and repeat steps 2 and 3 until row and column sums of the updated OD matrix converge to updated (target) values of total productions and attractions, as set by equations 4 and 5.

As noted by Ortuzar and Willumsen (2001), a major requirement of the growth factor model is that total updated productions and attractions are equal:

$$\sum_i \left(\tau_i \cdot \sum_j T_{ij} \right) = \sum_j \left(\tau'_j \cdot \sum_i T_{ij} \right) \tag{15}$$

> The outcome of trip distribution is the number of daily trips performed between zone pairs, for different trip purposes and/or other population characteristics.

Mode Choice

Mode choice is the third step of the FSM and introduces traveler options and tastes with respect to available travel modes. Essentially, mode choice aims at splitting trips to available modes of travel, by taking into account traveler preferences and the modes' particular characteristics. For that purpose, related models attempt to describe the way people make choices, with respect to their preferred travel mode; traveler's choice depends upon their socioeconomic characteristics and the attractiveness of each mode (Ortuzar and Willumsen, 2011).

Discrete choice models have been for years the primary modeling approach for mode choice in the FSM context; these models produce disaggregated results at the individual level. These are then aggregated following appropriate methods (for a detailed description of aggregation techniques, the interested reader is referred to Ortuzar and Willumsen (2011)). In this context, travelers evaluate available mode alternatives, compare the attractiveness of each model with respect to their tastes and decide upon the mode they will use. The attractiveness of each alternative is represented by its utility, a "theoretical construct," which attempts to quantify and rate a traveler's preference towards selecting a particular mode (Ortuzar and Willumsen, 2011). In this context, travelers aim at maximizing their utility, by selecting their preferred mode of travel. Of course, their final decision of travelers is a combination of their personal characteristics, the value they impute to different elements of the transportation system and the different way they perceive trip costs; some of these elements are captured by the utility function parameters.

The utility function consists of two components: the systematic utility component and the random disturbance. Its mathematical formulation is the following:

$$U_i = V_i + \varepsilon_i \tag{16}$$

where U_i is the utility of mode i, V_i is the systematic (measurable) component of the utility and ε_i is the random disturbance.

The systematic utility component is represented as a linear combination of variables, which correspond to different decision parameters (attributes) for travelers. These parameters are derived from the attributes of alternatives or the socioeconomic characteristics of individuals.

$$V_i = \lambda_1 X_{i1} + \lambda_2 X_{i2} + \ldots + \lambda_n X_{in} \tag{17}$$

where X_{in} represents the n attribute of the alternative i and λ_n is a coefficient indicating the contribution of each attribute to the overall utility.

Relevant attributes include trip characteristics such as travel time and travel cost, traveler characteristics (income, age, trip purpose, etc.) or even mode characteristics (parking availability, tolls, etc). On the other hand, random part ε_i represents particular preferences and characteristics of individuals, as well as modeling errors. Because of this random term, the mode whose utility is maximized cannot be determined with certainty; as such, the probability of an individual choosing mode i with respect to another mode j is estimated instead:

$$P_i = \Pr\left[(V_i + \varepsilon_i) > (V_j + \varepsilon_j)\right] \tag{18}$$

Multinomial Logit Models

MNL models relate the probability that a traveler chooses a given alternative from a set of alternatives to the utility of these alternatives according to the following equation:

$$P_n(i) = \frac{\exp\,(V_{ni})}{\sum \exp\,(V_{nj})} \tag{19}$$

where $P_n(i)$ = probability for traveler n to choose alternative i, C_n = set of alternatives for traveler n and V_{ni} = systematic utility component of alternative i for traveler n.

The MNL model for choice probabilities is based on the assumption that the utility of the alternative is a function of choice determinants, some unknown parameters and an additive, identically distributed Gumbel error term.

Equation 19 may be further transformed as follows:

$$P_n(i) = \frac{\exp\,(V_{ni} - V_{nm})}{1 + \sum_{j \in C_n \neq m} \exp\,(V_{nj} - -V_{nm})} \tag{20}$$

According to equation 20, choice probabilities are only affected by the difference between the utilities of all but one alternatives and the utility of the other alternative. The choice of the omitted alternative is arbitrary and different reference alternatives may be used for different variables.

MNL models are specified by determining relative utilities for alternatives; these are in turn a function of explanatory variables that describe each utility. These variables can be either generic (which have the same effect and parameter value on the utility of different modes) or alternative specific (which have different effects on the utility of different modes). Typically, explanatory variables are related to the traveler (e.g. trip purpose) or the alternative (e.g. trip cost) and can also be a function of the origin and destination (e.g. travel time between origin and destination).

An important property of the Multinomial Logit Model is the Independence from Irrelevant Alternatives (IIA) property. According to this property, the comparison between two alternatives is not affected by the existence of other available options. This can be shown by examining the ratio of the probabilities for two alternatives i and j.

$$\frac{P_i}{P_j} = \frac{\exp\,(\beta V_i)}{\exp\,(\beta V_j)} = \exp\,\left[\beta(V_i - V_j)\right] \tag{21}$$

It is clear that the ratio is independent from the other alternatives of the choice set. This property is convenient only in the case of different modes without strong similarities to a relatively distinct attractiveness to travelers (for example, private cars versus a bus network).

Nested Logit Models

There exist cases of either non-independent or decisions considering a chain of modes. Then, mode choice is not a straightforward process, but the individuals have to decide in more than one stage, especially when the similarities between the alternatives are important. For example, if a traveler has to select between public transport and private vehicles, the next question concerns the particular (public or private) mode that he or she will choose. In this context, the decision-making mechanism of the traveler is described by a two-step process. Under this approach the mode choice can be considered as an output of an hierarchical process. This multidimensional approach is a solution to problems derived from the IIA property, especially when alternatives are not independent, or when there are observed variations of perception or preferences between individuals. This model category is called Hierarchical Nested Logit Models (NL models) and a typical formulation of the structure of a NL model is provided at Figure 13.7. Interesting discussions on Nested Logit Models are offered by Ortuzar and Willumsen (2011) and Koppelman and Bhat (2006).

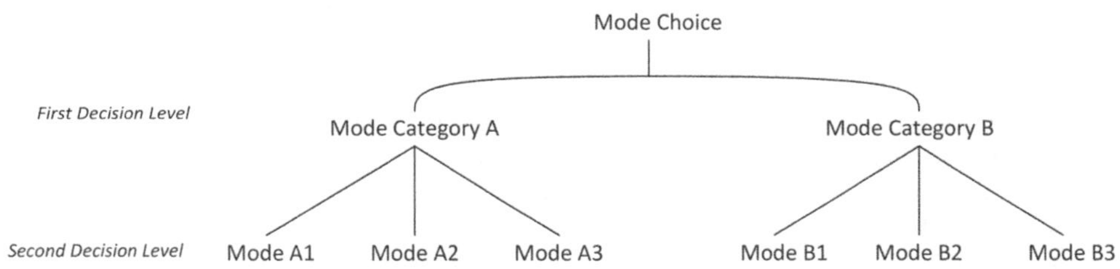

Figure 13.7 Hierarchical structure of an NL model

> The output of the third step is the mode share (percentage of traffic, transit, walk, etc.) attributed to each pair of zones. This is further applied to origin destination matrices, and separate O–D matrices per mode are derived.

Trip Assignment

Previous FSM steps focused on estimating and distributing demand among zones and modes; the last step deals with equilibrations, as it links demand to the transport system. Under trip assignment, optimal travel paths are determined between zones and passenger flows and other attributes are calculated for the links of the different transportation networks (traffic, transit, etc.). Required data for trip assignment include:

- Origin and Destination (O–D) matrices.
- The structure of the encoded transport network. The network is composed by links and nodes, while each link is characterized by some attributes, such as capacity, distance, etc.

The main output of the assignment process are network link flows, which can further be used for assessing network conditions, for estimating travel costs and for evaluating the functionality and serviceability of the transportation system in the future. However, calculation of link flows requires determining the paths (sequences of links), which will be used for traveling between zones. These paths (routes) are in turn directly related to traveler behavior. As such, trip assignment consists of two substeps: selection of paths between zones and loading trips to corresponding links.

Route Choice

As in mode choice, the basic principle, which guides route choice, is utility maximization (i.e. the path/route with maximum attractiveness). Generalized cost is usually exploited as a measure of route attractiveness. While various factors may affect generalized cost and route decision, usually, the generalized cost is considered to be composed by two elements: i) the monetary cost for the traveler, ii) the total travel time. An expression of the generalized cost is the following:

$$g_{ijr} = c_{ijr} + \tau t_{ijr} \ , \ \forall i,j,r \tag{22}$$

where g_{ijr} is the generalized cost of path *r* from the point *i* to the point *j*, c_{ijr} is the monetary cost of path *r* from point *I* to point *j*, t_{ijr} the total travel time and τ is a coefficient that converts the travel time in monetary cost. Coefficient τ reflects the perceived cost of time for individuals, so its value is not constant, but varies depending on the assignment model used. It should be noted that, despite the assumption of minimum generalized cost, individuals don't always select the same path between a particular origin-destination pair. Different perceptions of generalized cost, unobserved attributes of the alternative paths and varying individual priorities with respect to trip characteristics (transfers, comfort, congestion, etc.) can explain this behavior.

Path Determination

A required step for assignment involves the determination of the minimum cost (shortest) path between zones. For that purpose, two widely used, efficient algorithms proposed by Moore (1957) and Dijkstra (1959) are applied. The output of algorithms is a sequence of nodes which form the shortest paths originating from a zone to all others.

Traffic Assignment Models

The literature exhibits a variety of traffic assignment methods, ranging from the simple "All-or-nothing assignment" to complex, dynamic traffic assignment methods. This section discusses most popular methods (All-or-Nothing, User Equilibrium, Stochastic User Equilibrium and System Optimum).

All-or-Nothing Assignment

The "All-or-Nothing" traffic assignment method is a simple, applicable but not always effective approach, as its assumptions are not often realistic. According to this method, all trips between a pair of zones are allocated to a single, minimum cost path, while other paths are neglected. The method assumes that there is no congestion and that all travelers exhibit attributes with respect to path choice. Such an approach could be reasonable in cases of uncongested networks with limited path options between zones, and a path has always a comparative advantage among others.

User Equilibrium

User equilibrium-based assignment was first introduced by Wardrop (1952). According to Wardrop's first principle: "The journey times on all the routes actually used are equal, and less than those which would be experienced by a single vehicle on any unused routes." Considering that travel time reflects travel cost, "no individual can reduce his path cost by selecting an alternative route". Based on Wardrop's first principle, User Equilibrium Assignment (UE) is based on certain assumptions (Mathew and Krishna Rao, 2006). First, UE can handle network congestion. Second, all travelers have a similar perception of travel costs and full knowledge on the cost composition. Third, travel time on a link is a function of the link's flow. UE is formulated as the following non-linear mathematical programming model:

$$\min \sum_{a} \int_{0}^{q_a} t_a(q_a)\, dq \tag{23}$$

$$\sum_{k} f_k^{rs} = d_{rs}$$

$$\sum_{r}\sum_{s}\sum_{k} f_k^{rs}\, \delta_{ak}^{rs} = q_a \quad \forall a \in A$$

$$q_a \geq 0 \quad \forall a \in A$$

$$f_k^{rs} \geq 0 \quad \forall r \in R, \forall s \in S, \forall k \in K_{rs}$$

where t_a = cost at link a, q_a = flow at link a, f_k^{rs} = flow on path k from origin r to destination s, d_{rs} = demand for the pair r-s, A = set of links, R = set of all origins, S = set of all destinations, K_{rs} = set of routes between r and s.

An example traffic assignment output using the UE model is shown in Figure 13.8.

Figure 13.8 Example traffic assignment

System Optimum (Social Equilibrium)

System optimum is based on Wardrop's second principle, according to which all drivers behave (cooperate) in such a way so that the total network travel time is minimized. Such an approach is relatively unrealistic in most cases, unless network operators are able to manage traffic to particular routes, so that traffic is minimized. Optimal assignment for system optimum can be found, as in the case of User Equilibrium, through a non-linear mathematical model, where the aim of the objective function is the minimization of travel cost:

$$\text{Minimize } Z = \sum_a x_a \cdot t_a(x_a) \tag{24}$$

$$\sum_k f_k^{rs} = d_{rs} \quad \forall r,s$$

$$x_a = \sum_r \sum_s \sum_k \delta_{a,k}^{rs} \cdot f_k^{rs} \ \forall a$$

$$f_k^{rs} \geq 0 \quad \forall k,r,s$$

$$x_a \geq 0 \quad \forall a \in A$$

where t_a = cost at link a, x_a = flow at link a, f_k^{rs} = flow on path k from origin r to destination s, d_{rs} = demand for the pair r-s, A = set of links, R = set of all origins, S = set of all destinations, K_{rs} = set of routes between r and s.

Stochastic User Equilibrium (SUE)

SUE is an extension of User Equilibrium which incorporates particular stochastic characteristics. Under SUE all travelers do not have the same perception of travel cost and its composition. Each individual is motivated by a different cost perception, and therefore the existence of multiple routes between two points of origin and destination guides travelers to the selection of different, available paths. Wardrop's principle remains the same: in equilibrium conditions, no user can improve his/her travel cost by switching route. However, each user has a different perception of travel cost and therefore not all users select the same route, for a specific O-D pair. Thus, alternative paths are loaded and corresponding flow patterns are produced. According to Mathew and Krishna Rao (2006), this method can be implemented in uncongested traffic conditions, since it presents resistance to network changes and the travel cost of each link is independent from its corresponding flow.

Transit Assignment

Unlike traffic assignment, transit assignment refers to the movement of people rather than vehicles around a network. This implies that apart from actual usage of the transit system (in-vehicle part of a trip), travelers need to access the system, egress and transfer. In this context, the generalized cost for transit is a linear combination of different factors, including (a) in-vehicle travel time, (b) access, egress and waiting times, (c) transfers and (d) trip fare; each factor is attributed to a different weight, tailor made to fit prevailing conditions in the study area.

As in its traffic counterpart, transit assignment focuses on obtaining minimum cost paths between zones and on obtaining flows of transit lines. However, transit assignment is considerably more complex for a number of reasons (Ortuzar and Willumsen, 2011):

1. Paths include different elements, such as walking and transfer parts, in addition to in-vehicle trips. This practically adds to the complexity of obtaining minimum cost paths.
2. Fare structures are not always distance-based: A flat fare, multifare card or other, similar fare types may be used. Also, transfers may or may not be charged. As such, introducing fares in a generalized cost function is not always trivial.
3. The common lines problem: Often, transit lines share paths of the same corridor. In such cases, travelers may have the option of using a set of alternative lines for accessing their destination. As such, they will board the first arriving vehicle that belongs to this set of alternative lines.
4. Frequency-based or schedule-based services: Frequency-based services offer short headways and correspond to random passenger arrivals. On the contrary, schedule-based services allow passengers to access bus stops shortly before boarding and facilitate network schedule. Depending on the service type, waiting and transfer behavior and practice of travelers is obviously different.

The simplest among transit assignment procedures is the "All-or-Nothing" approach. More advanced methods (such as the EMME/2 Optimal Strategies, the TransCAD Pathfinder and the schedule-based method) consider multiple paths and assign trips proportionally to service frequencies (Spiess and Florian, 1989; Caliper, 2008). Other equilibrium-based methods consider the capacity of transit lines and crowding impacts on ridership, and allocate flows among paths under these limitations (Caliper, 2008).

> The output of the fourth step includes flows of passengers, vehicles or goods along network links, as well as prevailing paths between zones. Based on these values, additional information can be derived for the performance of a transportation system. For example, for highway networks, calculation of link travel times and speeds under prevailing flow conditions is possible, along with indicators such as volume to capacity ratio, veh-km and veh-h traveled. For transit assignment passenger flows can be used for determining service characteristics, transfers, and so on.

Model Validation

FSM validation is performed using observed data, preferably not used in any part of the modeling process. Its objective is to assess the accuracy and reliability of the model outcomes over its steps. For example, traffic assignment outputs (flows) are compared to observed traffic counts. The Root Mean Square Error (RMSE) and Percentage RMSE (%RMSE) are useful indicators for that purpose:

$$RMSE = \sqrt{\frac{\sum_{i=1}^{N}\left(Count_i - Model_i\right)^2}{N}} \tag{25}$$

$$\%RMSE = 100 \cdot \frac{RMSE}{\sum_{i=1}^{N} Count_i / N} \tag{26}$$

where i = network link, N = total number of links considered in a group of links, $Count_i$ = observed traffic counts for link i, $Model_i$ = assignment output flows for link i.

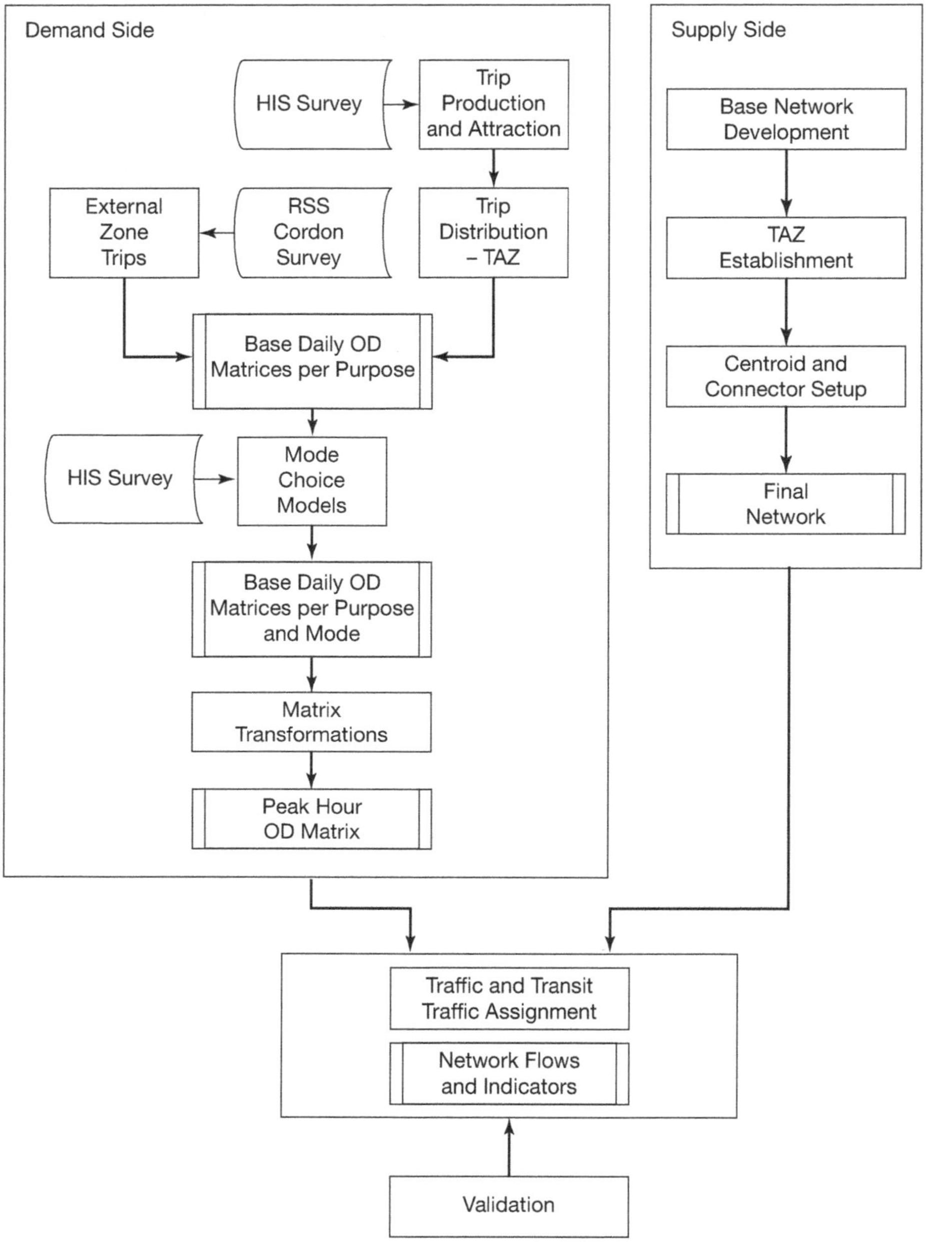

Figure 13.9 Example framework for developing a FSM

In general, lower %RMSE values indicate higher similarities between counts and model flows. RMSE and %RMSE measure the average error between the observed and modeled traffic volumes. They can be determined for different highway categories, or parts of the study area (a sub area) and so on. It is noted, however, that these indicators become less useful for aggregated cases (such as the whole study area) (Cambridge Systematics, 2010). For detailed, practice-oriented discussions on model validation in all FSM steps, the interested reader is referred to

NCHRP Report 716 (Cambridge Systematics et al., 2012) and FHWA *Travel Model Validation and Reasonableness Checking Manual* (Cambridge Systematics, 2010).

An Example Framework for Developing a FSM

Four-Step Modeling often becomes complex and requires: a) pre-processing a vast amount of data originating from surveys and census information; b) consideration of particular (often study specific) model details in each step of FSM, and c) intermediate OD matrix transformations. While major steps remain the same, each individual model may require different assumptions, pre-processing and so on.

Figure 13.9 presents an example framework for developing a Four-Step Model; the framework is not exhaustive but may form the basis for developing similar models.

According to Figure 13.8, the supply side includes all necessary steps for effectively representing the transportation network and activity system. Major highways are selected for introduction to the model, based on operational, density and geographical criteria, and main attributes such as direction, lane numbers, free flow speeds, free flow travel times and capacity values per direction are assigned to network links. Zones (TAZ) are defined next, according to network structure, geography and land use information. Finally, zone centroids and connectors are established—these are assigned to TAZ, while a number of them is used for managing external trips.

The demand side, on the other hand, focuses on developing a base peak hour origin-destination matrix. In detail, based on household surveys (HIS), trip productions and attraction models were developed for TAZ and a distribution (gravity) model is calibrated for estimating flows between TAZ, for different trip purposes (Home-based-work (HBW) and Home-based-other (HBO)). NHB trips are neglected for the sake of simplicity in representation. For estimating external, incoming and outgoing trips, roadside survey data, weighted with traffic counts are used. These were used for building base, daily origin destination matrices for different trip purposes. Next, HIS data are used for estimating mode choice models, for prevailing modes (private transport and public transport); multinomial logit model structures are used and models for both trip purposes are developed.

Using appropriate transformations, final, daily origin-destination matrices are constructed for different modes and are transformed to peak-hour matrices using information from traffic count surveys. From a practical perspective, a) OD matrices for different purposes are summed up for each mode, b) passenger OD matrices are converted to vehicle OD matrices for private transportation, and c) peak-hour OD matrices are derived. Supply and demand models are finally combined for performing traffic and transit assignment. Network flows and other indicators are extracted and validated using traffic counts.

References

Caliper (2008). *Travel Demand Modeling with TransCAD* version *5 User's Guide*. Caliper Corporation.

Cambridge Systematics (2010). *Travel Model Validation and Reasonableness Checking Manual*, 2nd ed., Washington, DC: Federal Highway Administration.

Cambridge Systematics, Vanasse Hangen Brustlin, Gallop Corporation, Chandra R. Bhat, Shapiro Transportation Consulting, LLC, Martin/Alexiou/Bryson, PLLC (2012). *Travel Demand Forecasting: Parameters and Techniques*. NCHRP Report 716, Washington, DC: National Academy of Sciences Press.

Dijkstra, E.W. (1959). Note on two problems in connection with graphs (spanning tree, shortest path). *Numerical Mathematics* 1, 269–271.

Florian, M., Gaudry, M., and Lardinois, C. (1988). A two-dimensional framework for the understanding of transportation planning models, *Transportation Research Part B*, 22B, 411–419.

Koppelman, F.S. and Bhat, C. (2006). *A Self Instructing Course in Mode Choice Modeling: Multinomial and Nested Logit Models*. Report prepared for the Federal Transit Administration, Washington, DC.
Manheim, M.L. (1979). *Fundamentals of Transportation Systems Analysis*, MIT Press, Cambridge, MA.
Martin, W.A. and McGuckin, N.A. (1998). *Travel Estimation Techniques for Urban Planning*. NCHRP Report 365, Washington, DC: National Academy of Sciences Press.
Mathew, T.V. and Krishna Rao, K.V. 2006. Introduction to Transportation Engineering. Civil Engineering – Transportation Engineering. IIT Bombay, NPTEL. Available at: www.cdeep.iitb.ac.in/nptel/Civil%20Engineering/ (accessed July 2014).
McNally, M.G. (2008). The four step model. In: *Handbook of Transport Modeling*, Eds. Henser, D.A., Button, K.J., New York: Pergamon, 35–52.
Meyer, M.D. and Miller, E.J. (2001). *Urban Transportation Planning: A Decision-oriented Approach*. New York: McGraw-Hill.
Miller, H. (2001). *Towards Consistent Travel Demand Estimation in Transportation Planning: A Guide to the Theory and Practice of Equilibrium Travel Demand Modeling*. Report prepared for the US Department of Transportation, Washington, DC.
Moore, E.F. (1957). The shortest path through a maze. *Proceedings of the International Symposium on the Theory of Switching*. Harvard University Press, Cambridge, MA.
O'Neill, W. A. (1991). Developing Optimal Transportation Analysis Zones Using GIS. *ITE Journal* – Institute of Transportation Engineers, 61 (12), 33–36.
Ortuzar, D.D.J. and Willumsen, L. (2011). *Modelling Transport*, Wiley, Chichester.
Spiess, H. and Florian, M. (1989). Optimal strategies: A new assignment model for transit networks. *Transportation Research Part B*, 23(1), 83–102.
Wardrop, J.G. (1952) Some theoretical aspects of road traffic research. *Proceedings of the Institution of Civil* Engineers, *Part II*, 1, 325–362.
Wright, P. (1996). *Highway Engineering*, 6th ed., Wiley, New York.

14

TRANSPORTATION AND LAND USE

Giovanni Circella[1] *and Francesca Pagliara*[2]

(1) GEORGIA INSTITUTE OF TECHNOLOGY; UNIVERSITY OF CALIFORNIA,
(2) UNIVERSITY OF NAPLES

Introduction

Many studies have investigated the complex relationships linking land-use patterns to travel behavior and the use of transportation. More compact, high-density urban areas with mixed land uses and better public transportation are found to encourage the use of non-automobile travel options, e.g. walking, biking and public transportation. At the other end of the spectrum, low-density suburban developments characterized by the separation of land uses, i.e. a very popular type of settlement in the United States as well as in other countries, are associated with a larger reliance on the use of private vehicles. Not only a topic of interest for transportation planners, the interactions between land use and transportation have been the object of research in many fields including urban and regional planning, community development, travel behavior, public transportation, physical activity, public health and environmental sustainability. In addition, the topic has generated important discussions in policy-making: if individuals' behavior can be influenced by land-use policies, coordinated planning strategies that favor more compact land-use patterns and efficient public transportation solutions will improve the liveability and environmental sustainability of cities.

According to forecasts from the United Nations Secretariat, more than 60 percent of the world population will live in urban areas by 2030. With a total world population expected to reach 8.2 billion in the same year, the pressure that this enormous mass of new urban dwellers will cause on transportation infrastructures will be unbearable. It will cause the conversion of large portions of land into urbanized areas and require vast investments in transportation infrastructures, raising concerns about public expenditures, as well as about traffic congestion and safety, environmental pollution, and land consumption.[1]

Empirical studies have provided robust evidence that some features of land use can affect travel behavior in an environmentally beneficial way. Increasing the understanding of the relationships between land use and travel behavior can assist planners in guiding future processes of urban development. It can help develop solutions that improve the performance of transportation, satisfy individuals' mobility needs and reduce the environmental externalities of transportation. Land-use policies have already been promoted as part of strategies to reduce the use of private vehicles and support the adoption of alternative modes of transportation. They are an important component of policy packages designed to promote smart growth and transit-oriented development, among other forms of integrated urban and transportation planning.

In the remainder of this chapter, we briefly summarize the research findings from recent studies that have investigated the relationships between land use and transportation, and have attempted to quantify the effects of the characteristics of the built environment on travel behavior. After that, we discuss how the complex interactions between land use and transportation, and the future development of cities, can be modeled through the adoption of integrated land use and transportation modeling approaches. Finally, we discuss the development of land-use policies that can increase liveability and environmental sustainability of cities, and reduce the environmental impacts of human settlements.

The Interactions among Land Use, Transportation, and the Environment

The current form of cities and metropolitan areas is the result of complex interactions among the numerous needs that influence the location of economic activities and residences. Living in cities allows individuals to interact with other components of a community in their work and social activities, saves on transportation costs, provides diversity in the availability of goods and services, and facilitates communication and social interactions. Central locations allow firms to benefit from agglomeration economies, minimize the transportation costs for people and goods, and ensure proper access to information and the needed connection with commercial partners, consumers and suppliers (O'Sullivan, 2011). Still, the organization of cities is not always efficient in all its dimensions, as demonstrated by the high congestion levels in many urban areas.

Urban economic theory has, since the nineteenth century, studied the development of cities and the relationships between rent and the distribution of residences and economic activities for competing socioeconomic groups within a region. According to location theories, residents and activities, if allowed to freely choose their location, will tend to maximize their utility and adjust their location in order to fulfill their needs. They trade their desire for space against transportation costs to the city center or other locations to which proximity is desirable.

The current urban form of a city or region is the result of many counteracting forces, social and economic factors, local geographical characteristics, land-use regulations and historical processes. Transportation has a fundamental role in shaping the urban form of cities, and in affecting the location of economic activities and residences. It modifies the relationships among destination accessibility, land values and space availability. European cities, and in general older cities that were largely built before or during the nineteenth century, mainly grew along public transportation corridors (at a time in which most trips were carried out walking or by public transportation). More recently, the larger availability of cars has affected the land-use patterns in cities and suburban areas that were developed during the second half of the twentieth century. It has contributed to shaping the urban form of most American cities along the elements of the road network, e.g. freeways and surface roads.

Transportation technology has the ability to modify the distance constraints that limit the mobility of goods and people, and the relationships between distance and destination accessibility. Over time, technological development increases travel speed through the introduction of new means of transportation and/or the improvement of road and rail infrastructures (cf. Muller, 2004). It also helps make travel cheaper, reducing the costs of moving goods and people over longer distances. The combined effect of these technological improvements is a contraction of the average travel times needed to reach specific destinations of interest, an effect known as time-space convergence (Janelle and Gillespie, 2004). This allows for an expansion of the available destinations that can be reached in a specific unit of time. However, time-space convergence patterns can be rather uneven in space and over time: for example, they mirror the investment patterns that privilege the accessibility of some places or regions over others, or the development of transportation services along dedicated corridors and radial axes of a city. Modern technologies allow for higher efficiency in the use of land, thus allowing for the expansion of cities beyond previous limits (cf. the concept of technology–land substitution, e.g. in Kim et al., 2009). They allowed for an increase in urban density, e.g. with the development of the first skyscrapers in the first half of the twentieth century, and provided better accessibility, reducing the friction of distance and favoring urban expansion in low-density areas in the second half of the twentieth century. Modern information and communication technology (ICT) can further lift space-time constraints, although disagreement exists on the extent of that relaxation (cf. Maeng and Nedović-Budić, 2008; and Janelle, 2012) and the resulting impact on urban form. ICT has been considered a potential substitute for physical trips, and from this perspective has been often promoted as a trip replacement strategy (cf. Salomon, 1998; Mokhtarian et al., 2005). Modern communication technologies could further contribute to the decentralization of cities (even if the futuristic visions of the "end of geography", cf. O'Brien (1992), or "death of distance", cf. Cairncross (2001), associated with the rise of ICT will probably never become reality). Cities will maintain a strong attractive power due to the benefits they provide. Still, some ICT-enabled options, e.g. telecommuting, provide increased flexibility in residential locations, and may act as a catalyst making possible a move that is prompted by other reasons, such as the interactions with other household members' needs, a relocation to places where lots of housing units are larger, cheaper or available in a larger quantity, or the desire to be located near vibrant parts of a city (cf. Circella and Mokhtarian, forthcoming). Therefore, ICT might facilitate an eventual shift to more decentralized and lower-density land-use patterns, which in turn affect mobility patterns (Mokhtarian, 2009).

Influence of Land Use on Mobility and Accessibility

Many studies have investigated the relationships between land use and travel behavior, and tried to estimate the impacts of specific features of the built environment on the use of transportation. Research findings tend to agree that denser urban forms, which feature mixed land uses and are served by better public transportation services, are associated with lower levels of dependence on the use of private vehicles and higher use of non-automobile means of transportation. The extensive reviews in Badoe and Miller (2000), Ewing and Cervero (2001), Leck (2006), National Research Council (2009), Ewing and Cervero (2010) provide detailed overviews of the studies in this field, and their evolution over time. Most studies included in these reviews refer to data from US cities, although the research findings are likely to be of wider application (Stopher and Stanley, 2014).

Cervero and Kockelman (1997) proposed the use of density, diversity and design as relevant measures of the characteristics of the built environment that affect the use of transportation. These indicators have been further enriched by additional land-use measures, in particular including destination accessibility and distance to transit, as the 5 Ds of development (cf. Cervero and Kockelman, 1997; Ewing and Cervero 2001, 2010). Additional measures of land-use features have also been proposed, including network connectivity, job-housing balance and mixed land use, among others. The effects of many of these variables are often similar, and not always easy to measure and to isolate from each other. In addition, research findings often vary depending on the specific local contexts, the data and the methodology that are used. The empirical findings from these studies suggest the following conclusions:

- Residential density is usually measured as a ratio of persons divided by land area, e.g. persons per square mile, or housing units divided by land area. The impact of residential density on travel behavior, and in particular on the use of private cars and vehicle miles traveled (VMT), has been the object of many studies. Early studies that used aggregate data for cities, counties, or metropolitan areas (e.g., Newman and Kenworthy, 1989, 1999; van de Coevering and Schwanen, 2006) found reductions in VMT, and an increase in the use of alternative modes of transportation, associated with an increase in residential density. These findings were confirmed by studies that used individuals' travel data and controlled for the contemporaneous impact of additional land use and socio-demographic variables (cf. Bento et al., 2005; Brownstone and Golob, 2009; Heres-Del-Valle and Niemeier, 2011; Kim and Brownstone, 2013). Land-use policies that result in higher densities are often mentioned in the suite of tools that can reduce VMT and greenhouse gas (GHG) emissions associated with transportation. These policies include combinations of changes in land-use regulations, such as relaxing the minimum lot size requirements, increasing the density of allowed development, and encouraging urban infill, or combinations of infrastructure, zoning, or public finance policies that, for example, focus development around transportation nodes (including transit stations).
- Employment density is usually measured as the ratio of the number of employees divided by land area (e.g. employees per square mile). The impact of employment density on VMT has been less studied than residential density (some studies, though, summed up employment and population to compute an overall "activity density" per areal unit). Early studies likely overestimated the impact of employment density because they did not control for the impact of other correlated factors. Most recent studies attribute a rather weak impact on transportation to employment density: for instance, the meta-analysis by Ewing and Cervero (2010) concluded that the effect of employment density on VMT is close to zero. Certainly, increasing employment density in cities seems to reduce the reliance of travelers on private vehicles, and it increases the use of alternative modes of transportation. This is mainly the effect of other characteristics of the built environment, such as street density, access to public transit, quality of the pedestrian environment, and job-housing balance, which are usually correlated with employment density, rather than an effect of density per se. Once these characteristics are accounted for, employment density has a weak, and often non-significant, effect on travel behavior (cf. Badoe and Miller, 2000; Leck, 2006; Ewing and Cervero, 2001, 2010).
- Diversity refers to the land-use mix, or mixed-use development, that results from the practice of accommodating more than one type of land-use function, including residential, office, retail, and personal services, as well as parks and open space within a building, a set of buildings, a block or a neighborhood. Land-use diversity can be promoted through a better balance of land uses through zoning that allows housing, retail establishments, and employment centers to exist in close proximity, as well as policies that encourage in-fill development and vertical mixing of uses within the same building. Various measures of land-use mix entropy, land-use dissimilarity and job-housing balance have been proposed in the literature to measure urban diversity. Because mixed-use neighborhoods offer a variety of employment, shopping, and recreational opportunities and access to many destinations within short distances of residences, they facilitate the use of non-automobile travel modes and shorten car trips (cf. Kockelman, 1997; Bento et al., 2005; Ewing and Cervero, 2010).
- Design includes many characteristics of the built environment that are expected to affect travel behavior. Several measures of design have been discussed in the literature to analyze

Figure 14.1 Regular street networks and improved accessibility lower VMT and favor the use of alternative transportation modes

Source: Susan Handy and Kristin Lovejoy, modified by the authors

the characteristics of different neighborhoods, including elements of street network characteristics and network connectivity, sidewalk coverage, density of pedestrian crossings, presence of street trees and characteristics of buildings, among other features that differentiate neighborhoods located in central, more walkable and dense locations from more auto-oriented developments. Many elements of the urban design are thought to create a sense of place, improve the enjoyment from local residents, and encourage walkability and non-automobile trips, thus contributing to reducing the use of cars (Cervero and Kockelman, 1997; Frank and Engelke, 2005; Ewing and Cervero, 2010).

- Destination accessibility measures the number of potential destinations, including jobs, stores, parks, schools, and other attractions that are reachable in a given unit of travel time (or distance). Destination accessibility is usually higher in more central locations and lower in suburban developments. This measure of the built environment depends on local land-use characteristics (in particular, on the mix of different land uses). The higher the destination accessibility, the lower the dependence on cars, as shown in more walkable, diverse and dense areas of cities (Ewing and Cervero, 2001, 2010; Ewing et al., 2011).
- Distance to transit measures the distance from the home or work location to the nearest public transit stop or station. The distance from public transportation stops is an important determinant of public transportation patronage. Usually, better public transportation services are found in more compact urban areas that have bus and rail stops conveniently located in walking distance from residential and commercial areas. Shorter distances to transit stops reduce the dependence on the use of private vehicles, and encourage the use of alternative modes of transportation (Frank and Engelke, 2005; Ewing and Cervero, 2010).
- Network connectivity describes the quality of the connections that link each location in a community with others, and is affected by the structure of the street network, including the patterns of streets and intersections, and the directness of these connections, which often differ by mode. It is often measured as the ratio of the number of four-way or three-way intersections to all intersections (or the ratio of the number of intersections to the number of street segments, e.g. "nodes" to "links"), the average block size, the intersection density (e.g. intersections per square mile) or street density (e.g. lane miles of street per square mile). Increased network connectivity reduces the distance between trip origins and destinations, and affects-trip frequency, destination choice, and mode choice. All else equal,

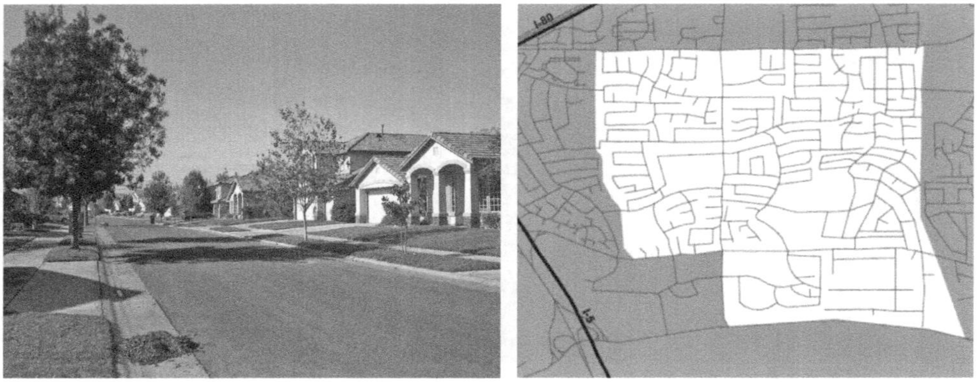

Figure 14.2 Suburban design is associated with a larger use of private cars and reduced access to public transportation

Source: Susan Handy and Kristin Lovejoy, modified by the authors

greater connectivity means shorter travel distances, and it encourages residents to walk or bike instead of driving (Cervero and Kockelman, 1997; Fan and Khattak, 2008; Zhang et al., 2012).

Estimating the impact of each reported characteristic of the built environment in isolation is not easy: many of these land-use features are correlated with the others, and their cumulative effects may differ when implemented as part of more complex policy packages. For example, residential and employment density is usually correlated with a large number of land-use traits, including mixed land uses, access to public transit, quality of the pedestrian environment, and proximity to regional employment or shopping centers. While density is easily measured, and it is often attributed most of the effects of land use on travel behavior, many researchers believe that policy attention should focus on a more holistic set of land-use characteristics (cf. Chatman, 2008). If multiple policy instruments are used together, e.g. mixing residential and commercial land use, improving metropolitan job accessibility and providing improved access to public transit services, their combined impact could be considerably larger than what would be obtained by only changing urban density. Increases in density, and other changes in the 5 Ds of development, should be considered as part of coordinated land-use plans rather than in isolation (National Research Council, 2009).

Many additional co-benefits can derive from land-use policies that encourage higher residential densities, concentrations of employment, shopping, and service destinations, and infrastructure and urban design that makes non-motorized travel modes, e.g. walking and bicycling, more attractive. Several land-use characteristics, including higher residential density, are associated with increased walking and biking (e.g. Boarnet et al., 2008, 2011), stronger social ties and a sense of belonging to the local community. Further, increases in non-motorized travel generate health benefits for the local population (Handy et al., 2002; Ewing et al., 2011).

Individuals might also choose to live in high-density settings with varied land uses because they seek to drive less and enjoy these types of settings. If this is true, they do not adopt these travel patterns as a direct effect of the built environment, but as a consequence of their personal attitudes and preferences. This residential self-selection effect may significantly reduce the effects of policies designed to reduce the use of private vehicles and incentivize alternative transportation modes. If residential self-selection is in place, these policies would reduce VMT by providing living places for persons who already seek to drive less. If there is a shortage of such places, building higher density neighborhoods would reduce VMT to the extent that it would modify the travel patterns of the persons choosing to live in these areas. Many studies have attempted to quantify the effects of residential self-selection, highlighting how it often accounts for a large portion of the impact that could otherwise be attributed to the built environment (cf. Cao et al., 2009).

If the impact of land-use characteristics on travel behavior has been the object of many studies, the reverse relationship, i.e. the impact of the characteristics of the transportation system on land-use development, has been less investigated in quantitative terms and through rigorous studies. Still, this is an important topic due to the enormous impact that transportation investments have on the accessibility and attractiveness of different areas. The characteristics of transportation have a fundamental role in affecting the local accessibility of places, and the location of residences and economic activities. Improved transportation accessibility significantly impacts economic development, as a result of the lower costs and easier movement of goods and passengers. It does impact job creation, in particular for specific types of jobs and economic sectors, and affects residential location through both a direct and indirect effect. Transportation accessibility modifies the attractiveness of an area, and affects rent and property values, influencing the location of residences and activities, and generating differential effects among different social classes. It might eventually lead to processes of gentrification, e.g. as the result of the opening of a new subway station or a freeway corridor.

Identifying the impacts of transportation accessibility on land-use development is not easy: in addition to the difficulty of isolating the effects of transportation from the impact of other variables (e.g. economic cycles and socioeconomic patterns), it is not easy to analyze behavioral changes of residents and demonstrate causality. This is in particular true for most studies that analyze cross-sectional datasets, and/or focus on short-term effects. The impacts of transportation accessibility on the location of activities and residences, and land-use development, take effect in the longer run and can be measured only after several years. This makes isolating the effects of this variable even more difficult, due to the concurring effects of several other factors affecting land-use development in the long run.

Land-Use Transportation Interaction Models

Several approaches have been proposed for modeling the complex interactions between land use and transportation. Land-use transportation models are usually composed by a land-use modeling component that runs in iteration with a transportation model. Wegener (2004) and more recently Pagliara et al. provide a detailed review of the state-of-the-art in this field, focusing in particular on the recent approaches that model respectively residential location choice (Pagliara et al., 2010) and employment location choice (Pagliara et al., 2012). Further discussion on the assumptions used by these models, and their application to urban modeling, is available in Simmonds et al. (2013). The remainder of this section presents a non-comprehensive review of some of these models, and briefly discusses their application to the simulation of future development scenarios in urban areas.

The first contribution to residential location modeling dates back to the work of Alonso (1960), who laid the foundations for the economic analysis by applying von Thunen's key "bid rent" idea to residential location. Employment location choice modeling has also played a central role ever since the start of applied urban models. The Lowry (1964) model is prototypal in this respect: it foresees the location of working residents based on the employment required by basic industries, and it forecasts the location of the service industry employment based on consumer demand. Some years later, Echenique et al. (1969) introduced physical constraints on location arising from buildings, town planning and the transportation network, with the objective of predicting the distribution of employment and population within a city according to the functional relationships among urban activities subject to those constraints.

Early models used Newton's gravity law to model the location of businesses and households. Then spatial interaction models based on entropy maximization were introduced (Wilson, 1970). Over the years, the modeling approach became progressively grounded in micro-economic theory. MEPLAN (Echenique, 2004) is an example of an integrated land use and transportation modeling framework that was developed at the beginning of the 1980s. Close integration of the MEPLAN framework with input–output tables (Leontief and Leontief, 1986), random utility theory for location choice (McFadden, 1974) and travel demand modeling (Ben-Akiva and Lerman, 1985) have produced practical tools that can be used to explore and assess urban land use and transportation development options. Three models, i.e. TRANUS (de la Barra, 1989), MUSSA (Martínez, 1996) and PECAS (Hunt and Abraham, 2003) follow the MEPLAN methodology, developing certain aspects of employment and household allocation. MEPLAN and TRANUS use a spatially disaggregated input–output structure to represent the behavior of industry sectors and household categories, and their interactions to simulate the spatial distributions of incremental production and floor space rents arising from an initial allocation of exogenously generated "basic" components. The PECAS model system develops and applies spatial economic models and includes a representation of the elements of residential choice behavior. It includes a computable general equilibrium structure for representing how activities (including industrial sectors, government and household categories) locate within the building space provided by developers, and how these activities interact with each other at a given point in time.

The IRPUD model (Wegener, 1982) was the first example of an employment location choice model that uses logit models based on the random utility theory. In recent years, several approaches that model businesses or jobs by micro simulation have been developed. The premise is that microsimulation is better suited to simulate location behavior and the interactions of different agents. The first example of micro simulation of jobs was developed for the California Urban Futures (CUF) model (Landis and Zhang, 1998). All potential uses, such as single-family, apartments, commercial, or industrial uses, bid for developable land. Khan et al. (2002) developed a microsimulation model for small and medium-sized businesses using a synthetic study area based on raster cells; discrete choice models are used to simulate location decisions determined by the price of floor space. Maoh and Kanaroglou (2012) developed a model simulating small- and medium-sized businesses with less than 200 employees for Hamilton, Canada. Finally, one of the few models simulating firms instead of jobs is the one contained in the Spatial Firm demographic Microsimulation (SFM) (cf. de Bok and Pagliara, 2012).

DRAM and EMPAL are respectively the household and employment allocation submodels in the Integrated Transportation and Land Use Package (ITLUP). Trip generation and distribution functions for the travel forecasting model are developed within DRAM, simultaneously with household location. Travel times from runs of the travel model are used to compute new activity distributions (Putman, 2014). The DELTA land-use economic modeling package (Simmonds, 2001) is divided into processes representing spaces and activities. The model processes simulate household formation and dissolution, employment growth or decline, location and property markets, and the employment status of individuals. DELTA adopts a detailed classification of households reflecting household composition, age of household members, working status of working-age adults, and the socioeconomic group to which the household belongs. The MUSSA model forecasts the expected location of agents, residents and firms in an urban area. It adopts a modified version of the bid-choice framework: it combines bid rent and discrete choice approaches to land markets by dealing simultaneously with both sides of an auction in an integrated framework.

The UrbanSim model system (Waddell, 2002) contains model components representing household and employment location choices, and real-estate development. Several applications of the UrbanSim model have been developed for major cities in the US and abroad. UrbanSim explicitly models individual agents in the land development process, including both the supply and demand sides of the real-estate market as well as the prices, which are adjusted dynamically. Recent applications of the UrbanSim model (Waddell, 2010) adopt a temporally and spatially detailed micro simulation approach, which is integrated in practical applications with modern activity-based travel demand models.

Integrated land-use transportation models are often used for the simulation of future development scenarios and for policy testing. Simplified models have also been developed, allowing for faster execution time and less demanding data input. For example, the Metropolitan Activity Relocation Simulator (MARS) (Pfaffenbichler 2003; Pfaffenbichler et al., 2008) is a dynamic land-use transportation interaction modeling framework based on the principles of systems dynamics and synergetic. MARS-based models have been developed for several cities in Europe and Asia, as well as in Brazil (Porto Alegre) and the US (Washington, DC). MARS includes two main submodels, a transportation model and a land-use model. The land-use model simulates the generation and allocation of new housing units and the location of workplaces for two main categories: production and services. Guzman et al. (2014) discuss the application of a MARS-based model for the analysis of future scenarios of development, and the evaluation of synergies associated with the simultaneous adoption of multiple transportation and land-use policies.

Land-Use Transportation Strategies and Urban Sustainability

The integration of land-use policies with transportation plans has received increased attention as a key strategy to improve the sustainability of cities and local communities, provide efficient transportation alternatives, and reduce the dependence on private cars. Integrated land-use transportation plans are a fundamental component of the widely promoted "three-legged" approach to mitigate climate change and reduce CO_2 emissions from transportation, with one leg related to vehicle fuel economy, a second to the carbon content of the fuel itself, and a third to the amount of driving or vehicle miles traveled (VMT) (cf. Ewing et al., 2008). Bringing residents closer to destinations and providing alternatives to driving may lead people to drive less (Handy and Clifton, 2001; Handy et al., 2005; Ewing and Cervero, 2010), and have positive effects on health and physical activity (Handy et al., 2002; National Research Council, 2005). Research evidence has confirmed that poor design of the built environment and urban sprawl is usually associated with low levels of walking, bicycling and transit use, lower levels of physical activity and social connections among residents, reduced mobility of senior members and individuals physically or psychologically impaired to drive, exposure to traffic-generated air pollutants, increased obesity and risks of cardiovascular or pulmonary disease, and the risk of injuries due to low transportation safety and traffic-related accidents (cf. Frank et al., 2008; Ewing et al., 2011, 2014).

Providing mixed-use, compact, walkable transit oriented development with easier access to mass transportation, through the improvement of existing public transit or, eventually, the introduction of new transportation options, e.g. bus rapid transit (BRT) or light rail, helps reduce the mode share of "driving alone" and GHG emissions from transportation. Further, it increases quality of life, reduces social exclusion, supports the development of liveable and vibrant communities, promotes healthy and active living, and increases citizens' physical activity, also as the result of the higher adoption of active transportation modes, as opposed to the sedentary life associated with car-oriented mobility patterns (cf. Handy et al., 2002; National Research Council, 2005; Ewing et al., 2008; Kooshian and Winkelman, 2011).

Many of these topics have become central to the research and planning agendas that promote the principles of smart growth, new urbanism, mixed-use development, transit-oriented development, healthy communities and in-fill development. Common to these movements is the idea that land-use policies can influence the use of transportation and promote more sustainable urban forms while decreasing the environmental externalities of human settlements (cf. Handy et al., 2005; Ewing et al., 2008). More compact cities with better accessibility and efficient public transportation are often considered the most efficient and desirable form of urban settlements (Cervero, 2001). Empirical findings from existing cities show that individuals travel shorter distances and rely less on the use of cars when they live in a neighborhood with compact urban form. Similarly, the expectation is that new urban designs with higher density, mixed land uses, grid-pattern street network, and proximity to commercial activities will contribute to decreasing automobile travel also in other areas, although less agreement exists on the ability to achieve similar results in already-built cities that feature low-density, auto-oriented development (Stopher and Stanley, 2014).

Policies inspired by the principles of smart growth seek to promote more balanced development of cities through the coordination of transportation and land-use policies, preserving open space, and promoting the redevelopment of already urbanized areas and in-fill development. Among the objectives of smart growth are the promotion of higher density, job-housing balance, easier access to transit, walkable streets and the preservation of open space, thus reducing the land consumption that would be caused by suburban development (Downs, 2001). The smart growth principles aim to create walkable neighborhoods, with mixed land uses and a broader range of alternative transportation options, capitalizing on the resources of existing communities and the advantages of compact building design (Duany et al., 2009).

Similar objectives are promoted by the New Urbanism movement, which calls for a return to traditional planning principles from the era before the massive adoption of automobiles. Since the early 1990s, the Congress of New Urbanism has been advocating for traditional planning and design solutions that focus on neighborhood centers, through the mix of land uses, the location of key attractors and centers of daily life, such as shops, libraries, churches, and public buildings in walking distance from residential units (Duany et al., 2001, 2009).

The importance of integrating the development of transportation infrastructures with land use is central to the implementation of policies for transit-oriented development (TOD). TOD aims at the development of high-density urban areas, with mixed land use and high quality of life, which favor the use of transit facilities. It combines the objectives of proximity to transit (within an easy walk) and development shaped by transit, promoting a pedestrian-oriented layout around major transit nodes (Dunphy et al., 2004). Among the benefits from TOD, residents are able to reduce car use while maintaining high levels of accessibility and mobility (Hess and Lombardi, 2004; Dunphy and Porter, 2006). In the short term, transit oriented development maximizes the benefits from transportation investments, supporting the increase in density, mixed land use and urban quality along the transportation corridors. In the long run, it contributes to a modal shift toward public transportation, and limits urban sprawl.

Policies that coordinate land-use development with the adoption of active modes and public transportation reduce urban sprawl and land consumption, improve energy efficiency, support the efficient use of local infrastructures and reduce infrastructure costs. Many efforts are underway to promote integrated land-use transportation policies as part of national and state transportation strategies, and to include them in regional and local plans adopted by metropolitan planning organizations and local governments. Smart-growth inspired plans have been promoted in geographically distant regions, often with similar objectives, e.g. that one of coordinating the location of new public transportation facilities with the development of mixed residential, commercial, and business activities, while providing a diverse set of transportation options and promoting healthy communities. Examples of such plans can be found in San Francisco, Sacramento, San Diego, Portland, New York, as well as in many TOD projects developed in rapidly growing Asian cities. The coordination of land use and transportation as part of a strategy to increase sustainability in urban development, reduce land and resource consumption, and improve public health has also become a priority in the political agendas of many European countries, and has entered the legislation of several US states, including Oregon, Washington and California.

In California, the landmark legislation set forth through the Sustainable Communities and Climate Protection Act (Senate Bill 375) in 2008 has introduced the requirement for Metropolitan Planning Organizations (MPOs) to adopt Sustainable Community Strategies in order to meet established targets for GHG emissions in 2020 and 2035, and as part of the effort mandated by the Assembly Bill 32 to reduce transportation GHG emissions through increasing vehicle efficiency, decreasing transportation demand, and shifting transportation demand to more environmentally sustainable transportation modes through new patterns of land use and transportation investment. Sustainable Community Strategies need to meet the transportation and housing needs of a region while ensuring an appropriate reduction in the environmental impact from transportation and an increase in the liveability of California's communities, identifying, among other things, the location of land uses, densities, areas to house future population, and the transportation network investments needed to serve these areas (cf. McFadden et al., 2014). Still, several uncertainties are associated with the outcomes from these policies and their ability to achieve the proposed targets of reduction in transportation-related emissions. In particular, the lack of coordination with economic tools (e.g. pricing, gas taxes or incentive programs) might significantly reduce the potential results from the adoption of these strategies (Chatman, 2008; Heres-Del-Valle and Niemeier, 2011).

Conclusions

If current trends continue, more than eight billion people will inhabit the Earth by 2030. Most of them will live in urban areas, and will travel using two billion vehicles (Sperling and Gordon, 2009). Such a scenario raises serious concerns about the environmental (un)sustainability of human settlements, and calls for immediate actions to influence the development of cities and reduce the environmental impacts of transportation. Research has demonstrated that land-use characteristics are able to affect the use of transportation in several ways. In particular, individuals who live in high-density and high-walkability neighborhoods usually depend less on the use of private vehicles for their mobility needs, take shorter trips and tend to walk, bike and use public transportation more often.

Compact urban development generates a number of co-benefits associated with the reduced use of private vehicles, including lower GHG emissions from transportation and reduced land consumption. Several approaches have been proposed to model the effects of land-use development on transportation and, vice versa, the impact of transportation investments on land-use development. Land-use transportation interaction models are useful tools for scenario testing and policy analysis; they can support decision-making providing information about long-term outcomes of policy implementation, and the potential synergies from the development of multiple policies.

The integration of land-use and transportation policies that aim to increase environmental sustainability in cities is central to the principles of smart growth, new urbanism, and transit-oriented development. Even with some differences in their main focus, these movements rely on a closer coordination of land-use development with transportation investments. They promote mixed land use, and support the development of public transportation. Integrated transportation land-use plans hold promise for more sustainable travel. However, the correct implementation of plans that coordinate land-use development with transportation investments requires strict cooperation and action alignment from local and regional governments. Further, the achievement of the objectives of sustainable transportation and reduced use of cars may be limited by political, fiscal, and other constraints affecting policy development and transportation investments.

Understanding the dynamics behind urban development, and predicting future travel patterns is not easy. For example, important changes have been recently observed in the development patterns of many American cities. A process of urban transformation is underway, featuring an increase in urban density and the regeneration of the central parts of cities, in the US as well as in other developed countries, and in contrast with the trend observed in the previous decades, which were dominated by increasing suburbanization of cities and the large use of private vehicles. Identifying the causes behind this change, and predicting its future evolution poses several challenges. The impact of economic cycles and fuel prices, the high levels of traffic congestion together with the changes in the household composition and personal lifestyles, e.g. the desire of a larger number of residents to relocate near the most accessible and vibrant parts of the cities, are all possible factors influencing these trends (Wachs, 2013). Further research is needed to investigate this and other related topics, in order to better understand the consequences of this change on urban development and future travel demand, and to assist planners in the definition of policies for sustainable, liveable and healthy communities, and efficient transportation for future generations.

Note

1. Land scarcity is already a binding constraint to the development of new infrastructures in many urban areas.

References

Alonso, W. (1960). A theory of the urban land market. *Papers in Regional Science*, *6*(1), 149–157.

Badoe, D. A. & Miller, E. J. (2000). Transportation–land-use interaction: Empirical findings in North America, and their implications for modeling. *Transportation Research Part D: Transport and Environment*, *5*(4), 235–263.

Ben-Akiva, M. E. & Lerman, S. R. (1985). *Discrete Choice Analysis: Theory and Application to Travel Demand*. Cambridge, MA: MIT Press.

Bento, A. M., Cropper, M. L., Mobarak, A. M., & Vinha, K. (2005). The effects of urban spatial structure on travel demand in the United States. *Review of Economics and Statistics*, *87*(3), 466–478.

Boarnet, M. G., Greenwald, M., & McMillan, T. E. (2008). Walking, urban design, and health toward a cost-benefit analysis framework. *Journal of Planning Education and Research*, *27*(3), 341–358.

Boarnet, M. G., Joh, K., Siembab, W., Fulton, W., & Nguyen, M. T. (2011). Retrofitting the suburbs to increase walking: Evidence from a land-use-travel study. *Urban Studies*, *48*(1), 129–159.

Brownstone, D. & Golob, T. F. (2009). The impact of residential density on vehicle usage and energy consumption. *Journal of Urban Economics*, *65*(1), 91–98.

Cairncross, F. (2001). *The Death of Distance: How the Communications Revolution is Changing Our Lives*. Boston, MA: Harvard Business School Press.

Cao, X., Mokhtarian, P. L., & Handy, S. L. (2009). Examining the impacts of residential self-selection on travel behaviour: A focus on empirical findings. *Transport Reviews*, *29*(3), 359–395.

Cervero, R. (2001). Efficient urbanisation: Economic performance and the shape of the metropolis. *Urban Studies*, *38*(10), 1651–1671.

Cervero, R. & Kockelman, K. (1997). Travel demand and the 3Ds: Density, diversity, and design. *Transportation Research Part D: Transport and Environment*, *2*(3), 199–219.

Chatman, D. G. (2008). Deconstructing development density: Quality, quantity and price effects on household non-work travel. *Transportation Research Part A: Policy and Practice*, *42*(7), 1008–1030.

Circella, G. & Mokhtarian, P. L. (forthcoming). Impact of information communication technology on transportation. In S. Hanson & G. Giuliano (Eds.), *Geography of Urban Transportation* (5th ed.). New York: The Guilford Press.

de Bok, M., & Pagliara, F. (2012). Choice set formation in microscopic firm location models. In F. Pagliara, M. de Bok, D. Simmonds, & A. Wilson (Eds.), *Employment Location in Cities and Regions* (pp. 263–281). Berlin: Springer.

De la Barra, T. (1989). *Integrated Land Use and Transport Modelling: Decision Chains and Hierarchies*. Cambridge: Cambridge University Press.

Downs, A. (2001). What does 'smart growth' really mean? *Planning*, *67*(4), 20–25.

Duany, A., Plater-Zyberk, E., & Speck, J. (2001). *Suburban Nation: The Rise of Sprawl and the Decline of the American Dream*. New York: Farrar, Straus & Giroux.

Duany, A., Speck, J., & Lydon, M. (2009). *The Smart Growth Manual*. New York: McGraw-Hill Education.

Dunphy, R. T., Cervero, R., Dock, F. C., McAvery, M., Porter, D. R., & Swenson, C. J. (2004). *Developing Around Transit: Strategies and Solutions that Work*. Washington, DC: Urban Land Institute.

Dunphy, R. T. & Porter, D. R. (2006). Manifestations of development goals in transit-oriented projects. *Transportation Research Record: Journal of the Transportation Research Board*, *1977*(1), 172–178.

Echenique, M. (2004). Econometric models of land use and transport. In D. Hensher et al, Oxford, pp. 185–202. (Ed.), *Handbook of Transport Geography and Spatial Systems* (Vol. 5, pp. 185–202). Oxford: Elsevier.

Echenique, M., Crowther, D., & Lindsay, W. (1969). A spatial model of urban stock and activity. *Regional Studies*, *3*(3), 281–312.

Ewing, R., Bartholomew, K., Winkelman, S., Walters, J., & Chen, D. (2008). *Growing Cooler: The Evidence on Urban Development and Climate Change*. Washington, DC: Urban Land Institute.

Ewing, R. & Cervero, R. (2001). Travel and the built environment: A synthesis. *Transportation Research Record: Journal of the Transportation Research Board*, *1780*(1), 87–114.

Ewing, R. & Cervero, R. (2010). Travel and the built environment: A meta-analysis. *Journal of the American Planning Association*, *76*(3), 265–294.

Ewing, R., Meakins, G., Bjarnson, G., & Hilton, H. (2011). Transportation and land use. In A. L. Dannenberg, H. Frumkin, & R. J. Jackson (Eds.), *Making Healthy Places* (pp. 149–169). Washington, DC: Island Press.

Ewing, R., Meakins, G., Hamidi, S., & Nelson, A. C. (2014). Relationship between urban sprawl and physical activity, obesity, and morbidity—Update and refinement. *Health & Place*, *26*(0), 118–126.

Fan, Y. & Khattak, A. J. (2008). Urban form, individual spatial footprints, and travel: Examination of space-use behavior. *Transportation Research Record: Journal of the Transportation Research Board*, *2082*(1), 98–106.

Frank, L. D. & Engelke, P. (2005). Multiple impacts of the built environment on public health: Walkable places and the exposure to air pollution. *International Regional Science Review*, *28*(2), 193–216.

Frank, L. D., Kerr, J., Sallis, J. F., Miles, R., & Chapman, J. (2008). A hierarchy of sociodemographic and environmental correlates of walking and obesity. *Preventive medicine*, *47*(2), 172–178.

Guzman, L. A., de la Hoz, D., & Circella, G. (2014). Evaluation of synergies from transportation policy packages using a social welfare maximization approach: A case study for Madrid, Spain. *Case Studies on Transport Policy*.

Handy, S., Cao, X., & Mokhtarian, P. (2005). Correlation or causality between the built environment and travel behavior? Evidence from Northern California. *Transportation Research Part D: Transport and Environment*, *10*(6), 427–444.

Handy, S. L., Boarnet, M. G., Ewing, R., & Killingsworth, R. E. (2002). How the built environment affects physical activity: Views from urban planning. *American Journal of Preventive Medicine*, *23*(2), 64–73.

Handy, S. L. & Clifton, K. J. (2001). Local shopping as a strategy for reducing automobile travel. *Transportation*, *28*(4), 317–346.

Heres-Del-Valle, D. & Niemeier, D. (2011). CO_2 emissions: Are land-use changes enough for California to reduce VMT? Specification of a two-part model with instrumental variables. *Transportation Research Part B: Methodological*, *45*(1), 150–161.

Hess, D. B. & Lombardi, P. A. (2004). Policy support for and barriers to transit-oriented development in the inner city: Literature review. *Transportation Research Record: Journal of the Transportation Research Board*, *1887*(1), 26–33.

Hunt, J. D. & Abraham, J. E. (2003). Design and Application of the PECAS Land Use Modelling System. Paper presented at the 8th International Conference on Computers in Urban Planning and Urban Management, Sendai, Japan.

Janelle, D. G. (2012). Space-adjusting technologies and the social ecologies of place: Review and research agenda. *International Journal of Geographical Information Science*, *26*(12), 2239–2251.

Janelle, D. G. & Gillespie, A. (2004). Space–time constructs for linking information and communication technologies with issues in sustainable transportation. *Transport Reviews*, *24*(6), 665–677.

Khan, A. S., Abraham, J. E., & Hunt, J. D. (2002). Agent-based Micro-simulation of Business Establishments. Paper presented at the Congress of the European Regional Science Association (ERSA), Dortmund, Germany.

Kim, J. & Brownstone, D. (2013). The impact of residential density on vehicle usage and fuel consumption: Evidence from national samples. *Energy Economics*, *40*, 196–206.

Kim, T. J., Claus, M., Rank, J. S., & Xiao, Y. (2009). Technology and cities: Processes of technology-land substitution in the twentieth century. *Journal of Urban Technology*, *16*(1), 63–89.

Kockelman, K. M. (1997). Travel behavior as function of accessibility, land use mixing, and land use balance: evidence from San Francisco Bay Area. *Transportation Research Record: Journal of the Transportation Research Board*, *1607*(1), 116–125.

Kooshian, C. & Winkelman, S. (2011). *Growing Wealthier: Smart Growth, Climate Change and Prosperity*. Washington, DC: Center for Clean Air Policy.

Landis, J. & Zhang, M. (1998). The second generation of the California urban futures model. Part 1: Model logic and theory. *Environment and Planning B*, *25*, 657–666.

Leck, E. (2006). The impact of urban form on travel behavior: A meta-analysis. *Berkeley Planning Journal*, *19*, 37–58.

Leontief, W. W. & Leontief, W. (1986). *Input-output Economics*. New York: Oxford University Press.

Lowry, I. S. (1964). A model of metropolis. *RAND Research Memorandum RM-4035-RC*. Santa Monica, CA: The RAND Corporation.

Maeng, D.-M. & Nedović-Budić, Z. (2008). Urban form and planning in the information age: Lessons from literature. *Spatium*, *17–18*, 1–12.

Maoh, H. & Kanaroglou, P. (2012). Modelling firm failure: Towards building a firmographic microsimulation model. In F. Pagliara, M. de Bok, D. Simmonds, & A. Wilson (Eds.), *Employment Location in Cities and Regions* (pp. 243–261). Berlin: Springer.

Martinez, F. (1996). MUSSA: Land use model for Santiago city. *Transportation Research Record: Journal of the Transportation Research Board*, *1552*(1), 126–134.

McFadden, D. (1974). Conditional logit analysis of qualitative choice behavior. In P. Zarembka (Ed.), *Frontiers in Econometrics* (pp. 105–142). New York: Academic Press.

McFadden, A. J., Circella, G., Haydu, B., & Linesch, N. J. (2014). Impact of Proposed Land Use and Transportation Investments on Future Travel Patterns in California. *Transportation Research Record: Journal of the Transportation Research Board*, *2430*, 207–215.

Mokhtarian, P. (2009). If telecommunication is such a good substitute for travel, why does congestion continue to get worse? *Transportation Letters*, *1*(1), 1–17.

Mokhtarian, P. L., Salomon, I., & Choo, S. (2005). Measuring the measurable: Why can't we agree on the number of telecommuters in the US? *Quality and Quantity, 39*(4), 423–452.

Muller, P. (2004). Transportation and urban form. In S. Hanson & G. Giuliano (Eds.), *The Geography of Urban Transportation* (4th ed.). New York: Guilford Press.

National Research Council (2005). *Does the Built Environment Influence Physical Activity? Examining the Evidence*. Washington, DC: The National Academies Press.

National Research Council (2009). *Driving and the Built Environment: The Effects of Compact Development on Motorized Travel, Energy Use, and CO2 Emissions*. Washington, DC: The National Academies Press.

Newman, P. & Kenworthy, J. (1999). *Sustainability and Cities: Overcoming Automobile Dependence*. Washington, DC: Island Press.

Newman, P. G. & Kenworthy, J. R. (1989). *Cities and Automobile Dependence: An International Sourcebook*. Brookfield, VT: Gower Publishing.

O'Brien, R. (1992). *Global Financial Integration: The End of Geography*. New York: Council on Foreign Relations Press.

O'Sullivan, A. (2011). *Urban Economics*. New York: McGraw-Hill Education.

Pagliara, F., de Bok, M., Simmonds, D., & Wilson, A. (2012). *Employment Location in Cities and Regions: Models and Applications*. Heidelberg: Springer.

Pagliara, F., Preston, J., & Simmonds, D. (2010). *Residential Location Choice: Models and Applications*. Heidelberg: Springer.

Pfaffenbichler, P. (2003). The strategic, dynamic and integrated urban land use and transport model MARS (Metropolitan Activity Relocation Simulator). Vienna: Institut für Verkehrsplanung und Verkehrstechnik der Technischen Universität Wien.

Pfaffenbichler, P., Emberger, G., & Shepherd, S. (2008). The integrated dynamic land use and transport model MARS. *Networks and Spatial Economics, 8*(2–3), 183–200.

Putman, S. H. (2014). *Integrated Urban Models Volume 2: New Research and Applications of Optimization and Dynamics* (reprint ed.). New York: Routledge.

Salomon, I. (1998). Technological change and social forecasting: The case of telecommuting as a travel substitute. *Transportation Research Part C: Emerging Technologies, 6*(1), 17–45.

Simmonds, D. (2001). The objectives and design of a new land-use modelling package: DELTA. In G. Clarke & M. Madden (Eds.), *Regional Science in Business* (pp. 159–188). Berlin: Springer.

Simmonds, D., Waddell, P., & Wegener, M. (2013). Equilibrium versus dynamics in urban modelling. *Environment and Planning B: Planning and Design, 40*(6), 1051–1070.

Sperling, D. & Gordon, D. (2009). *Two Billion Cars: Driving Toward Sustainability*. New York: Oxford University Press.

Stopher, P. & Stanley, J. (2014). *Introduction to Transport Policy: A Public Policy View*. Cheltenham: Edward Elgar Publishing.

van de Coevering, P. & Schwanen, T. (2006). Re-evaluating the impact of urban form on travel patterns in Europe and North America. *Transport Policy, 13*(3), 229–239.

Wachs, M. (2013). Turning cities inside out: Transportation and the resurgence of downtowns in North America. *Transportation, 40*(6), 1159–1172.

Waddell, P. (2002). UrbanSim: Modeling urban development for land use, transportation, and environmental planning. *Journal of the American Planning Association, 68*(3), 297–314.

Waddell, P. (2010). Modeling residential location in UrbanSim. In F. Pagliara, J. Preston & D. Simmonds (Eds.), *Residential Location Choice* (pp. 165–180). Berlin: Springer.

Wegener, M. (1982). Modeling urban decline: A multilevel economic-demographic model for the Dortmund region. *International Regional Science Review*, 7(2), 217–241.

Wegener, M. (2004). Overview of land-use transport models. *Handbook of Transport Geography and Spatial Systems, 5*, 127–146.

Wilson, A. G. (1970). *Entropy in Urban and Regional Modelling*. London: Pion.

Zhang, L., Hong, J. H., Nasri, A., & Shen, Q. (2012). How built environment affects travel behavior: A comparative analysis of the connections between land use and vehicle miles traveled in US cities. *Journal of Transport and Land Use, 5*(3), 40–52.

15

PUBLIC TRANSPORTATION SYSTEMS

Partha Chakroborty and Vinod Vasudevan

INDIAN INSTITUTE OF TECHNOLOGY

Introduction

Public transportation includes any type of transportation that is available to the public for safe and efficient movement for trip purpose satisfaction. In its broadest sense, it includes for-hire individual transport means like taxis (cabs) to rail rapid-transit systems like urban metro-rails or greater-metropolitan area surface rails that transport a large number of people. In this chapter, the term "public transportation system" is used to mean for-hire group transport systems.

As public transportation systems are for hire, it is important to understand issues related to the cost of public transportation and pricing of these services. Pricing of any service ultimately depends on the customers' willingness to pay for such services and hence is influenced by what is important to the paying public and how they view the quality of service. Public transportation systems (PTS) cater to a group of typically unrelated trip makers and therefore rely on collecting trips in space and time. Such agglomeration of trips is possible if 1) the schedule of the PTS is such that trip makers can use the system by minimally modifying their departure times, and 2) the routes are such that accessing the system from their points of origin and destination is easily possible. Hence, it is important to study how public transport networks and schedules can be developed so that they are successful in catering to a large number of trip makers. Discussions on network design and scheduling, economics of public transportation, and measurement of quality of service form the subject matter of later sections. Today, PTS around the world cover a wide range in terms of the size of the cabin used to transport people, the type of routes and schedules they use, the right of way they enjoy, the kind of fuel used for propulsion, and the technology used for guidance. Cabin sizes range from units that can carry three to four people and run as individual units to those that carry a hundred or more and generally are linked to form a train. There are public transportation systems that run on fixed routes but variable schedules to those that run on fixed routes and schedules.

Public transportation systems use a wide range of right of way (ROW) categories. Some share their right of way with others on the street (Category C) while others have a completely separate right of way (Category A); there are some that use a partially separate right of way (Category B). Systems that use ROW Category C offer lesser travel time reliability to users and typically have lower operating speeds and safety levels than those that use other ROW categories.

On the other hand, systems using ROW Category A have higher operating speeds, improved safety levels (since control of units can be centralized) and better travel time reliability. However, providing ROW Category A is costlier and requires large tracts of land.

PTS often use cleaner fuel like electricity or compressed natural gas (CNG). The choice of fuel also impacts the operating characteristics of the system. PTS use both guided as well as steered technologies. Transit systems using guided technologies (like vehicles running on rails) can be allowed to move at higher speeds than those using steered technologies which invariably rely on visual acuity and alertness of the driver to move safely.

Before closing this introductory section and moving on, the topics of multimodal public transportation and the evolution of public transportation with urban size need to be touched on. Any transportation system can be described by two basic attributes: the level of accessibility it provides and the speed at which it operates. It is also well understood that any system that operates at high speed offers a lower level of accessibility (for example, an access controlled road allows greater travel speeds and rail rapid transit systems have fewer stops than a bus system, etc.). Any trip, however, requires both accessibility and (ideally) greater travel speed. However, transit systems, like any other transportation system, cannot achieve both speed and accessibility simultaneously. Thus, a transit system that wants to cater to a large number of people and wants to reduce their in-vehicle travel time must necessarily be multimodal. For users with average-to-long trip lengths in such a system, the trip will consist of both high accessibility, low speed systems and low accessibility, rapid transit systems. Hence, the planning of transit systems should ideally be done as a multimodal system. Various issues such as i) access/egress, ii) connectivity, iii) frequency balance and schedule coordination, and iv) fare structure and ticketing, etc. need to be integrated in such planning. The reader may refer to Phanikumar et al. (2013), Mishra et al. (2012) and Krygsman et al. (2004) for further understanding of various issues related to multimodal public transportation systems.

Another area that a student of PTS must be conversant with is the relationship between city size and public transportation systems. As a settlement grows, the public transportation facilities must evolve with it. Towns and cities around the world that traditionally ignored public transportation during their growth and instead relied on private modes, today face problems and pose a considerable challenge to sustainable development. There are other urban settlements where the public transportation system did not move out of the small-cabin, fixed-route, variable-schedule para-transit systems that today face serious congestion and pollution. Vuchic (1981) gives a succinct description of how settlement size and public transportation systems need to evolve together. Of course, just as settlement size has an impact on public transportation facilities, the latter also influences how a settlement grows. For example, policies that encourage private transportation modes often create large suburban, low-density settlements around a major urban center from where most people commute to work, whereas cities that have mass rapid transit systems developing alongside the settlements have reasonably dense suburban growths along the transit lines.

As stated earlier, discussions on network design and scheduling, economics of public transportation and measurement of quality of service form the subject matter of the next three sections.

Public Transportation Network Design and Scheduling

In this section the problem of designing a public transportation network is discussed. A network, in its generic sense, is a structure that consists of nodes and links. The links are entities that connect nodes and nodes are points where something that is traversing the network can either start or end its journey, or change from one link to another. Any transportation system can be naturally represented as a network. Unlike the road network, the public transportation (transit) network can connect an ordered pair of nodes through multiple links (representing transit routes) and the links only exist when a transit unit (say, a bus or a train) traverses it. That is, at a given time a user may not be able to use a link if at that time there are no transit units on that link.

Hence, the public transportation (or transit) network design problem consists of two main parts (Chakroborty, 2003): i) the development of transit routes and ii) development of schedules (a statement that says when these routes are alive or useable). In the literature, this problem is often referred to as the transit network design and scheduling problem. Of course, there are various other classifications of the tasks involved in the design and planning of transit networks. One of the most often quoted (see for example, Guihare and Hao, 2008) classification of tasks is by Lampkin and Saalmans (1967); a similar classification is also given in Ceder and Wilson (1986). In this, the problem is divided into a sequence of five phases: i) design of routes, ii) determination of frequency of operation on these routes, iii) development of schedules that take into account transfers, iv) scheduling/assignment of buses/rakes to the bus/train operations on the various routes, and v) scheduling of crew to the buses/trains. The last two activities help operationalize the running of transit units as per the schedule on developed routes and are often referred to as operational planning. The first three activities relate to strategic planning. Also, it is generally assumed that these activities are performed in a sequence with the output of the previous activity becoming an input to the next.

Such a sequential procedure, especially in the strategic planning phase, is not entirely justified as the demand for public transportation depends on the quality of service it provides, which in turn depends on the route set and the schedule of operation. Ideally, therefore, the routing and scheduling phases (including frequency setting) should be done concurrently. Yet, as Guihaire and Hao (2008) point out, "due to the exceptional complexity of the process, this global approach appears intractable in practice."

Before discussing some of the work that has been done on the transit network design problem, another important feature that has received lesser attention from researchers is brought to the fore. As can be imagined, the entire design process strongly depends on how demand will change on changing design features of the system. For example, questions such as how the demand will change if the route set is changed or the frequency is altered are important to the design process. Further, the change in demand can happen at two levels: i) demand may move to or from other modes, and ii) demand may shift from one route to another. Typically, the design process, while evaluating various designs, assumes that the demand for the mode remains constant and only intra-mode changes occur based on the quality of the routes. Even the way such changes in demand are taken into account during the demand assignment process is often influenced by considerations of reducing formulation complexity. Guihaire and Hao (2008) also highlight some aspects of this. In short, while a variety of algorithms exist for the strategic planning phases, the way demand is handled remains a weakness.

There are a large number of studies on modal split and these can be integrated into the network design process. Unfortunately, there is little work on the intra-mode choice process of users. For example, studies on why one traveler picks one route over another while going from their origin to destination has not been studied extensively.

In the rest of the section some literature covering the five phases are described. In the description, frequency setting and timetabling are discussed under the Schedule Development phase and the vehicle assignment and crew assignment are discussed under the Operational phase. Further, this description is not meant to be a comprehensive review of the literature but is aimed at indicating to the reader the kind of work that has gone on and where gaps exist. For a comprehensive review of the strategic planning phases one may refer to Guihaire and Hao (2008). For a review of the operational planning phases one may refer to Bunte and Kliewer (2009) and Steinzen et al. (2010).

Route Design Phase

In this phase, given the passenger demand, the underlying spatial characteristics (like the urban activity zones and road network for bus routes) and link travel time information, the task is to determine a set of routes along which transit units (buses or trains) will ply. A route consists of two terminal stops and a series of intermediate points that it touches. The ordered set of points from the starting stop to the end stop should be such that every link (joining two consecutive points) is meaningful, i.e., it should be possible to go from the first point of the pair to the second point of the pair without touching any of the other points in that route. (Sometimes when routes are constructed as a juxtaposition of links, which connect two points directly, it must be ensured that the start point of a subsequent link is the same as the end point of the previous link.)

Typically, there are constraints on the number of routes (since a service provider may not have the resources to provide even the minimum level of service on each of the routes if routes become numerous) in a given area and the maximum route length. The maximum route length requirement stems from one or more of the following reasons: i) restrictions on working hours of the crew, ii) energy related restrictions on the distance a vehicle can travel before refuelling or recharging, iii) the maximum time a vehicle can remain away from the depot due to servicing related restrictions, and iv) restrictions on minimum service levels in terms of frequency of operations.

Although individually a route may be good or bad depending on how well and how much of the demand it caters to, ideally, routes should not be evaluated in isolation. Rather, an entire set of routes serving a given geographical area (like a city) should be evaluated as a unit. This is so because routes compete as well as cooperate with each other to transport people in a given area. A good route set should i) satisfy most, if not all, of the existing demand, ii) require only a small percentage of the users to transfer in order to complete their journey, and iii) offer low travel time, including waiting time, in-vehicle time and transfer time (where applicable), to most of those who use it (Chakroborty and Dwivedi, 2002).

The problem of optimal route design is a discrete combinatorial problem with an objective that cannot be evaluated through a function. The difficulties posed by the nature of the problem have restricted its solution methodologies to primarily heuristics and meta-heuristics. Although there have been some attempts to look at the problem analytically (for example, Holroyd (1967), Byrne and Vuchic (1972), Byrne (1975), Guan et al. (2006)), most of these methods optimize some parameters like route spacing and route length for idealized networks.

On the other hand, a large amount of work has been done on developing routes using heuristic and meta-heuristic methods. The methods range from heuristics (for example, Sonntag (1977)) where routes are successively deleted from a starting set (often connecting each origin-destination pair in the demand matrix) to the use of meta-heuristics like genetic algorithms (for example, Chakroborty and Dwivedi (2002)), ant colony algorithms (for example, Yang et al. (2007) and Yu et al. (2012)), and recently particle swarm optimization (for example, Kechagiopoulos and Beligiannis (2014)). These different approaches use different objectives that aim to capture the goodness of a route set and use different procedures to ensure that routes are meaningful. For a reasonably recent and good review of the network design problem, one may refer to Guihare and Hao (2008).

Related to the route design problem is the issue of stopping policy and stop spacing. Transit units may be required to stop at all the stops or they may stop only when there is an alighting or boarding demand at a stop, or they may stop anywhere along the route where passengers may want to board or alight. In the last case there are no defined stop locations; some small-cabin para-transit systems follow this type of policy. On the other hand, high-speed rail rapid-transit facilities use all-stop stopping policy. Kikuchi and Vuchic (1982) present a simple analysis that relates demand on a line to the ideal stopping policy choice. Wirasinghe and Ghoneim (1981) also talk about stopping policy.

Stop spacing impacts the accessibility characteristics as well as in-vehicle travel time of a route. If the number of stops (per unit route length) is many, then the accessibility to the route improves but the in-vehicle travel time increases (due to frequent stoppages); the effect is just the opposite if the number of stops are few. Accessibility and in-vehicle travel time impact the demand for a route. More stops also increase the turn-around time of a transit unit thereby requiring the operator to deploy a greater number of vehicles for the same frequency of service. This increases the operator cost. Hence, selecting the number of stops optimally is important.

Surprisingly, as also pointed out by Ibeas et al. (2010), "less work has been published about optimal spacing and location of bus stops" although the problem has been studied from the early 1900s, especially in Europe (for a review of early literature in this area, refer to Vuchic and Newell (1968)). Vuchic and Newell (1968) look at the problem of interstation spacing where passengers are assumed to travel to a single destination. Later, Wirasinghe and Ghoneim (1981) relax this assumption but analyze only stop spacing on a single route. Kikuchi and Vuchic (1982) integrate the study of stopping policy and stop spacing. In most of the past work the impacts of stop spacing on demand for the route as well as for the mode have not been adequately addressed. Ibeas et al. (2010) present a heuristic method to develop optimal stop spacing for a network of routes while considering the effect of stop spacing on demand. Ibeas et al. (2010) also present a reasonable overview of the past work in this area. Some studies have looked at the problem of optimal stop spacing and frequency of operation jointly (for example, Medina et al. (2013)). In a somewhat different approach to the issue of stop locations in an area, Murray (2003) developed a procedure to optimally choose zones to be served by stops while keeping the total number of stops in the area constant. Murray (2003) views stops as access points to the public transport facility as a whole rather than to specific routes.

Schedule Development Phase

The route design phase is concerned with the spatial aspects of a PTS and aims at developing routes along which transit units will run and at what locations they will stop to pick up passengers or drop them off. The schedule development phase is concerned with the temporal aspects of the system. In this phase, given 1) the route network and stop spacing (since it impacts travel time on a route), 2) the demand (by different periods of the day, week, etc.), and 3) the fleet size, the task is to determine the timetable for each route at every stop. The objective is to ensure minimal waiting time for users while meeting the concerns of the operator in terms of fleet (and crew) deployment. The timetable can either be presented as arrival and departure times of transit units of various routes at a given stop or for a given route at all its stops.

Traditionally, often the schedule has been developed in two stages. First, the frequency of operations on various routes is to be determined and next the timetable is developed using these frequencies. But, as Guihare and Hao (2008) also point out, the frequency setting phase can be skipped. The frequency setting phase assumes importance in only those studies where the routes are developed along with the frequencies. Ideally, the spatial and temporal parts of the PTS design should be done concurrently since both impact the quality of service and ultimately the demand. The use of frequency as a representative of the temporal part of the system (rather than the timetable) makes the combined process more tractable. It may be pointed out that, as mentioned in Guihare and Hao (2008), some argue against such concurrent treatment since the spatial layout of routes and stops tend to be the more permanent features of PTS.

Lampkin and Saalmans (1967) in their classic paper concentrate on developing the routes and frequencies in a sequential process. Similar sequential processes for developing frequencies and routes were developed by others also. Ideally, however, the process should not be sequential. Hasselstrom (1979) and later Van Nes et al. (1988) were among the first to attempt a solution to the joint route and frequency determination problem. Van Nes et al. (1988) posed the problem as that of simultaneously determining routes and frequencies such that the number of direct trips is maximized for a given fleet size. A heuristic solution approach was prepared for this problem. In a more recent study Cipriani et al. (2012) present a heuristic and a parallel genetic algorithm to develop a set of routes with associated frequencies for an urban transit system. The interested reader may refer to Guihare and Hao (2008) and Farahani et al. (2013) for a good review of the current literature in this area.

Even though the frequency setting phase is important in certain cases, a PTS ultimately requires a timetable or time schedule. Timetabling becomes all the more important when headways are not small and the percentage of users transferring is not negligible. Over the years a lot of work has been carried out on various aspects of timetabling and the interested reader may refer to Guihare and Hao (2008) for a good review. Approaches to solving the timetabling problem include mixed integer programming (for example, Chakroborty et al. (1995, 1998) and Wong et al. (2008)) and various heuristic and meta-heuristic methods (for example, Chakroborty et al. (1995, 1998, 2001)).

Operational Phase

At the end of the earlier phases the transit network gets its spatial and temporal skeleton. The transit service at this stage has a set of routes and the times at which these routes will be active or usable. In order to operationalize the service so that it can actually move people and goods, transit units or vehicles need to operate on the routes as per the determined schedule. In order for this to happen vehicles (such as buses or rakes) and crew members (such as drivers, ticket collectors, etc.) need to be assigned to each and every trip (a trip is a run on a route). Often the vehicle scheduling (assignment) problem is solved first and is followed by the crew scheduling problem.

Any scheduling (assignment) problem gives rise to an optimization problem that attempts to minimize the cost to the operators and users while satisfying all the constraints posed by the problem. In this case, since all the user-related requirements are taken care of in the earlier two phases and through constraints the objective function typically reflects only the operator cost. The cost is generally a function of the number of vehicles used, the total distance these vehicles travel over and above the distance they cover while servicing the routes (often referred to as deadheading distance), etc. The constraints depend on the kind of transit system being designed. However, there are certain constraints that exist in some form in any vehicle and crew scheduling problem. These are:

1. For each trip exactly one vehicle should be assigned.
2. The sequence of trips assigned to a vehicle should be feasible (i.e., it should be possible for the vehicle to be available at the origin node of every trip before the start of that trip).

3. The crew requirement should be such that it is less than the maximum number available (for each crew type) and meets the maximum duty time (distance) requirements between every pair of relief points for each crew type and various other labour laws.

Often an additional requirement that every vehicle return to its original depot is stipulated. Although the problem of vehicle and crew scheduling has been studied extensively, the issue of maintenance activities being scheduled together with trips has not received adequate attention. There are various transit systems where vehicles go through a fixed servicing/maintenance schedule. Only a few yards exist where such activities are possible. Generally, these yards are also depots and/or starting or ending stops of trips; but not all depots or stops are maintenance yards. In these cases the vehicles need to be scheduled to trips in such a way that they get enough lay-off time at one of the possible maintenance yards while ensuring that the assignment is optimum. The method of using maximum route length constraints (see Bunte and Kliewer (2009) and Li (2013)) in order to handle maintenance is not adequate for such systems. Vaidyanathan et al. (2008) realizes this issue in a locomotive assignment problem and proposes a sequential method.

Before leaving this section it must be pointed out that a large body of literature exists in this area. Large numbers of modeling approaches, including single and multi-commodity models, set-partitioning models and time-space network models, have been attempted. The interested reader should refer to Bunte and Kliewer (2009) and Li (2013). "The crew scheduling problem is usually formulated as a set-partitioning or covering problem" (Steinzen et al. 2010). The first instance of viewing the crew and vehicle assignment problems in a combined manner is by Freling (1997). For a recent study on the combined problem one may refer to Steinzen et al. (2010).

Economics of Public Transportation

Public transportation systems require huge capital investment and therefore, the economics of such systems assume importance in its planning. One of the major differences of public transit, compared to other service commodities, is the inverse relationship between its demand and cost; as more users choose public transit, the average user cost decreases. At the same time, an increased user base enables the service providers to make improvements at minimal additional cost to the users. These improvements could include higher frequency (hence reduced delay), improved quality of buses, comfortable bus shelters, improved information system, efficient integration of various components of the system, just to name a few. This section discusses the following related issues: 1) the benefits of PTS, 2) the costs of PTS, and 3) the pricing of PTS.

Benefits of PTS

Estimating the benefits from a PTS is a difficult task and as Litman (2014) points out, most existing studies tend to overlook or underestimate several indirect benefits. The common benefits that an efficient PTS brings include mobility options for all, improved convenience and comfort for users, equity benefits, improved operating efficiency of the overall transportation systems, improved security, mobility benefits to the physically challenged, the elderly, young travellers, and those without personal vehicles, and increased public fitness and health due to reduction in air and noise pollution. One of the benefits, which is often ignored is the equity benefits. Since many users (especially the economically weaker sections of the society, elderly or young people) do not have any other travel options, it is important to give sufficient weight to such benefits. Choice riders (or potential future users) also benefit from PTS through reduced traffic congestion, savings in vehicle ownership, infrastructure and parking facility costs, and improved traffic safety. Double-counting benefit is another common mistake that many economists do. Therefore, it also needs to be avoided.

Costs of PTS

There are various elements to the PTS costs. Some of them are briefly discussed here.

1. *Capital cost*: It includes the cost of creating the infrastructure including land procurement.
2. *Operating and maintenance cost*: It includes costs associated with labor, fuel, and vehicle, line, and yard maintenance costs.
3. *Travel time cost*: It includes access time, egress time, waiting time, riding time, and time taken during transfers. These times are converted to monetary values; during the conversion the unit cost of the riding time is typically taken as the lowest followed by the waiting and access/egress time; transfer time is generally assigned the highest value.
4. *Fare*: It is an out-of-pocket cost that often influences user behavior significantly.
5. *Government subsidies*: Since most of the PTS cannot sustain its service by only depending on the revenue generated from users, the government offsets actual costs with subsidies of various kinds.

The first four costs are direct costs; the first two are borne by the operator and the next two are borne by the user. The last cost is passed on to the taxpayer.

Pricing

Pricing is one of the most critical elements of the public transit system, which defines its success and failure. A higher fare would enable a better performing transit system with low societal costs. However, it has a negative relationship with ridership, making pricing a complicated problem. Identifying an efficient pricing scheme is a challenging task and requires a fine balance between ridership and performance of the service. Willingness-to-pay is one of the methods used to address this issue.

Willingness-to-pay (WTP) towards improving the quality of transit service is an important policy tool employed to understand the user perspective on an existing or proposed feature of the system. Although there are several ways to determine WTP, the methods that are commonly used are: bidding game, payment card, discrete choice, and discrete choice with follow-up (for example Kroes and Sheldon (1988) and Dissanayake et al. (2008)). Examples of studies to understand how much users are willing to pay for improvements in travel time, information systems, crowding, reliability, and safety may be found in studies by Li et al. (2010), Chaturabong et al. (2011), Ibeas et al. (2011), and Tirachini et al. (2013).

Quality of Service

Quality of service plays a critical role in the success of a PTS. As the PTS quality improves, it is expected to attract new users and retain the existing users even after they have other choices. However, like the assessment of quality of any service, evaluation of the quality of service of a PTS is a tricky task. Since a PTS has several stakeholders, quality of service definitions could vary for each of these groups, such as users, operators, motorists, and decision makers. However, in this section, the focus is mainly on users.

While some of the existing studies have used indicators such as travel time, waiting time, and frequency of service, many others have used more qualitative indicators such as cleanliness, comfort, and value of information. These indicators are evaluated to assess the quality of service of the system considered. Most of the commonly used parameters can be grouped into four major categories:

- travel time related;
- service related;
- customer service related;
- safety and security related parameters.

Travel time-related parameters are the most critical ones that impact the quality of service of a PTS. Although some other alternatives exist, while evaluating service quality, converting various travel time-related parameters to a corresponding in-vehicle value of time (VoT) is the most common practice followed (Concas and Kolpakov (2009; Balcombe (2004) and Wardman (2004)). Unit rates of bus VoT are typically higher than that in rail mode because of the perception that the riding quality of buses is inferior to that of rail (Abrantes and Wardman (2011); Fosgerau et al. (2010)).

Service-related parameters include adherence to schedule, travel comfort, passenger environment, access facilities, information systems, parking facilities, shuttle and feeder services. Similarly, bus stop amenities such as better shelter, improved lighting and comfortable seats induce positive VoT values (Balcombe (2004)). Studies show that for rail-based PTS, the walk-access distances of more than 0.8 km were acceptable to many users as compared to 0.4 km for bus based PTS (O'Sullivan and Morrall (1996) and Coffel et al. (2012)). In cities with reasonable bicycle trips, bicycle access trips are the most for access trips between 1.6 and 2 km and in these cases, the parking and on-vehicle bicycle carrying facilities play an important role in defining quality of service (Park and Kang (2008)).

The main parameters related to customer services are cleanliness, crowding, travel comfort, ease of ticketing, and amenities. A consolidated ticketing system improves the service quality of the participating public transit systems. Amenities at a transit stop that impact quality of service include seating, shelters, lighting, information, and vending machines.

Safety and security of public transportation are extremely important factors in deciding how the user chooses a mode of transport. While safety includes vehicle accident risks and passenger accident risks, cities with high pedestrian related crashes could improve quality of service by initiating efforts to improve safety of its users, who are predominantly pedestrians. Security can be improved by making the transit system including the stops well lit and making several emergency phones available to the users.

Although quality of service from the users' perspective is discussed in this section, it is important to note that other stakeholders, such as operators, motorists and decision makers, have different parameters to evaluate quality of service. These are discussed in detail in TCRP Report 165 (TRB (2014)).

Assessment of Quality of Service

Quality assessment from the users' perception is used for identifying 1) the users' level of satisfaction, and 2) the relative importance of service attributes to the users. Level of Service (LOS) is generally used to measure users' perception about a service. Mfinanga and Ochieng (2006) introduced LOS scores. In this concept, service levels of different public transport service attributes are converted to an equivalent travel time. Another team of researchers (Correia et al., 2008) developed a method based on the users' perception of actual service levels. In this method, different levels of LOS are predefined and responders, through random on-board surveys, are asked a few questions to identify if the existing definition needs modification or not. It is important to note that although LOS is ideal from the users' perspective, it could mean a wasteful use of resources from the operator's perspective. LOS definitions vary among different services and modes, and these LOS definitions cannot be transferred between them. For example, the definitions of LOS for fixed-route, fixed-schedule service would be very different from that of a demand-responsive transit service. Some recent efforts (TRB (2014)) consider user perspectives in developing LOS scores. Along with the direct quality determinants of a public transportation system such as availability, accessibility, on-time performance, information, customer care, comfort, safety and security, it is important to consider other measures such as transit stop quality, pedestrian facilities including safe crossing facilities, bicycle facilities, and access-related facilities also before finalizing quality of service.

References

Abrantes, P.A.L. and Wardman, M.R. (2011). "Meta-analysis of UK values of travel time: An update." *Transportation Research Part A*, Vol. 45, pp. 1–17.

Balcombe, R. (2004). "The demand for transport: A practical guide." TRL Report 593, TRL, Wokingham. Accessed on April 4, 2014 at www.demandforpublictransport.co.uk/TRL593.pdf

Bunte, S. and Kliewer, N. (2009). "An overview on vehicle scheduling models." *Public Transportation*, Vol. 1, pp. 299–317.

Byrne, B.F. (1975). "Public transportation line positions and headways for minimum user and system cost in a radial case." *Transportation Research*, Vol. 9, pp. 97–102.

Byrne, B.F. and Vuchic, V.R. (1972). "Public transportation line positions and headways for minimum cost." Proceedings of the Fifth International Symposium on Traffic Flow and Transportation, New York, pp. 347–360.

Ceder, A. and Wilson, N.H.M. (1986). "Bus network design." *Transportation Research Part B*, Vol. 20, pp. 331–344.

Chakroborty, P. (2003). "Genetic algorithms for optimal urban transit network design." *Computer-Aided Civil and Infrastructure Engineering*, Vol. 18, pp. 184–200.

Chakroborty, P. and Dwivedi, T. (2002). "Optimal route network design for transit systems using genetic algorithms." *Engineering Optimization*, Vol. 34, pp. 83–100.

Chakroborty, P., Deb, K., and Sharma, R.K. (2001). "Optimal fleet size distribution and scheduling of urban transit systems using genetic algorithms." *Transportation Planning and Technology*, Vol. 24, pp. 209–226.

Chakroborty, P., Deb, K., and Srinivas, B. (1998). "Network-wide optimal scheduling of transit systems using genetic algorithms." *Computer-Aided Civil and Infrastructure Engineering*, Vol. 13, pp. 363–376.

Chakroborty, P., Deb, K., and Subrahmanyam, P.S. (1995). "Optimal scheduling of urban transit systems using genetic algorithms." *Journal of Transportation Engineering*, Vol. 121, pp. 544–553.

Chaturabong, P., Kanitpong, K., and Jiwattanakulpaisarn, P. (2011). "Analysis of cost of motorcycle accidents on Thailand by willingness-to-pay method." Transportation Research Records, *Journal of Transportation Research Board*, Issue 2239, pp. 56–63.

Cipriani, E., Gori, S., and Petrelli, M. (2012). "Transit network design: A procedure and an application to a large urban area." *Transportation Research Part C*, Vol. 20, pp. 3–14.

Coffel, K., Parks, J., Semler, C., Ryus, P., Sampson, D., Kachadoorian, C., Levinson, H.S., Schofer, J. (2012). "Guidelines for providing access to public transportation stations." TCRP Report 153, Transportation Research Board, Washington, DC.

Concas, S. and Kolpakov, A. (2009). "Synthesis of research on value of time and value of reliability." Report BD54946. Florida Department of Transportation, Tallahassee, FL. Accessed March 20, 2014 at www.nctr.usf.edu/pdf/77806.pdf

Correia, A.R., Wirasinghe, S.C., and A.G., Barros (2008). "Overall level of service measures for airport passenger terminals." *Transport Research Part A*, Elsevier Publications, Vol. 42, pp. 330–346.

Dissanayake, D., Umar, R.S., Faudzi, M., Bentotage, S., Widyastuti, H., and Herijanto, W. (2008). "Inter-regional valuation of road casualties and exploration of social attributes to road casualty reduction in Southeast Asian countries, Thailand, Malaysia and Indonesia." Indonesia, International Cooperative Research Activity.

Farahani, R.Z., Miandoabchi, E., Szeto, W.Y., and Rashidi, H. (2013). "A review of urban transportation network design problems." *European Journal of Operations Research*, Vol. 229, pp. 281–302.

Fosgerau, M., Hjorth, K., and Lyk-Jensen, S.V. (2010). "Between-mode-differences in the value of travel time: Self-selection or strategic behaviour?" *Transportation Research Part D: Transport and Environment*, Elsevier Publications, Vol. 15, No. 7, 2010, pp. 370–381.

Freling, R. (1997). "Models and techniques for integrating vehicle and crew scheduling." Unpublished doctoral dissertation, Erasmus University of Rotterdam, The Netherlands.

Guan, J.F., Yang, H., and Wirasinghe, S.C. (2006). "Simultaneous optimization of transit line configuration and passenger line assignment." *Transportation Research Part B*, Vol. 40, pp. 885–902.

Guihaire, V. and Hao, J-K. (2008). "Transit network design and scheduling: A global review." *Transportation Research Part A*, Vol. 42, pp. 1251–1273.

Hasselstrom, D. (1979). "A method for optimization of urban bus route networks." Technical Report. Volvo Transportation Systems. Goteborg.

Holroyd, E.M. (1967). "The optimal bus service: A theoretical model for a large uniform urban area." Proceedings of the Third International Symposium on the Theory of Traffic Flow, New York, pp. 308–328.

Ibeas, A., Cecin, P., and dell'Olio, F. (2011). "Willingness to pay for improving service quality in multimodal area." *Transportation Research Part C*, Elsevier Publications, Vol. 19(6), pp. 1060–1070.

Ibeas, A., dell'Olio, L., Alonso, B., and Sainz, O. (2010). "Optimizing bus stop spacing in urban areas." *Transportation Research Part E*, Vol. 46, pp. 446–458.

Kechagiopoulos, P.N. and Beligiannis, G.N. (2014). "Solving the urban transit routing problem using a particle swarm optimization based algorithm." *Applied Soft Computing*, Vol. 21, pp. 654–676.

Kikuchi, S. and Vuchic, V.R. (1982). "Transit vehicle stopping regimes and spacings." *Transportation Science*, Vol. 16, pp. 311–331.

Kroes, E.P. and Sheldon, R.J. (1988). "Stated preference method: An introduction." *Journal of Transportation Economics and Policy*, Vol. 22, pp. 11–25.

Krygsman, S., Dijst, M., and Arentze, T. (2004). "Multimodal public transport: An analysis of travel time elements and the interconnectivity ratio." *Transport Policy*, Vol. 11, pp. 265–275.

Lampkin, W. and Saalmans, P.D. (1967). "The design of routes, service frequencies, and schedules for a municipal bus undertaking: A case study." *Operation Research Quarterly*, Vol. 18, pp. 375–397.

Li, J.-Q. (2013). "Transit bus scheduling with limited Energy." *Transportation Science* (article in advance).

Li, Z., Hensher, D.A., and Rose, J.M. (2010). "Willingness to pay for travel time reliability in passenger transport: A review and some new empirical Evidence." *Transportation Research Part E*, Elsevier Publications, Vol. 46, pp. 384–403.

Litman, T. (2014), "Evaluating public transit benefits and costs—best practices guidebook." Victoria Transport Policy Institute, Victoria, BC.

Medina, M., Giesen, R., and Muñoz, J.C. (2013). "Model for the optimal location of bus stops and its application to a public transport corridor in Santiago, Chile." Transportation Research Record No. 2352, pp. 84–93.

Mfinanga, D.A. and Ochieng, M.O.A. (2006). "Development of a model for assessing urban public transport level of service in cities of developing nations." *African Journal of Science and Technology*, Science and Engineering Series, Vol. 7, pp. 33–35.

Mishra, S., Welch, T.F., and Jha, M.K. (2012). "Performance indicators for public transit connectivity in multi-modal transportation networks." *Transportation Research Part A*, Vol. 46, pp. 1066—1085.

Murray, A.T. (2003). "A coverage model for improving public transit system accessibility and expanding access." *Annals of Operations Research*, Vol. 123, pp. 143–156.

O'Sullivan, S. and Morrall, J. (1996). "Walking distances to and from light-rail transit stations." Transportation Research Record, *Journal of the Transportation Research Board*, Vol. 1538, pp. 19–26.

Park, S. and Kang, J. (2008). "Factors that influence walking and biking to the station: Modeling commuter rail user's access mode choice." Proceedings of the TRB 87th Annual Meeting, Transportation Research Board, Washington, DC.

Phanikumar, P., Parida, M., and Swami, M. (2013). "Performance evaluation of multimodal transportation systems." *Procedia—Social and Behavioral Sciences*, Vol. 104, pp. 795—804.

Sonntag, H. (1977). "Linienplanung im ffentlichen personennahverkehr." Ph.D. thesis, Technische Universitt, Berlin.

Steinzen, I., Gintner, V., Suhl, L., and Kliewer, N. (2010). "A time-space network approach for the integrated vehicle- and crew-scheduling problem with multiple depots." *Transportation Science*, Vol. 44, pp. 367–382.

Tirachini, A., Hensher, D.A., and Rose, J.M. (2013). "Crowding in public transport systems: Effects on users, operation and implication for the estimation of demand." *Transportation Research Part A*, Elsevier Publications, Vol.53, pp. 36–52.

TRB (2014). "Transit capacity and quality of service manual." TCRP Report 163, 3rd ed., Transportation Research Board, Washington, DC.

Vaidyanathan, B., Ahuja, R.K., and Orlin, J.B. (2008). "The locomotive routing problem." *Transportation Science*, Vol. 42, pp. 492–507.

Van Nes, R., Hamerslag, R., and Immers, B.H. (1988). "Design of public transport networks." Transportation Research Record No. 1202, pp. 74–83.

Vuchic, V.R. and Newell, G.F. (1968). "Rapid transit interstation spacings for minimum travel time." *Transportation Science*, 2(4), pp. 303–339.

Vuchic, V.R. (1981). "Urban public transportation: Systems and technology." Prentice-Hall, NJ.

Wardman, M. (2004). "Public transport values of time." *Transport Policy*, Elsevier Publications, Vol. 11, pp. 363–377.

Wirasinghe, S.C. and Ghoneim, N.S. (1981). "Spacing of bus stop for many to many travel demand." *Transportation Science*, Vol. 15, pp. 210–221.

Wong, R.C.W., Yuen, T.W.Y., Fung, K.W., and Leung, J.M.Y. (2008). "Optimizing timetable synchronization for rail mass transit." *Transportation Science*, Vol. 42, pp. 57–69.

Yang, Z., Yu, B. and Cheng, C. (2007). "A parallel ant colony algorithm for bus network optimization." *Computer Aided Civil and Environmental Engineering*, Vol. 22, pp. 44–55.

Yu, B., Yang, Z.-Z., Jin, P.-H., Wu, S.-H., and Yao, B.-Z. (2012). "Transit route network design-maximizing direct and transfer demand density." *Transportation Research Part C*, Vol. 22, pp. 58–75.

16

VEHICLE ROUTING AND SCHEDULING

Xingyin Wang,[1] *Maria Battarra,*[2] *Bruce Golden,*[1] *and Edward Wasil*[3]

(1) UNIVERSITY OF MARYLAND, (2) UNIVERSITY OF BATH, (3) AMERICAN UNIVERSITY

Introduction

The Vehicle Routing Problem (VRP) was introduced by Dantzig and Ramser (1959) as the truck dispatching problem. They presented the problem of distributing gasoline from a bulk terminal to a set of service stations. The VRP, as defined by Dantzig and Ramser (1959), modeled a distribution problem where goods had to be delivered from a central depot to a set of customers or demand points. The VRP differs from the Traveling Salesman Problem (TSP) (Applegate et al., 2006) because a fleet of vehicles is stationed at the depot. Each vehicle can deliver goods to customers, but the goods carried by a vehicle cannot exceed its capacity. The VRP requires that we generate a route for each vehicle (a closed circuit starting and ending at the depot) such that the demand of each customer is satisfied and the transportation costs are minimized.

This pioneering work of Dantzig and Ramser introduced a problem of much larger scope and impact than the gasoline dispatching application. The truck dispatching problem modeled, with an acceptable degree of accuracy at that time, one of the most crucial aspects in distribution. Clearly, effective transportation planning is a key to success for a business or organization. Delivering goods in time while minimizing costs is critically important. Therefore, the potential savings due to effective VRP solvers has motivated academics and practitioners to develop high-quality transportation planning algorithms. Over the last five decades, major efforts from academia and industry have provided powerful algorithms for the basic VRP and its variants.

The prominence of the VRP in the field of operations research is due to the practical relevance of the problem and its complex combinatorial structure. The VRP is a generalization of the TSP. When the fleet is composed of a single vehicle, the VRP reduces to a TSP. The VRP is closely related to the bin packing problem (see Martello and Toth (1990)). When no distribution costs are incurred, the VRP reduces to the problem of assigning customers to vehicles so that the customer demands do not exceed the vehicle capacity. Clearly, the VRP is a challenging problem with its complexity following from two hard combinatorial problems.

The first attempts to solve the VRP used construction heuristics. Clarke and Wright (1964) developed one of the first greedy heuristics for the VRP. They manually solved instances with 13 and 30 customers. Due to its performance, simplicity, and flexibility, the Clarke and Wright algorithm is one of the most popular and successful VRP construction heuristics. Gaskell (1967) extended the Clarke and Wright algorithm to construct all routes sequentially or in parallel and considered alternative criteria in a more complex heuristic.

In the first 10 years after the publication of the article by Dantzig and Ramser (1959), very few articles were devoted to the VRP. However, there were three important contributions made in 1969. Christofides and Eilon (1969) proposed the first VRP branch-and-bound algorithm (solving instances with up to 13 customers exactly). They compared the Clarke-and-Wright heuristic to the 3-optimal tour method. In addition, three new instances were introduced with up to 100 customers. Newton and Thomas (1969) solved a school bus routing problem by solving the TSP over all customers and then partitioning the tour into a set of feasible routes. Pierce (1969) surveyed the literature on truck dispatching problems and referred to both the TSP and VRP.

It is interesting to point out that the authors of articles on the VRP published in the 1960s seemed well aware of the extensions and broad applicability of the VRP. For example, Hausman and Gilmour (1967) described an application of the VRP to oil and fuel distribution over an extended time horizon. Schruben and Clifton (1968) emphasized that routing costs could be reduced by 8 percent to 12 percent in the farming industry using heuristic solution methods instead of manual methods. They highlighted the possible savings achieved by combining delivery and pickup services, the fact that only trucks with special equipment or special characteristics may be able to serve certain customers, and the need to visit customers within time windows.

Beginning in the mid-1970s, the number of articles on the VRP that were published in academic journals increased dramatically. In Figure 16.1, we give the number of articles published on vehicle routing in four journals—Journal of the Operational Research Society (JORS), Networks, Computers & Operations Research (C&OR), and European Journal of Operational Research (EJOR). Vehicle routing researchers from North America and Europe have published extensively in these journals. The attention of researchers and practitioners toward this class of problems has tended to increase over the years and these four journals have been the primary outlets for publishing work on the VRP.

The methodological milestones achieved in the academic world up to the late 1990s have been surveyed in the book edited by Toth and Vigo (2002). Exact algorithms based on branch-and-cut schemes were able to solve problems with up to 100 customers due to the dynamic separation of effective cuts, and they outperformed branch-and-bound schemes. A growing interest in column generation algorithms for the VRP with time windows was also reported in Toth and Vigo (2002). In addition, construction heuristics have been widely studied and their performance has been carefully assessed. Interest in heuristic methods has been directed toward metaheuristic algorithms that use systematic techniques to escape local optima including simulated annealing, tabu search, and genetic algorithms.

Finally, key features of real-world VRPs are identified in Toth and Vigo (2002). Problems with time windows, backhauls, pickups and deliveries, site dependencies, and route duration constraints are considered and solved using a variety of methodologies. It was pointed out that the gap between the types of VRPs modeled in academia and the types of VRPs encountered in practice, where many additional features are needed to model realistic applications, seemed to be quite large.

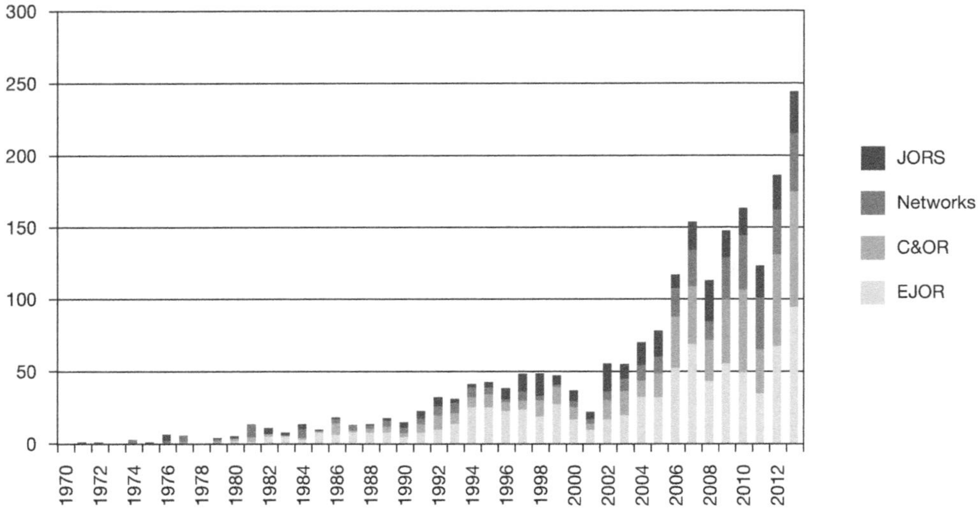

Figure 16.1 Number of articles published in *JORS*, *Networks*, *C&OR*, and *EJOR* from 1970 to 2013

The book by Golden et al. (2008) provided an indication of the rapid progress of research and practice in the VRP community. Faster and cheaper computers, new technologies (e.g., GIS and GPS, RFID), more demanding transportation regulations (e.g., driving time regulations) and a growing competition among quick delivery services (e.g., online shopping and home-delivery services) extended the boundaries of the VRP applications even further.

The methodologies that were used to solve the VRP became more effective and, at the same time, the variants of the VRP were greatly enriched. Real-world characteristics of distribution problems were now being considered by academics and algorithms were being developed to effectively solve more realistic routing variants.

Today, exact algorithms based on branch-and-cut-and-price are capable of solving instances with as many as 200 customers to optimality. Metaheuristic frameworks can solve complex vehicle routing variants to near-optimality in minutes of computing time. New and innovative variants have been proposed recently and the VRP community is now providing ways to solve these challenging distribution problems.

In addition to the books by Toth and Vigo (2002) and Golden et al. (2008), there have been other, influential book-length surveys (compilations) on vehicle routing published since 1971. We refer the reader to Table 16.1.

Although there are many state-of-the-art algorithms that have been developed by the academic community, relatively few have been implemented by real-world distribution companies. Knowledge transfer has been limited and opportunities for discussion between industry and academia have been scarce. Today, there is a communication void to be filled. Academics and practitioners would benefit from regular, open discussions. Practitioners and commercial software developers could improve their solution methods, while academics could be more responsive to new, real-world routing features and constraints. We believe that a common terminology and a clear and concise overview of the state-of-the-art on the VRP would facilitate communication between industry and academia.

Table 16.1 Book-length surveys on vehicle routing

Title	*Reference*
Distribution Management: Mathematical Modelling and Practical Analysis	Eilon, Watson-Gandy, and Christofides (1971)
Routing and Scheduling of Vehicles and Crews: The State of the Art	Bodin, Golden, Assad, and Ball (1983)
Vehicle Routing: Methods and Studies	Golden and Assad (1988)
Fleet Management and Logistics	Crainic and Laporte (1998)
Arc Routing: Theory, Solutions, and Applications	Dror (2000)
The Vehicle Routing Problem	Toth and Vigo (2002)
The Vehicle Routing Problem: Latest Advances and New Challenges	Golden, Raghavan, and Wasil (2008)
Arc Routing: Problems, Methods, and Applications	Corberán and Laporte (2014)
Vehicle Routing: Problems, Methods, and Applications	Toth and Vigo (2014)

This chapter aims to complement the existing literature on the VRP. It is our goal to help bridge the gap between academics and practitioners, so that practitioners and software developers might more easily access the latest VRP algorithms developed by academics.

We start by classifying VRP variants (i.e., extensions of the VRP that model real-world features) and by providing a few key references for each variant. This should allow practitioners to identify real-world features using the same terminology found in the academic literature and to have up-to-date references to the relevant published articles.

We highlight recent trends in the academic literature including new variants that have been proposed and studied, as well as algorithms that provide the best results. We present examples of commercial and open source VRP software, discuss their features, and mention characteristics of good VRP software. The latter discussion takes into account algorithmic aspects and marketing considerations.

Problem Classification

The basic capacitated VRP is modeled on a symmetric graph (i.e., the distance between any two vertices in the network is the same in both directions), all vehicles have the same capacity, and there is no maximum duration restriction on the length of a route.

In the last 10 years, effective exact and metaheuristic algorithms have been proposed in the literature. Trends can be identified among the techniques that proved to be the most effective. We briefly describe these techniques and highlight the main characteristics.

Among exact procedures for the VRP, integer linear programming algorithms based on branch-and-cut-and-price are the state-of-the-art methods. Typically, in these methods, the problem is modeled using a set partitioning formulation (i.e., a variable assumes the value one if the corresponding route is traversed by one of the vehicles and zero otherwise). Sophisticated separation algorithms dynamically generate those routes capable of improving the solution quality. These algorithms are embedded in branch-and-bound frameworks and valid inequalities are typically added to the formulation in a dynamic fashion (see Lübbecke and Desrosiers (2005), Pessoa et al. (2008), and Poggi and Uchoa (2014) for more details). Currently, branch-and-cut-and-price algorithms are solving VRP instances to optimality with more than 200 customers in hours of computing time (Martinelli et al. (2014)). Heuristic algorithms are usually more flexible and can provide high-quality solutions in reasonable computing times. Heuristics are frequently used to solve real-world problems, even when commercial solvers based on branch-and-cut-and-price algorithms are available (an example is the software described in Ceselli et al. (2009)).

The effectiveness of metaheuristic techniques relies on a good balance between two techniques.

1 Intensification techniques. These are methods that aim at improving solution quality. Typically, local search techniques explore neighborhood solutions. The computational burden of local search can be mitigated by effective solution evaluation techniques, such as the concatenation method proposed in Kindervater and Savelsbergh (1997), using neighborhoods with limited complexity (the number of solutions explored is at most quadratic in the size of the problem), or exploring large neighborhoods in effective ways (see the survey of Ahuja et al. (2002) for very large neighborhood search techniques).
2 Diversification techniques. These techniques allow searches to escape from local optima. If intensification deeply explores a specific area of the search space (e.g., all moves in the neighborhood of an initial solution), diversification investigates alternative areas in the solution space. Tabu search algorithms accept non-improving and non-tabu moves (Cordeau and Maischberger (2012)). Iterative local search performs random moves in the solution space (Subramanian et al. (2013)). Population-based metaheuristics, such as genetic algorithms, path relinking algorithms, and scatter search, consider individuals in the population with different characteristics. Another diversification technique involves including infeasible solutions in the search (these solutions are usually penalized).

In the past 10 years, several metaheuristic frameworks have provided consistently high-quality results for many variants of the basic VRP including the tabu search heuristic of Cordeau et al. (1997), the adaptive large neighborhood search of Pisinger and Røpke (2007), the hybrid algorithm of Subramanian et al. (2013), and the hybrid genetic algorithm of Vidal et al. (2014).

Among VRP variants, we distinguish between well-studied VRP variants and newer problems that have emerged in the past 20 years. First, we comment on the well-studied variants.

1 Asymmetric VRP (AVRP). The distance between two vertices in the network varies depending on the travel direction. Asymmetric distances are more realistic for urban transportation applications, while distances and travel times could be approximated to be symmetric in long-haul transportation problems.

2 Open VRP (OVRP). Vehicles are not required to return to the depot at the end of their routes. The objective is to minimize the total routing cost, but variants, that minimize the number of vehicles, are also presented in the literature (Li et al. (1992)). The OVRP is suitable for applications in which an external distribution company is providing delivery services, so that the vehicles are not expected to return to the depot at the end of the work day.

3 Distance Constrained VRP (DVRP). In this variant, each route is bounded by a maximum duration (Almoustafa et al. (2013)). The DVRP is suitable to model the constraint on the maximum duration of a driver's work day.

4 Multi-Depot VRP (MDVRP). Multiple depots are available with a fixed number of vehicles assigned to each depot. A vehicle must end its route at the depot from which it started.

5 Heterogeneous VRP (HVRP). Vehicles in the fleet have different capacities and different costs. Fixed costs can be assigned when a vehicle is used. Costs can be proportional to the distance traveled and the type of vehicle used. An overview of HVRP variants is presented by Baldacci et al. (2008).

6 VRP with Time Windows (VRPTW). A customer expects a delivery or collection service within a specified time window. A time window is hard whenever a customer must be visited within the specified time slot. It is soft whenever it is preferred that a customer be visited within the time slot. In the latter case, customer dissatisfaction is typically modeled as a penalty term in the objective function.

7 Site-Dependent VRP (SDVRP). This is an HVRP in which each customer can be visited by only a subset of the vehicle types. The SDVRP is often studied in conjunction with time window constraints.

8 Pickup and Delivery VRP (PDVRP). Customers can ask for a pickup, a delivery, or both. Goods may originate from and be destined for the depot for all customers. Goods may be shipped from a specific customer to another customer. A generic commodity may be collected from any customer and delivered to any other customer. Additional constraints may require that all goods have to be delivered before there are pickups (backhaul problem) or allow the possibility of dropping goods at intermediate customers along the route (preemptive problems). Battarra et al. (2014) and Doerner and Salazar-Gonzalez (2014) provide overviews of the PDVRP.

9 Multi-Trip VRP (MTVRP). Vehicles are allowed to return to the depot multiple times during a work day to replenish or deliver their loads. The duration of the work day is bounded.

10 Multi-period VRP or Periodic VRP (PVRP). Customers may need to be visited multiple times within a time horizon and provide preferences among visiting patterns (e.g., delivery takes place on Monday and Wednesday or Tuesday and Friday).

The classical VRP variants are challenging combinatorial problems, but they do not always reveal the full complexity of real-world distribution problems. In the last 20 years, academics have proposed and studied new VRP variants motivated by practice. Now we provide a few examples of new variants (we point out that our intention here is not to be exhaustive).

11 VRPs considering driving regulations. Route planning considers real-world regulations about a driver's workload (e.g., breaks and maximum working day duration). Regulations vary from country to country. A metaheuristic algorithm has been proposed by Goel and Vidal (2014).

12 Green VRP. The objective involves minimizing gas emissions.

13 Time dependent and speed optimization VRP. In the future, VRP software may not only provide an ordered sequence of customers to visit (and corresponding paths), but also the best travel speed on each street. Routing costs can decrease if driver speed is optimized on each route segment. Recent articles solving VRPs over a congested network include Dabia et al. (2011) and Norstad et al. (2011) for a maritime routing application.

14 VRP with loading constraints. Route planning decisions can be integrated with vehicle load planning decisions. Algorithms would generate feasible vehicle routes and feasible loading patterns.

15 Split Delivery VRP. Vehicles may take advantage of visiting a customer multiple times.

16 Dynamic VRP, Stochastic VRP, and Robust VRP. Data that are necessary to define routing solutions are incomplete or uncertain. Data are revealed (or acquired) over time in dynamic routing problems where unplanned customers request a visit during the work day. In stochastic and robust problems, statistical knowledge about the data is available. In stochastic models, the decision maker is either considering expected realizations for the uncertain data or accepts risky solutions only with a controlled degree of risk. Robust solutions are achieved when the worst solution in an uncertainty set is considered where an uncertainty set contains likely values to be attained by the uncertain parameters.

17 Multi-objective VRP. There are multiple objectives such as minimize routing cost and minimize the number of vehicles. Typically, the objective function components are weighted according to their relative importance when weights can be easily estimated. More sophisticated techniques aim at defining the Pareto frontier, i.e., the set of solutions that are not dominated by any other in terms of any of the objectives. This can be useful in practice, because it provides alternative solutions in which at least one of the objective components is at its best. For example, Gulczynski et al. (2011) consider balancing the workload among drivers across routes in their study of the PVRP.

18 Two-echelon VRP. One of the techniques to reduce congestion in city centers involves placing logistic centers on the outskirts of large cities. Large trucks deliver goods from those centers while smaller vehicles deliver goods in the city center. Algorithms allocate goods to the logistic centers, route goods to the logistic centers, and then to final destinations.

19 Inventory routing. Customers do not place delivery orders, but the distribution company is responsible for replenishing each customer's inventory before it runs out of stock. This complex variant responds to the needs of a supply chain logistics system.

20 Prize-collecting VRPs. This problem models applications in which not all potential customers are served, but only a most convenient subset. Many variants have been proposed that quantify the trade-off between serving additional customers (thereby collecting the profit) and increasing the length of a route (thereby incurring higher routing costs).

21 Consistent VRP. In order to improve customer satisfaction, regular customers must be visited by the same driver at roughly the same time of day. These additional constraints

Table 16.2 Recent Survey Articles in Vehicle Routing

Topic	*Authors*	*Publication outlet*	*Year*
Vehicle routing in general	Marinakis and Migdalas	*Operational Research*	2007
	Laporte	*Naval Res. Log.*	2007
	Laporte	*Trans. Science*	2009
	Eksioglu et al.	*Computers & IE*	2009
Arc routing	Hertz	book chapter	2005
	Wøhlk	book chapter	2008
	Corberán and Prins	*Networks*	2010
Periodic vehicle routing	Campbell and Wilson	*Networks*	2014
Inventory routing	Andersson et al.	*Computers & OR*	2010
	Bertazzi and Speranza	*EURO J. Transport and Logistics*	2012
	Bertazzi and Speranza	*EURO J. Transport and Logistics*	2013
	Coelho et al.	*Trans. Science*	2014
Multi-objective vehicle routing	Jozefowiez et al.	*EJOR*	2008
Location-routing	Nagy and Salhi	*EJOR*	2007
	Prodhon and Prins	*EJOR*	2014
Green vehicle routing	Lin et al.	*Expert Systems with Applications*	2014
	Demir et al.	*EJOR*	2014
Split delivery	Archetti and Speranza	*ITOR*	2012
Synchronization in vehicle routing	Drexl	*Trans. Science*	2012
Static pickup and delivery	Berbeglia et al.	*TOP*	2007
Dynamic vehicle routing	Pillac et al.	*EJOR*	2013
Time windows	Bräysy and Gendreau	*Trans. Science*	2005
	Bräysy and Gendreau	*Trans. Science*	2005
School bus routing	Park and Kim	*EJOR*	2010
Prize/profit collecting VRPs	Laporte and Martin	*Networks*	2007
	Vansteenwegen et al.	*EJOR*	2011
Routing and loading	Iori and Martello	*Yugoslav J. of OR*	2013
Rich vehicle routing	Drexl	*Logistics Research*	2012
Exact algorithms	Baldacci et al.	*4OR*	2007
Local search	Funke et al.	*J. of Heuristics*	2005
Evolutionary algorithms	Potvin	*INFORMS JOC*	2009

increase the routing cost, but may simplify the loading and unloading operations, possibly reduce customer service time, and increase customer satisfaction.

22 Close-enough VRP. In this application, a vehicle is not required to visit each customer. It is sufficient to get close enough to each customer location. This variant has been proposed for applications like meter reading. In this setting, a driver is able to access data within a radius of 500 or so feet from the location of the meter. Getting close enough allows for significant routing cost savings and reduction in the service time at the customers.

23 Humanitarian relief routing. This problem has been proposed to model the delivery of emergency supplies to people affected by a disaster. This application differs from standard distribution problems in that the priority is to provide the fastest and most effective service to all people in need, rather than reducing routing costs. Fairness is taken into account, i.e., the level of guaranteed service should be as similar as possible among all those affected by the disaster. We refer the reader to Campbell et al. (2008) and Panchamgam et al. (2013).

Since 2000, there are many survey articles on specific vehicle routing topics that have been published. In Table 16.2, we list 31 survey articles.

VRP Research over Time

The vehicle routing problem continues to attract extensive academic research. Work on the vehicle routing problem with time windows, period vehicle routing, and the inventory routing problem has been ongoing since the 1980s (see Figures 16.2, 16.3, and 16.4). Research on green vehicle routing (see Figure 16.5) and humanitarian logistics and relief vehicle routing (see Figure 16.6) has only recently begun to emerge.

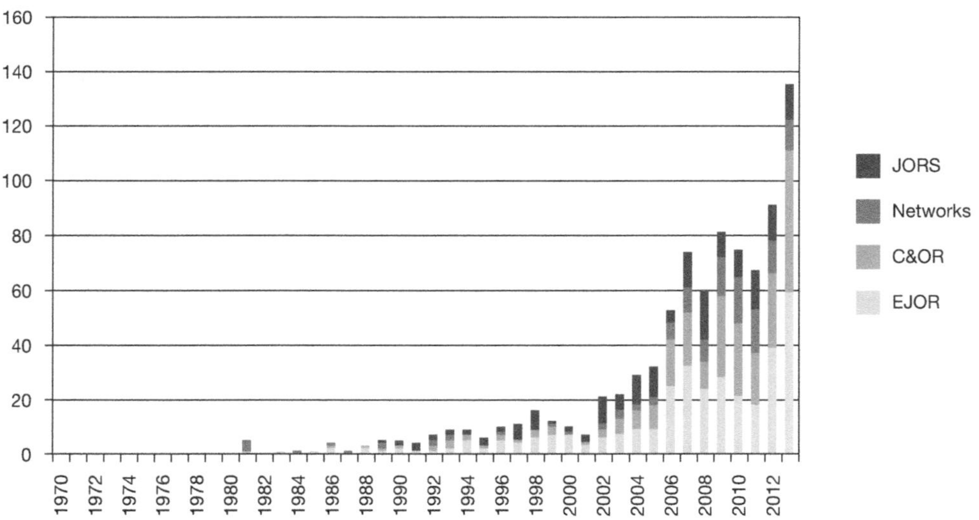

Figure 16.2 Articles on vehicle routing with time windows

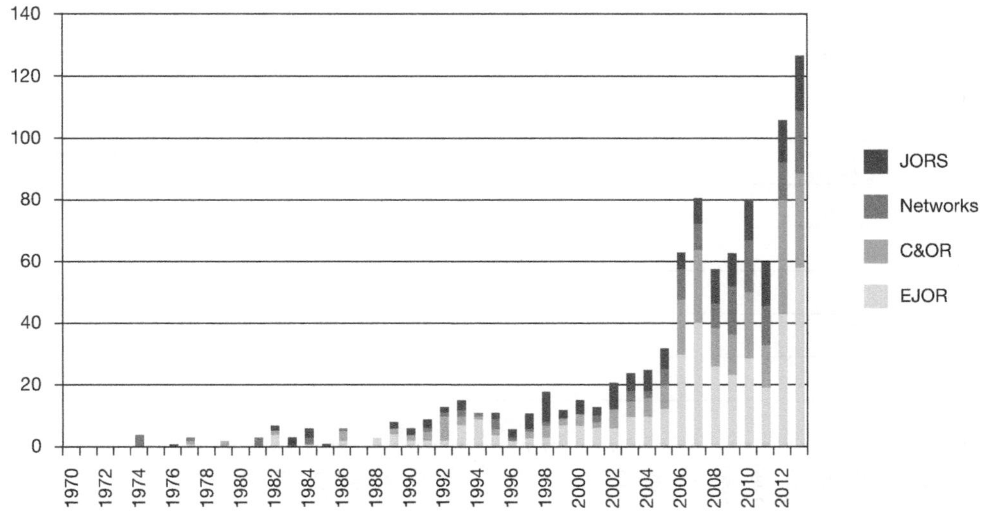

Figure 16.3 Articles on period vehicle routing

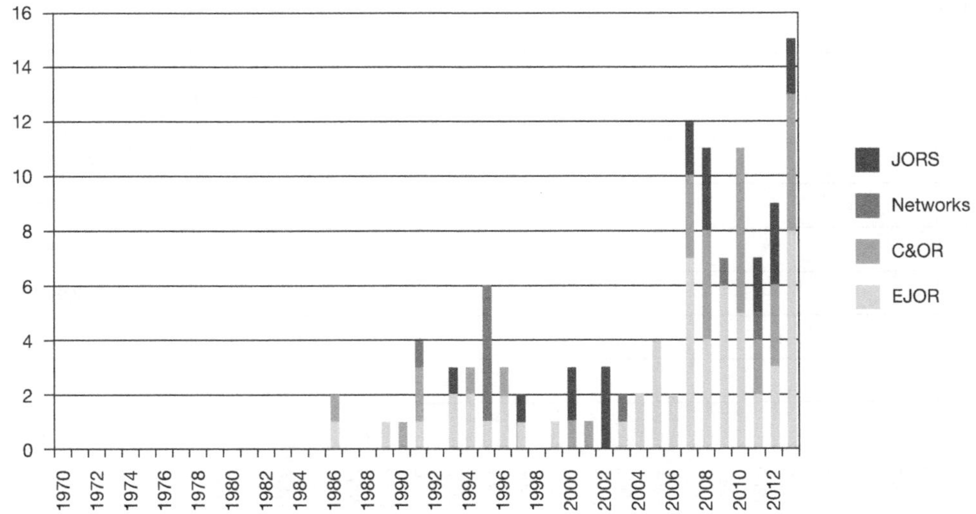

Figure 16.4 Articles on inventory routing

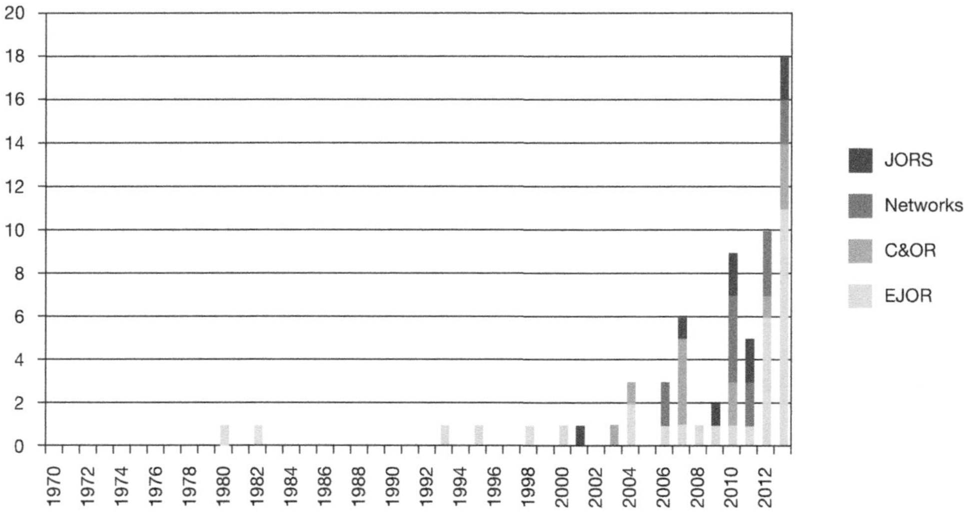

Figure 16.5 Articles on green vehicle routing

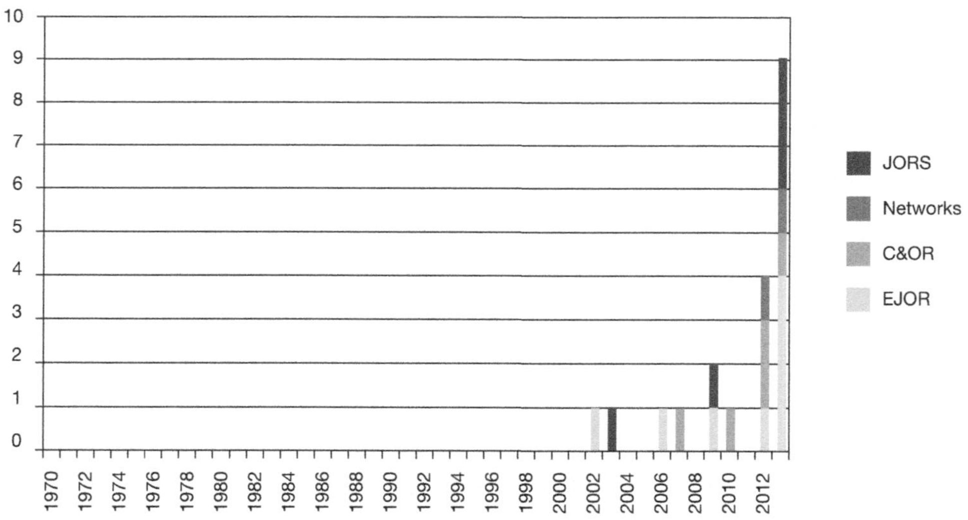

Figure 16.6 Articles on humanitarian logistics and relief vehicle routing

Commercial Vehicle Routing Software

Today, commercial VRP solvers are sophisticated software packages. A commercial solver can provide high-quality routing solutions for a large variety of problem variants. A solver needs to integrate with new devices and to interface easily and effectively with data visualization and manipulation tools. In this section, we focus on the characteristics that a VRP algorithm should have to be effective in VRP commercial software. We discuss additional components of a software package that are desirable.

In our discussion, we refer to general purpose commercial solvers and not to algorithms developed specifically for a real-world distribution problem. The latter type of algorithm is usually developed in the context of consultancy agreements, and provides a one-time response to the specific needs of a company. This software solution may not be easily extended or maintained and may lack in integration with other software or technological tools. The algorithms that are part of the software are high performing and respond specifically to the customer's needs.

Table 16.3 Commercial Software Companies

Vendor	*Year founded*	*Country*	*Website*
Descartes	1981	Canada	www.descartes.com
DNA Evolutions GmbH	Unknown	Germany	www.dna-evolutions.com
ESRI	1969	USA	www.esri.com
KKT srl	2011	Italy	www.routist.com
MapMechanics	1987	USA	www.truckstopsrouting.com
MJC2	1990	UK	www.mjc2.com/distribution-logistics-software.htm
Netakil	2011	Turkey	logvrp.com
Optrak Distribution	1988	UK	www.optrak.com
ORTEC	1981	Netherlands	www.ortec.com
Paragon Software Systems, Inc.	1991	UK	www.paragonrouting.com
Roadnet Technologies	1983	USA	www.roadnet.com
RouteSmart	1980	USA	www.routesmart.com
RouteSolutions	2004	USA	www.routesolutions.com
Runzheimer International	1933	USA	www.runzheimer.com
TMW Systems	1983	USA	appian.tmwsystems.com
WM Logistics	2000	USA	wmlogistics.wm.com

Many of the algorithms presented in the academic literature focus on the classical VRP or variants of the VRP with additional features. This allows academics to compare the performance of their algorithms on benchmark instances and better assess how the algorithm's components contribute to solve a specific problem variant. In contrast, commercial software needs to be sufficiently flexible to handle most of the real-world features arising in distribution problems.

Partyka and Hall (2014) identified crucial characteristics to assess a general purpose VRP software application. Some characteristics are related to the performance and requirements of an algorithm, such as solution quality, maximum number of customers, trucks, and depots that the software can handle, specific hardware, and speed of computation. The flexibility of the software is usually as important as the quality of the VRP algorithm. According to Partyka and Hall (2014), the main features that arise in routing practice are: modeling arc routing and node routing problems, solving periodic problems, considering pickup and delivery services, modeling incompatibilities between truck/driver and sites to be visited, incompatibilities between segments of the network and vehicles, taking into account driving regulations, using a heterogeneous fleet of vehicles, different types of services (e.g., taxis, buses, emergency services), and the presence of clusters of customers.

The authors specified the most desirable characteristics of the software as a package. VRP packages should easily integrate both data management software (e.g., spreadsheets and databases) and GIS maps. The software should be able to run on standard PCs (with different operating systems), tablets, and smartphones. It should be able to run online applications. The software may integrate with online data services and collect real-time traffic information to generate the most suitable routing solutions depending on congestion. A key output component of software is the set of instructions for each driver. Important considerations are the possibility of monitoring the real-time positions of trucks and the integration of information collected from bar code readers. These types of considerations become important in real-world applications.

Often, VRP software providers are asked to customize the package to a company's specific needs and provide technical support. The type of licensing policies, the package price, and the support are important relevant characteristics of VRP commercial software.

The survey by Partyka and Hall (2014) was based on interviews with 15 VRP software providers, who agreed to participate in a questionnaire about their software, their vision of the market directions, and future needs. In Table 16.3, we provide information about these plus one additional company. The picture arising from this survey is a market that continues to grow, though facing many challenges. Integration with devices and software, the need for effective routing solutions in minutes of computing time, and the possibility of dynamic re-optimization of the routing decisions with respect to traffic information or driver feedback seem to be priorities in the VRP industry. Customers are expecting punctual and fast deliveries within short time windows. They would like real-time information about deliveries. Clearly, the bar has been raised in the last few years for software providers. Commercial VRP packages must be integrated with sophisticated tools to be competitive. The market seems to be responding positively to the advances in the VRP industry. Customers are aware of their distribution environment and they know that good VRP software can provide substantial cost savings and improve customer satisfaction.

Free Vehicle Routing Software

In this section, we provide a list of free software and software libraries that are available to solve the VRP. We present information that is available with respect to the solvers. We do not compare performances of the software.

5.1 *SYMPHONY* . This solver aims at providing an exact solution to the capacitated VRP using a parallel branch, cut-and-price framework (Ralphs et al. (2003); Ralphs (2003)). The code is maintained at the COR@L Lab at Lehigh University and is released under the Eclipse Public License (version 5.4 and later). It compiles under all major C compilers. The source code (precompiled binaries for various platforms) and the user manual can be downloaded at projects.coin-or.org/SYMPHONY.

5.2 *CVRPSEP* . This is a collection of C/C++ routines that provide optimal solutions for the capacitated VRP. It implements a branch-and-cut algorithm in which the valid inequalities include rounded capacity inequalities and hypotour inequalities (Fukasawa et al. (2003); Lysgaard et al. (2004)). The software is released under the Common Public License and can be downloaded at www.hha.dk/lys/CVRPSEP.htm .

5.3 *VRPH* . An extensive well-documented C/C++ library of local search heuristics (2-opt, 3-opt, Oropt, one point move, two point move) for the capacitated VRP with and without time windows (Groër et al. (2010)). There are several solvers in this library that use an initial solution procedure (a variant of the Clarke-Wright algorithm (Yellow (1970)) or the sweep algorithm (Gillett and Miller (1974)) and one of the several metaheuristic frameworks including simulated annealing (Kirkpatrick et al. (1983)) and the record-to-record travel (Dueck (1993)). An application accepts a text file as input and outputs the solution as a text file. An interface with the PLPlot library produces colored plots. The library is compliant with all major compilers. It is hosted by Coin-OR and is released under the Common Public License (CPL). Additional information is available at sites.google.com/site/vrphlibrary. VRPH can be downloaded at coinor.org/download/source/VRPH .

5.4 *VRP Spreadsheet Solver.* This Excel workbook provides a unified platform to represent, solve, and visualize the results of VRPs. It solves the capacitated closed or open VRP with time windows using large neighborhood search (Shaw (1998); Pisinger and Røpke (2007)). It also solves the team orienteering problem with time windows. The solver can either accept distance and driving time data from manual input into the spreadsheet or retrieve data from Bing Maps based on street addresses. It is capable of displaying solutions using a scatter chart. The spreadsheet solver and a comprehensive user manual can be downloaded at verolog.deis.unibo.it/vrp-spreadsheet-solver.

5.5 *jsprit.* This open source toolkit is written in Java. It implements the ruin-and-recreate principle of Schrimpf et al. (2000) and targets several VRP variants (multi-depot VRP, VRP with time windows, backhauls, pickups and deliveries, heterogeneous fleet, dial-a-ride problem). It is hosted by github and released under the GNU Lesser General Public License. Several examples of building problem instances and reading VRP instances into text files can be viewed at github.com/jsprit/jsprit/wiki. The software has plotting capabilities as well. The software can be downloaded at github.com/jsprit/jsprit.

5.6 *Open-VRP* . This is a framework to solve VRPs (capacitated, time windows, with pickup and deliveries) written in Common Lisp's CLOS. The user can build solution procedures or adapt the tabu search algorithm found in the software. The program accepts problem instances via text file and can plot the solution. It is released under the Lisp Lesser GNU Public License and can be downloaded at github.com/mck-/Open-VRP.

5.7 *OptaPlanner* . This program targets general planning problems and has a VRP solver for the capacitated VRP and VRP with time windows. The program is well documented and has a user friendly graphical interface. It is written in Java and runs on any Java virtual machine. The program is released under the Apache Software License and can be downloaded at http://www.optaplanner.org .

5.8 *vroom*. This is a Java library for the VRP that includes jMSA, which is a framework for dynamic vehicle routing (Pillac et al. (2012)), and Technicians, which is a framework for the technical routing and scheduling problem (Pillac et al. (2013a)). It is released under GNU General Public License and vroom can be downloaded at victorpillac.com/vroom.

5.9 *Windows VRP Solver*. This is a stand-alone Windows application for the capacitated VRP. The software has the randomized Clarke-Wright algorithm and improvement heuristics including 2-opt, OR-opt, and swap. The solver accepts manual input or text files giving the coordinates and demands of the customers, and outputs the solution in both graphical and text formats. It can be downloaded at coral.ie.lehigh.edu/ larry/software/vrp-solver .

Conclusions

In this chapter, we provided a survey of the most common techniques, problem variants, and research trends for the vehicle routing problem. We described commercial and open source software for modeling and solving the VRP.

The VRP continues to be a popular area for research and practice. New variants motivated by challenging real-world applications and technological advances are posing important research questions. Collaboration among academics, practitioners, and software providers will be essential to successfully answering these questions. We hope that our survey helps to point them in the right direction.

References

Ahuja, R. K., Ergun, Ö., Orlin, J. B., and Punnen, A. P. (2002). A survey of very large-scale neighborhood search techniques. *Discrete Applied Mathematics* 123(1), 75–102.

Almoustafa, S., Hanafi, S., and Mladenovic, N. (2013). New exact method for large asymmetric distance-constrained vehicle routing problem. *European Journal of Operational Research* 226(3), 386–394.

Andersson, H., Hoff, A., Christiansen, M., Hasle, G., and Løkketangen, A. (2010). Industrial aspects and literature survey: Combined inventory management and routing. *Computers & Operations Research* 37(9), 1515–1536.

Applegate, D., Bixby, R., Chvatal, V., and Cook, W. (2006). *The Traveling Salesman Problem: A Computational Study*. Princeton University Press, Princeton.

Archetti, C., and Speranza, M. G. (2012). Vehicle routing problems with split deliveries. *International Transactions in Operational Research* 19(1–2), 3–22.

Baldacci, R., Battarra, M., and Vigo, D. (2008). Routing a heterogeneous fleet of vehicles. In: B. Golden, S. Raghavan, and E. Wasil (Eds.), *The Vehicle Routing Problem: Latest Advances and New Challenges*, Springer, New York, 3–27.

Baldacci, R., Toth, P., and Vigo, D. (2007). Recent advances in vehicle routing exact algorithms. *4OR* 5(4), 269–298.

Battarra, M., Cordeau, J.-F., and Iori, M. (2014). Pickup-and-delivery problems for goods transportation. In: P. Toth, and D. Vigo (Eds.), *Vehicle Routing: Problems, Methods, and Applications*. SIAM, Philadelphia, 161–191.

Berbeglia, G., Cordeau, J.-F., Gribkovskaia, I., and Laporte, G. (2007). Static pickup and delivery problems: A classification scheme and survey. *TOP* 15(1), 1–13.

Bertazzi, L., and Speranza, M. G. (2012). Inventory routing problems: An introduction. *EURO Journal on Transportation and Logistics* 1(4), 307–326.

Bertazzi, L., and Speranza, M. G. (2013). Inventory routing problems with multiple customers. *EURO Journal on Transportation and Logistics* 2(3), 255–275.

Bodin, L., Golden, B., Assad, A., and Ball, M. (1983). Routing and scheduling of vehicles and crews: The state of the art. *Computers & Operations Research* 10(1), 63–212.

Bräysy, O., and Gendreau, M. (2005). Vehicle routing problem with time windows, Part I: Route construction and local search algorithms. *Transportation Science* 39(1), 104–118.

Bräysy, O., and Gendreau, M. (2005). Vehicle routing problem with time windows, Part II: Metaheuristics. *Transportation Science* 39(1), 119–139.

Campbell, A., Vanderbussche, D., and Hermann, W. (2008). Routing for relief efforts. *Transportation Science* 42(2), 127–145.

Campbell, A., and Wilson, Jill. (2014). Forty years of periodic vehicle routing. *Networks* 63(1), 2–15.

Ceselli, A., Righini, G., and Salani, M. (2009). A column generation algorithm for a rich vehicle-routing problem. *Transportation Science* 43(1), 56–69.

Christofides, N., and Eilon, S. (1969). An algorithm for the vehicle-dispatching problem. *Journal of the Operational Research Society* 20(3), 309–318.

Clarke, G., and Wright, J. (1964). Scheduling of vehicles from a central depot to a number of delivery points. *Operations Research* 12(4), 568–581.

Coelho, L., Cordeau, J.-F., and Laporte, G. (2014). Thirty years of inventory routing. *Transportation Science* 48(1), 1–19.

Corberán, Á., and Laporte, G. (Eds.). (2014). *Arc Routing: Problems, Methods and Applications*. SIAM, Philadelphia.

Corberán, Á., and Prins, C. (2010). Recent results on arc routing problems: An annotated bibliography. *Networks* 56(1), 50–69.

Cordeau, J.-F., Gendreau, M., and Laporte, G. (1997). A tabu search heuristic for periodic and multi-depot vehicle routing problems. *Networks* 30(2), 105–119.

Cordeau, J.-F., and Maischberger, M. (2012). A parallel iterated tabu search heuristic for vehicle routing problems. *Computers & Operations Research* 39(9), 2033–2050.

Crainic, T., and Laporte, G. (Eds.). (1998). *Fleet Management and Logistics*. Springer, New York.

Dabia, S., Røpke, S., Van Woensel, T., and De Kok, T. (2011). Branch and cut and price for the time dependent vehicle routing problem with time windows. *Transportation Science* 47(3), 380–396.

Dantzig, G., and Ramser, J. (1959). The truck dispatching problem. *Management Science* 6(1), 80–91.

Demir, E., Bektas, T., and Laporte, G. (2014). A review of recent research on green road freight transportation. *European Journal of Operational Research* 237(3), 775–793.

Doerner, K.F., and Salazar-Gonzalez, J.-J. (2014). Pickup-and-delivery problems for people transportation. In: P. Toth, and D. Vigo (Eds.), *Vehicle Routing: Problems, Methods, and Applications*. SIAM, Philadelphia, 193–212.

Drexl, M. (2012). Synchronization in vehicle routing—A survey of VRPs with multiple synchronization constraints. *Transportation Science* 46(3), 297–316.

Drexl, M. (2012). Rich vehicle routing in theory and practice. *Logistics Research* 5(1), 47–63.

Dror, M. (Ed.). (2000). *Arc Routing: Theory, Solutions, and Applications*. Springer, New York.

Dueck, G. (1993). New optimization heuristics: The great deluge algorithm and the record-to-record travel. *Journal of Computational Physics* 104(1), 86–92.

Eilon, S., Watson-Gandy, C., and Christofides, N. (1971). *Distribution Management: Mathematical Modelling and Practical Analysis*. Griffin, London.

Eksioglu, B., Vural, A., and Reisman, A. (2009). The vehicle routing problem: A taxonomic review. *Computers & Industrial Engineering* 57(4), 1472–1483.

Fukasawa, R., de Aragão, M. P., Reis, M., and Uchoa, E. (2003). Robust branch-and-cut-and-price for the capacitated vehicle routing problem. In: *Proceedings of The International Network Optimization Conference* 106, 231–236.

Funke, B., Grünert, T., and Irnich, S. (2005). Local search for vehicle routing and scheduling problems: Review and conceptual integration. *Journal of Heuristics* 11(4), 267–306.

Gaskell, T. (1967). Bases for vehicle fleet scheduling. *Journal of the Operational Research Society* 18(3), 281–295.

Gillett, B. E., and Miller, L. R. (1974). A heuristic algorithm for the vehicle-dispatch problem. *Operations Research* 22(2), 340–349.

Goel, A., and. Vidal, T. (2014). Hours of service regulations in road freight transport: An optimization-based international assessment. *Transportation Science* 48(3), 291–412.

Golden, B., and Assad, A. (1988). *Vehicle Routing: Methods and Studies*. North-Holland, Amsterdam.

Golden, B., Raghavan, S., and Wasil, E. (Eds.), (2008). *The Vehicle Routing Problem: Latest Advances and New Challenges*. Springer, New York.

Groër, C., Golden, B., and Wasil, E. (2010). A library of local search heuristics for the vehicle routing problem. *Mathematical Programming Computation* 2(2), 79–101.

Gulczynski, D., Golden, B., and Wasil, E. (2011). The period vehicle routing problem: New heuristics and real-world variants. *Transportation Research (Part E)* 47(5), 648–668.

Hausman, W., and Gilmour, P. (1967). A multi-period truck delivery problem. *Transportation Research* 1(4), 349–357.

Hertz, A. (2005). Recent trends in arc routing. In: M. Golumbic, and I. Hartman (Eds.), *Graph Theory, Combinatorics and Algorithms*, Springer, New York, 215–236.

Iori, M., and Martello, S. (2013). An annotated bibliography of combined routing and loading problems. *Yugoslav Journal of Operations Research* 23(3), 311–326.

Jozefowiez, N., Semet, F., and Talbi, E. (2008). Multi-objective vehicle routing problems. *European Journal of Operational Research* 189(2), 293–309.

Kindervater, G. A. P., and Savelsbergh, M. W. P. (1997). Vehicle routing: Handling edge exchanges. In: E. H. L. Aarts and J. K. Lenstra (Eds.), *Local Search in Combinatorial Optimization*, Princeton University Press, Princeton, 337–360.

Kirkpatrick, S., Gelatt, C. D., and Vecchi, M. P. (1983). Optimization by simulated annealing. *Science* 220 (4598), 671–680.

Laporte, G. (2009). Fifty years of vehicle routing. *Transportation Science* 43(4), 408–416.

Laporte, G. (2007). What you should know about the vehicle routing problem. *Naval Research Logistics* 54(8), 811–819.

Laporte, G., and Martin, I. R. (2007). Locating a cycle in a transportation or a telecommunications network. *Networks* 50(1), 92–108.

Li, C. L., Simchi-Levi, D., and Desrochers, M. (1992). On the distance constrained vehicle routing problem. *Operations Research* 40(4), 790–799.

Lin, C., Choy, K. L., Ho, G. T. S., Chung, S. H., and Lam, H. Y. (2014). Survey of green vehicle routing problem: Past and future trends. *Expert Systems with Applications* 41(4), 1118–1138.

Lübbecke, M. E., and Desrosiers, J. (2005). Selected topics in column generation. *Operations Research* 53(6), 1007–1023.

Lysgaard, J., Letchford, A. N., and Eglese, R. W. (2004). A new branch-and-cut algorithm for the capacitated vehicle routing problem. *Mathematical Programming* 100 (2), 423–445.

Marinakis, Y., and Migdalas, A. (2007). Annotated bibliography in vehicle routing. *Operational Research* 7(1), 27–46.

Martello, S., and Toth, P. (1990). *Knapsack Problems: Algorithms and Computer Implementations*. Wiley, New York.

Martinelli, R., Pecin, D., and Poggi, M. (2014). Efficient elementary and restricted non-elementary route pricing. *European Journal of Operational Research* 239(1), 102–111.

Nagy, G., and Salhi, S. (2007). Location-routing: Issues, models and methods. *European Journal of Operational Research* 177(2), 649–672.

Newton, R., and Thomas, W. (1969). Design of school bus routes by computer. *Socio-Economic Planning Sciences* 3(1), 75–85.

Norstad, I., Fagerholt, K., and Laporte, G. (2011). Tramp ship routing and scheduling with speed optimization. *Transportation Research Part C: Emerging Technologies* 19(5), 853–865.

Panchamgam, K., Xiong, Y., Golden, B., Dussault, B., and Wasil, E. (2013). The hierarchical traveling salesman problem. *Optimization Letters* 7(7), 1517–1524.

Park, J., and Kim, B. (2010). The school bus routing problem: A review. *European Journal of Operational Research* 202(2), 311–319.

Partyka, J., and Hall, R. (2014). VR delivers the goods. *OR/MS Today* 41(1), 40–46.

Pessoa, A., Uchoa, E., and Poggi de Aragão, M. (2008). Robust branch-cut-and-price algorithms for vehicle routing problems. In: B. Golden, S. Raghavan, and E. Wasil (Eds.), *The Vehicle Routing Problem: Latest Advances and New Challenges*. Springer, New York, 297–325.

Pierce, J. (1969). Direct search algorithms for truck-dispatching problems. *Transportation Research* 3(1), 1–42.

Pillac, V., Gendreau, M., Guéret, C., and Medaglia, A. L. (2013). A review of dynamic vehicle routing problems. *European Journal of Operational Research* 225(1), 1–11.

Pillac, V., Guéret, C., and Medaglia, A. (2013a). A parallel matheuristic for the technician routing and scheduling problem. *Optimization Letters* 7(7), 1525–1535.

Pillac, V., Guéret, C., and Medaglia, A. L. (2012). An event-driven optimization framework for dynamic vehicle routing. *Decision Support Systems* 54(1), 414–423.

Pisinger, D., and Røpke, S. (2007). A general heuristic for vehicle routing problems. *Computers & Operations Research* 34(8), 2403–2435.

Poggi, M., and Uchoa, E. (2014). New exact algorithms for the capacitated vehicle routing problem. In: P. Toth, and D. Vigo (Eds.), *Vehicle Routing: Problems, Methods, and Applications*. SIAM, Philadelphia, 59–86.

Potvin, J. (2009). State-of-the-art review—Evolutionary algorithms for vehicle routing. *INFORMS Journal on Computing* 21(4), 518–548.

Prodhon, C., and Prins, C. (2014). A survey of recent research on location-routing problems. *European Journal of Operational Research* 238(1), 1–17.

Ralphs, T. K. (2003). Parallel branch and cut for capacitated vehicle routing. *Parallel Computing* 29(5), 607–629.

Ralphs, T., Kopman, L., Pulleyblank, W., and Trotter, L. (2003). On the capacitated vehicle routing problem. *Mathematical Programming* 94 (2–3), 343–359.

Schrimpf, G., Schneider, J., Stamm-Wilbrandt, H., and Dueck, G. (2000). Record breaking optimization results using the ruin and recreate principle. *Journal of Computational Physics* 159(2), 139–171.

Schruben, L., and Clifton, R. (1968). The lockset method of sequential programming applied to routing delivery and pickup trucks. *American Journal of Agricultural Economics* 50(4), 854–867.

Shaw, P. (1998). Using constraint programming and local search methods to solve vehicle routing problems. In: M. Maher, and J. F. Puget (Eds.), *Principles and Practice of Constraint Programming – CP98. Vol. 1520 of Lecture Notes in Computer Science*. Springer, Berlin, Heidelberg, 417–431.

Subramanian, A., Uchoa, E., and Ochi, L. S. (2013). A hybrid algorithm for a class of vehicle routing problems. *Computers & Operations Research* 40(10), 2519–2531.

Toth, P., and Vigo, D. (Eds.), (2014). *Vehicle Routing: Problems, Methods, and Applications*. SIAM, Philadelphia.

Toth, P., and Vigo, D. (Eds.), (2002). *The Vehicle Routing Problem*. SIAM, Philadelphia.

Vansteenwegen, P., Souffriau, W., and Oudheusden, D. (2011). The orienteering problem: A survey. *European Journal of Operational Research* 209(1), 1–10.

Vidal, T., Crainic, T.G., Gendreau, M., and Prins, C. (2014). A unified solution framework for multi-attribute vehicle routing problems. *European Journal of Operational Research* 234(3), 658–673.

Wøhlk, S. (2008). A decade of capacitated arc routing. In: B. Golden, S. Raghavan, and E. Wasil (Eds.), *The Vehicle Routing Problem: Latest Advances and New Challenges*. Springer, New York, 29–48.

Yellow, P. C. (1970). A computational modification to the savings method of vehicle scheduling. *Journal of the Operational Research Society* 21(2), 281–283.

17

AIRPORT PLANNING AND DESIGN

Alexandre G. de Barros

UNIVERSITY OF CALGARY

The Airport System

Since airports are a place of modal interchange, the airport system is bounded by the interfaces with the ground and air transportation systems. Figure 17.1 illustrates the typical elements of an airport system and its interfaces with the other systems. Not all airports will have all elements shown in 17.1. For example, an airport may not have dedicated cargo facilities. General aviation airports may not have a passenger terminal building. The runway is arguably the one essential element in an airport system.

Master Planning

Airports are generally built to last for as long as there are aircraft in need to take off and land. They generally take up a vast amount of land and—since they mostly serve large urban areas—need to be close to an urban area. As that urban area grows, so does the airport size requirement. The vast majority of the world's large airports started as small airfields and were expanded as demand for air transportation grew. If the airport site and its surroundings are not properly protected, the airport's future ability to expand to serve the demand may be seriously compromised.

In addition to the ability to expand the airport to meet demand, the ability to develop an optimal airfield layout to meet the needs of said demand may also be compromised by decisions made in the past. For instance, the construction of a passenger terminal building and its respective support infrastructure at a certain location will preclude the construction of a runway at that location or even within a certain distance from it.

To avoid these problems in the future development of the site, an airport master plan is essential. The purpose of the master plan is to provide an outlook of the future airport layout to guide the decisions made in the present. A complete master plan should illustrate what the airport layout is expected to be at the end of the planning horizon, and the expansion stages that will take it there. It is important to note that any planning exercise is based on conditions that are forecast at the time of such exercise, which are in turn largely based on existing conditions. As existing conditions change with time, so do the forecasts, and so should the master

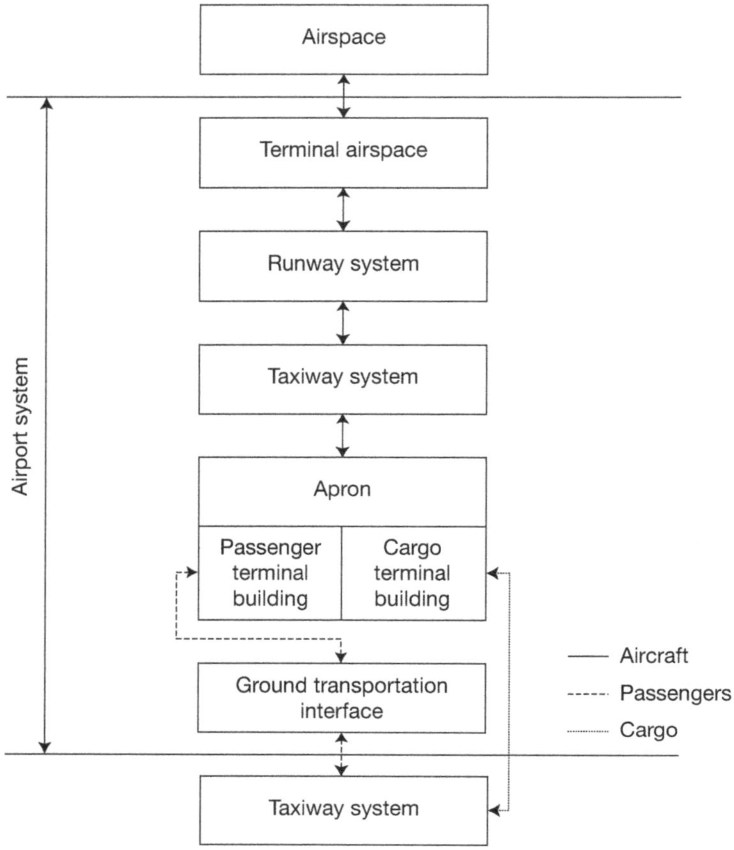

Figure 17.1 The airport system
Source: Adapted from Horonjeff et al., 2010).

plan. Thus the master plan needs to be updated frequently. Typical planning horizons for an airport master plan range between 20 and 40 years, with updates every 3–5 years.

Airfield Design Standards

The International Civil Aviation Organization (ICAO) is a United Nations specialized agency that sets international standards and recommended practices for aviation. Each ICAO state member uses these standards to develop their own local regulations. Throughout this chapter reference will be made to the ICAO standards for airfield design. Although most countries adopt the ICAO standards almost entirely in their regulations, local requirements may differ. The reader is strongly advised to check local regulations before proceeding with any airport planning exercise.

The ICAO airfield design standards are based on the aircraft categorization shown in Table 17.1. The categorization is based on the aircraft's wingspan, wheel track and reference field length. The latter is the aircraft's required runway length for take-off under standard weather conditions and at its maximum take-off weight, serving as proxy for aircraft mass.

Table 17.1 Aircraft categorization for airfield design

Aerodrome code number	*Aircraft reference field length (m)*	*Aerodrome code letter*	*Aircraft wingspan (m)*	*Outer main gearwheel span (m)*
1	<800	A	<15	<4.5
2	800–<1200	B	15–<24	4.5–<6
3	1200–<1800	C	24–<36	6–<9
4	≤1800	D	36–<52	9–<14
		E	52–<65	9–<14
		F	65–<80	14–<16

Source: ICAO (2004)

The Runway System

The single most important element of an airport is the runway. The runway is a linear strip of land used for aircraft take-off and landing runs. A well-designed runway must:

- have a straight, leveled, finished surface to allow aircraft to run at high speeds;
- support the weight of aircraft without significant deformation over the runway lifespan;
- provide sufficient drainage to prevent the accumulation of water on the runway surface;
- be of sufficient length to allow aircraft to take-off and land safely;
- provide sufficient width to allow for lateral aircraft deviations during take-off and landing;
- be free of obstacles that could pose a hazard to aircraft during take-off and landing, not only on the runway itself but also on its surroundings and on climb and approach paths along its extended centerline;
- provide sufficient visual guidance to pilots, especially during low-visibility conditions such as night-time and deteriorated weather;
- be oriented such that crosswinds are minimized.

All requirements above depend on aircraft characteristics such as size, weight, engine type and performance capabilities. Therefore, a critical part of the planning process is the choice of a design aircraft, which will determine the requirements for the runway design. The choice of the design aircraft will depend on market and traffic forecasts, and is normally done as part of the master planning exercise.

Runway Orientation

The ability of aircraft to fly comes from the lift generated by the air moving at high speeds over and under the wings—the airspeed. Wind speeds and direction are, therefore, critical in the choice of runway orientation as they affect both airspeed and handling.

Headwinds will add to the aircraft's airspeed, whereas tailwinds will subtract from it. As an example, an aircraft flying at a ground speed of 250 km/h with a headwind of 20 km/h will have an airspeed of 270 km/h, whereas the same aircraft at the same ground speed but with a tailwind of 25 km/h will be flying at an airspeed of 225 km/h. For this reason, take-offs and landings should always be performed with either headwinds or calm winds. The direction the wind is blowing from will then determine the direction of take-off and landing operations on a runway.

Table 17.2 Maximum recommended crosswinds for airport design

Reference field length (RFL)	*Maximum crosswind*
<1200 m	19 km/h (10.5 knots)
1200–1499 m	24 km/h (13 knots)
>1500 m	37 km/h (20 knots)

Source: ICAO (2004)

Winds rarely blow in the exact same direction as the runway, however. Most often the wind will be blowing at an angle with the runway, generating both parallel and perpendicular wind components, as seen in Figure 17.2. The parallel component will determine the direction in which take-offs and landings need to occur. However, the wind's component perpendicular to the runway—known as crosswind—pushes the aircraft laterally, away from the runway centerline. Strong crosswinds may severely affect aircraft operations. Table 17.2 shows typical crosswind limits for aircraft based on their reference field lengths.

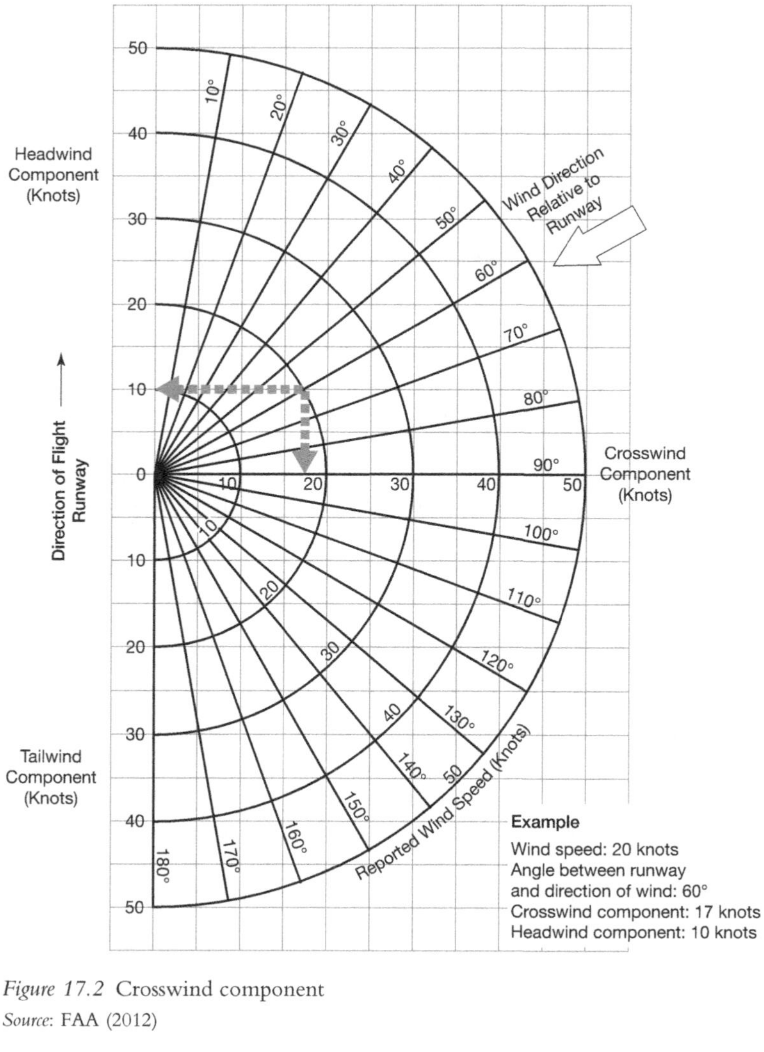

Figure 17.2 Crosswind component
Source: FAA (2012)

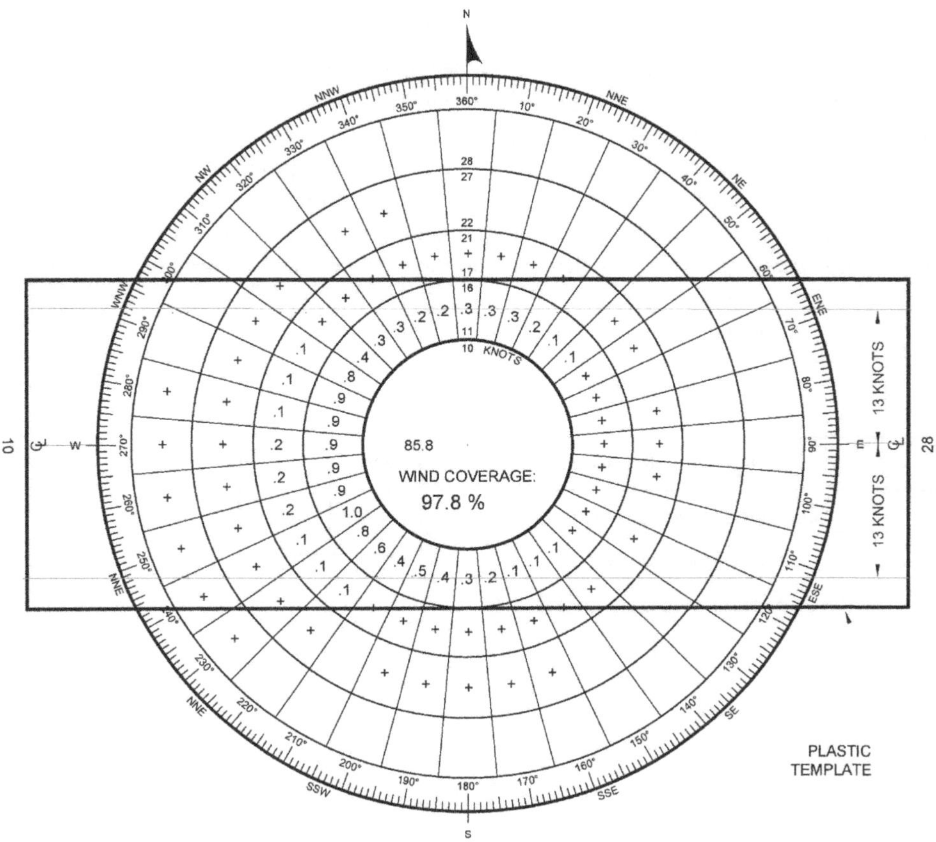

Figure 17.3 Wind rose and runway template
Source: FAA (2012)

The orientation of a runway may be limited by site constraints such as airport property land constraints and the presence of obstacles in the airport surroundings such as buildings and terrain. If such constraints are not present or can be neglected, then the runway should be oriented as to minimize the effect of crosswinds on aircraft operations. A weather analysis using a wind rose such as the one in Figure 17.3 can be used to determine the availability of a runway for a given aircraft operation, i.e. the percentage of time the crosswinds are within the limit. In Figure 17.3, the heading represents the direction from which the wind is blowing; the distance from the center represents the wind speed; and the numbers represent the percentage of time the wind is blowing in that speed-heading range. Using a rectangular template centered on the wind rose and with a half-width equivalent to the maximum crosswind allowed, it is possible to determine the availability of a runway by adding up the numbers that fall under the runway template.

The optimal runway availability may vary from airport to airport and needs to be determined based on local factors. If the minimum availability cannot be achieved in any direction, then it may be necessary to build two or more runways with different orientations.

Runway Length

The runway must be long enough to allow the design aircraft to take off and land safely. Not all of the runway length—hereby denoted *field length* (FL)—needs to be full-strength pavement (FS). A portion of the field length can be a rectangular area called *clearway* (CL) centrally located about the runway's extended centerline that is free of objects or protrusions and under airport control. There may also be a *stopway* (SW), an area beyond the full-strength pavement with pavement strength just enough to support the weight of an aircraft during an aborted take-off. Figure 17.4 illustrates these runway elements and the distances that will need to be evaluated.

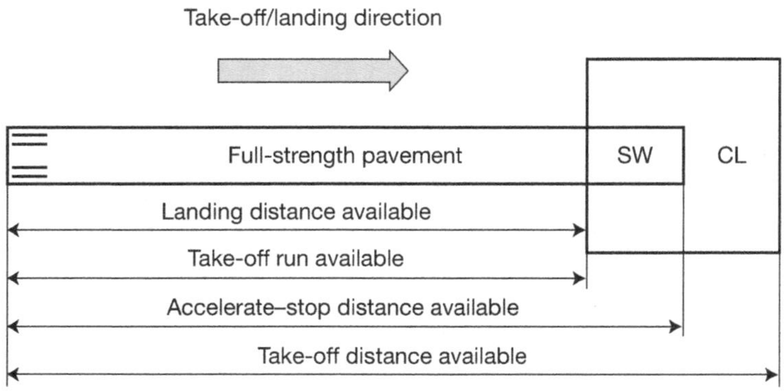

Figure 17.4 Distances comprising runway length

In the evaluation of the runway length, four scenarios must be considered:

1. Take-off with all engines.
2. Take-off with one engine down at $V_{1.}$
3. Aborted take-off at $V_{1.}$
4. Landing.

V_1, the decision speed, is set for each aircraft in its operations manual. It is the speed beyond which the pilot can no longer abort take-off and must proceed. Should an engine failure or any other problem deemed to be a hazard to the flight occur before reaching V_1, the pilot will cut engine thrust and apply the brakes, bringing the aircraft to a full stop. After V_1, the pilot must continue with the take-off. Since the average longitudinal slope may vary with the direction of the landing/take-off run, the lengths above must be calculated for each runway direction.

Case 1: Take-Off with All Engines

ICAO (2001) establishes that on take-off the aircraft must be able from stop to reach a height of 10.5 m (35 ft) above the ground over the runway's field length. The distance travelled during take-off is denoted as $D35_1$. To allow for variations in aircraft and pilot performances, a 15 percent safety margin is added. Therefore, the take-off distance required for the design aircraft with all engines is:

$$TOD_1 = 1.15 \cdot D35_1 \qquad (1)$$

Not all of TOD_1 needs to be full-strength pavement. Up to half the difference between TOD_1 and 115 percent of the lift-off distance (LOD_1, the distance from full stop to lift-off) can be a clearway—as long as it does not exceed 300 m. In summary,

$$CL_1 = \min\left[300 \text{ m}, (TOD_1 - 1.15 \cdot LOD_1)/2\right] \tag{2}$$

where CL_1 is the clearway length.

Therefore, the take-off run required for all-engines take-off is:

$$TOR_1 = TOD_1 - CL_1 \tag{3}$$

The required full-strength pavement is:

$$FS_1 = TOR_1 \tag{4}$$

and the required field length is:

$$FL_1 = TOD_1 \tag{5}$$

Case 2: Take-Off with One Engine Failure at V_1

V_1 is the lowest possible aircraft speed during take-off at which an engine failure can occur and the aircraft can still take off. In that case, the aircraft still needs to be able to reach a height of 10.5 m (35 ft) above the runway. However, since the aircraft will now have a lower thrust from V_1, it will travel a longer distance from stop to 10.5 m (35 ft) above the ground ($D35_2$). Engine failures during take-off are extremely rare, and the probability that one will occur at exactly V_1 is extremely low. For these reasons there is no margin of safety applied to the take-off distance, which will hence be:

$$TOD_2 = D35_2 \tag{6}$$

and the clearway length allowed, also with no safety margins, is:

$$CL_2 = \min\left[300 \text{ m}, (TOD_2 - LOD_2)/2\right] \tag{7}$$

where LOD_2 is the lift-off distance with one engine down at V_1. The take-off run, the full-strength pavement and field length requirements are respectively:

$$TOR_2 = TOD_2 - CL_2 \tag{8}$$

$$FS_2 = TOR_2 \tag{9}$$

$$FL_2 = TOD_2 \tag{10}$$

Case 3: Aborted Take-Off at V_1

V_1 is the highest possible speed at which the pilot can abort take-off by cutting engine thrust and applying the brakes. Should that happen, the aircraft needs to be able to come to a full stop safely within the runway length. The total distance travelled by the aircraft from full stop to V_1 to full stop again is called the accelerate-stop distance (DAS). Since this is a rare event, no safety margin is applied and the field length requirement is:

$$FL_3 = DAS \tag{11}$$

Also, not all of *DAS* needs to be on full-strength pavement. The difference between *DAS* and the full-strength pavement provided for landing and the other take-off cases can be a stopway as defined earlier in this chapter. The stopway length is:

$$SW = \max\left[FS - DAS, 0\right] \tag{12}$$

Case 4: Landing

ICAO (2001) defines the landing distance as the distance travelled by an aircraft from 15 m (50 ft) above the runway threshold to full stop. It also establishes that this distance should be no more than 60 percent of the available full-strength pavement for landing. Therefore:

$$FL_4 = FS_4 = LD/0.6 \tag{13}$$

Overall Runway Length Requirement

The final runway length must satisfy all requirements for the four cases described above. In summary, the field length and the full-strength requirements are respectively:

$$FL = \max\left[FL_1, FL_2, FL_3, FL_4\right] \tag{14}$$

$$FS = \max\left[FS_1, FL_2, FS_4,\right] \tag{15}$$

The stopway length is given in equation 12. The clearway length is:

$$CL = FL - FS \tag{16}$$

Number of Runways and Airfield Layout

Small airports and some midsized ones will likely not need more than one runway to meet demand. The capacity of a single runway—defined as the number of take-off and landing movements it can handle per unit of time—is, however, limited by a number of factors, including (but not limited to):

- air traffic control (ATC) requirements for separation between aircraft;
- weather conditions—wind speed and direction, visibility, ceiling, precipitation;
- aircraft mix;
- proportions of take-offs and landings;

Table 17.3 Minimum separations between parallel runways with instrument approaches

Runway operations	*Minimum separation between runway centerlines*
Independent parallel approaches	1,035 m
Dependent approaches (minimum separation between aircraft approaching to the two runways required)	905 m
Independent parallel departures and/or segregated operations (one runway used only for landings, the other used only for departures)	760 m

Source: ICAO (2004)

- location and design of the runway exits;
- ATC technology and performance;
- noise constraints.

Typical single runway capacities range between 30 and 60 movements/hour. The capacity of multi-runway systems will depend heavily on the layout of the system. Parallel runways may be considered independent if there is sufficient separation between their centerlines, in which the capacity of an *n*-runway system is *n* times the capacity of a single runway. Table 17.3 shows the minimum separation between parallel runways established by ICAO for different kinds of operations.

In multi-runway systems where two or more runways are laid out at an angle—intersecting or not—the capacity of the system will depend on the operational constraints imposed by the layout. The US Federal Aviation Administration (FAA, 1983) provides runway capacity estimates for a variety of runway layouts. Although the FAA estimates can be used for high-level planning, a more detailed analysis is recommended for more accurate evaluations. Both analytical tools and computer simulation can be used to model in detail the specifics of the runway system to evaluate its capacity.

Obstacle Clearance

To ensure the safe operation of aircraft, airport facilities must be designed to provide space that is free of obstacles and hazards. Since aircraft operations near the ground extend beyond the airport site itself, ensuring an airspace free of obstacles and hazards is a task that involves not only the airport operator but also the local and provincial authorities that regulate and enforce land use restrictions in the airport surroundings. These restrictions include—but are not limited to—limitations on the heights of buildings, towers and even terrain in the proximity of the airport, as well as within the airport site.

ICAO (2004) defines a number of imaginary three-dimensional surfaces in and around the airport site that should not be penetrated by obstacles. These surfaces, illustrated in Figure 17.5, are called obstacle limitation surfaces (OLS). They also establish minimum dimensions for ground facilities such as runways, taxiways and aprons, as well as minimum separations between them. The dimensions of the surfaces depend on the aircraft that the airport is certified to handle.

For the purpose of determining the surfaces' required dimensions, aircraft are classified both by size and manoeuvrability. Size will determine the parameters related to ground facilities such as runway widths and taxiway-to-taxiway separations, whereas maneuverability will be the main driver for the surface dimensions. Generally, the larger the aircraft, the larger the separation requirements between the aircraft and ground facilities. Also, the more kinetic energy an aircraft has—a function of its mass and speed—the more difficult it is to maneuver and the more space it needs in case of deviation from the intended flight path. Other parameters such as weather conditions and the type of operation—whether visual or by instruments—will also affect those restrictions.

Two issues arise from the approach used in current standards. First, they were developed many decades ago, assuming aircraft performance parameters that may not be up to date, as aircraft

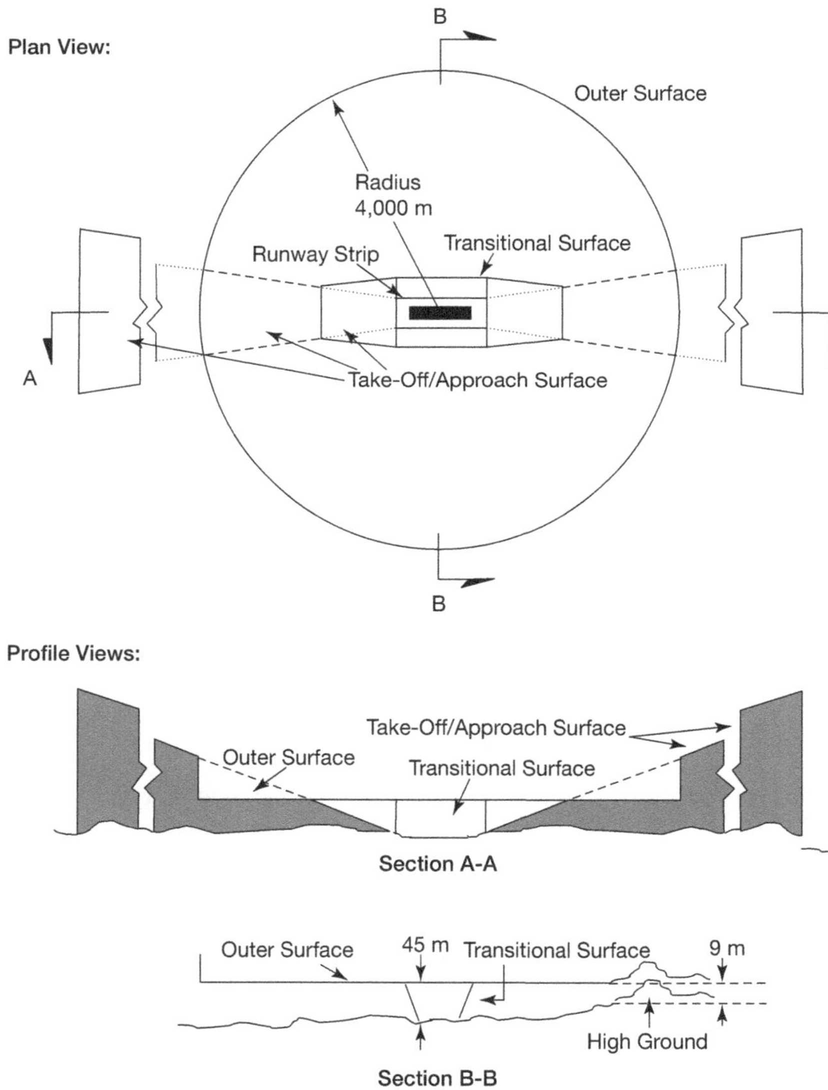

Figure 17.5 Obstacle limitation surfaces
Source: Adapted from Transport Canada (1993)

control systems have evolved significantly in the last few decades. Second, the discrete approach used in the categorization of aircraft means that where the design aircraft is marginally within a given category the airport will need to follow the standards for that category. For example, code E airports must be designed for aircraft with a wingspan of up to 65 m, whereas code F airports must be able to accommodate wingspans of up to 80 m. There is no provision for airports designed for an aircraft with a wingspan between 65 m and 80 m. An airport designed for such aircraft will need to meet the requirements for code F. This "overdesign" may be impractical and uneconomical at many airports. Although most countries allow for deviations that can accommodate cases like this, clearly there is room for a continuous or quasi-continuous approach.

This issue is of paramount importance for the increasingly common situation where airports are inserted in densely developed urban areas. Most commercial airports in operation were built a safe distance away from the city to avoid this issue. Mid-sized and large airports are, however, centers of business that tend to attract urban development in their vicinities. Furthermore, the recent urbanization trend observed in most countries has led to rapid urban sprawl in many cities, to the point where many of them have reached and even engulfed their airport—which used to be located far away from the city. As a result, some airports are now a severe limitation to the urban development. Careful coordination between airport authorities and local municipalities is essential to establish appropriate land use zoning in the airport vicinities.

The Taxiway System

Taxiways are the paths that connect the runways to the apron. Since taxiing occurs at very low speeds (typically 25 km/h or less), taxiway width and obstacle clearance need not be as stringent

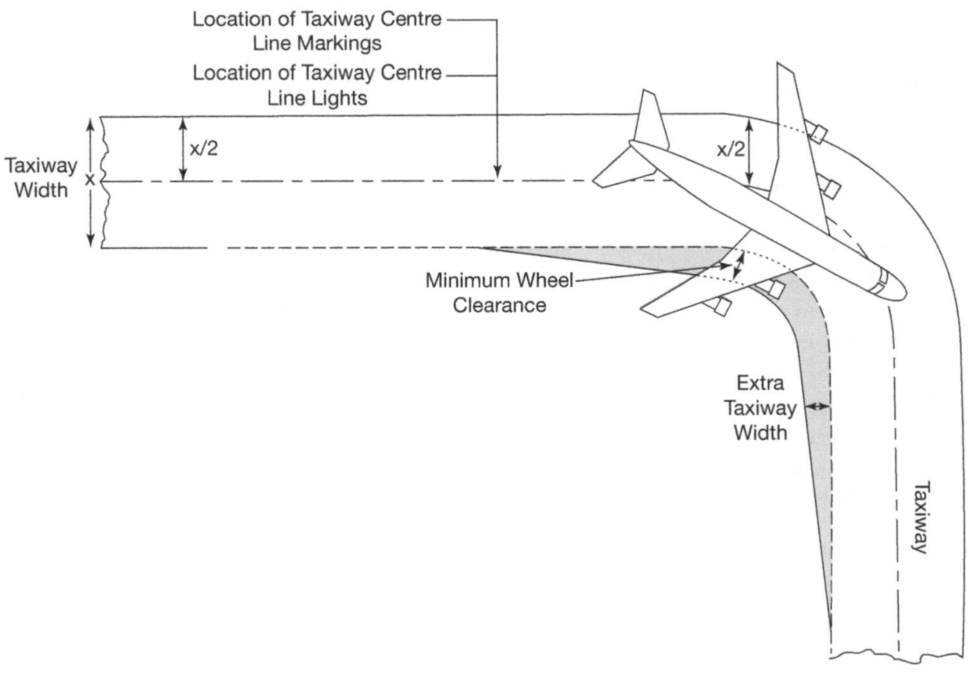

Figure 17.6 Taxiway fillet
Source: ICAO (2004)

as the runway's. The taxiway must accommodate the design aircraft's wheel track, with some extra width to accommodate small deviations while taxiing. Taxiway shoulders may also be required to avoid jet blast erosion. At taxiway intersections, pavement fillets are usually required to allow for aircraft to negotiate the curves safely without wandering off the pavement. Figure 17.6 shows an example of a taxiway fillet. ICAO (2004) sets standards for taxiway widths, shoulder and separation to other taxiways.

Apron

The apron is the area where aircraft parking stands are located. It is the place where the actual interface between air and ground transportation modes begins or ends. Passenger and cargo loading and unloading, as well as aircraft fueling and servicing, all happen at the apron. Clearly, it is not just aircraft movements that drive the design of the apron—it is also the traffic of ground vehicles, equipment and people. The apron should be designed to allow the safe movement of all these entities.

Number of Apron Stands

The apron should provide sufficient stands to meet demand during peak periods. A high-level estimate of the average number of stands required during a given period is:

$$G = A(T + s) \tag{17}$$

where G is the number of stands, A is the aircraft arrivals rate in aircraft/hour, T is the average aircraft turnaround time (from arrival to departure) in hours, and s is the buffer time between consecutive aircraft on the stand in hours. The buffer time allows for both outbound and inbound aircraft manoeuvers and is generally set at 15 minutes (0.25 h) for planning purposes.

The aircraft turnaround time varies widely with aircraft size. Commuter turboprops with up to 50 seats can have turnaround times as short as 20 minutes, whereas long-haul wide-bodied aircraft may take up to two hours to service. Because of these wide variations, it may be necessary to evaluate the number of stands separately for each aircraft size or category. Also, since the peaks for each aircraft size or category may not coincide in time, designing the apron with the flexibility to accommodate different aircraft sizes at different times may provide significant savings in apron area.

More detailed analysis of apron stand requirements and operations is usually required at large airports. Bandara and Wirasinghe (1990) propose a cost optimization model that attempts to balance the cost of the apron against the cost of delays caused by insufficiency of available stands during peak periods. De Barros and Wirasinghe (2004) also propose a cost optimization model that takes flexible apron design and stage construction into account. Other techniques may be employed such as gating analysis—where a mock flight schedule is used to assign stands to individual flights in the schedule—and discrete-event simulation.

Passenger Terminal Building

The passenger terminal building (PTB) is part of the interface between the air and ground transportation modes for passengers starting or ending their air trips at the airport. The PTB may also provide an interface between flights, allowing for passenger transfers between them. Figure 17.7 illustrates the passenger flows and processing functions that occur in the PTB.

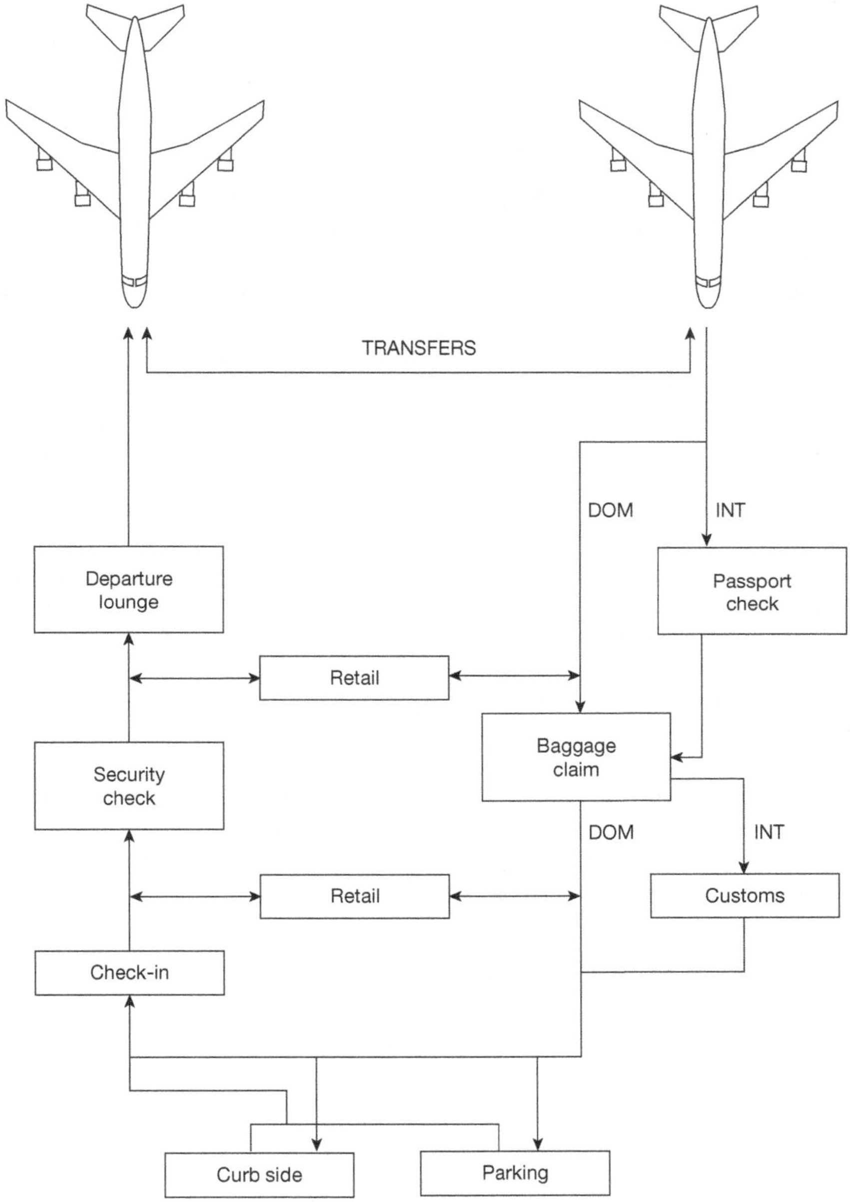

Figure 17.7 Passenger flows in the terminal building

Check-In

Payload weight and weight distribution on board the aircraft are critical in aircraft operations. The need to correctly and accurately measure and balance payload weight generates the need for passenger and baggage pre-processing. For that reason, passengers cannot simply move directly from their ground transportation vehicle to the aircraft. They need to first check in with the airline and hand in their baggage. The check-in process for passengers without baggage may be performed at a check-in counter, an automated self-service kiosk, or remotely via the Internet. Passengers with baggage to check in may also check in remotely but will need to stop at a baggage drop-in counter at the PTB before proceeding for boarding.

Security Check

Once checked in with the airline, passengers proceed through a security check. The purpose of the security check is to ensure that passengers do not bring objects that could pose a threat or a hazard to the integrity of the aircraft and its occupants. For maximum efficiency, the security checkpoint is normally located at the entrance to the boarding gate area, with one checkpoint being used for access to a group of gates. Each checkpoint is comprised of a number of security lanes, each equipped with a hand-luggage screening machine and a passenger screening gantry or machine. As of the writing of this book, X-ray is the most common screening technology for hand luggage, whereas metal detection and millimeter wave machines are used for passengers. Screening technology is constantly evolving and may not be the same as today years from now.

Boarding Area

Most passengers are likely to arrive early at the boarding gate. An area must then be provided at the boarding gate for passenger accommodation while they await boarding. This area is commonly known as the departure room, boarding room, departure lounge, or boarding area. The size of this area, as well as the number of seats provided, will depend on the aircraft size and on how early passengers arrive at the gate.

The boarding gate is the actual interface between the PTB and the apron. Contact gates will normally be equipped with one or more boarding bridges connecting the building directly to the aircraft door. At small PTBs, passengers may be required to walk between the building and the aircraft. Aircraft may also be parked at remote stands, i.e. away from the PTB. In that case, the boarding area will lead to a bus/transporter loading area, from where passengers will be transported by ground vehicle to the aircraft door.

Baggage Claim

Passengers who are ending the air portion of their trip at the airport must get off the aircraft either through the boarding bridges or through the stairs on the apron, from where they can either walk or be transported to the PTB. Once at the PTB, they proceed to the baggage claim area.

Typically, the baggage claim area comprises one or more baggage claim devices. These devices consist of baggage conveyors that move in a closed loop, such that baggage loaded onto them are constantly displayed until their owner collects them.

Border Control

After clearing security, departing international passengers may need to be screened by border control. Some countries such as the United States and Canada do not screen outbound international passengers. Outbound border control usually comprises counters and/or booths, each occupied by one border control agent who checks each passenger's travel documents required by local and international law for travel.

Departing domestic passengers need no screening and can proceed directly to the gate.

Arriving international passengers are typically required to clear immigration and customs before they can even be allowed contact with domestic passengers. In many countries, arriving international passengers must be physically segregated from domestic and even departing international passengers until they have cleared border control. In that case, a sterile corridor fully dedicated for the flow of arriving international passengers between the gates and passport control must be provided. Only after clearing immigration are these passengers allowed to proceed to baggage claim.

Many countries also have customs control, in which government agents may check the passengers' baggage for articles that are either not allowed in the country or subject to taxation. In this case, the international baggage claim area must be fully segregated from domestic and customs control is located after baggage claim.

Transfer Passengers

Ideally, passengers connecting between flights will not need any processing at the airport and will be allowed to proceed directly to their departure gate from their arrival gate. This is typically the case with passengers connecting between domestic flights with the same airline. Inter-airline passengers will need to go through the full arrivals procedure—including baggage claim—and check in for their next flight as if they were originating passengers.

The procedure for connections between an international flight and a domestic flight will vary from country to country and even from airport to airport. Typically, passengers arriving on an international flight will need to clear customs and immigration. They will then need to recheck their baggage with the airline and proceed to security check for access to their departure gate. Baggage recheck counters may be provided at customs exit to facilitate the connection.

PTB Configuration

The layout of the PTB should aim to facilitate passenger movements between the aircraft gates and the landside for originating and terminating passengers, and between gates for transfer passengers. These are often conflicting goals and some trade-off may be required.

Originating passengers need to be checked in before departure, whereas arriving passengers may need to collect their checked baggage before proceeding to ground transportation. Passenger and baggage check-in on departure can be done with check-in counters and self-service kiosks. Baggage claim on arrival is most often done with baggage conveyors where the baggage is deposited by the airline and displayed to the passengers for collection. At small sized airports, it is possible and often more economical to place these two functions side by side at ground level. At large airports, the large size of these areas requires a stacked configuration, with arrivals at ground level and departures at the upper level. This vertical layout minimizes both the building footprint and the distances passengers need to walk to and from the gates. This building is commonly referred to as the central processor.

The gate area—also commonly referred to as the terminal's airside—is where the passenger boarding and unloading happens. This portion of the terminal building is often of linear shape, although other shapes are possible. The actual layout of the PTB depends on the proportions of originating/terminating and transfer passengers.

Originating/terminating passengers must walk or otherwise be transported between the gate and the central processor. Ideally, the PTB configuration will minimize this distance. A linear configuration as shown in Figure 17.8 has the potential to minimize that distance if the processor can be spread along the landside face of the building and multiple security checkpoints can be

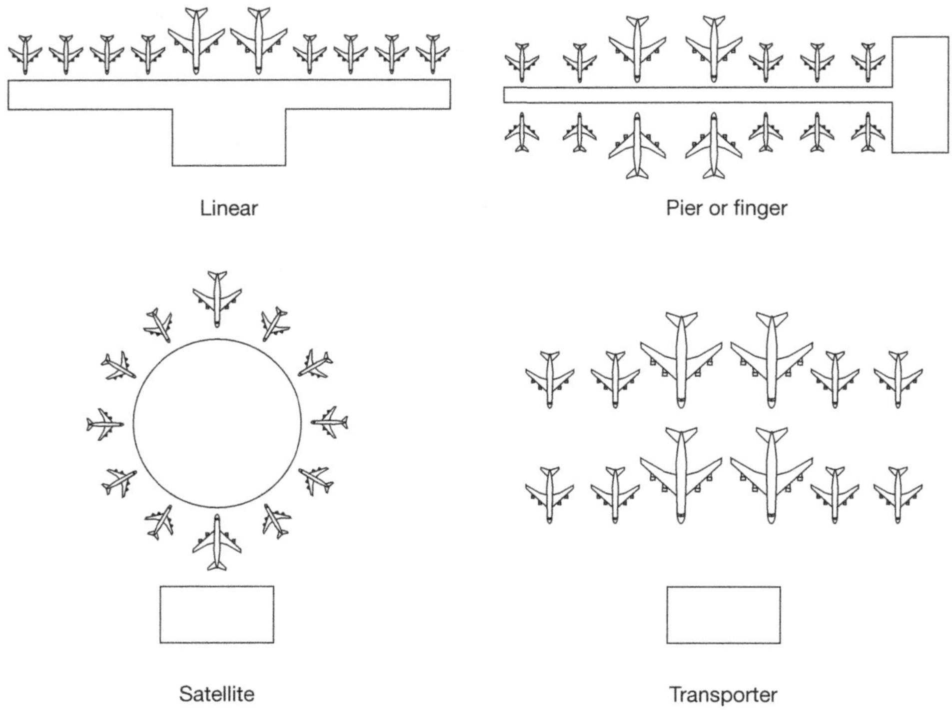

Figure 17.8 Basic terminal configurations

provided. This can prove to be very difficult and inefficient for a PTB with a large number of gates, as the building can be quite long. Munich is an example of an airport with linear PTBs.

For PTBs with high transfer rates, a layout that minimizes the distances to be overcome between gates is preferable. A pier shape with aircraft parked on both sides of the pier will cut the average distance between gates by half with respect to a linear building. Since aircraft are parked on both sides, the central processor must be either attached to one of PTB ends or to its center portion. For a larger number of gates, it may be necessary to break the piers into several parallel ones with tunnels and/or automated people movers connecting the piers to the central processor and to each other. Atlanta and Denver are examples of parallel-pier configurations.

One or more piers attached to a common central processor is a configuration known as pier-fingers. Bandara and Wirasinghe (1992) have shown that, for a PTB with a high intra-pier transfer rate, this configuration provides the shortest average walking/moving distance considering both transfers and originating passengers. Amsterdam/Schiphol, San Francisco, Toronto and São Paulo/Guarulhos are examples of pier-finger PTBs.

In the transporter configuration, aircraft are parked remotely from the central processor and transporters (buses) are used to transport passengers between the PTB and the aircraft stand. The level of service to passengers provided by this configuration is lower but the operational flexibility may be a valuable feature at airports with wide variations in demand profile over time. For example, airports with high seasonal demand can accommodate a large number of aircraft operations during peak season while still avoiding having large facilities being unused for large portions of the year.

These base configurations are often combined to meet specific, local requirements. Figure 17.9 shows an example of a combination of a pier-satellite configuration.

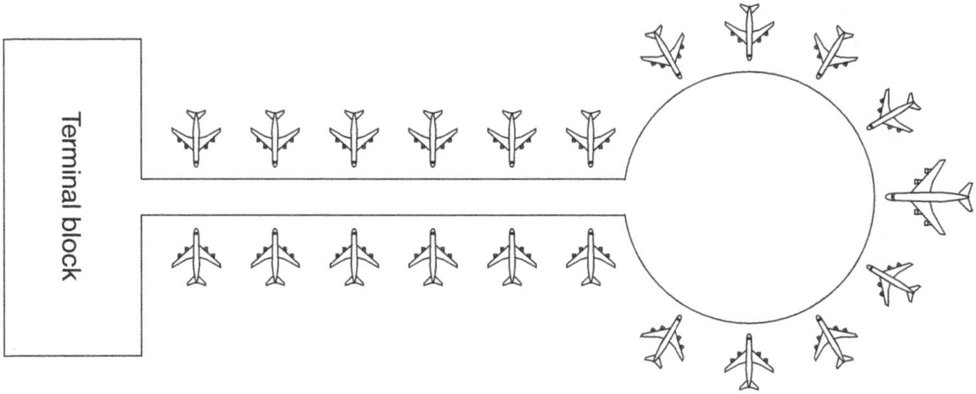

Figure 17.9 Pier-satellite combination

Environmental Impact of Airports

Aircraft Noise

Low-flying aircraft are a significant source of ambient noise on the ground. Departing aircraft have their engines working at or close to full power. On final approach to land at an airport, the aircraft's engines are operating at low power; however, the aircraft's flying path is at a much lower angle to the ground—typically 3°–5°—which means it flies close to the ground for a longer time, exposing a wider area to its noise. The noise generated by aircraft in the airport's vicinity can severely affect its neighbors, to the point where certain noise-sensitive land uses such as residential areas and hospitals may need to be restricted (de Neufville and Odoni, 2003; Horonjeff and McKelvey, 1994).

Actual levels of aircraft noise vary significantly with aircraft and engine size and type, and arrival and take-off procedures. Also, the actual level of discomfort noise causes on people will depend not only on actual noise levels, but also on its duration and frequency. These factors can vary with time: the airport's traffic levels can grow or shrink; airlines may change their fleet, introducing new aircraft with different characteristics; aircraft technology also changes, with newer aircraft being generally quieter than older ones. These variations make planning and determining adequate land use zoning schemes very difficult and are a significant source of uncertainty in urban planning.

In an attempt to facilitate urban planning in the airport's vicinity, several methods have been developed to estimate noise levels in the vicinity of an airport (FAA, 1983, 1999). Each method produces a set of noise contours that mark the zones within which the threshold levels for discomfort are exceeded. In principle, such noise contours should be used by local authorities to regulate the land use in those areas, preventing residential developments wherever the predicted noise levels exceed standard thresholds for acceptable discomfort.

In the US, guidance on noise issues is provided by the Code of Federal Regulations, better known as FAR Part 150 and by the FAA (1983). This regulation is based on the Day–Night Level (DNL) method for measuring aircraft noise, which takes into account noise levels, frequency and duration of exposure, as well as time of day (night or daytime). According to FAR Part 150, DNL measures over 65 dB(A) can cause significant discomfort to people and need to be addressed through land-use restrictions and noise-abatement procedures.

However, there has been significant disagreement on whether this threshold is adequate, as many people report a significant discomfort even where DNL measures are below 65 db(A) (Ward, 2010). Similar issues are faced in Australia, where the Australian Noise Exposure Forecast (ANEF) is used to draw the noise contours and the threshold of 20 ANEF is used as the limit for acceptable noise (Freestone and Baker, 2010); and in Canada, which uses an annoyance prediction system called Noise Exposure Forecast (NEF) (Transport Canada, 2005). Regardless of the noise level threshold used, measures can be taken to reduce noise in the airport vicinity. Netjasov (2012) lists ICAO's "balanced approach" measures recommended for noise reduction: in-flight noise abatement procedures, such as special arrival and take-off trajectories and recommended flying techniques; engine run-up restrictions; use of preferential runways; airport curfews; noise charges; auxiliary power unit operating restrictions; noise level limits above which airlines may be subject to charges or fines; restrictions to aircraft that do not meet specific noise requirements; operating quotas; noise budget restrictions; sound insulation for residences affected by aircraft noise; purchase assurance for homeowners located within the airport noise contours; overflight easements; zoning laws; real-estate/property disclosure laws; acquire land for noise compatibility; determining the number of people residing within specific noise contours; airport noise contour overlay maps. Netjasov (2012) also notes that novel noise management techniques are currently being applied by airports around the world, including optimal arrival and departure trajectories, active role of air traffic controllers, and economic approaches to noise regulation.

References

Bandara, S. and Wirasinghe, S.C. (1990). Airport gate position estimation for minimum total cost—approximate closed form solution. *Transportation Research B* 24, 287–297.

Bandara, S. and Wirasinghe, S. C. (1992). Walking distance minimisation for airport terminal configurations. *Transportation Research A* 26, 59–74.

FAA (1983) *Noise Control and Compatibility Planning for Airports*. Advisory Circular AC 150/5020–1, Federal Aviation Administration, Washington, DC.

FAA (2012) *Airport Design*. Advisory Circular 150/5300–13A. Federal Aviation Administration, Washington, DC.

Freestone, R. and Baker, D. (2010). Challenges in land use planning around Australian airports. *Journal of Air Transport Management* 16, 264–271.

Horonjeff, R., McKelvey, F.X., Sproule, W.J., Young, S.B. (2010) *Planning & Design of Airports*. McGraw Hill, New York.

ICAO (2001) *Annex 6 to the Convention on International Civil Aviation—Operation of Aircraft—Part I—International Commercial Air Transport—Aeroplanes*. International Civil Aviation Organization, Montreal.

ICAO (2004) *Annex 14 to the Convention on International Civil Aviation—Aerodromes—Volume I—Aerodrome Design and Operations*. International Civil Aviation Organization, Montreal.

Netjasov, F. (2012) Contemporary measures for noise reduction in airport surroundings. *Applied Acoustics* 73, 1076–1085.

Transport Canada (1993) *Aerodrome Standards and Recommended Practices*. TP-312E, 4th ed. Ottawa, Canada.

Transport Canada (2005) *Aviation—Land Use in the Vicinity of Airports*. TP 1247E. Ottawa, Canada.

Ward, S. (2010) *Enhancing Airport Land-use Compatibility in Airport Areas*. Planning Advisory Service Report 562, ABI/INFORM Trade & Industry.

18

THE ATC (AIR TRAFFIC CONTROL) SYSTEM

Milan Janić

DELFT UNIVERSITY OF TECHNOLOGY

Introduction

In general, it can be said that the air transport system consists of three main components: airports, ATC (Air Traffic Control) system, and airlines. The first two represent the system's infrastructure with the main attribute capacity as capability to handle aircraft/flight demand under given conditions. These flights are scheduled by the last/third system's component—airlines. Handling by ATC implies their monitoring and controlling while being at and in between their origin and destination airports.

- *Airports* consist of landside and airside area. The airport landside area includes the airport ground access systems/modes, vehicles' parking and maneuvering areas, and the airport passenger and cargo terminal buildings. The airport airside area consists of the runways with adjacent airspace, taxiways, and apron/gate complex.
- *The ATC (Air Traffic Control) system* consists of the facilities and equipment, and operating rules and procedures. The facilities and equipment are the radio-navigational aids on the ground (beacons) and in the air (satellites), the ground and air-based surveillance equipment, avionics; and computing facilities processing the flight (and supportive) data for different planning and operational purposes (Graves, 1998; Horonjeff and McKelvey, 1994).
- *Airlines* operate the fleets consisting of different aircraft types (size, seat capacity, etc.). These are scheduled over the air route networks generally having hub-and-spoke and point-to-point spatial configuration influencing the volume and intensity of demand for the ATC system in a given airspace.

This chapter focuses on the ATC (Air Traffic Control) system. In addition to this introductory section, it consists of four other sections. Section 2 describes the main components and operations of the ATC system. Section 3 presents demand, capacity, and aircraft/flight delays in the European and US ATC system as their main operational performances. Section 4 deals with modelling of the capacity of ATC system expressed by the ATC controller's workload. The last section contains some conclusions.

1. The ATC (Air Traffic Control) System

The ATC (Air Traffic Control) system is established over some designated airspace to provide safe, effective, and efficient air traffic.That implies accommodating aircraft/flights at the system's demand without conflicts (safety), significant delays (effectiveness), and additional costs (efficiency). Such a system is characterized by the airspace, components, automation, separation rules, and performances.

Airspace

In order to fulfil the above-mentioned objectives, the ATC divides a large controlled airspace into smaller parts at two levels. First, the airspace is divided into airport zones, terminal areas, and low and high altitude en-route areas. Second, each of the above-mentioned areas is subdivided into smaller parts—the ATC sectors, assigned to one or a team of ATC controllers. Figure 18.1 shows a simplified scheme of such airspace division (Janić, 2000).

- *Airport zone* is established around an airport with a usual radius of 2–2.5 nm (nm: nautical mile) and altitude of 2,000 ft (600 m) MSL (Middle Sea Level) to enable control and management of aircraft landing and taking off.
- *Terminal airspace* spreads around and above the airport zone. It is established around a single busy or several airports of the same airport system with a radius of 40 to 50 nm and altitude up to FL (Flight Level) 100 (10,000 ft or 3,000 m—Class B airspace; each FL is determined by 10^3 ft (1 ft (0.305 m)) (ICAO, 2007). The aircraft fly through this area along the prescribed arrival and departure trajectories, defined either by radio-navigational facilities, the ATC radar vectoring, and/or RNAV (Area Navigation) and RNP (Required Navigation Performance).

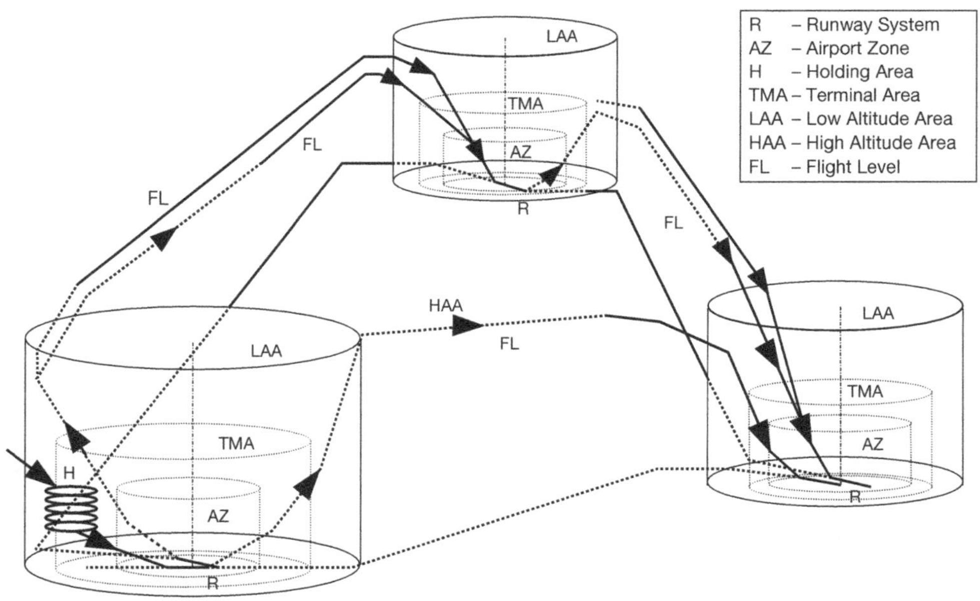

Figure 18.1 Simplified scheme of the ATC system's airspace
Source: Janić (2000)

- *Low altitude airspace* spreads around and above the terminal airspace between the altitudes between FL100 and FL180 (or FL245) MSL (Class E airspace) (Horonjeff and McKelvey, 1994; ICAO, 2007). The aircraft use this airspace for the climbing and descending phase of flights.
- *High-altitude airspace* covers the altitudes between FL180 (or FL245) and FL600 MSL (Class A airspace). The aircraft use this airspace for cruising phase of their flights (ICAO, 2007).

2. Components

The main components of the ATC system are: (1) communication equipment; (2) navigational facilities; (3) surveillance equipment, both with the components on the ground, at aircraft, and in the (air)space; (4) air traffic management system; and (5) the ATC staff, including the ATC controllers and ultimately the aircraft pilots.

1. *Communication equipment* includes communication channels for transmission of information between pilots and ATC controllers (VHF/UHF air/ground voice and non-voice communication links); particular ATC's control units; and the ATC and the outside environment. The forthcoming communication systems under development within European (EU) SESAR and US NextGen initiatives imply mostly automated non-voice transmission of information throughout the ATC system (www.sesarju.eu/; www.faa.gov/nextgen/).
2. *Navigational facilities* include ground aids and space satellites both used for the aircraft primary navigation. The ground aids are VOR, DME, VOR/DME, and RNAV system (use the information about distance and azimuth by VORTAC stations as input); the over-water en-route OMEGA, LORAN-C; Doppler Navigation System; Inertial Navigation System (INS); the terminal airspace/airport ILS (Instrumental Landing System) and MLS (Microwave Landing System), and the airport navigational equipment, including the approach lighting systems, slope indicators, the surface detection equipment, etc. The forthcoming navigational procedures are to be primarily based on RNAV and RNP, which will enable flying along the most preferred routes thus contributing to increased flexibility of using airspace, and consequently higher efficiency and effectiveness of flights.
3. *Surveillance equipment* consists of the primary and the beacon (secondary) radars enabling the ATC controllers to monitor and control separation between aircraft and make appropriate decisions, which are sent to pilots by the A/G communication channel link (Horonjeff and McKelvey, 1994). The forthcoming ADS-B (Automatic Dependent Surveillance) in combination with CDTI (Cockpit Display Traffic Information) system will enable the aircraft/pilots' self-monitoring of their mutual separation, thus partially relieving the ATC controllers' workload (www.sesarju.eu/; www.faa.gov/nextgen/).
4. *ATM (Air Traffic Management)* system enables performing a two-level ATC control function: a) safe separation between aircraft according to the prescribed rules; and b) safe, efficient, and effective movement of traffic flows under given conditions. The ATM system's subfunctions are AS (Airspace Management), ATS (Air Traffic Services), and ATFM (Air Traffic Flow Management) (Janić, 2000).
    - *AM (Airspace Management)* focuses on the availability of airspace for the aircraft/flight optimal trajectories. For example, in the currently very busy North Atlantic airspace, where the traffic flows are kept on tracks (airways), the ATC applies the minimum longitudinal separation between aircraft on the same flight level(s) of 10 minutes,

vertical separation of 2,000 ft, and the lateral separation between tracks (airways) of 60 nm. By reducing the vertical separation to 1,000 ft (also thanks to the ADS-B) for the aircraft flying at the altitudes FL290–FL410, six additional cruising levels can be provided. In addition, reducing the lateral separation from the current 60 to 30 nm, which will be possible thanks to RNAV, proportionally more additional tracks (airways) can be opened. Consequently, the airspace capacity could be doubled compared to its current counterpart (www.sesarju.eu/; www.faa.gov/nextgen/).

- *ATS (Air Traffic Services)* is a core component of the ATM function continuously providing conditions for safe, efficient, and effective movement of air traffic flows.
- *ATFM (Air Traffic Flow Management)* balances the air traffic flows as demand and the available ATC system capacity aiming at preventing long ground/airport and air/airspace aircraft/flight delays, and possible safety implications. The ATFM can be applied to the strategic and tactical level. At the former, it provides planning of handling aircraft/flights respecting their preferred four-dimensional trajectories several hours and/or several months in advance. At the tactical level, ATFM sets up the acceptance rates for the specific en-route points and/or destination airports accompanied by the allocation of departure slots (EEC, 2007, 2008; Odoni, 1987). The tactical ATFM operates on line, implying updating of the aircraft/flights position regularly in the very short time spots (1–3 minutes).

5. *The ATC staff* includes the ATC controllers, and ultimately the pilots. The former are responsible for monitoring and controlling air traffic in the airspace of their jurisdiction. They are usually organized as teams of two or three people with a precise division of tasks and responsibilities. Pilots participate in the ATC processes by following instructions, in addition to their own, and guidance from the ATC controllers. These mainly relate to allocated four-dimensional trajectories and separation from other aircraft/traffic.

Automation

The more intensive automation of the ATC system's operations and processes has recently started through the European SESAR (Single European Sky ATM Research) and US NextGen (Next Generation) research and development programmes (Erzberger, 2004; www.sesarju.eu/; www.faa.gov/nextgen/). Their main objective is to develop a new generation ATM (Air Traffic Management) system, which will be able to cope with growing demand while at the same time increase safety and improve flight fuel/energy/cost efficiency, and consequently the overall environmental and social friendliness of air transportation.

The main ATC components to be developed with slight variations in both above-mentioned programs depend on the phase of flight as given for the US NextGen program in Table 18.1.

The components in Table 18.1 should enable:

- Improved access to different information about the state of airspace through common network during the flight planning.
- More precise planning of the actual departure times, including improvements in the departure performances by using RNAV- and RNP-supported multiple departure paths, particularly over the metropolitan areas.
- Improved surveillance enabling reduction of the aircraft separation minima during the climbing and cruising phase of flight (the responsibilities will be shared between the ATC controllers and pilots, thanks to ADS-B and CDTI).

Table 18.1 Components of the US ATC NextGen development initiative

FLIGHT PLANNING

Ground ATC facilities/equipment
- Common Support Services-Weather (GSS-Wx)
- Data Communication (Data Comm.)
- En Route Automation Modernization (ERAM)
- Modernized Aeronautical Information Management system (AIM)
- NextGen Weather Processor (NWP)
- System Wide Information Management (SWIM)
- Terminal Flight Data Manager (TFDM)
- Traffic Flow Management System (TFMS)

PUSH BACK, TAXI, AND DEPARTURE

Ground ATC facilities/equipment
- Automatic Dependent Surveillance Broadcast (ADS-B) ground stations
- Airport Surface Detection Equipment—Model X (ASDE X)
- GSS-Wx
- Data Comm.
- Integrated Departure and Arrival Coordination System
- Modernized AIM
- NWP
- Satellite Based Augmentation System (SBAS)
- Standard Terminal Automation Replacement System (STARS)
- SWIM. TFDM. TEMS

Avionics
- ADS-B In and out with associate displays like Cockpit Display of Traffic Information (CDTI)
- Area Navigation (RNAV) and Required Navigation Performance (RNP)
- Data Comm.

CLIMB AND CRUISE

Ground ATC facilities/equipment
- ADS-B ground stations
- Advanced Technologies and Oceanic Procedures
- GSS-Wx, Data Comm., ERAM, NVP,
- Time-Based Flow Management (TBFM)
- TFMS

Avionics
- ADS-B In and out with associate displays CDTI
- Data Comm. Including integration with Flight Management System (FMS)
- Future Air Navigation System in oceanic airspace
- RNAV and RPN

DESCENT AND APPROACH

Ground ATC facilities/equipment
- ADS-B ground stations
- ASDE-X
- Data Comm.
- GSS-Wx
- NWP, SBAS
- STARS enhancements, TBFM, TEDM, TFMS

Avionics
- ADS-B In and out with associate displays like CDTI
- Enhanced Flight Vision System (EFVS)
- Data Comm.
- RNAV and RNP, Vertical Navigation (VN)

LANDING, TAXI, AND ARRIVAL

Ground ATC facilities/equipment
- ADS-B ground stations
- ASDE-X
- CSS-W
- Data Comm.
- Ground-Based Augmentation System (GBAS)
- Integrated Departure and Arrival Coordination System
- Modernized AIM
- SBAS, STARS enhancements, WIM, TBFM, TFDM, TFMS

Avionics
- ADS-B In and out with associate displays like CDTI
- EFVS, Data Comm., GBAS

Source: FAA, 2013; www.faa.gov/nextgen/

- Planning well in advance arrivals and departures along the optimal single and/or multiple 4D trajectories (supported by RNAV and RNP) between the end/beginning of cruising and the runway landing/departure phase of flight, respectively, in terms required safety, fuel consumption (and related emissions of GHG (Green House Gases)), and noise. These trajectories will also end/begin at the runways depending on the current and predicted congestion and delays. As well, during the arrivals, ground maneuvering, and departures, independently on the weather conditions (good or bad), both ATC controllers and pilots will be able to monitor the aircraft position relative to the surrounding traffic on the associated screens (CDTI).
- Reducing the aircraft/flight separation minima thanks to the surveillance improved precision.
- Transmitting most of the information/data automatically, thus reducing the quantity and intensity of the A/G voice communications independently on the content (routine, exceptional, emergency, weather, etc.), and consequently increasing the ATC controller's capacity at the same workload.

The European SESAR has the very similar components as those in Table 18.1 and the quantitative targets as follows (www.sesarju.eu/):

- Decreasing the ATM-related charges for the airspace users for about 50 percent.
- Reducing the impacts of aircraft/flights on the environment for about 10 percent.
- Increasing the ATC system capacity for about 200 percent, which will enable reducing both ground and airborne delays.
- Improving the safety level by 10 percent.

The ATC Separation Rules

The main task of the ATC system (i.e., ATC controller) is to provide, by using the above-mentioned facilities and equipment, safe separation between aircraft/flights. In the airport zone and terminal airspace, the ATC horizontal distance-based separation rules are applied between the landing aircraft on the same runway depending on the sequences in order to protect trailing aircraft from the wake vortices of leading aircraft. These are 2.5, 3, 4, 5, and 6 nm. The ATC time-based separation rules of 1 and 2 min are applied depending on the aircraft taking-off sequence from the same runway. In addition, only one aircraft can occupy the runway at a time (ICAO, 2007; Janić, 2012).

In the low- and high-altitude airspace, both ATC vertical and horizontal time- or distance-based separation rules are applied. The vertical separation rules of 1,000 ft are applied to the aircraft/flights cruising at FLs up to FL290 (29,000 ft) and 2,000 ft above. In some controlled airspaces, the vertical separation of 1,000 ft is applied up to FL410 (41,000 ft) and 2,000 ft above. In addition, the vertical separation can be applied between two climbing/descending aircraft when one reach required FL before the other arrives on its previous FL. Otherwise, the ATC horizontal separation rules between these and the aircraft cruising on the same route and FL are applied.

The distance-based longitudinal separation rule is 15 nm (28 km). In case of the radar/ADS-B/CDTI surveillance, the minimum horizontal separation rules between the aircraft/flights on the same FL is 5 nm. The minimum time-based separation rules are 3, 5, 10, and 15 min, depending on the air-route configuration, navigational aids used, and traffic situation (for example, crossing routes at different angles/courses, climbing and descending along the same route, etc.). The minimum lateral separation rules between routes of the aircraft flying on the same FL if RNP is applied need to be at least four times greater than the required limits for accuracy (ICAO, 2007; Janić, 2012).

Performances

The ATC system performances can be categorized as operational, economic, environmental and social given the infrastructure (airspace/ATC sectors and airports, i.e., air route network) and the above-mentioned supportive facilities and equipment.

- The operational performances are demand, capacity, and congestion and related delays of aircraft/flights as an outcome from the dynamic interaction of the first two.
- The economic performances can be expressed in two ways: i) the efficiency of ATC system expressed by its charges of the users–aircraft/flights; and ii) the efficiency of performed flights expressed by the cost of deviation of the actual from the optimal/planned flight paths trajectories due to any reason (for example, airspace congestion, bad weather, and/or failures of the facilities and equipment, etc.).
- The environmental performances can be expressed by an extra fuel consumption and related emissions of GHG due to the airside delays comprising those before landings, flying at the fuel non-optimal cruising FLs, and rerouting due to the ATC system's and other air traffic and non-air traffic reasons.
- The social performances are expressed by the effects of measures for reducing the impact of air traffic on the overall population and individual users/air passengers on board. These impacts are noise around airports and air traffic accidents and incidents, i.e., safety, which from some perspectives can also be considered as the operational performance.

3. Analysis of the Operational Performances of the ATC System

Demand

The ATC demand is expressed by the number of flights requesting to pass through the airspace under jurisdiction of a given ATC unit during the specified period of time (hour, day, month, season, year). The volume/number and intensity of flight demand are mainly driven by the overall airline activities satisfying the passenger and freight/cargo air transport demand.

The ATC systems such as, for example, European EUROCONTROL and US FA (Federal Aviation Administration) usually express their demand in terms of the number of flights to be served during the specified period of time as shown on Figure 18.2.

As can be seen, the demand in the US airspace has been and is expected to be about three times greater than that in Europe, which is mainly due to the difference in the size of both airspaces. In addition, the forecast for both areas indicates that a steady growth of the number of IFR flights could be expected over the forthcoming decades (2013-2032/2035) (FAA, 2011; EEC, 2013).

Capacity

The ATC capacity is defined as the maximum number of aircraft/flights which can be handled during the specified period of time under given conditions. Usually, the capacity of airspace including the aircraft/flight trajectories, i.e. the air route network and the capacity of

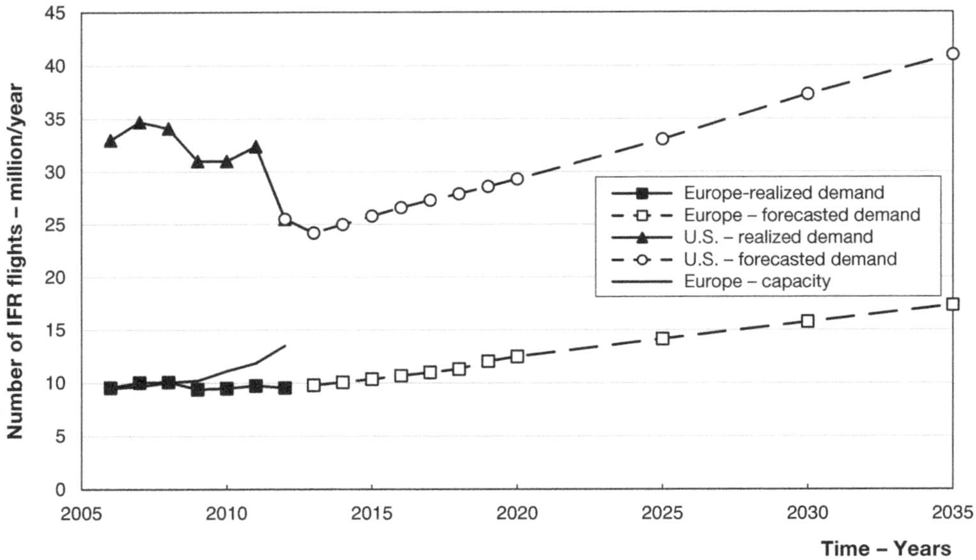

Figure 18.2 Past, present, and future development of IFR aircraft/flight demand in the European and US airspace

Source: EEC, 2013; FAA, 2011

corresponding ATC unit(s)/sector(s) are considered (usually separately). In both cases, given conditions are defined by constant demand for service during the specified period of time (the "ultimate capacity"), the average acceptable delay of aircraft/flights before getting service (the "practical capacity"[1]), and sustainability of the capacity over the specified period of time (for few continuous hours) ("sustainable" capacity). An additional important condition for all above-mentioned definitions is the required level of safety.

The capacity is usually determined for a given ATC sector under the jurisdiction of the ATC controller (or associated team) as the maximum number of aircraft/flights safely handled during a given period of time (one hour) under conditions of constant demand and limited acceptable ATC controller's workload. Alternatively, this capacity can be expressed by the maximum number of aircraft/flights simultaneously monitored in the sector.

For example, in the European airspace, the ATC sector capacity typically varies between 40 and 70 aircraft/flights per hour imposing the ATC controller's workload of 70 percent or more (i.e., active working time of 42 min/h or longer). This capacity/ATC controller workload is mainly influenced by the airspace configuration, the military use of airspace, the air traffic complexity, etc. In addition, the maximum number of simultaneously monitored aircraft/flights in the sector can be about 25 (EEC, 2003; Janić, 2000). By summing up the sectors' capacities and assuming that all sectors operate continuously (i.e. 24 h/day during 365 days/year), the annual ATC system capacity can be obtained as shown in Figure 18.2.

Delays

The ATC delays occur whenever the demand exceeds the available capacity of the ATC airspace (sector/unit). They can be stochastic when the generally lower demand temporally exceeds the capacity. These delays are deterministic when the demand exceeds the capacity for the longer

a) Average delay per delayed flight

b) Proportions of delayed flights

c) Average delays in en-route airspace

d) Average en-route flight delays in reaching required FL (European airspace–summer 2009)

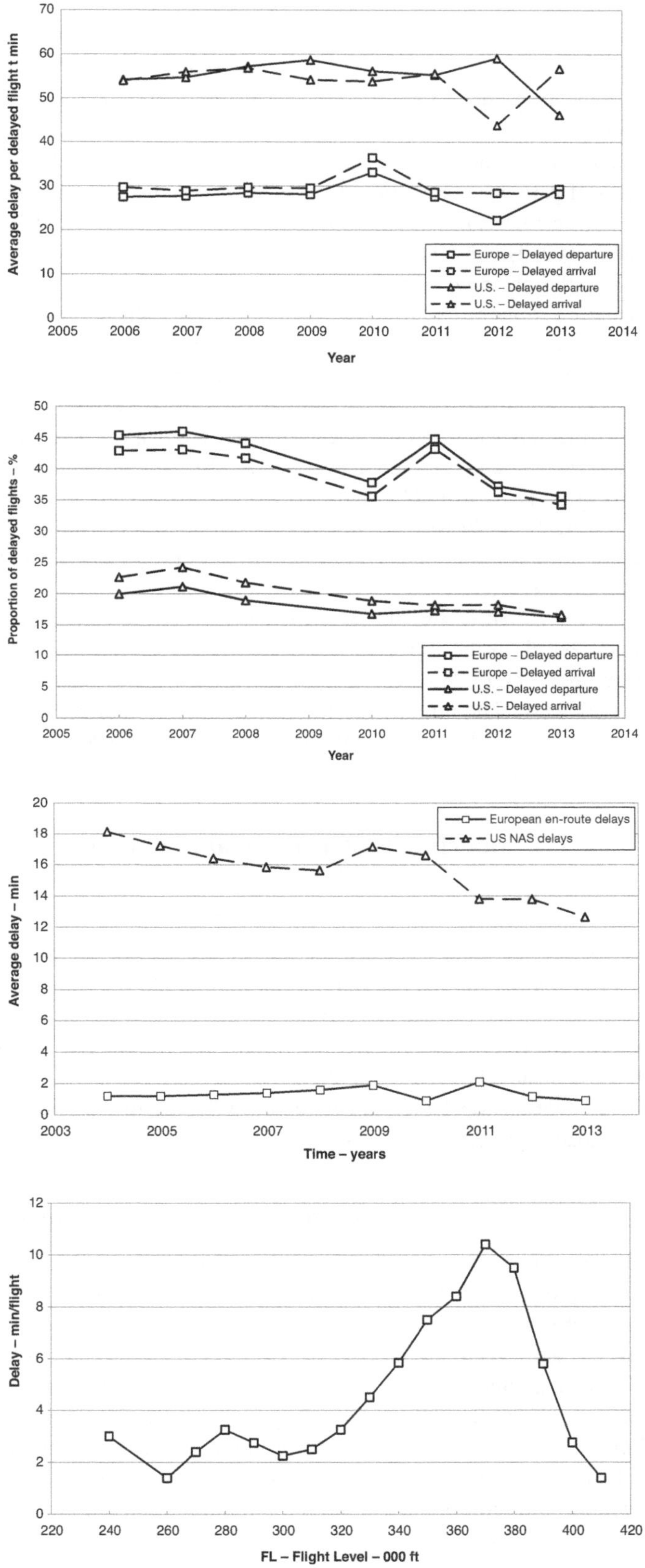

Figure 18.3
Development of ATM delays in European and US airspace
Source: EEC, 2009; 2013; Griffiths, 2013; www.transtats.bts.gov/

period of time (for example, for a few hours). The ATC delays generally can take place individually and/or in different combinations before the aircraft/flight departures and/or landings, entering particular ATC sector(s), and before reaching the optimal/planned cruising altitude (FL). Consequently, these delays are labelled accordingly. Figure 18.3 shows development of particular categories of the ATC delays in the European and US ATC system.

Figure 18.3a shows that the average arrival and departure delays, with slight variations, have been relatively constant in the European (around 30 min) and US (around 55 min) ATC systems during the observed period. In the US, the arrival delays have been slightly longer than their departure counterparts, while in Europe it has been the opposite. In addition, Figure 18.3b shows that the proportions of delayed arrival and departure flights have decreased over the observed period in both systems. At the same time, they have been much lower in the US (around 15–25 percent) than in the European (35–45 percent) ATC system. The mutual relations of proportions of arrivals and departures have been analogous to that of the average delays in both ATC systems. The average departure and arrival delays and their proportions have also been dependent on each other in both ATC systems.

Figure 18.3c shows that the average en-route delays have also decreased during the observed period, but as much higher in the US than in the European ATC system (12–18 min and up to 2 min), respectively.

Figure 18.3d shows the average delay of an aircraft/flights while waiting to reach the requested FL in the European ATC system. As can be seen, these delays amounting to about 4 min to reach FL280 have increased to about 10 min with the requested FL increasing up to FLs 360–390. The number of aircraft/flights has also increased in line with the increase in the average delays. This means that about 100,000 to 150,000 flights have been waiting for up to about 4 minutes to reach FLs240–320 and FLs400–410, and 500,000 flights to reach FLs360–390.

4. Modeling the Operational Performances of the ATC System

General

The above-mentioned performances of different components of ATC system can be modeled using analytical, simulation, and optimization models. In general, the capacity of airspace, ATC sector based on the ATC controller's workload, the A/G communication link, air route network, and ATFM operations and processes have been under focus (Bertsimas and Patterson, 2000; Janić and Tosic, 1991; Odoni, 1987; Péter, T. and Szabó, Z., 2012; Vranas, 1994). Here, the analytical model for estimating the ATC sector capacity based on the ATC controllers workload is presented (Janić, 1997).

Model of the ATC Sector Capacity

The models for estimating the ATC sector ("ultimate") capacity are commonly based on estimation of the ATC controller's workload generated by executing the main ATC tasks/activities such as monitoring air traffic on the radar and other (CDIT) screens, decision-making, and executing the control tasks, all under given conditions (EEC, 2003; Janić, 1997). Monitoring screens enables the ATC controller to memorize the current air traffic situation in the sector by using his/her internal "mental" model and store it into the short-term memory. This is then used together with information exchanged with pilots via the A/G/ communication link and that with the air traffic controllers from neighboring ATC sectors for creating and carrying out control tasks. It is commonly assumed that the ATC controller cannot simultaneously monitor screens and carry out control tasks, which appears crucial for modeling the ATC sector capacity in the given context. Assuming that the ATC controller operates as at the channel serving demand consisting of the control and monitoring tasks, his/her workload can be expressed by the workload coefficient as follows:

$$\rho = \rho_m + \rho_c = 1.0 \qquad (1)$$

where

ρ_m is the workload coefficient due to monitoring screens; and
ρ_c is the workload coefficient due to performing control tasks.

Equation 1 implies the ATC controller's maximum workload during a given active period of time (for example, 42 min of 1 hour (EEC, 2003)).

The workload coefficient ρ_m in equation 1 can be estimated as:

$$\rho_m = \frac{t_1^*}{t_1^* + t_2^*} \qquad (2)$$

where

t_1^* is the average time of one observation of traffic situation on the screens; and
t_2^* is the average time between successive observation of traffic situation on the screens.

The average time t_1^* in equation 2 is estimated as follows:

$$t_1^* = \lambda_c t_f p_c H \tau_{01} + \frac{(\lambda_c t_f p_c)^2 - (\lambda_c t_f p_c)}{\tau_{02}} \qquad (3a)$$

where

λ_c is the intensity of air traffic flow entering the sector (aircraft/h);
t_f is the aircraft average flying time through the sector (min);
H is the entropy the ATC controller is faced with while observing the screens (bits);
τ_{01} is the time for processing a unit of information obtained from observing the screens (s/bit),
τ_{02} is the time for establishing an appropriate relationship between any pair of targets (aircraft) on the screens (s); and
p_c is the proportion of targets (aircraft) under one observation on the screens ($p_c \leq 1.0$).

The time t_2^* in equation 2 is estimated as follows:

$$t_2^* = \frac{3\varepsilon_0 m (t_e + t_b)}{\Phi^{-1}(0.5 - (p_0/2))} \qquad (3b)$$

where

m is the number of control tasks carried out and causing change of the actual traffic situation as compared to its memorized counterpart;

t_e is the average time of carrying out a control task (s);
t_b is the time of inventorying/remembering traffic situation in the short-term memory (s);
ε_0 is the maximum tolerable error during comparison of two pictures of traffic situation from two successive observations of the screens (this is assumed to be the random variable with Gauss (normal) probability distribution),
p_0 is the probability that the actual error is greater than the maximum tolerable value ε_0 set up in advance,
Φ^{-1} is the inverse Laplace's function.

The workload coefficient ρ_c in equation 1 can be estimated as:

$$\rho_c = \lambda_c L t_e \tag{4}$$

where

L is the number of control tasks performed for one aircraft while in the sector.

The other symbols are as the previous equations.

By a relatively simple mathematical manipulation of equations 1, 2, 3a and 3b, 4, and 5, the polynomial of third degree with λ_c as the unknown variable is obtained as follows:

$$a_0 + \sum_{n=1}^{3} a_n \lambda_c^n = 0.5 \tag{5}$$

where

$$a_0 = \frac{6\varepsilon_0 m\left(t_e + t_b\right)}{L t_e t_f^2 p_c^2 \tau_{02}^2 \left[\Phi^{-1}\left(0.5 - p_0/2\right)\right]}$$

$$a_1 = \frac{6\varepsilon_0 m\left(t_e + t_b\right)}{t_f^2 p_c^2 \tau_{02}^2 \left[\Phi^{-1}\left(0.5 - p_0/2\right)\right]}$$

$$a_2 = \frac{2\left(H\tau_{01} - 0.5 p_c \tau_{02}\right)}{p_c^2 \tau_{01} \tau_{02}}$$

$$a_3 = 1.0$$

The real root of equation 5 represents the ATC sector capacity λ_c. After determining this capacity, the number of aircraft simultaneously in the ATC sector can be estimated as:

$$N(\lambda_c) = \lambda_c t_f \tag{6}$$

where all symbols are as in the previous equations.

The model is applied to a hypothetical ATC control sector and traffic scenarios there. Some of the results are shown in Figure 18.4.

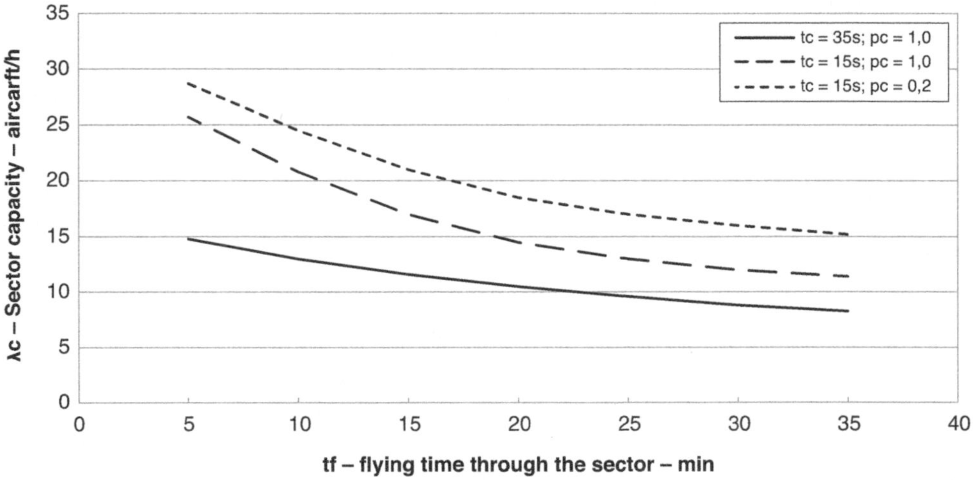

Figure 18.4 Relationship between the ATC sector capacity and the aircraft flying time through the sector, duration of control tasks, and tactics of monitoring screens

Source: Janić (1997)

As can be seen, the ATC sector capacity decreases more than proportionally with increasing of the aircraft flying time through the sector, i.e. the time under the ATC control. It increases with shortening of time for preparing and executing the control tasks, and the number of targets (aircraft) under one observation on the screens.

Consequently, the reserves for increasing the ATC sector capacity are contained in shortening the preparation and execution of control tasks and tactics for monitoring screens. The former will be achieved by the above-mentioned automated exchange of flight data and related information including reducing the number of tasks (and communications) for establishing and maintaining the aircraft separation. Thanks to ADS-B and CDTI, a part of these tasks will be

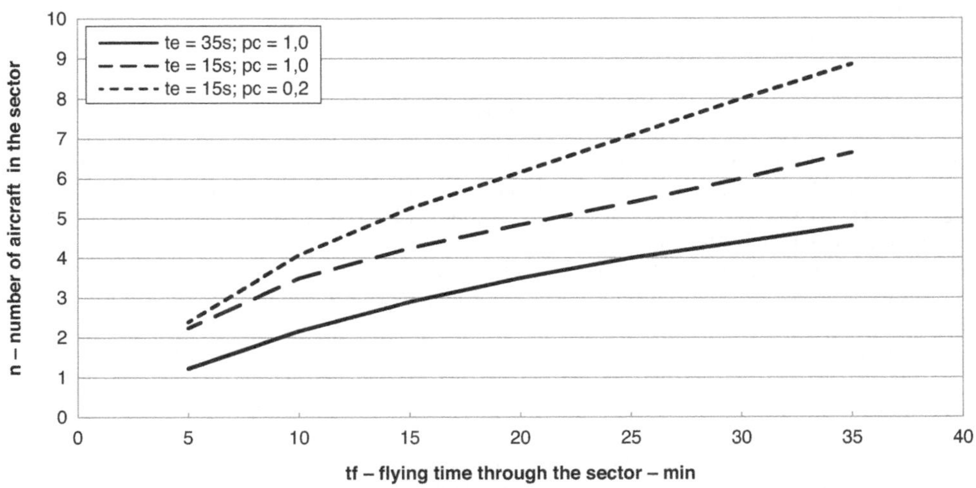

Figure 18.5 Relationship between the number of aircraft in the ATC sector, flying time through the sector, and the sector capacity

allocated to pilots. The latter will be achieved by reducing the entropy in recognizing particular targets (aircraft) and needed frequency of observations of the screens. It is to be expected that more or less "passive" monitoring will dominate in the above-mentioned automated ATC system, which will reduce the ATC controller's workload per single aircraft and consequently increase the number of handled aircraft, i.e., the ("ultimate") sector capacity, under given conditions.

Further, the corresponding number of aircraft simultaneously monitored in the sector will increase at a decreasing rate with the increase of aircraft flying time through the sector and the ATC sector capacity, as shown in Figure 18.5.

5. Conclusions

This chapter has dealt with the description, analysis, and modeling of the ATC (Air Traffic Control) system as the main component of the air transport system together with airports and airlines. The airspace, main components, particularly current and prospective state of automation, and a range of performances of the system are described. This is followed by an analysis of the ATC system main operational performances such as demand, capacity, and the outcome of their dynamic interaction—aircraft flight/delays. In addition, the modeling of the ATC system's operational performances has been illustrated by an analytical model of the ATC sector capacity based on the air traffic controller workload.

The general findings indicate that the contemporary ATC system is and will increasingly be faced with the medium- to long-term growing demand requiring provision of the sufficient capacity in order to maintain aircraft/flight delays and other performances (particularly safety) under prescribed limits (targets). The solutions are contained in the forthcoming increasing level of the ATC system's automation currently underway within different programs such as, for example, the European SESAR and US NextGen research and development initiative to be implemented by the year 2020.

Note

1 This is also called "effective" capacity expressed by the traffic volume handled by an ATC unit with an average en-route flight delay of 1 min (EEC, 2007).

References

Bertsimas, D. and Patterson, S. S. (2000) "The Traffic Flow Management Rerouting Problem in Air Traffic Control: A Dynamic Network Flow Approach," *Transportation Science*, Vol. 34, No. 3, pp. 239–255.

EEC (2003) Pessimistic Sector Capacity Estimation, EEC Note No. 21/03, project COCA, EUROCONTROL, Brussels.

EEC (2007) Capacity Assessment & Planning Guidance: An Overview of the European Network Capacity Planning Process, Edition, September 2007, EUROCONTROL, Brussels.

EEC (2008) European Medium-Term ATM Network Capacity Plan Assessment 2009–2012, DMEAN—Dynamic Management of the European Airspace Network Edition, September, EUROCONTROL, Brussels.

EEC (2009) CODA Digest: Delays to Air Transport in Europe, Cooperative Network Design, EUROCONTROL, Brussels.

EEC (2013) Challenges of Growth 2013: Task 4: European Air Traffic in 2035, European Organization for the Safety of Air Navigation (EUROCONTROL), Brussels.

Erzberger, H. (2004) "Transforming the NAS: the Next Generation Air Traffic Control System", 24th ICAS (International Congress of the Aeronautical Sciences), September, Yokohama.

FAA (2011) FAA Aerospace Forecast: Fiscal Years 2012–2035, US Department of Transportation, Aviation Policy and Planning, Federal Aviation Administration, Washington, DC.

FAA (2013) NextGen Implementation Plan, Office of the NextGen, Federal Aviation Administration, US Department of Transportation, Washington, DC.

Graves, D. (1998) *UK Air Traffic Control: A Layman's Guide*, 3rd ed., Airlife Publishing, Shrewsbury.

Griffiths, P. (2013) SES Performance Targets and Achievements, EUROCONTROL, Brussels.

Horonjeff, R. and McKelvey, X. F. (1994) *Planning & Design of Airports*, 3rd ed., McGraw Book Hill Company, New York.

ICAO (2007) Air Traffic Management (Doc. 4444), 15th ed., International Civil Aviation Organization, Montreal.

Janić, M. (1986) "A Practical Capacity Model of an Air Route Network", *Transportation Planning and Technology*, Vol. 11, pp. 87–103.

Janić, M. and Tosic, V. (1991) "En Route Sector Capacity Model", *Transportation Science*, Vol. 25, No. 4, pp. 299–307.

Janić, M. (2000) *Air Transport System Analysis and Modelling: Capacity, Quality of Services, and Economics*, Gordon & Breach Science, Amsterdam.

Janić M. (2012) "Modelling the Effects of Different Air Traffic Control (ATC) Operational Procedures, Separation Rules, and Service Priority Disciplines on Runway Landing Capacity," *Journal of Advanced Transportation* (wileyonlinelibrary.com). DOI: 10.1002/atr.1208

Odoni, A. R. (1987) "The Flow Management Problem in Air Traffic Control," in *Flow Control of Congested Networks*, NATO ACI Series, Vol. F38, Springer-Verlag, Berlin.

Péter, T. and Szabó, K., (2012), "A New Network Model for the Analysis of Air Traffic Networks," *Transportation Engineering*, Vol. 40, No. 1, pp. 39–44.

Vranas, B. M., Bertsimas, D., and Odoni, A.R. (1994) "Dynamic Ground-Holding Policies for a Network of Airports," *Transportation Science*, Vol. 28, No. 4, pp. 275–291.

Websites

www.eurocontrol.int/articles/air-traffic-control-services
www.faa.gov/nextgen/
www.sesarju.eu/
www.transtats.bts.gov/

19

RAILWAY PLANNING AND OPERATIONS

Nigel G. Harris

THE RAILWAY CONSULTANCY, LONDON

Introduction

Key Attributes of the Mode

Good railway planning should attempt to take into account the inherent attributes of the mode, which include:

- effectively being what Professor Felix Schmid at the University of Birmingham has termed a "guided missile," constrained by track layout;
- a limited network leading to a finite number of possible options to optimize;
- having a critical interface between wheel and rail, and friction at only about 11 percent of the value of road vehicles, which puts limits on gradients (typical maxima 1 percent for heavy freight and only 4 percent for suburban);
- having low friction leading to moderate acceleration and long braking distances, but energy efficiency if momentum can be maintained;
- being environmentally efficient, with the potential for any pollution to be emitted elsewhere, under more controlled conditions;
- as a public mode of transport, placing safety highly, and therefore leading to risk-averse industry behaviour such as over-engineering;
- a tendency toward high structural strength, leading to high weight and higher centrifugal forces, limiting curvature (with appropriate compensating super-elevation) to c.1km at 100kph and c.7km at 300kph (Lichtberger, 2005);
- being economically efficient with large volumes and/or medium-long distances (variants of the mode (e.g. metro, tram) are better for shorter distances);
- having fixed costs, with unsure economies of scale but significant economies of density and diseconomies associated with complexity (e.g. multiple traction systems or gauges);
- having low marginal costs, which means that the costs of carrying one extra unit of traffic can be very low, provided that this does not require additional capacity beyond that needed for peak traffic levels.

Because many competing modes are largely subsidized in a hidden fashion (e.g. externalities not paid for), rail is not always able to achieve some of these economies; for instance, where domestic air services are allowed to compete directly with rail, they may merely rob it of passengers who could be carried at marginal cost.

Objectives

Railways are also built and operated to satisfy a wide range of objectives, which can determine what is planned. These may include:

- maximizing punctuality or other measures of service or customer satisfaction;
- maximizing traffic;
- maximizing profit;
- maximizing time saved to users and/or non-users;
- maximizing road congestion relief;
- maximizing access to employment;
- maximizing economic activity, whether at the local or national level;
- minimizing the environmental impact of transport;
- providing a strategic network (e.g. for military purposes).

All of these are legitimate objectives, but problems can arise in two situations: first, when railways are (as is quite common) required to achieve multiple but mutually conflicting objectives (e.g. maximizing traffic and punctuality, or maximizing social benefit within a profitability target (as was the case with London Underground in the 1990s). The other occasion when railway objectives come into play is when there are rapid changes of direction, e.g. in Britain, the Serpell Report of 1982 and the Northern Rail Review of 2007 both threatened a switch from socially to financially based objectives.

Having established the relevant objectives, policy decisions can then be evaluated and detailed plans developed. A well-managed process (part sequential, part parallel and part involving appropriate feedback loops (as options are found to be suboptimal) needs to be followed, prior to the actual act of construction or implementation.

Demand Planning

Once the background has been set, however, transport planning theory is quite clear: railways should seek to minimize generalized cost for all potential customers (the detail here is set out for a passenger railway, but is not dissimilar for freight):

$$\min \sum_{i=1}^{n} \sum_{j=1}^{n} \sum_{p=1}^{n} \left(P_{ij} \times G_{ij} \right)$$

where

P_{ij} = the number of passengers between i and j, summed across all passenger types p; and

G_{ij} = the generalized cost of travel from i to j (between n origins and destinations).

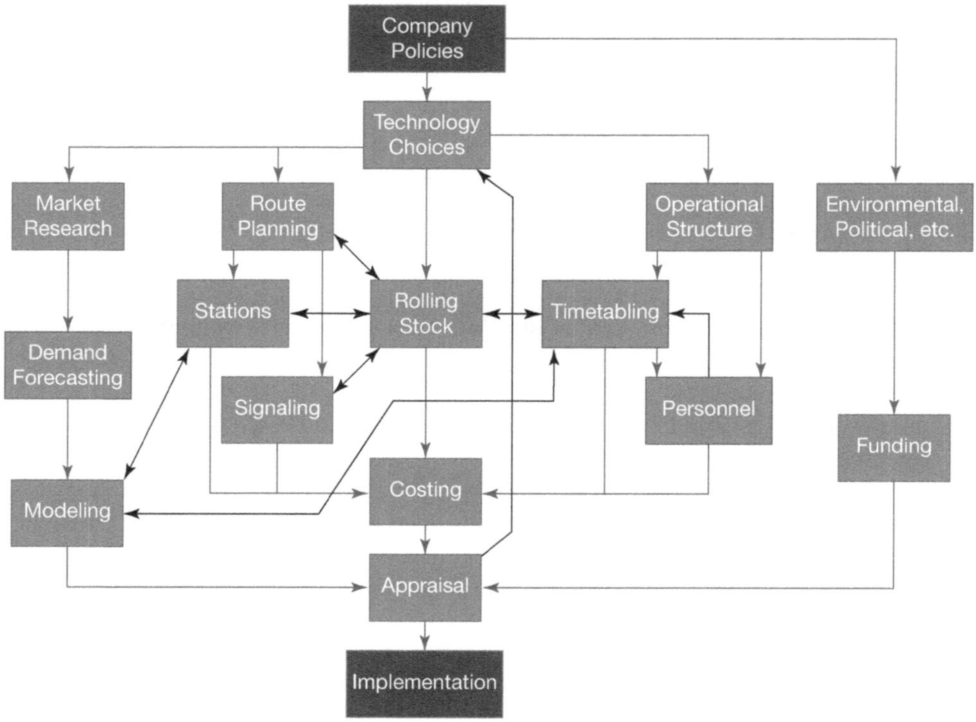

Figure 19.1 The planning process

Generalized cost is a concept developed in the 1960s (Warner, 1962), discussed in the 1970s (Grey, 1978) and generally adopted as normal practice from the 1980s (e.g. Pienaar, 1997). It may be considered in layman's terms as a basket of hassle; in economists' terms, it is a measure of disutility, defined in a form such as

$$GC = F + b_1 \cdot A + b_2 \cdot W + b_3 \cdot R + n \cdot I + b_0 \quad (1)$$

where:

A = access (and egress) time
W = waiting time
R = running (in-vehicle) time
$b_1 - b_3$ are weighting parameters
F = fare (which needs to be converted into a time equivalent using a Value of Time)
$n \cdot I$ = number of interchanges (or possibly alightings)
b_0 = other unmeasured or quality factors.

Research from the literature (e.g. Cascetta et al., 2011) typically shows parameter values of b_1 and b_2 to be about twice that of b_1,[1] reflecting the more onerous conditions (e.g., physical effort, weather) associated with access and the uncertainty and waste of time elements of waiting. But railways could, and should, plan to minimize all the parts of generalized cost, and should take account of those elements to which passengers are most sensitive. That requires attention not only to the parameter weightings, but also to those elements of the journey which are greatest in that market segment. For urban journeys, access to stations and waiting for trains can be onerous, so regional transport authorities such as the British PTEs have opened new stations and increased frequencies. For long-distance journeys, journey time and fares are proportionately more important, so providing wifi and snacks enable people to use their journeys better (effectively reducing the journey time), while cheap offers (ideally linked to less popular trains) can reduce the disincentive of expensive fares.

The true application of generalized cost theory to railway planning has some important implications for the mode, implications which are sometimes forgotten, e.g. in the marketing desire to compete against airlines for end-to-end traffic. This can add hugely to the inconvenience of the service for intermediate traffic and, since the correct objective function is the minimization of market size *and* generalized cost, may lead to the wrong answer. For instance, advertising four-hour journeys between London and Glasgow calling only at Preston is irrelevant to anyone living in some of the larger towns and cities en route, e.g. Milton Keynes or Carlisle. Moreover, this all assumes that time-savings are linearly consistent, an issue with which transport economists continue to grapple: perhaps (e.g. in terms of journey opportunities) the real value of saving one hour to 1,000 intermediate passengers is actually greater than saving one minute to 60,000 passing through?

Fares Policy

The extent to which sufficient income can be derived for a railway operation to cover all its costs is dependent on the level and structure of fares and charges, and how users may respond to changes in these. The latter is determined by the elasticity of demand, where

$$e_i = \frac{\text{\% change in demand}}{\text{\% change in fares}}$$

although this type of response is equally valid to any of the elements of generalized cost. Fares' elasticities vary by fares level (hence country), journey purpose and overall spend. Values are negative (because higher fares reduce demand) and typically between 0 and –1; a value of 0 indicates a demand completely inelastic (i.e. unresponsive) to changes in fares, whereas a value of –1 provides an interesting tipping point where revenues remain the same through a change in fares, as reductions in demand exactly compensate for increases in fare levels (and vice versa). Typical values are shown in Table 19.1 below (see also Litman, 2011). Note that off-peak values are typically twice those in the peak, and that values over the longer term (e.g. three years) may be as much as three times the initial values, as passenger behaviour changes (e.g. by moving house).

Table 19.1 Typical elasticity values

Situation	*Fares elasticity*	*Reference*
Commuting, Australia	–0.3	Booz Allen Hamilton, 2008
Urban transit, France	–0.3 to –0.6	Bressonet al., 2003
Business to/from London, Britain	–0.5 to –0.9	Wardman and Shires, 2003
Suburban rail, Britain	–0.5 to –1.0	Paulley et al., 2006
Inter-urban leisure, Britain	–1.1 to –1.5	Wardman and Shires, 2003

One might wonder why an issue which is clearly economic in nature is included in a chapter on railway planning, but the temporal and unstorable nature of demand is critical here. Railway demand is highly peaky: on commuter railways, hourly peak demand can be ten times that of a typical off-peak hour. This results in poor asset use in the peaks, with train crew, train sets, tracks, platforms, and even complete stations effectively required only for a few hours per day. Although there can be huge social benefits in providing these services (enabling road congestion-reducing travel to work to provide economic gain, and concise (as opposed to sprawling) urban form), these operations are generally loss-making (contrary to popular belief).

Discouraging passengers to travel outside peak periods is therefore a critical part of railway planning, whether this be through fares policy or wider social policies, e.g. introducing flexible working hours, or providing concessionary travel only at off-peak times.

Freight

For freight, many of the principles are similar. Bulk freight (exploiting economies of scale) can generate significant profits from long and/or heavy trains, where 1 km long and 10,000 tonnes would not be considered particularly unusual parameters. But the number of traffic origins and destinations which can support such volumes is limited to mining, heavy industry (e.g. steel) and major ports. Elsewhere, rail's Achilles heel is, once again, the limitations of its network and access to it; as soon as goods have to be brought to rail terminals, the extra interchange penalty (here, a transshipment) becomes significant, especially as railfreight service frequencies are often low (e.g. daily), so road competition can have quite a head start. More recently, however, environmental concerns, the price of diesel fuel, and difficulties in finding and paying lorry drivers for multi-day trips are beginning to tip the balance in favour of railfreight, in situations where commercially minded companies are allowed to thrive.

Infrastructure Planning

This overview of railway planning deliberately discussed high-level objectives first and commercial considerations second. But railway planning also works from the physical end too, leading to important trade-offs.

Route Planning

Underlying economic geography matters immensely when planning railway routes because of the capital-intensive nature of railways and their constraints. Even highly populated rural areas can be difficult to serve with a complex network, while linear valleys (e.g. those of South Wales) are much easier. Here, a single route can minimize access to many potential customers.

Physical geography is important too. With a tunnelled route perhaps ten times as costly as an at-grade route, the treatment of obstacles has a serious impact on capital cost, although tunnelled routes can be shorter, less steep and involve less land-take (and hence potentially less public controversy). Climate change is also requiring engineers to consider the implications of excess rainfall (e.g. in Germany in 2012), or coastal defence (e.g. the washing-away of the British Great Western main line at Dawlish in early 2014).

When considering deviations to a potential route, the benefit of shorter access to local users needs to be balanced against the ongoing (time, maintenance and energy) disbenefits of the longer route. A 1 km diversion to serve a huge interchange such as London's Clapham Junction can be worthwhile, but Brunel's Great Western Railway realized that it could not serve all the relevant towns in Devon and Cornwall without using branch lines. Short branch lines are rarely cost-effective, however: operating costs can be relatively high (turn-round times can be relatively high) while passengers can drive to, and freight even be put on a conveyor belt to, the mainline interchange.

However, proposals for a maglev system in Britain demonstrated that increases in line-haul speed permit routes to be more circuitous while still remaining competitive. At 400 kph, a London–Scotland route could have been routed via Birmingham, Manchester and Leeds, leading to a considerable saving in capital costs compared to the branched solution adopted by HS2. Moreover, the longer route would also have enabled regional traffic between those three major cities to be carried on London–Scotland route services, thereby increasing load factors.

Route planners also need to consider the range of traffic objectives. Connecting city centres together may be good for business traffic, but commuters require both suburban and city-centre stations. The most profitable lines (such as London's Thameslink) link together a wide range of major traffic centres (e.g. office locations, airports, suburban shopping centres, dormitory and seaside towns) in a "chain of pearls" of dense trip ends likely to generate traffic around the clock.

Wider economic objectives are increasingly seen as a benefit from rail schemes, enabling increased employment catchment areas, and shorter business-to-business travel times. These "agglomeration economies" were first quantified by Graham (2007). But one has to be careful with logic, and not double-count benefits between different projects (some schemes may relatively disbenefit urban areas distant from them—see Muir (2013)).

Station Spacing

Access to the railway is more important than often thought: not only is it weighted higher by users, but there are perceptual benefits of stations too: out-of-town stations do not have the same presence. The "beacon" function of the Paris metro "M" should not be underestimated. In order to minimize station access time, positioning stations on street corners, or "double-ending" them with multiple entrances are both valuable techniques.

As railways are a mass transit mode, connections by other modes of public transport (e.g. bus, tram) are to follow primary concerns about access on foot. Multimodal interchanges can provide huge benefits to railways, in terms of maximizing their catchment area, but park-and-ride sites (while valuable in places) can deter the potential user by surrounding the station with huge car parks.

Line Planning

Railways are a genuine system, so capacity is multifaceted, including infrastructure, rolling stock, timetabling and even wider demand matters, e.g. fares policy. In general, however, the capacity of a railway line is determined by signalling except in inner urban areas (where the time spent in station stops is critical). Daily line capacity may be described through Scott's equation, which can be written:

$$C = \frac{(24 - w) \times 60}{T + \tau}$$

where:

C = sectional capacity in number of trains per day in either direction;
W = any engineering "white" period during which the line is to be closed;
T = time taken in minutes by slowest train to cover longest block section; and
t = time taken in minutes to close the line, obtain line clear and start the next train.

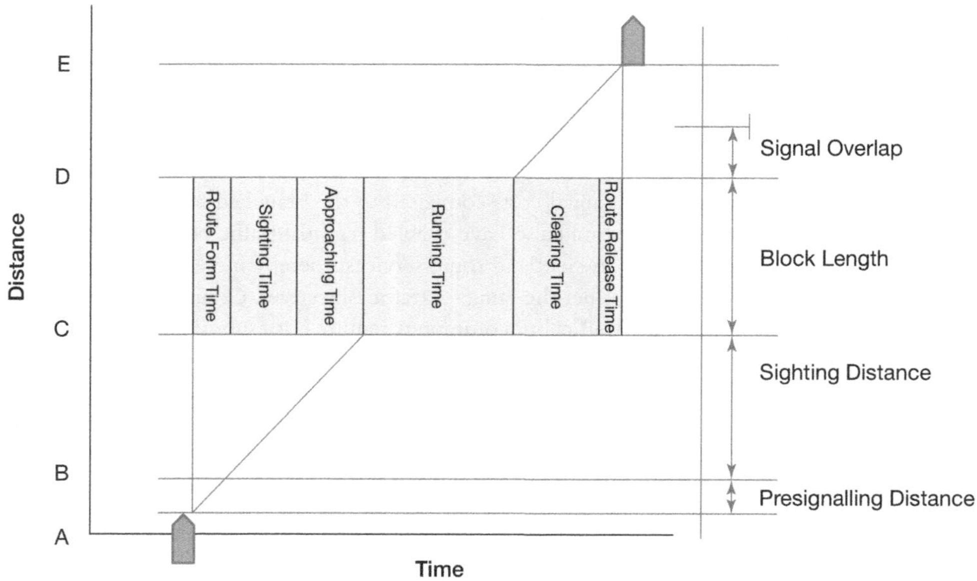

Figure 19.2 Blocking time of a train through a line section

Reducing the time T can be achieved by faster or better accelerating rolling stock, but the standard signalling solutions would aim to reduce block length. This increases maintenance and operations costs, but at least (unlike in the early days) electronic signalling can be undertaken both distantly and automatically. It should also be remembered that the effective occupation of any particular piece of track is greater than the actual time a train is on it, since the line must be cleared for its entry well prior to its arrival (if the train is not to be stopped), while a safety margin is applied to ensure that a train is indeed well clear of any fouling point (see Figure 19.2). The detailed application of these principles to different European standards can be found in Pachl (2008).

Track Layout Planning

Insufficient attention is sometimes paid to track layouts. The aim of having multiple tracks is to disaggregate trains by speed category, thereby enabling a considerable increase in capacity (see Figures 19.3a and b).

Standard double junctions may cost more to install and maintain, but enable the parallel working of trains. Those extra costs may therefore be justified (in terms of capacity enhancement) and be classified as "golden assets", on whose successful operation the railway depends commercially.

Track layouts within stations also matter. Although partly determined by train service quantities and passenger volumes, the requirements for simple (e.g. cross-platform) interchange, the segregation of passengers of different types (e.g. the needs of commuters and InterCity passengers are different) and the physical geography of the site also impact. More platforms are needed for a given number of trains at termini, since turn-round times (8–60 minutes, depending upon journey duration) are generally greater than normal (0.5–5-minute) station stop times.

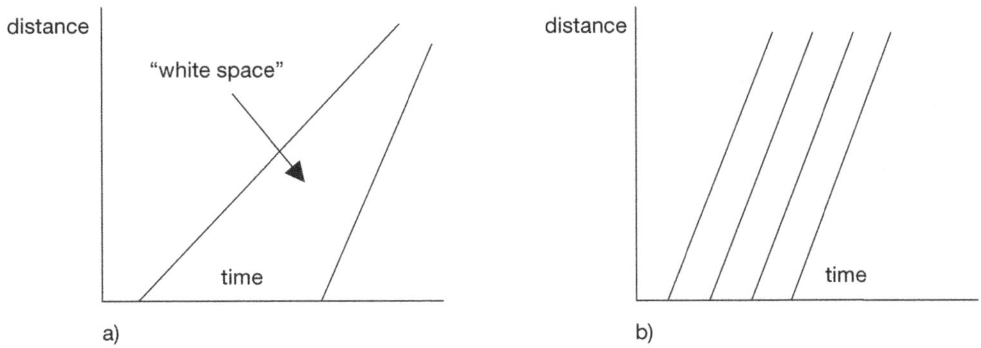

Figure 19.3 Train graphs for services with (a) differing and (b) similar characteristics

Station Planning

As well as good practices in the fields of architecture, maintainability, urban, structural and interior design (Ross, 2000), station planning needs to take into account key attributes of the rail mode. Stations need to be planned both individually and together (along a line), so that passengers' needs (e.g. ticket purchase, interchange) can be addressed where appropriate and passenger flow can be spread out. For instance, along a line of route, entrances to platforms should be at different places along the train, to even out the load between different carriages; this is one of the few design failures of the otherwise excellent Victoria Line in London, a failure which is behind the current £500 million scheme to enlarge Victoria Station through provision of a new northern access.

There are four main functions of stations: operational (e.g. boarding and alighting); support (e.g. ticket issuing); beacon (advertising the railway's point of access within the local community), and retail. Unfortunately, an over-emphasis on the last of these can give it an unhelpful pre-eminence, actually prejudicing the more important functions. Network Rail is to be praised for removing some retail kiosks at London's Waterloo Station; even though this reduced their income, they realized that passenger flow was being affected.

The different elements of a station should enable a prospective passenger to undertake a logical sequence of activities: for instance, ticket issuing is better done before providing information about which platform is required. Minimizing the distance between street and platform is a key target. However, there are always constraints in designing stations, whether these are externally driven or the result of railway requirements. Examples of external impacts include the foundations of other buildings, utilities (e.g. water pipes and telecoms cables), and even the need to build around ancient monuments (e.g. at Pino Suarez, Mexico City). Railway safety regulatory authorities usually require platforms to be straight and level, as well as having minimum widths (e.g. 3 m for a single-sided platform, but 5 m for a double-sided one). Platform requirements (e.g. their layout) are partly determined by the train service: individual platforms may be dedicated for particular purposes, either to help passengers (e.g. for airport express services) or to reflect track geography.

The location of stations within the urban environment is also important, as they need to serve areas of high trip end density. Along a line, this should include both key employment destinations and other traffic generators, including high-rise residential development, shopping centres and sport stadia. This enables the railway to maintain decent off-peak traffic levels, a key to its commercial success. The minimization of access time to the station is good practice, so locating stations at street corners, or providing them with multiple entrances, is more cost-effective in serving an area than the provision of two closely spaced stations. Because access is so heavily weighted within the journey, escalators can be used to cover horizontal distance, enabling multiple constraints to be overcome (e.g. entrance on a street corner leading to a location 50 m away where platforms can be wide and straight enough).

Maintaining good passenger flow through stations is not only an element of good customer service, it can be an essential part of safe railway operation. The basis of most pedestrian planning work is the seminal paper by Fruin (1971), and his level-of-service categories are still used to determine acceptable levels of crowding. This is especially important in terms of railways and metros running underground in urban areas, when the key design criteria is the ability to evacuate the station within a given time-period (e.g. 4 minutes to a place of safety, 6 minutes to the street). Designs need to be reworked until it can be shown that the most crowded conditions (levels E and F) are unlikely, except for a few seconds, as in these circumstances pedestrian flow reduces towards zero as densities rise. The use of dynamic and artificial intelligence-based modelling has helped planners visually understand their designs, and important constraints such as ticket gatelines, crossing movements and a sheer lack of capacity (e.g. Daamen and Hooghendoorn, 2004).

In practice, however, matters are slightly more subtle. As passengers do not walk right next to the walls of passageways, "edge effects" may be observed, with the effective passageway width up to 50 cm less on either side than its physical width. This problem is exacerbated where there are sharp changes in direction, where considerable space may go unused on the outer side of corners. Passengers are known to move more slowly up and down stairs, but these are often created to be narrower than the surrounding passageways, leading to inevitable queues. Passenger flow is also more bunched leading away from platforms (with large numbers of passengers following train arrivals) but more spread out on entering stations, but few railways acknowledge this in their allocation of capacity.

Figure 19.4 Unequally split stairway, Tokyo

Figure 19.5 Container terminal, Saudi Arabia

Station Stop Management

For railways in urban areas, line capacity is determined by train (capacity, braking) characteristics and station stop times as much as signalled headway. "Function" time (door opening and closing, train despatch) must be minimized through good design and procedures, while passenger flow time is affected by a wide range of factors including train, passenger flow and platform characteristics (Harris et al., 2014). These latter include width, distance to exits and minimizing the gap between the train and the platform. Careful management of the situation is needed where capacity is fully-utilized; this may include control of passenger entry to the platform or to the busiest ("critical") doors of the train (as at Town Hall station, Sydney).

Terminal Planning

Similar principles apply to the planning of freight, as opposed to passenger, terminals. The separation of road and rail movements, sufficient space to store traffic units (which, unlike passengers, do not move themselves), and the relegation of supporting functions (such as offices) to the periphery of the site reflect analogous principles. Access to locations of high passenger trip ends is replaced by the need for easy access to major industrial sites and centres, ideally reached directly by trunk road. Reachstackers and/or gantry cranes are appropriate for loading containers, but hoppers and conveyor belts may be more appropriate for bulk commodities such as coal. However, it is important to understand the required shape of railfreight terminals: freight trains are particularly long and thin, so yards and terminals may need to be 2 km long but only 50 m wide.

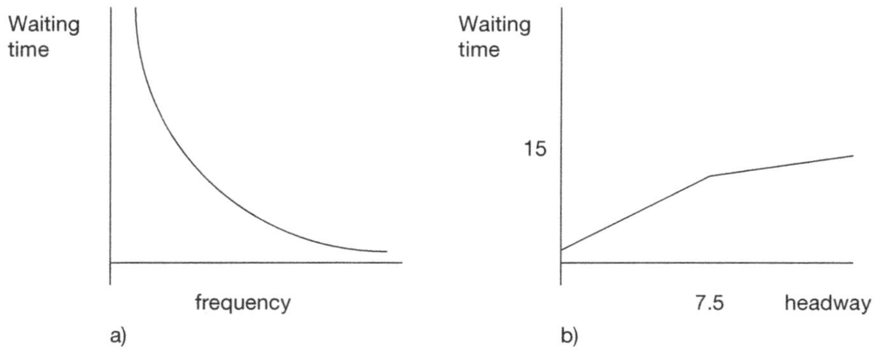

Figure 19.6 Impact of (a) frequency and (b) headway on passenger waiting times

Operational Planning

The first stage in planning a train service is to estimate the number of trains required to carry the (passenger or freight) load, bearing in mind the level of traffic peakiness, as discussed earlier. However, a key issue is then to decide whether or not to match the time-of-day profile of the traffic or to provide a standard (e.g. hourly) service. The former of these appears to be more customer-focused, but this may not necessarily be the case. The reason for this is that waiting times are not a linear function of train service frequency, but fall exponentially, as shown above.

This means that, as frequencies increase, the proportionate reduction in waiting time falls. This is despite a kink in the relationship between waiting time and headway (the time between successive departures), as passengers usually aim to catch a specific train at lower frequencies while turning up at random (therefore enduring waits of half the headway) at higher frequencies. The kink point relates to frequencies of about four trains per hour, which is therefore a common target for policy-makers in urban rail environments.

The expected waiting time for headways greater than 15 minutes is a function similar to

$$W = \frac{15}{2} + \frac{(h-15)}{4}$$

where

W = waiting time; and
h = headway.

Rail planners should bear in mind that, even if they wish to aim for a specific train, some passengers are unable to do so, being constrained by other transport services (e.g. connections) or lifestyle constraints (e.g. the start/end times of work, leisure events etc.). Railways can mitigate the impact of these by providing (e.g. eating or shopping) facilities at stations which reduce any unproductive time spent waiting.

Providing a standard-hour service may therefore appear to under-provide for the peak and over-provide for the offpeak, but the principle of equation (1) needs to be remembered, while simplicity is also a valid objective, as well as gen cost minimization. The same principles apply to the consideration of traffic across two branches: even if demand is split 2:1, running an equivalent level of service may still be the best answer.

Timetable Planning

Timetable planning is the art of trading off the minimization of generalized cost with the minimization of resource use within the physical constraints of the network while trying to maintain a relatively simple solution. For instance, the mere act of stopping a train costs money (typically, a few euros per stop), so it is undesirable to stop a train if no one is going to get on or off—but service simplicity is helpful for staff and the public alike. Request stops lead to inconsistent times, and are most used in peak periods when there is most pressure on the service anyway.

Of course, there are different ways of providing service, for instance using through trains or shuttle services. At this point, we encounter another of the many trade-offs: passengers' preferences for no interchange need to be balanced against their preference for punctual service. Longer-distance trains, running through from one place to another, far away, are more likely to be delayed. But passenger preferences are not the same around the world: railways with a rigorous culture of holding connections have passengers who are less deterred by changing trains, as they have increased confidence that it will work.

A specific example of a network planned on connections is that of Switzerland. Their timetable is based on "Taktfahrplan" principles, which include:

- standard departure times throughout the traffic day/week from both ends of the line;
- standard stopping patterns;
- where multiple services interwork, departures across both/all services to be evenly-spaced;
- times for all services at the major network nodes to be around the hour (xx:00)/half-hour (xx:30), or, failing that, xx:15 and xx:45, thereby enabling the maximum number of connections to be made easily.

These principles lead to a reduction in generalized cost across the network, and are undoubtedly one of the reasons why Switzerland has one of the highest per capita rail-trip rates in the world (*c.* 50 trips per person per annum). However, it does have significant capital requirements: for instance, platform utilization is rather poor: many platforms may be needed for only a few minutes per hour. It may not be practical for other railways to convert wholly to this system, if land constraints prevent station expansion to accommodate extra platforms—although the Swiss have generally provided sufficient funds to enable the construction of underground facilities if necessary. Nevertheless, elements of this should and can be implemented in all rail networks.

Because passenger trains are large, of relatively fixed size (additional capacity comes in the form of (several) extra coaches) and hence expensive, balancing loads between trains becomes important, especially as passengers dislike crowding (values of b_3 in equation 1 commonly increase by 10 percent). Passengers also dislike intermediate stops: a rule of thumb is that trains should stop no more than ten times. In an urban area, this leads to the development of a hierarchy of services: inner suburban, outer suburban and long distance. The exact stopping pattern should

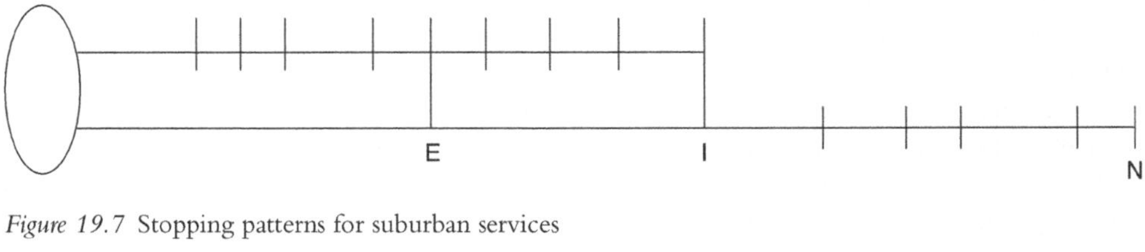

Figure 19.7 Stopping patterns for suburban services

reflect traffic demands, but this is a juggling act to which passengers respond in their own interest. Consider the following example of stopping patterns out of a major city:

Having outer-suburban services calling at all stations would be a poor policy, because those trains would become overcrowded, being available to all passengers; in contrast, "short-turning" trains at I would be relatively poorly loaded. The service indicated may balance demand better between trains, although passengers from stations E and I will tend towards the outer-suburban trains, because these are faster. Removing the stop at E would obviously transfer those passengers to the slower service, but would instead make the outer-suburban service faster, encouraging even more people from station I to use it. This can be a serious source of contention in urban areas, exacerbated by the fact that development density and political affiliation can vary by distance from the city centre, leading to accusations of favoritism by the railway.

Resource Constraints

All the above has to take into account limited numbers of train sets and tracks, and raises the issue of how much slack to leave. Train sets are expensive but need maintenance: efficient railways normally manage to get about 92 percent of electric trains and 85 percent of diesel trains into service in peak periods.

Maintenance aside, the number of train sets required to operate a train service is determined by the formula:

$$N = \frac{2(T + t)}{h}$$

where:

T = travel time from one end of the line to the other
t = turnround time at the terminus
h = headway.

Note that values have to be rounded up to the nearest integer.

Consideration of appropriate turn-round times needs to take into account operational requirement (e.g. passenger alighting/boarding, changing drivers, restocking catering vehicles, cleaning), the level of robustness required (so that any lateness can be recovered) and the length of the preceding journey (since this is a primary driver of delays).

Unfortunately, the preceding formula implies that train service efficiency is affected by network geography, itself a function of where towns are located, which the railway cannot influence directly. However, the railway can make adjustments to improve its efficiency. The Swiss approach (funded through public referendum–driven investment programmes such as Bahn 2000) is to invest to reduce journey times between key nodes to values which are operationally efficient, e.g. 54 minutes. An alternative approach is to adjust service patterns.

Suppose a line from A to B has a journey time of 61 minutes and an hourly service is to be operated. This would use 3 train sets, even though each of them would only run for 122 minutes every 180 (67 percent). Missing out a couple of stops, each saving perhaps 3 minutes, could reduce the journey time to 55 minutes, but could leave stations without service. Alternative solutions include "interworking" this service with others. Where this interworking involves services on the same side of a major centre, it is called "cross-forming" (e.g. from A, B and C to X, Figure 19.8a), in contrast to the "through-running" of services on the other side of that node (e.g. from D to E via Y – Figure 19.8b). Both types of interworking worsen punctuality,

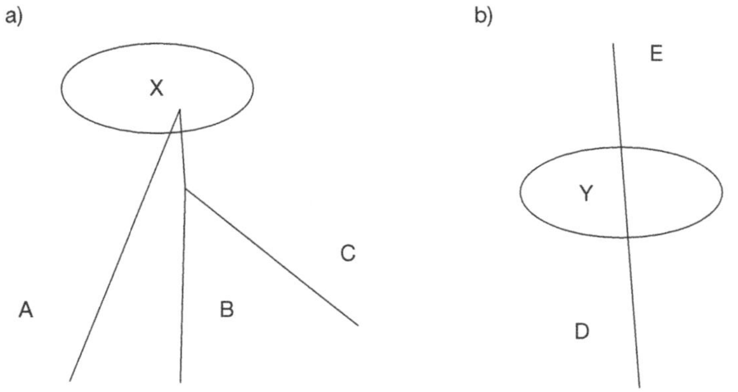

Figure 19.8 Interworking: (a) cross-forming; (b) through-running

because delays from one part of the network are imported into another. If demand for both services is not equal, it can lead to larger trains operating on routes which only need lower capacity. However, with trains costing over £1 million per carriage to buy, the rationale for reducing cost is strong, and this can be reinforced by other constraints, such as a lack of platforms at a terminus station, requiring trains to be despatched as soon as reasonably practical, rather than waiting the time of the next departure to the place they have just arrived from.

On the other hand, through-running offers an additional key benefit: by definition, it provides through services, removing an interchange, and therefore stimulating demand. Unfortunately, as the impact of this is greatest for shorter journeys, this can lead to localized overcrowding on longer distance services, e.g. in Britain, between Derby and Bristol on CrossCountry trains from Newcastle to Plymouth via Birmingham.

This is another classic example of railway planning demonstrating trade-offs. Service complexity and reliability are being exacerbated, in an attempt to reduce costs. The relative value of this trade-off depends upon the type of service: rural services are more likely to be cost-driven, while higher earning intercity services are more likely to be able to sustain more assets, since the commercial consequences of train set unavailability are higher (especially if there are customer service guarantees).

As well as constraints regarding the number of trains available, there are further constraints associated with the network, e.g. the number of tracks. The UIC (2013) recommends using 75–80 percent of line capacity on an ongoing basis, leaving the remainder for service recovery. However, line capacity is impacted by a range of issues, including not just the signalled headway on plain line, but by junctions, station stops and the mix of traffic. To understand this, we need to remember that train graphs (see Figure 19.3) reflect the *planned* position of the *front* of each train.

When slow and fast trains are run alternately (as in Figure 19.3a), considerable track capacity is wasted, reflected in "white space" being shown on the graph. Where trains of a similar type are operated consecutively (as in Figure 19.3b), a higher utilization of line capacity can be achieved—this is the norm for metro operation. Equalizing the speed of trains along a line is therefore helpful, especially when it is impractical to provide more capacity: Eurotunnel, for instance, require high-speed Eurostar services to slow down through the Channel Tunnel between England and France in order to minimize their impact on the capacity of a tunnel whose other traffics (primarily car shuttles) operate at a lower speed.

"Flighting" is a name variously used to describe a range of techniques designed to minimize the amount of white space on a train graph. This can be achieved either by running several services with similar characteristics together, or by having a group of fast, then semi-fast, then slow services. Without this, the effective capacity of a double-track line signalled to three-minute headways can be as low as eight trains per hour, if services are truly mixed. Multiple tracks enable the segregation of traffics, and therefore provide significant increases in capacity. However, it should be noted that it is the *difference* in the *time* taken over the critical line section which determines capacity, and a mix of semi-fast passenger and non-stop freight can work well. It is truly high-speed trains which reduce capacity significantly, and which are the reason for the construction of separate lines such as HS2 in Britain. The actual capacity of lines can be derived from a given timetable using the compression method (see e.g. Gibson et al., 2002).

Ensuring Punctuality

Of course, services planned on paper do not always work on the day, and railway planning must take into account the operational robustness of services. A key measure here is the ratio of secondary (consequential or knock-on) delay relative to the primary (input) delay, which may be caused by a defect with infrastructure or rolling stock.

Operational robustness may be measured by the ratio of secondary:primary delay, where average values of <1:1 would be sought (even if specific incidents led to higher values). A wide range of factors affect operational robustness (including weather conditions and variability between drivers) and these need to be countered by a range of management measures, some of which are within the purview of the planning process. Timetabling, in particular, needs to recognize the need for appropriate recovery margins, although excessive slack tends to lead to capacity which is apparently overused, while not encouraging proactive staff behaviour.

Simulations are commonly used to undertake the increasingly complex trade-offs required in operational planning across ever-larger networks, taking into account the probability of train positions, given input timetables and delay distributions. This is important in understanding what happens in practice, rather than on paper: for instance, Figure 19.9 (after Medeossi et al., 2011)

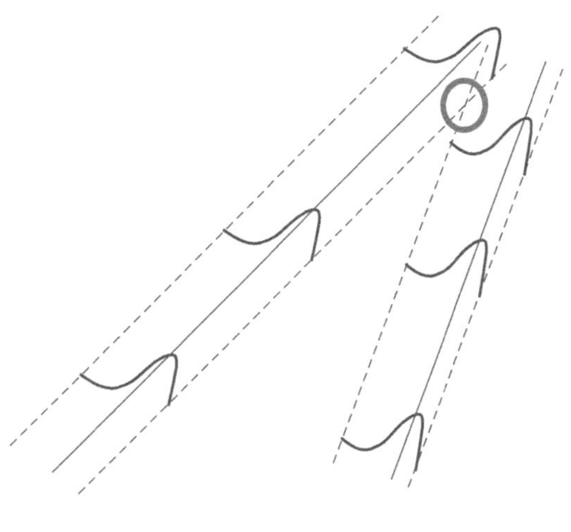

Figure 19.9 Interaction of trains during perturbations

shows a static image of the possibility of two trains conflicting in practice (based on distributions of their likely position), while a simple comparison of the expected positions (shown in the solid lines) would not suggest a problem. For a comprehensive treatment of this area, see Hansen and Pachl (2008).

Conclusions

Ideally, railway planning should follow a sequential flow of disaggregating thoughts from concept to detail, in a process which is normally iterative. While this author would advocate a commercially driven approach, railway planning is characterized by a wide range of trade-offs, e.g. between capital and operating expenditure, capital expenditure and demand, operating expenditure and service reliability, end-to-end speed and enabling easy access to train services, etc. Guidance on these should be derived by applying generalized cost theory and its implications, paying attention to the whole journey, from door to door, and not just considering that part of the journey which happens to fall on the railway. While those and other principles are undoubtedly the starting-point, detail matters on the railway, at least once initial capacity has been used up.

Note

1 b_1 is conventionally set to 1. Note even higher values of 2.5 for walking down stairs and 4 for walking up stairs reflecting the greater difficulty of these activities.

References

Booz Allen Hamilton (2008) "(Sydney) City Rail Fares Elasticities", report for Independent Pricing and Regulatory Tribunal (IPART), Australia.

Bresson, G., Dargay, J., Madre, J.-L., and Pirotte, A. (2003), "The Main Determinants of the Demand for Public Transit: A Comparative Analysis of Great Britain and France", ESRC Transport Studies Unit, University College London.

Cascetta, E., Papola, A., Pagliara, F. and Marzano, V. (2011) "Analysis of Mobility Impacts of the High-Speed Rome-Naples Link", *Journal of Transport Geography*, 19, pp. 635–643.

Daamen, W. and Hooghendoorn, S. P. (2004) "Pedestrian Traffic Flow Operations on a Platform: Observations and Comparisons with Simulation Tool SimPed", *COMPRAIL IX*, pp. 125–134.

Fruin, J.J. (1971) "Pedestrian Planning and Design". Updated in Bowman, B.L. et al. (1989) "Handbook on Planning, Design and Maintenance of Pedestrian Facilities", US Department of Transportation, Virginia, USA.

Gibson, S., Cooper, G. and Ball, B. (2002) "The Evolution of Capacity Charges on the UK Rail Network", *Journal of Transport, Economics and Policy*, 36, pp. 341–354.

Graham, D. J. (2007) "Agglomeration, Productivity and Transport Investment", *Journal of Transport, Economics and Policy*, 41, pp. 317–343.

Grey, A. (1978) "The Generalised Cost Dilemma", *Transportation*, 7, pp. 261–280.

Hansen, I. A. and Pachl, J. (Eds.) (2008) *Railway Timetable and Traffic*, Eurailpress, Hamburg (228pp.).

Harris, N. G., Graham, D. J, Anderson, R. J. and Haojie, L. (2014) "The Impact of Urban Rail Boarding and Alighting Factors", TRB, 3rd Annual Meeting, Washington, DC.

Lichtberger, B. (2005) *Track Compendium*, Eurailpress, Hamburg, Germany (634pp.).

Litman, T. (2011) "Transit Price Elasticities and Cross-Elasticities", Victoria Transport Policy Institute paper, Australia (24pp.).

Medeossi, G., Longo, G. and de Fabris, F. (2011) "A Method for Using Stochastic Blocking Times to Improve Timetable Planning", *Journal of Rail Transport Planning & Management*, 1 pp. 1–13.

Muir, G. (2013) "HS2 and the Dialogue of the Deaf", *Passenger Transport*, 33, pp. 26–29.

Pachl, J. (2008) *Railway Operation and Control*, 2nd ed., VTD Rail Publishing, Mountlake Terrace, WA (268pp.).

Paulley, N. et al. (2006) "The Demand for Public Transport: The effects of fares, quality of service, income and car ownership", *Transport Policy*, 13, pp. 295–306.

Pienaar, W. J. (1997) "Modelling Transit Generalised Commuter Costs for use in Urban Planning", *Transport Built Environment*, 30, pp. 307–315.

Ross, J. C. (2000) *Railway Stations: Planning, Design and* Management, Architectural Press, Oxford (350pp).

UIC (2013) "Capacity", leaflet 406 (2nd ed.).

Wardman, M. and Shires, J. (2003) "Review of Fares Elasticities in Great Britain", University of Leeds Institute of Transport Studies, Working Paper 573.

Warner, S. L. (1962) "Stochastic Choice of Mode in Urban Travel: A Study in Binary Choice", Northwestern University, Evanston, IL.

20

INLAND WATERWAY TRANSPORTATION

Paul Schonfeld

UNIVERSITY OF MARYLAND

Introduction

Inland waterways are navigable bodies of water contained within land masses and include rivers, lakes and canals. Thus, inland waterway transportation may compete with overland transportation modes, such as roads, railroads, airlines and pipelines. In some cities such as Venice, Amsterdam and Bangkok, dense networks of canals provide a major form of local urban transportation. Canals may also provide shortcuts between seas for oceangoing ships when a narrow isthmus can be crossed by a canal, such as the Suez, Panama, Kiel and Corinth Canals.

Inland waterways have been widely used for transporting both freight and persons since prehistoric times. Their main advantage over competing transportation modes has usually been their relatively low cost per transported unit, e.g. per ton kilometer. That relatively low cost is primarily attributable to the relatively low resistance of vehicles (e.g. ships, boats, rafts) moving through water at low speeds. A related advantage is the ability of many waterways to support relatively large vehicles. This not only provides large transportation capacity but also economies of vehicle scale which help reduce costs. One further advantage is the pre-existence of natural waterways which may not have to be built or expensively maintained. While considerable expenditures have been made to improve some natural waterways, the economies of using pre-existing waterways are very significant. In some cases even vehicles are dispensable. For example, logs can be tied together into rafts and floated downstream or towed upstream.

One obvious disadvantage of waterway transportation is its relatively low speed compared to competing modes. The main reason is that the resistance and energy consumption of floating vehicles is very slight at low speeds but increases highly non-linearly with speed. Another reason is that travel paths on rivers and lakes may be more circuitous than for roads, railroads and pipelines. (Rivers in relatively flat plain regions tend to be especially windy.) Low speeds may be economically suitable for moving freight, especially bulk cargoes with relatively low values per weight or volume unit (such as petroleum products, grains, coal, minerals, construction materials, fertilizer, garbage), but are rarely suitable for moving passengers. For moving passengers on waterways at high speeds it may be desirable to use unconventional vehicle technology, such as hydrofoils and air cushions, which reduce high-speed resistance by raising the vehicles partly or fully above the water surface.

Another significant disadvantage of waterways in regions with cold winters is that they freeze for part of the year and become unavailable for transportation. That reduces the utilization of vehicles and facilities, reduces the reliability of deliveries, and may require large stored inventories or some backup form of transportation.

Historical Development

Natural waterways can compete very strongly with other freight transportation modes when they are available in suitable form. However, they are quite unevenly distributed over the world and those suitable for economic transportation operations are even scarcer. Nevertheless, it is notable that existing waterways have very strongly influenced the development of human settlements and economic activities, resulting in actual transportation flow patterns that are quite well aligned with those waterways.

When waterways are not naturally available they can be created artificially by building canals or by flooding some regions. The latter option has been used in many regions, for instance by impounding water behind a dam, but has significant disadvantages including loss of useful land, expensive dams, lack of connectivity with other waterways, induced earthquakes, landslides and various adverse environmental impacts. The construction of canals may be very expensive if modern large vessels must be accommodated. Historically, some very long canals have been built in Europe, Asia and North America before long-distance railroads were practical. The Grand Canal of China, completed in approximately 600 CE and linking the Yellow and Yangtse Rivers, was 1,776 km long. Long canals in the Western world, such as the Canal du Midi in France linking the Mediterranean to the Atlantic (240 km, with 101 locks, built in 1661–1681) and the Erie Canal in the US (584 km, with 83 locks, built in 1817–1825) were designed for relatively small and shallow draft boats which were pulled by animals from parallel tow paths. Much wider and deeper canals are needed for modern vessels large enough to compete with trucks or rail vehicles. Such canals are usually too uneconomical to build over long distances; they are built to connect natural waterways which are rarely more than 100 km apart.

Extensive waterway networks have developed over time in habitable regions of the world that already had navigable rivers. Early civilizations and large empires in Egypt, Mesopotamia and China developed along navigable rivers, which provided water for irrigation as well as relatively easy transportation to connect and unify large regions. The many navigable rivers of Europe and Asia have served as major transportation arteries for millennia. When traffic rose sufficiently and societies were well enough organized to build and defend them, canals were built to link rivers and seas, e.g. in France, Britain, Germany, Sweden, Russia, China, USA and various other countries. Eventually, a very extensive network of waterways was developed in Europe, as illustrated in considerable detail in a map provided by the United Nations Economic Commission for Europe (UNECE) at www.unece.org/fileadmin/DAM/trans/main/sc3/European_inland_waterways_-_2012.pdf. In this European network it is notable, for instance, that the Rhine River is connected with the Danube, Oder, Elbe, Rhone and Seine, among other rivers, and that the Volga is connected to the Black Sea, the Baltic and the White Sea.

A map of major navigable waterways in the United States is shown in Figure 20.1. The main waterways in the US include the Mississippi River and its tributaries (mainly Missouri, Illinois, Ohio, Tennessee and Arkansas), the Great Lakes, the Atlantic and Gulf Coast Intracoastal Waterways, and the Columbia River. It should be noted that numerous other canals were built before the twentieth century in both Europe and the United States and some are still used by small recreational boats. However, most of those were abandoned for commercial cargo transportation because they were too narrow and shallow for economical use.

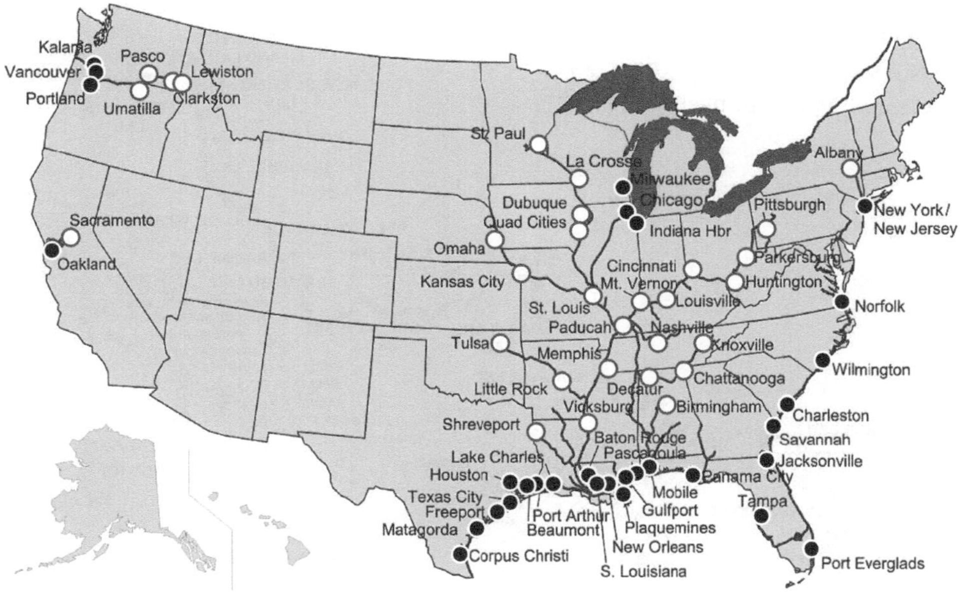

Figure 20.1 US waterways map

Source: US Army Corps of Engineers—Institute for Water Resources (www.iwr.usace.army.mil/Portals/70/docs/portswaterways/rpt/June_20_U.S._Port_and_Inland_Waterways_Preparing_for_Post_Panamax_Vessels.pdf)

Improvement of Natural Waterways

Natural waterways can be and sometimes are considerably improved artificially in order to accommodate larger vessels, reduce travel distances or enhance their reliability. Thus, windy rivers in relatively flat regions may receive short cuts between their curves, in order to reduce travel distances. Shallow waterways may be deepened by dredging to allow their use by larger vessels. Obstructions such as rocks, trees and sunken ships may be cleared. In hillier regions rivers may be too fast-flowing and turbulent when snows melt in the spring and too shallow in dry seasons the rest of the year. The typical solution to both problems, i.e. turbulence or shallow water, is to build a series of dams along the river, as illustrated in Figure 20.2 for the Upper Mississippi River between Minneapolis and St. Louis. By creating a stepped series of pools with calm water, sufficient water depth may be insured almost always (except during severe and long-lasting droughts). However, that creates the problem of moving the vessels across the dams.

Three main technical solutions have been employed for moving the vessels across the dams.

Locks

These structures provide the means of lifting or lowering vessels (locks are practically always usable in both directions) by changing the water level in a lock chamber enclosed between two gates. Their operation is shown in Figure 20.3. The vessels can enter the chamber from either direction when one gate is closed and the other is open. After the vessel enters the chamber both gates are closed and the water level is raised or lowered, along with the vessel, by opening and closing the right valves.

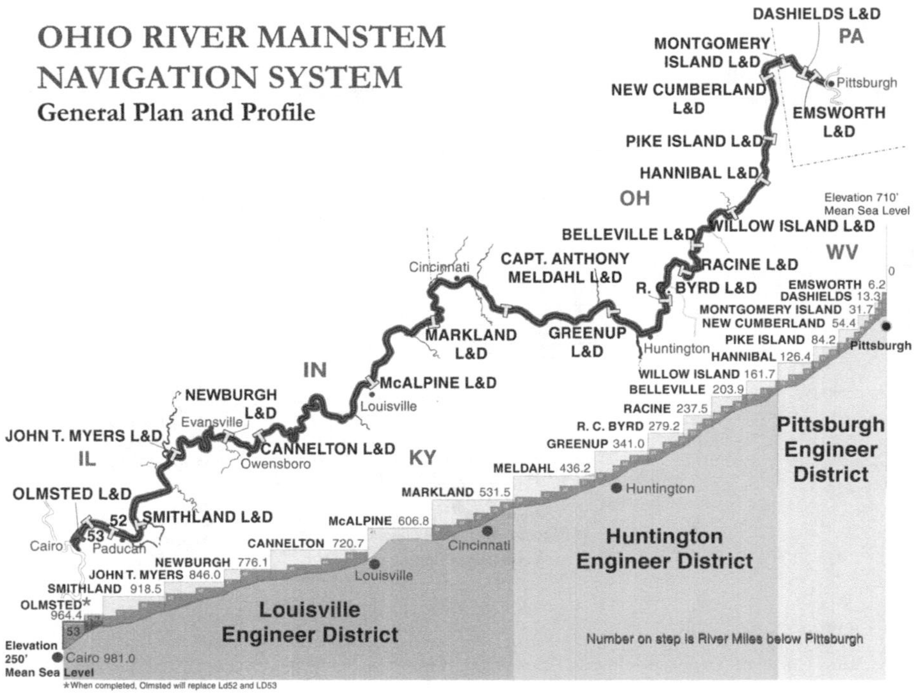

Figure 20.2 Series of dams along the Ohio River

Source: US Army Corps of Engineers (www.lrl.usace.army.mil/Missions/CivilWorks/Navigation/LocksandDams.aspx)

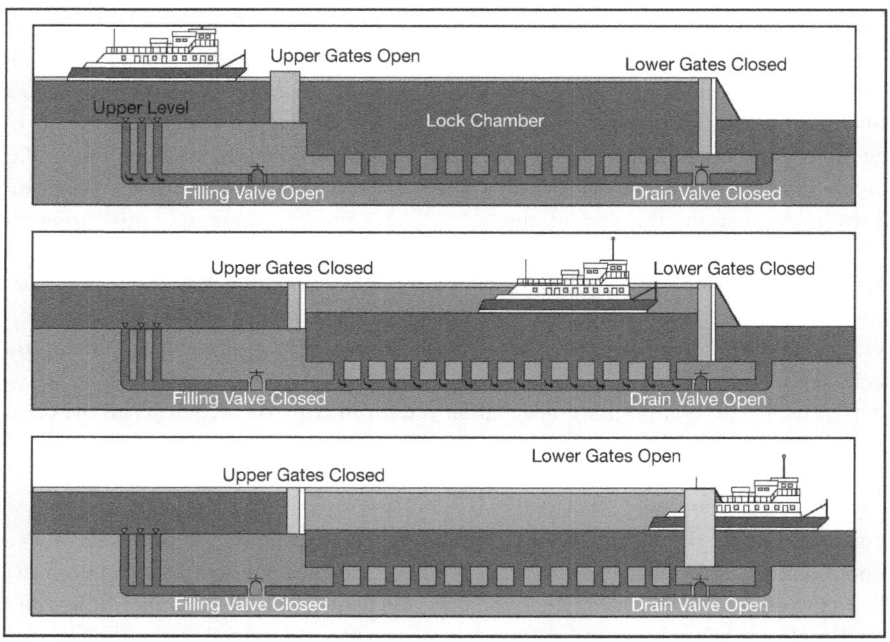

Figure 20.3 Lock operations

Source: US Army Corps of Engineers (www.lrl.usace.army.mil/Portals/64/docs/Ops/Navigation/lockthru.jpg)

Figure 20.4 Towboat pushing 15 barges into the chamber of lock and dam 12 on the Mississippi River

Source: US Army Corps of Engineers (www.mvr.usace.army.mil/Portals/48/docs/CC/FactSheets/Miss/Mississippi%20River%20Locks%20and%20Dams%20(2012).pdf)

Figure 20.5 Double lockage of large tow through single chamber lock

Source: US Army Corps of Engineers (www.mvr.usace.army.mil/Portals/48/docs/CC/FactSheets/Miss/Mississippi%20River%20Locks%20and%20Dams%20(2012).pdf)

Figure 20.6 Lock and dam 27 along the Mississippi River
Source: US Army Corps of Engineers (www.mvr.usace.army.mil/Portals/48/docs/CC/FactSheets/Miss/Mississippi%20River%20Locks%20and%20Dams%20(2012).pdf)

Multiple vessels can be "packed" in the chamber simultaneously. Thus, in Figure 20.4 a towboat is shown pushing 15 barges into one lock chamber, from the upstream end. (The closed downstream gate is angled to resist the water pressure from the higher level upstream.) All 16 vessels can fit very tightly in this lock chamber and be lowered together.

In Figure 20.5 a tow that is too large to fit into the lock chamber is broken into two "cuts" of nine and seven barges, which are served separately and then reassembled into one tow after the "double lockage" operation is completed.

Figure 20.6 shows a 16-barge tow approaching from downstream the larger of two chambers of Lock 27 along the Mississippi River. (The towboat is pushing barges into the chamber.) Other barges are shown docked at a port just upstream of the lock. The lock is located within the St. Louis Metropolitan Area on the Chain of Rocks Canal, which bypasses the Mississippi River, providing a non-turbulent shortcut. Downstream (south) of this lock the Mississippi River has no other locks and accommodates tows with up to 48 barges in 8 × 6 formations.

Inclined Planes

Such facilities move vessels uphill or downhill in an enclosed tub supported by wheels on rails, as shown in Figure 20.7. This figure shows the inclined plane at Saint-Louis Arzviller, on the Marne-Rhine Canal in France. This inclined plane is notable for having replaced (in 1969) a much older series of 17 locks, thus speeding up considerably the movements of vessels through the canal.

Figure 20.7 Vessels being moved on an inclined plane at Saint–Louis Arzviller, France
Source: Photo by Patrick Giraud (http://fr.wikipedia.org/wiki/Fichier:Arzviller_plan_incline_01.jpg)

Ship Lifts

Such a facility also lifts or lowers vessels in an enclosed tub, but vertically instead of on an inclined plane, as shown in Figure 20.8. The ship lift illustrated is located in the Trent–Severn waterway near Peterborough, Ontario. It was completed in 1904 and has dual parallel lifts which lift or lower boats by 65 ft.

It should be noted that additional ways of lifting or lowering ships exist. Both inclined planes and vertical elevators have been built in which the ships are moved dry, without being enclosed in a water tub. Small vessels may also be lifted or lowered by large cranes.

Of the above types of solutions, locks have been by far the most commonly used and are exclusively used on modern waterways in the United States. They range from very small locks used to lift or lower small pleasure boats to those as large as the Poe Lock between Lake Superior and Lake Huron which was rebuilt in 1968 (366 m long, 33.5 m wide, 9.75 m deep) to handle ships carrying up to 72,000 tons of cargo. Much larger new locks are being completed for the Panama Canal.

As the lock dimensions, including the lift per lock increase, fewer locks are needed along a given waterway. For instance, around the 1950s, 52 dams with locks along the Ohio River between Pittsburgh and the Mississippi River were replaced with only 19 new ones.

Currently, the 35 m lift of the Carrapatelo Lock on the Douro River in Portugal is the highest in the world. Such tall lock chambers require strong and expensive structures. The use of high dams and associated high lift locks can decrease the number of locks along a river and the total lockage delays but may still increase the overall cost of the (fewer) locks and dams while flooding larger areas behind those dams.

Figure 20.8 Peterborough ship lift on the Trent River, Canada
Source: Photo by Big Iron (http://commons.wikimedia.org/wiki/File:PeterboroughLiftLock23.jpg)

A series of adjacent locks without intervening waterways, sometimes called lock staircases, may be the most economical solution when large elevation differences must be overcome. For instance, for moving ships across China's Three Gorges Dam along the Yangtse River (China Three Gorges Corporation www.ctgpc.com/benefifs/benefifs_a_5.htm), where the elevation difference between the upstream and downstream levels can seasonally reach 113 m, two parallel lock staircases, each with five locks in series have been provided. The lock chambers are 280 m long, 35 m wide and 5 m deep. They can handle ships of up to 10,000 tons. The transit time through a lock staircase is about four hours. The two series provide redundancy and allow lock closures for maintenance purposes, although normally each series handles traffic in one direction. In addition to the two lock staircases a very tall ship lift with a basin of 120 × 18 × 3.5 m for smaller vessels of up to 3,000 tons is also provided at this dam, as can be seen on the left side of Figure 20.9. The combined capacity of the locks and ship lift is estimated to be approximately 100 million tons per year.

Figure 20.9 Model of parallel lock staircases and ship elevator at the Three Gorges Dam
Source: Photo by Sharon Nardo (http://commons.wikimedia.org/wiki/File:Three_Gorges_Dam,_model_of_bridge_at_the_reception_center.jpg)

Waterway Operations and Maintenance

Various types of vessels operate on inland waterways. These include ships, towed or pushed barges, towboats, tows and recreational vessels. On the relatively few inland waterways that are deep enough, such as the Great Lakes in the United States, the ships used may be comparable in size to most ocean ships. As shown in Figure 20.1, the Great Lakes are actually connected to the Atlantic Ocean by the St. Lawrence Seaway, which consists of highly modified s ections of the St. Lawrence River and the completely artificial Welland Canal that bypasses the Niagara Falls. Ocean ships displacing up to approximately 27,000 tons can pass through the Seaway, but even larger ships are used on the Great Lakes and cannot get out. Such captive ships may be optimized for local conditions, such as expected wave sizes and icing in winters, but largely resemble large ships used on oceans.

On rivers, the channel dimensions, such as depth, width and curvature, restrict the allowable dimensions of vessels. The vessel dimensions may also be limited by the dimensions of locks or ship lifts which they are expected to use. Over time, the ship dimensions may co-evolve with those of locks. Thus, on the Mississippi River and its tributaries in the United States the widths of most lock chambers have standardized over time at 110 ft while the widths of barges have standardized at 35 ft, thus allowing three side-by-side barges (= 3 × 35 = 105 ft) to fit into lock chambers. This can be seen in Figure 20.2. The 5 ft difference between the 110 ft lock chamber and 105 ft barge formation helps reduce resistance and ease maneuvering difficulties when barges are moved into and out of locks. The standard barge lengths are 195 ft. Since the chamber lengths on the Mississippi River System have also become standardized at 600 ft or 1,200 ft, the chambers can accommodate 3 × 3 or 3 × 6 barges, respectively. Similar co-evolution and standardization of vessel sizes and lock chambers, but reaching different dimensions, can be observed on various other river systems.

For moving large quantities of cargo on rivers, it is usually economical to separate cargo carriage and propulsion into different vessels. Thus, towboats are used to propel unpowered barges which carry the cargo, somewhat analogously to trains with distinct locomotives and unpowered cars or tractor–trailer combinations on highways, and for similar reasons: The unpowered barges may spend considerable periods getting unloaded and loaded in ports, or simply storing cargo. During these times the towboats may be gainfully employed elsewhere. In the US such combinations of barges and one towboat are called "tows". While barges were originally towed by towboats, in the US and some other parts of the world they are now mostly pushed by powered vessels which are still called "towboats". On relatively narrow rivers or canals one towboat may push one or two barges. On wider rivers, especially in the United States, it is found economical to push each large rigid formation of barges with one towboat. Such formations can reach up to 48 barges (6 wide × 8 long) on the lower Mississippi (south of St. Louis, where there are no locks to constrain tow dimensions). There are two major advantages to moving standard rectangular barges (195 × 35 ft) in rigid tow formations: 1) the wetted wall area of barges, and hence resistance, is reduced when they are tied side-by-side and 2) rigid tows are much easier than non-rigid formations to control and guide around river bends, into narrow lock chambers or between bridge piers.

Where they exist, locks (or, far more rarely, ship lifts or inclined planes) restrict the capacity of waterways and impose considerable delays on ship movements through them. In periods when the vessel arrival rate exceeds the service capacity of locks, queues of vessels are formed. In some seasons with heavy traffic vessels may wait in queues for days before getting through the locks. Long waits in queues may also occur when lock service is interrupted due to scheduled maintenance, accidents and other problems. The costs of such delays may be reduced by increasing lock capacity (with more or larger chambers), by speeding up lock-operating cycles and by various policies for controlling lock operations. Such policies may include grouping vessels in lock chambers in ways that better utilize chamber capacity, giving priority to certain vessels (passenger ships, larger tows, vessels carrying more valuable or time-sensitive loads), scheduling appointments or reserving places in queues, controlling vessel speeds toward locks, operating one-directional platoons of vessels through multiple lockage cycles, determining when to switch service to the opposite direction and coordinating the operations of nearby locks (Ting and Schonfeld, 2001).

Dams and the associated locks that permit navigation across them represent major infrastructure investments whose construction cost can exceed a billion dollars per location. Their operation and maintenance costs can also be very high. Furthermore, major maintenance of locks may require long closures which may greatly disrupt normal services or completely interrupt services for extended periods unless parallel chambers exist that can be kept open. The service disruptions and delay costs may be alleviated if closures are planned and announced to waterway users well in advance, so that alternate forms of transportation may be arranged, inventories can be built up to outlast the closure periods, and vessels can escape from being trapped on the wrong side of a closed lock. It should be noted that large-scale waterway networks, such as that of the Mississippi Basin in the US, are largely "tree" networks in which alternate waterway paths are usually not available. Thus, the distinction between preventive maintenance and corrective maintenance is especially important for infrastructure maintenance on waterways. Preventive maintenance may be cyclically scheduled based on expected deterioration rates of various lock components (e.g., walls, gates, pumps) as well as the results of periodic inspections. Some thorough inspections may require drying out the lock chambers, which can also result in substantial disruptions and delays to traffic. The US Army Corps of Engineers, which is the agency responsible for operating and maintaining the navigable waterways in the US, generally prefers to schedule preventive maintenance in winter months, when some waterways freeze and their traffic stops. Of course, such scheduling is limited by the availability of maintenance resources, which should also be efficiently employed in other seasons.

Considerable dredging of the navigable channels on waterways may also be required to maintain the required depths and widths and counteract the effects of silting. For example, in fiscal year 2012 the Corps of Engineers spent $1.22 billion to dredge 182 million cubic feet from US waterways. Of that dredged quantity, 84 percent was required simply to maintain existing channels rather than improve them or dredge new ones. A substantial industry with specialized personnel, techniques and equipment is involved in dredging activities. The expenses of maintaining navigable channels as well as locks may become financially unsustainable for some waterway sections. The operators must then decide which sections of waterways to improve, maintain or abandon. Many previously used waterways, mostly used by relatively few and small vessels, have been abandoned in various parts of the world. The reasons for abandonment have usually been economic or financial, but environmental considerations have also been important in some cases (such as the Kissimmee Waterway in Florida).

Capital investment alternatives for waterways (e.g. to improve lock capacities and service times, or channel dimensions) tend to be interrelated. The benefits of various proposed projects depend considerably on which other projects are also implemented and at what times. For instance, some improvements might mostly shift the capacity bottlenecks and delays elsewhere. The costs of projects may also be interrelated to some extent, especially when the costs of mobilizing resources and those due to lock closures are considered. Maintenance and rehabilitation projects are also interrelated in their costs and benefits. Budget and other resource considerations also complicate the planning and scheduling for both capital improvement and maintenance projects. Further descriptions and solution methods for dealing with such problems are provided in Wang and Schonfeld (2008).

It should be noted that the costs of waterway infrastructure such as dams may be justified by various other benefits beyond waterway transportation, including flood control, irrigation for agriculture, water supply for other purposes, hydroelectric power and recreation. Thus, evaluations and decisions regarding waterway investments may be complicated by numerous other benefits and environmental considerations, especially when the waterways serve multiple nations or jurisdictions.

Inland Ports

Ports on waterways range from very simple piers for docking one boat at a time to very extensive facilities for loading, unloading and servicing a wide range of vessels. The largest such ports are typically found at the mouths of navigable rivers, such as Rotterdam at the mouth of the Rhine and Hamburg at the mouth of the Elbe; such ports are major transshipment terminals where freight is transferred between large ocean ships and smaller vessels that can be accommodated on inland waterways.

Figure 20.10 shows a panoramic view of the Port of Duluth, Minnesota, which handles more tonnage (about 38 million tons/year, mainly iron ore, grains and coal) than any port on the North American Great Lakes. About 1,000 ships/year visit this major freshwater port which is

Figure 20.10 Aerial view of the Port of Duluth
Source: Courtesy of Dennis O'Hara/Northern Images Photography

Figure 20.11 Coal-loading dock at the Port of Duluth
Source: Courtesy of Dennis O'Hara/Northern Images Photography

located far inland (2,513 km from the Port of Quebec near the estuary of the St. Lawrence River). Historically, it has been especially used for loading iron ore on ships which deliver that cargo to steel mills around the Great Lakes and return to Duluth empty. One of its coal-loading docks is shown more closely in Figure 20.11.

Along most rivers the usable vessels are considerably smaller. Ports are mostly small (often with a single dock) and specialized in a narrow range of commodities, such as grain, coal, minerals, chemicals or construction materials.

Economics of Waterways and their Competitors

Freight transportation by waterway is economically competitive with overland transportation modes such as rail and truck when reasonably deep natural waterways are available for a substantial fraction of the shipment distance and the commodities being moved are sensitive to cost rather than travel time. This mainly characterizes non-perishable bulk commodities with a low ratio of value to weight or volume, such as coal, minerals, grains. Some higher value freight, such as chemicals and containers, can also be economically transported in some situations, although competition from railroads and trucks is likely to be stronger. Pipelines are typically the strongest competitors of waterways for the transportation of large quantities of fluid commodities (especially oil products); they can be fairly easily built over hilly or mountainous terrain where waterway transportation is seriously disadvantaged. Details on the quantities of various commodities transported on US waterways as well as much additional information on US waterways may be found in The US Waterway System—Transportation Facts and Information: www.navigationdatacenter.us/factcard13.pdf. Additional information on waterway operations is available on other websites of the US Army Corps of Engineers and of PIANC, which are listed in the References section.

Ground transportation modes may complement as well as compete with waterways. They can connect ports on waterways to origins and destinations remote from waterways. They can also provide backup to waterways during freezing seasons or when waterway service may be interrupted by low water, turbulence, lock closures, or accidents.

Since waterways are almost invariably operated by government agencies, the economies of carrier size are considerably lower than on railroads which provide their own infrastructure, and more similar to those of trucks operating on publicly provided roads. Thus, relatively small waterway operators can compete effectively, although some operators are quite large in the US. Operators may specialize in operating towboats or barges, although many operate both vessel types. A considerable fraction of the vessels is owned by major industrial shippers.

Surprisingly, there has been a resurgence of passenger travel on inland waterways, but this mainly constitutes leisure cruising rather than transportation of persons who must hurry. For instance, one company advertises riverboat cruises on the Nile, Yangtse, Mekong, Irrawaddy and numerous European rivers. Some hydrofoil vessels have provided fast passenger services since the 1960s on Russian rivers, when those rivers don't freeze.

Summary

Inland waterways provide economical freight transportation in many parts of the world where suitable natural waterways exist. This includes most of Europe, much of the Far East and North America, and several other regions. The potential for expanding operations seems considerable in Equatorial Africa and especially in South America. Even in regions blessed with good natural waterways the cost of improving the waterways for better economic performance and reliability, and then maintaining the resulting infrastructure, can be very large. The economic and environmental advantages of waterways in handling large freight flows depend considerably on local geographic characteristics and the magnitude of traffic flows. These advantages must be balanced against the considerable costs providing new waterways or even preserving existing ones.

References

China Three Gorges Corporation website: www.ctgpc.com/benefifs/benefifs_a_5.htm accessed October 12, 2014.

Hadfield, C. *World Canals: Inland Navigation Past and Present*, Newton Abbot: David & Charles, 1986.

Institute for Water Resources, National Waterways Study—A Framework for Decision Making, Final Report NWS-83-1, US Army Corps of Engineers, January 1983.

Institute for Water Resources, Compendium on Water Transportation Reliability: Lock Congestion and Lock Queues, IWR Report 93-R-9, US Army Corps of Engineers, April 1993.

McCartney, B., George, J., Lee, B.K., Lindgren, M. and Neilson, F. *Inland Navigation: Locks, Dams and Channels*, American Society of Civil Engineers, 1998.

PIANC website: www.pianc.org/ accessed October 12, 2014.

Ting, C.J. and Schonfeld, P. "Control Alternatives at a Waterway Lock", *Journal of Waterway, Port, Coastal and Ocean Engineering*, 127–2, March/April 2001, pp. 89–96.

United Nations Economic Commission for Europe, Map of the European Inland Waterways Network. Available at: www.unece.org/fileadmin/DAM/trans/main/sc3/European_inland_waterways_-_2012.pdf, accessed November 12, 2014.

US Army Corps of Engineers, Waterborne Commerce Statistics Center. Available at: www.iwr.usace.army.mil/About/TechnicalCenters/ *WCSCWaterborneCommerceStatisticsCenter.aspx*, accessed October 12, 2014.

US Army Corps of Engineers, The US Waterway System—Transportation Facts & Information. Available at: www.navigationdatacenter.us/factcard/factcard.htm, accessed October 12, 2014.

US Army Corps of Engineers, US Waterway Data. Available at: www.navigationdatacenter.us/data/datawcus.htm, accessed October 12, 2014.

Wang, S. and Schonfeld, P. "Scheduling Waterway Projects with Complex Interrelations", *Transportation Research Record 2062*, October 2008, pp. 59–65.

21
MARITIME TRANSPORTATION

Harilaos N. Psaraftis

TECHNICAL UNIVERSITY OF DENMARK

1. Introduction

The overall literature in the broad area of maritime transportation is immense, covering a vast array of topics, ranging from the economics of the various market segments to techno-economic aspects of ship design, from shipping network design to legal-regulatory aspects of the markets, from ship routing and scheduling to safety and security, and from the economic impact of air emissions to port and terminal management, to name just a few. It is clearly impossible to cover all this material in this chapter, and the reader is referred to a number of seminal works, for instance Stopford (2009) and Talley (2012) for ship economics, and Christiansen et al. (2007) for optimization in maritime transport, among others.

Rather, the purpose of this chapter is to focus only on some selected issues that have become important in maritime transportation. To do so, the rest of this chapter is structured as follows. Section 2 discusses basic criteria. Section 3 discusses the optimization of a ship's economic performance, with main focus on speed optimization. Finally, section 4 presents this chapter's conclusions and discusses areas where further research is necessary.

The material of this chapter is mainly based on work of the author and his colleagues that can be found in Psaraftis et al. (2012), Gkonis and Psaraftis (2012), and Psaraftis and Kontovas (2013, 2014), among others.

2. Basic Investment Assessment Criteria

Ships being investments like any other, the traditional set of criteria used in evaluating investments in any industry can also apply to ships. However, special considerations are warranted, so as to capture the special characteristics of the environment in which ships are operating during their lifetime.

Net Present Value

The traditional Net Present Value (NPV) criterion is defined as follows:

$$NPV = \sum_{t=0}^{N} \frac{I(t) - C(t)}{(1+i)^t} \tag{1}$$

where

N = lifetime of ship (years)
$I(t)$ = income generated by ship in year t ($t = 0, \ldots, N$)
$C(t)$ = expenditure spent on ship in year t ($t = 0, \ldots, N$)
i = ship owner's cost of capital (assumed constant)

$I(t)$ and $C(t)$ are supposed to include everything that will go into or out of the ship owner's pocket, during the entire ship construction, operation and scrapping cycle. These include a down-payment for the ship's contract, loan amounts received and paid, interest, taxes, all operating expenses paid for by the ship owner, amounts received when the ship is scrapped, and obviously charter revenues.

According to the *NPV* criterion, the ship or the combination of ships that yields the maximum possible *NPV* is the best alternative to the ship owner. However, the apparent simplicity of this criterion is deceptive, mainly due to the significant uncertainties associated with all elements in the *NPV* formula. For instance:

1. The income stream $I(t)$ is subject to the vagaries of the markets in which the ship will operate through its lifetime, and also depends on the way the ship will be utilized (which markets or trades it will serve, what cargoes it will carry).
2. The cost stream $C(t)$ is also uncertain, as it depends on variables such as fuel prices, which are not known with certainty, but also on uncertainties as regards repairs, maintenance, and other costs. It also depends on how the ship will be utilized (what routes it will serve, what cargoes it will carry, and at what speed- see also next section).
3. Last, but not least, the cost of capital i may not be known throughout a ship's lifetime (let alone a constant), not to mention that N itself is also unknown.

Yet, and in spite of such uncertainties, the *NPV* criterion is being widely used and can be useful as a criterion in maritime transportation.

Required Freight Rate

A criterion much related to *NPV* that is used very often in maritime transport is the Required Freight Rate (*RFR*), defined as follows.

If we assume that $I(t) = FX(t)$, where $X(t)$ is the cargo carried by the ship in year t and F is a constant freight rate, then the *RFR* of a ship is defined as the freight rate F for which the *NPV* associated with the ship throughout its lifetime is zero.

According to the *RFR* criterion, the ship whose *RFR* is the lowest among alternative ships is the best.

RFR is a widely used criterion to compare ships from a cost-effectiveness viewpoint. Yet, it suffers from at least the same deficiencies and limitations as *NPV*. For instance, being a ratio, it cannot say anything about the scale of the investment. Yet, it is known that larger ships will typically have a lower *RFR* than smaller ships due to economies of scale, making this criterion biased in favor of larger ships. Also, a constant freight rate F is very seldom (if at all) experienced by any type of ship, an exception being a ship engaged in a very long-term charter in tramp trades. Last, but not least, trying to predict $X(t)$ is no less difficult than trying to predict the market throughout a ship's lifetime.

In spite of all these, *RFR* is very frequently used as a criterion, mainly to compare alternative designs or alternative investment choices. It is thus very important to avoid some pitfalls, if *RFR* is to be used at all. The following considerations are important in that respect:

1. Make sure to compare ships of similar size. It would not make much sense to compare the *RFR* of a 500 TEU feeder with that of Maersk's 'Triple-E' 18,000 mega ships.
2. In addition to *RFR*, try also to compute the *NPV*s of the various alternatives. *NPV* is a more sound criterion, as a ship owner is more interested in what money he or she will make through a ship's lifetime than what value a particular ratio will take on.
3. Make a thorough sensitivity analysis.

Internal Rate of Return

In addition to *NPV* and *RFR*, some additional criteria are being used in maritime transportation. Among those, one can mention the Internal Rate of Return (*IRR*), defined as the interest rate in (1) which produces an *NPV* of zero.

According to this criterion, the ship that has the highest *IRR* among ship alternatives is the best.

The following caveats are worthy of note if this criterion is used:

a. As with *RFR*, *IRR* is also a ratio, and, as such, ignores scale. Therefore, we have to make sure we compare ships of similar size.
b. The root of the equation $NPV(i) = 0$ may not be unique, and if this is the case *IRR* is ambiguous.

3. Optimizing Shipping Economic Performance

The broader spectrum of problems related to the optimization of a ship's economic performance involves optimizing the operation of a fleet of ships, and/or, by a further extension, of the entire intermodal chain, including ports.

The Broader Picture

If one looks at this broader picture, it would be possible to examine one or a combination of problems from the following generic list (which is non-exhaustive):

- Optimal ship speed
- Optimal ship size
- Routing and scheduling
- Fleet deployment
- Fleet size and mix
- Weather routing
- Intermodal network design
- Modal split
- Transshipment
- Queuing at ports

- Terminal management
- Berth allocation
- Supply chain management.

Many of these problems methodologically fall under the disciplines of operations research—management science—transportation logistics problems, whose objective is to optimize one or more aspects of the operation of a ship, a fleet, or a supply chain system.

The literature on these problems is rapidly growing. See, for instance, among others, Christiansen et al. (2007), for surveys of ship routing and scheduling problems, Brown et al. (1987) for the scheduling of crude oil carriers, Fagerholt (2004) for vessel fleet scheduling, Thompson and Psaraftis (1993) for local search methods, Rana and Vickson (1991) for routing of containerships, and Agarwal and Ergun (2008), and Alvarez (2009), for optimization approaches in liner shipping. For terminal management problems, a branch-and-cut procedure is reported by Moccia et al. (2006) and the so-called "double cycling" procedures for loading and unloading are described by Goodchild and Daganzo (2006). For a review of the operations research literature of problems related to container terminal management the reader may refer to Stahlblock and Voss (2007), among others.

It should be said that, in spite of the growing literature, many of these problems are typically treated in isolation from one another, even though many of them are mutually connected. For instance, the problem of optimal loading of a containership is connected with the problem of optimal yard management, which in turn is connected with the problem of routing straddle carriers in the terminal, the problem of assigning cranes to ships when they berth, the berth allocation problem, the problem of determining optimal queuing strategies, the problem of selecting which port should be a transshipment port, the network design problem, and so on. All of these problems impact, directly, or indirectly, the economics of ships (including the basic criteria of section 2), but no very good way to treat them in a holistic way is yet known.

In the following section, we focus on one of the above categories of problems, the class of ship speed optimization problems. This class of problems has become increasingly important for a variety of reasons.

Ship Speed Optimization

Determining the optimal speed of ships is not new in the literature. A first observation is that most of the models (at least implicitly) assume that fuel costs are being borne by the ship owner. In the tramp shipping market (served by tankers, dry bulk carriers, product carriers, and gas carriers) this is the case if the ship is on spot charter. It is known that the predominance of charter party contracts are time charters, in which fuel costs are borne by the charterer. Even though most models assume the ship owner as the party that bears the costs, including fuel, the related optimization problem is typically cost minimization rather than profit maximization. This is tantamount to assuming that revenue for the service is fixed. This is not the case, however, in most instances and thus some of the models that optimize speed do not capture the trade-off between a higher speed to make more profit-earning trips per unit time and the impact of such higher speed on costs (mainly on fuel).

For problems in the liner market (served by containerships and roll-on/roll-off ships), a similar situation pertains. Ship owners who run liner services using their own ships want to maximize profits and the same trade-offs are at play. However, also using chartered ships to provide liner services is not uncommon, as a liner company typically employs a mix of owned and chartered vessels.

Maersk's new 18,000 TEU 'Triple-E'[1] containerships have a design speed of 17.8 knots, down from the 22–25 knots range that has been the industry's norm, and will emit 20 percent less CO_2 per container moved as compared to the *Emma Maersk*, previously the world's largest container vessel, and 50 percent less than the industry average on the Asia–Europe trade lane (Maersk, 2013).

For the so-called "industrial" types of problems, in which a company (for instance, an oil company) uses its own ships to move its own cargoes, again the usual objective is cost minimization. What is not often mentioned, however, is that these companies always have the option to enter the market, by either offering excess capacity at the prevailing spot rate, or hiring extra capacity at such rate. Thus, in a boom period it may make sense for an oil company to run its ships at a higher speed, so as to offer the excess capacity obtained to the spot market. Of course, that this opportunity exists does not necessarily mean it will be used.

There are many ways of classifying related papers and models in the literature. A first-order classification involves grouping references into two major categories:

1. those that present models in which emissions are not considered, and
2. those that present models in which emissions are considered (together with other considerations).

A finer-grain taxonomy classifies the literature of the previous two sections according to the following parameters (see also Psaraftis and Kontovas (2013)).

Optimization criterion: The main variants here are cost (to be minimized) and profit (to be maximized). Other variants include fuel consumption, transit time, or others. To be sure, some models in the literature are not cast as optimization problems. In these papers we set "cost", "profit", or other, depending on what the model described by the paper tries to measure.

What is the shipping market/context of the problem: This may be tankers, bulk carriers, containerships, or other ship types. It may even involve the whole commercial fleet.

Who is the decision maker: By this we mean who decides what the ship speed should be. This can be the ship owner or the charterer. For weather-routing problems, it is typically the ship's master. An attempt to designate who is the decision maker is made even if the model is not an optimization model.

Fuel price an explicit input? Yes, if fuel price is explicitly included as one of the explicit inputs of the problem, no otherwise.

Freight rate an input? Yes, if freight rate (spot, or other) is explicitly included as one of the explicit inputs of the problem, no otherwise. There are also models that compute that rate as an equilibrium rate depending on supply and demand.

Fuel consumption function: It could be cubic, non-linear, linearized, general or unspecified.

Optimal speeds in various legs: Whether or not the model computes optimal speeds for each leg of the route (versus a single optimal speed).

Optimal speed as function of payload: Whether or not the model can compute the optimal speed as a function of how full or empty the ship is.

Logistical context: This could be a fixed route scenario, a ship routing and scheduling problem, a fleet deployment problem, or other.

Size of fleet: One ship, or many ships.

Adding more ships an option: This is so if adding (or subtracting) ships is an option so as to maintain constant throughput.

Inventory costs included: Yes, if cargo-carrying (inventory) costs are included in the model, no otherwise.

Emissions considered: Yes or no.

Modal split considered: Yes, if model calculates the split among alternative and competing modes of transport as a function of problem inputs.

Ports included in formulation: Yes, if port times, costs, congestion, port emissions or other port-related variables are included in the model.

The full taxonomy is presented in Psaraftis and Kontovas (2013).

Table 21.1 below lists a limited sample of related papers in the literature, including some more recent ones, covering various logistical contexts, with respect to whether or not (a) fuel consumption is a function of payload, (b) fuel price is an explicit or implicit input, (c) freight rate is an input, and (d) cargo inventory cost is taken onboard. Various combinations of (a) to (d) can be seen. For more details, see Psaraftis and Kontovas (2014).

Slow steaming is governed by some fundamentals. The first fundamental is something that many papers in the literature seem to ignore: ships do not trade at predetermined speeds. Those who pay for the fuel, that is, a ship owner whose ship trades on the spot market, or a charterer if the ship is on time charter, may want to choose the ship speed as a function of (a) fuel price and (b) market spot rate. In periods of depressed market conditions, as is the typical situation these days, ships tend to slow steam. The same is the case if bunker prices are high. Conversely, in boom periods or in case fuel prices are low, ships tend to sail faster.

An exception to the case that the ship owner or the charterer can freely choose an optimal speed for the ship is in case the ship is on spot charter and speed is prescribed in the charter party contract, either explicitly (speed is, say, 15 knots) or implicitly (pickup and delivery dates are prescribed). In spot charters (rental of the ship for a single voyage) the fuel is paid for by the ship owner. Agreeing on a prescribed speed in the charter party involves in most cases only the laden part of the trip, with the owner free to choose his speed on the ballast return leg. The speed that is agreed upon for the laden leg may or may not be the speed that the ship owner would have freely chosen if no explicit agreement were in place. If it is higher, the ship owner may ask for a higher rate than the prevailing spot rate, understanding, of course, that in this case he may lose the customer to a competitor ship, with which the charterer can obtain more favorable terms. For a discussion of possible distortions and additional emissions that can be caused by charter party speed agreements, see Devanney (2011).

The second fundamental is perhaps not immediately obvious. This is that even though the owner's and time charterer's speed optimization problems appear at first glance different, the optimal ship speed for both problems turns out to be the same. A proof is in Devanney (2010) for a rudimentary scenario of a ship hauling cargo from port 1 to port 2 and returning to port 1 on ballast (empty), and goes roughly as follows.

Table 21.1 Sample of papers according to criteria (a) to (d)

Papers (listed chronologically)	*Logistical context*	*Fuel consumption: a function of payload?*	*Fuel price: explicit or implicit input?*	*Freight rate: an input?*	*Cargo inventory: cost taken onboard?*
Ronen (1982)	Tramp fixed route	No	Explicit	Yes	No
Perakis and Jaramillo (1991)	Liner fleet deployment	No	Explicit	Yes	No
Devanney (2007)	World oil network model	Only for laden and ballast conditions	Explicit	Equilibrium spot rate computed	Yes
Notteboom and Vernimmen (2010)	Container fixed route	No	Explicit	No	No
Norstad et al. (2011)	Tramp pickup and delivery	No	Implicit	No	No
Cariou and Cheaitou (2012)	Container fixed route	No	Explicit	No	Yes
Gkonis and Psaraftis (2012)	Tanker fixed route	Only for laden and ballast conditions	Explicit	Yes	Yes
Hvattum et al. (2013)	Fixed route	No	Implicit	No	No
Fagerholt and Ronen (2013)	Bulk ship routing, mixed chartering scenario	No	Implicit	Only for spot cargoes[2]	No
Psaraftis and Kontovas (2014)	Fixed or flexible route	For any loading condition	Explicit	Yes	Yes

Source: Psaraftis and Kontovas (2014)

For a given ship, a ship owner in the spot market should operate at a speed that maximizes profit per day. Then his speed optimization problem is the following:

$$\max_{v} \left\{ \frac{sC}{d/v} - pf(v) - E \right\} \tag{2}$$

where

s	=	spot rate received by the owner (in \$/tonne)
C	=	ship's cargo capacity (in tonnes)
d	=	roundtrip distance (in nautical miles)
v	=	sailing speed in nautical miles per day[3]
p	=	bunker price (in \$/tonne)
$f(v)$	=	daily fuel consumption function at speed v (tonnes/day), and
E	=	operating expenses borne by the ship owner other than fuel costs, including crew wages, insurance, etc. (in \$/day).

In the above scenario, time in port has been ignored, although including it is a straightforward extension. Also the function $f(v)$ is assumed to be the same in both directions (laden and ballast), although having different functions and different speeds on each leg is also a straightforward extension.

For a time charterer who has chartered the same ship, and who is the effective owner of the vessel during the period of the contract (also known in shipping parlance as the "disponent owner") faces the following problem:

$$\min_{v} \left\{ s\left(R - \frac{Cv}{d}\right) + pf(v) + T \right\} \tag{3}$$

where

R	=	how much cargo needs to be moved (tonnes/day)
T	=	time charter rate paid to the owner (\$/day)

Equation 3 above assumes that any difference between the cargo capacity required by the time charterer (R) and what the chartered ship can provide if sailing at speed v (Cv/d) can be chartered in the spot market at a spot rate of s. If the difference ($R - Cv/d$) is positive (meaning that the chartered ship sailing at speed v cannot fully satisfy the charterer's needs), then additional capacity is chartered in at a rate of s, assuming the spot chartered ship sailing at the same speed v. If this difference is negative (meaning that there is spare capacity in the time chartered ship), then that spare capacity can be chartered out at the same spot rate s.

It can be seen easily that problems (1) and (2) are mathematically equivalent. In fact, in (1) the term E does not depend on speed and can be discarded from the objective function, leading to

$$\max_{v} \left\{ \frac{sC}{d/v} - pf(v) \right\} \tag{4}$$

In (3), one can separate the term ($sR + T$) which does not depend on speed and thus can be discarded as well. What is then left is:

$$\min_{v} \left\{ pf(v) - s\frac{Cv}{d} \right\} \qquad (5)$$

Problems (4) and (5) are essentially the same, thus leading to the same optimal speed.

Factoring out the spot rate s, both problems can be rewritten as follows:

$$\min_{v} \left\{ \frac{p}{s}f(v) - \frac{Cv}{d} \right\} \qquad (6)$$

Equation 6 shows that for both problems, a key determinant parameter of the speed optimization problem is the *non-dimensional ratio* $\rho = p/s$ of the bunker price divided by the spot rate, since for a given ship and route the optimal speed will be the same as long as ρ remains constant. Higher ρ ratios will generally induce lower speeds than lower ρ ratios. This corresponds to the typical behavior of shipping lines, which tend to slow steam in periods of depressed market conditions and/or high fuel prices and go faster if the opposite is the case.

A third fundamental of slow steaming is that fuel consumption (and hence fuel costs and emissions) depend non-linearly on ship sailing speed. The simplest model is to assume one type of fuel consumed on the ship, available at a known price of p (in \$/tonne). Then the daily *at sea* fuel cost of a ship sailing from port i to port j is equal to $pf(v_{ij}, w_{ij})t_{ij}$, where $f(v_{ij}, w_{ij})$ is the ship's daily fuel consumption at sea (in tonnes/day), a known function of the ship's speed v_{ij} and payload w_{ij} from i to j, and t_{ij} is the ship's sailing time from i to j, given by the ratio (d_{ij}/v_{ij}), sailing distance divided by speed. Function f depends on many ship parameters, such as type and size of power plant, including main and auxiliary engines, geometry of ship hull, propeller design, and other parameters (weather conditions for instance). It can even be defined for $w_{ij} = 0$ (ship going on ballast). *In port* fuel costs are proportional to overall total port residence time, and these depend on per day fuel consumption of the ship's auxiliary engines while in port. In case the ship uses different fuels for its main engine and auxiliary engines (for instance, Heavy Fuel Oil (HFO) and Marine Diesel Oil (MDO) respectively), total fuel cost is the summation of all relevant fuel types.

The fact that function f can be a complex function which may not even be defined in closed form does not prevent us from considering some modeling approximations. A usual approximation is that function f is equal to $A + Bv_{ij}^{n}$ with A, B and n input parameters such as $A \geq 0$, $B > 0$ and $n \geq 3$. Another approximation is that for a given speed, f is proportional to $(w_{ij} + L)^{2/3}$, where L is the weight of the ship if empty plus fuel on board and consumables (modified admiralty formula, see also Barass (2005)). A combination of these two approximations can also be considered. Most papers in the literature assume a cubic function, that is, $A = 0$ and $n = 3$ and no dependency on payload. $n = 3$ is usually a good approximation for tankers and bulk carriers and for the range of typical operational speeds of these vessels. A basic drawback of a cubic function is that it is invalid for very low speeds. In fact, this function gives zero fuel consumption at zero speed, which is not the case in practice, as a ship, even stationary, consumes some fuel. Another drawback of a cubic function is that it may not be a good approximation for some ship types, containerships being the most notable example. For these ships, exponent n can be 4 or 5 or conceivably even higher.

Figure 21.1 shows how optimal speeds in the laden and ballast leg for a modern Very Large Crude Carrier (VLCC) operating from the Persian Gulf to Japan conditions may vary as a function of fuel price and market rate spot rates are expressed in terms of World Scale (WS) equivalents. For a certain tanker route, WS is defined as 100 times the ratio of the prevailing spot rate on that route divided by the "base rate" on that route (see Stopford, 2009).

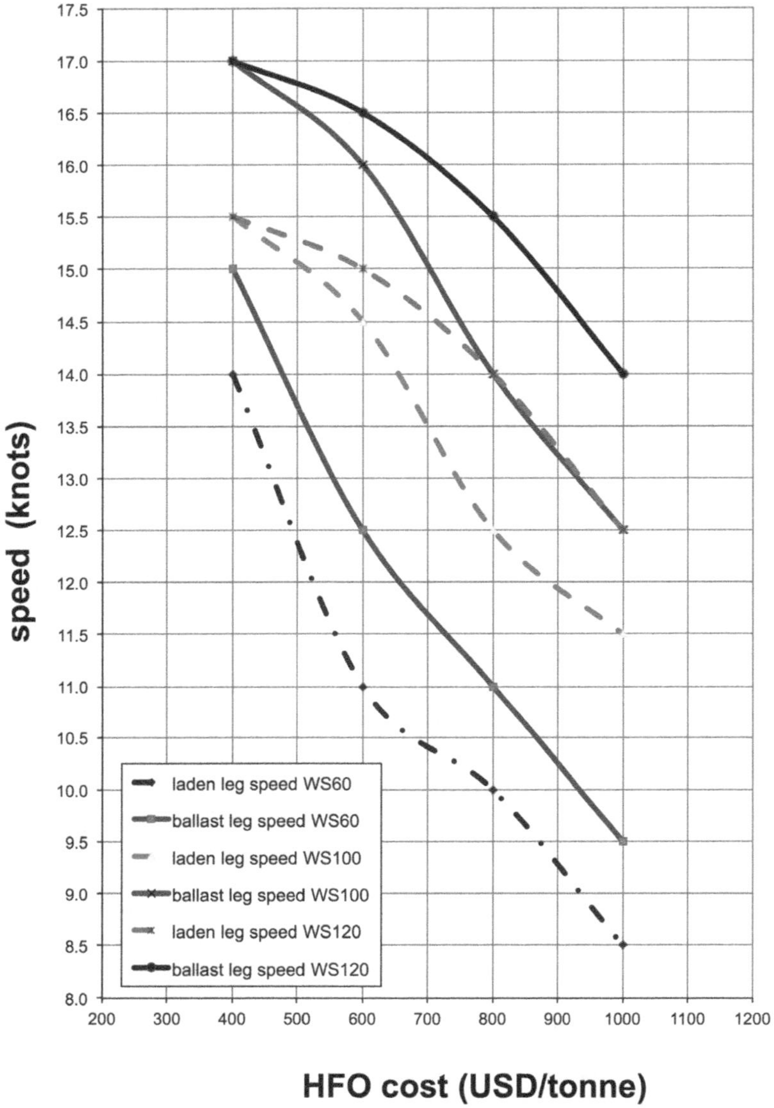

Figure 21.1 Optimal VLCC laden and ballast speeds as functions of fuel price and spot rate (spot rates are in WS)

Source: Gkonis and Psaraftis (2012)

Figure 21.2 shows the results if in-transit cargo inventory costs are factored into the cost equation. These costs on a per day and per tonne basis are equal to *PR*/365, where *P* is the value of the cargo and *R* is the cargo owner's cost of capital.

One can observe that optimal ballast speeds are typically higher (by 1 to 1.5 knots) than optimal laden speeds, except if cargo inventory costs are accounted for, in which case laden speeds can be higher than ballast speeds (depending on fuel price). In practice, however, many tankers sail faster on the laden leg than on the ballast leg, which is suboptimal. The reason for this is more likely to be attributed to charter party speeds than inventory costs (Devanney, 2011).

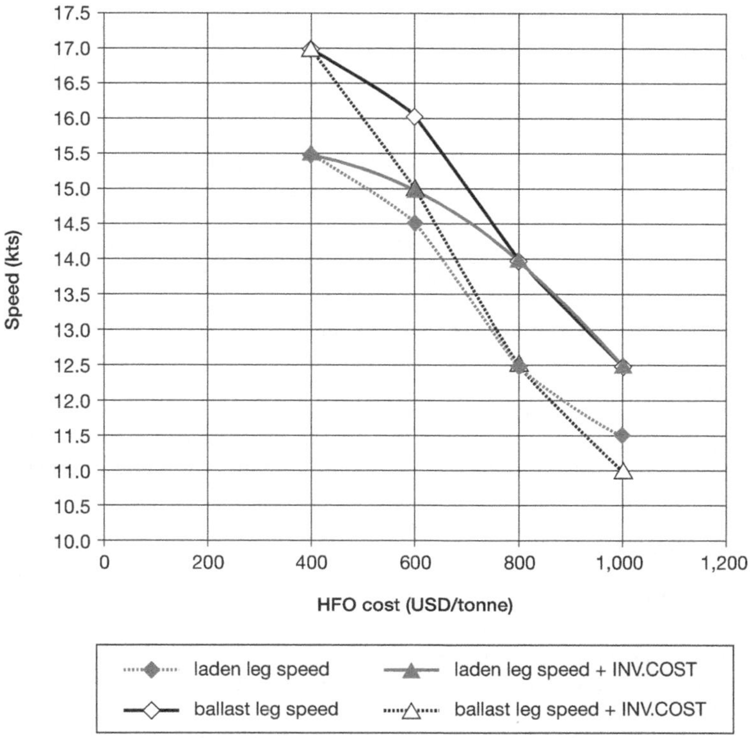

Figure 21.2 Optimal VLCC laden and ballast speeds with and without inventory costs
Source: Gkonis and Psaraftis (2012)

Combined speed-routing scenarios have been examined, among other things, in Psaraftis and Kontovas (2014).

4. Conclusions

This chapter has attempted to highlight a rather limited number of issues in maritime transportation that, in our opinion, will be important in the years ahead. We believe that research in these and related areas will further grow and will be useful from a practical perspective. With environmental concerns as regards air pollution from ships and on how to reduce it, it is imperative to improve the economic performance of ships to the greatest possible extent. Environmental criteria were not examined explicitly in this chapter, but as ship air emissions are directly proportional to fuel burned, it is clear that minimizing fuel consumption would also minimize emissions. To what extent this would also produce a win–win solution vis-à-vis other criteria is a subject that should be looked at. For some considerations on this subject, one may take a look (among others) at Corbett et al. (2010) and Psaraftis and Kontovas (2010).

Notes

1 Triple-E stands for Economy of scale, Energy efficiency and Environmentally improved performance.
2 A known revenue is assumed for each of the spot cargoes, implying a freight rate for them.
3 This is 24 times the speed in knots. We use this unit to avoid carrying the number 24 through the calculations.

References

Agarwal, R. and Ergun, O. (2008) Ship scheduling and network design for cargo routing in liner shipping, *Transportation Science*, 42 (2), 175–196.

Alvarez, J.F. (2009) Joint deployment and routing of a fleet of container vessels, *Maritime Economics and Logistics*,11 (2), 186–208.

Barrass, C.B. (2005). *Ship Design and Performance for Masters and Mates*, Butterworth-Heinemann, UK.

Brown, G.G., G.W. Graves, D. Ronen (1987) Scheduling ocean transportation of crude oil, *Management Science*, 33 (3), 335–346.

Cariou, P. and A. Cheaitou (2012) The effectiveness of a European speed limit versus an international bunker-levy to reduce CO_2 emissions from container shipping, *Transportation Research Part D*, 17, 116–123.

Christiansen, M., K. Fagerholt, B. Nygreen, D. Ronen (2007) Maritime transportation. In Barnhart, C. and Laporte, G. (Eds.) *Transportation* volume of *Handbooks in Operations Research and Management Science*, North Holland, Amsterdam.

Corbett, J., Wang, H. and Winebrake, J. (2010) The effectiveness and costs of speed reductions on emissions from international shipping, *Transportation Research Part D: Transport and Environment*, 14, 593–598.

Devanney, J.W. (2007) Solving elastic transportation networks. Center for Tankship Excellence (online). Available at: www.c4tx.org

Devanney, J.W. (2010) Impact of fuel prices on VLCC spot rates, *3rd Symposium on Ship Operations, Management and Economics*, SNAME Greek Section, Athens, October.

Devanney, J.W. (2011) The impact of charter party speeds on CO_2 emissions, Center for Tankship Excellence publication. Available at: www.c4tx.org

Fagerholt, K. (2004) A computer-based decision support system for vessel fleet scheduling—experience and future research. *Decision Support Systems*, 37(1), 35–47.

Fagerholt, K. and D. Ronen (2013) Bulk ship routing and scheduling: solving practical problems may provide better results, *Maritime Policy and Management*, 40 (1), 48–64.

Gkonis, K.G. and H.N. Psaraftis (2012) Modelling tankers' optimal speed and emissions, archival paper, SNAME Transactions, 120, 90–115.

Goodchild, A.V. and C.F. Daganzo (2006) Double cycling strategies for container ships and their effect on ship loading and unloading operations, *Transportation Science*, 40, 473–483.

Hvattum, L. M., I. Norstad, K. Fagerholt, and Laporte, G. (2013) Analysis of an exact algorithm for the vessel speed optimization problem, *Networks*, 62 (2), 132–135.

Maersk (2013) Building the world's biggest ship. Available online at: www.maersk.com/innovation/leadingthroughinnovation/pages/buildingtheworldsbiggestship.aspx

Moccia, L., J.-F. Cordeau, M. Gaudioso, and G. Laporte (2006) A branch and cut algorithm for the quay crane scheduling problem in a container terminal,*Naval Research Logistic*, 53, 45–59.

Norstad, I., K. Fagerholt, and G. Laporte (2011) Tramp ship routing and scheduling with speed optimization, *Transportation Research Part C*, 19, 853–865.

Notteboom, T.E. and Vernimmen, B. (2009) The effect of high fuel costs on liner service configuration in container shipping, *Journal of Transport Geography*, 17, 325–337.

Perakis, A.N. and D.I. Jaramillo (1991) Fleet deployment optimization for liner shipping. Part 1: Background, problem formulation and solution approaches. *Maritime Policy and Management*, 18 (3), 183–200.

Psaraftis, H.N. and Kontovas, C.A. (2010) Balancing the economic and environmental performance of maritime transportation, *Transportation Research* 15D, 458–462.

Psaraftis, H.N. and C.A. Kontovas (2013) Speed models for energy-efficient maritime transportation: a taxonomy and survey, *Transportation Research Part C: Emerging Technologies*, 26, 331–351.

Psaraftis, H.N. and C.A. Kontovas (2014) Ship speed optimization: concepts, models and combined speed-routing scenarios, *Transportation Research Part C*, 44, 52–69.

Psaraftis, H.N., D.V. Lyridis, and C. A. Kontovas (2012) The economics of ships, Chapter 19 of *Blackwell Companion to Maritime Economics*, W. K. Talley (Ed.) Wiley-Blackwell, Chichester.

Rana, K. and R.G. Vickson (1991). Routing containerships using Langrangean relaxation and decomposition, *Transportation Science*, 25(3).

Ronen, D. (1982) The effect of oil price on the optimal speed of ships, *Journal of the Operational Research Society*, 33, 1035–1040.

Stahlblock, R. and Voss, S. (2007) Vehicle routing problems and container terminal operations–an update of research. *The Vehicle Routing Problem: Latest Advances and New Challenges*. B. Golden, S. Raghavan and E. Wasil (Eds.), Springer, New York.

Stopford, M. (2009) *Maritime Economics* (2nd ed.) Routledge, London.

Talley, W.K. (Ed.) (2012) *Blackwell Companion to Maritime Economics*, Wiley-Blackwell, Chichester.

Thompson,P.M. and H.N. Psaraftis (1993) Cyclic Transfer Algorithms for Multi-Vehicle Routing and Scheduling Problems, *Operations Research* 41 (5), 935–946.

22

LOGISTICS AND SUPPLY CHAIN MANAGEMENT

Issues and Challenges for Modern Production Systems

Riccardo Manzini,[1] *Ricardo Accorsi,*[1] *Mauro Gamberi,*[2] *Alberto Regattieri,*[1] *and Matteo Mario Savino*[3]

(1) BOLOGNA UNIVERSITY, (2) UNIVERSITY OF PADUA, (3) UNIVERSITY OF SANNIO

Introduction

Logistics and supply chain management (SCM) are two of the most important disciplines in industrial engineering. They support managers of industrial and service sectors to make a success of their production systems operating in complex environments. Despite the common belief, logistic issues are crucial and strategic in the manufacturing industry (e.g. automotive industry, production of automatic machines, tile industry), in the processing industry (e.g. chemical industry and food industry) and in the service tertiary sector including the management of hospitals, research institutions (e.g. universities and departments), banks, etc. These sectors exemplify complex industrial/service systems operating worldwide in different countries and continents and involving several typologies of stakeholders (e.g. producers, logistic providers, wholesalers, final consumers).

Logistics deals with the integration of the flow of materials and the flow of information. The first traditionally moves from the source of raw materials to the final consumers passing through production plants, distribution centers, wholesalers and retailers. The reverse flow of information moves from the points of demand to the suppliers, across the supply chain stages. The adequate integration of these flows pushes and pulls goods throughout the pipeline to meet demand, quantity, quality and due-date constraints. The service level (SL) represents a renowned metric for the so-called "logistic efficiency" that includes a set of operative key performance indicators (KPIs).

Nowadays, the evaluation of the SL of supply chains passes also through the measurement and control of the quality, safety and sustainability of products, activities and the whole system according to an holistic perspective. A significant example of the level of complexity and criticalities to be faced is the food supply chain (FSC). The logistics of food has to guarantee high level of product quality at the place of consumption (1) combined with high levels of efficiency of the production/distribution system (2), which accounts for costs and requires the proper management of labor and activities in the presence of safety and security issues. Indeed, additional pillars to address the complexity of the food production/distribution system are food safety (3) and levels of sustainability (4). Safety, genuine and sustainable products from the place of origin to the place of consumption is the new challenge of the food industry and the FSC (Manzini and Accorsi, 2013).

Logistics supports decision-making both at the so-called "soft level", e.g. purchasing, inventory and fulfillment decisions, and at the "hard level", which deals with the infrastructure, resources and equipments required by storage, material handling, packaging, loading, transportation. Supply chain management (SCM) is "the integration of key business processes from end-user through original suppliers, that provides product, service, and information that add value for customers and other stakeholders" (Lambert et al. 1998). According to Sahin and Robinson (2002), supply chain "consists of the network of supplier/vendors, manufacturers, distributors, and retailers interconnected by transportation, information and financial infrastructure" whose role is "to provide value in terms of products and services" to each actor. The definition of supply chain management stated in this chapter highlights the role of logistics as a crucial enabling factor for the proper management of all the activities and operations that lead products from suppliers to consumers.

Strategic, Tactical and Operational Levels

Contributions in the supply chain planning literature discriminate between the strategic level on the one hand, and the tactical and operational levels on the other (Manzini et al. 2008). The strategic level deals with the decisions of the adequate configuration of the logistic infrastructural in terms of location and establishment of production/distribution nodes, where capacity, costs, and equipments of the facility are given. Conversely, the most important tactical and operational decisions include the inventory management (i.e. the level of inventory of the storage system), the stock reorder policy, the stock put-away operations, the order picking and product replenishment, the delivery schedule and tactical distribution decisions within the supply chain. In order to merge a multidimension and comprehensive decision-making, Shen (2005) highlights the importance to integrate the major strategic, tactical and operation cost drivers into the decision model. Recent contributions on the integration are presented and applied by Manzini et al. (2014) and Manzini (2012). One of the key goals of the recent research contributions is the integration of logistic problems and decision-making at different levels as a support for management in industrial and service companies. Figure 22.1 illustrates an exemplifying conceptual framework of the supply-chain planning (Manzini and Bindi, 2009), with a focus on some issues and decisions belonging to the strategic design, the tactical planning, and the operational planning and scheduling. Decisions are classified into three sets of logistics issues from the production stage to the supply stage:

1. *Purchase and production*, e.g. suppliers' selection, production facilities location, multifacilities material requirement planning, etc.
2. *Distribution and warehousing*, e.g. distribution network configuration, transportation modes decisions, storage capacity determination, etc.
3. *Supply*, e.g. customer demand assignment, vehicle loading and routing, etc.

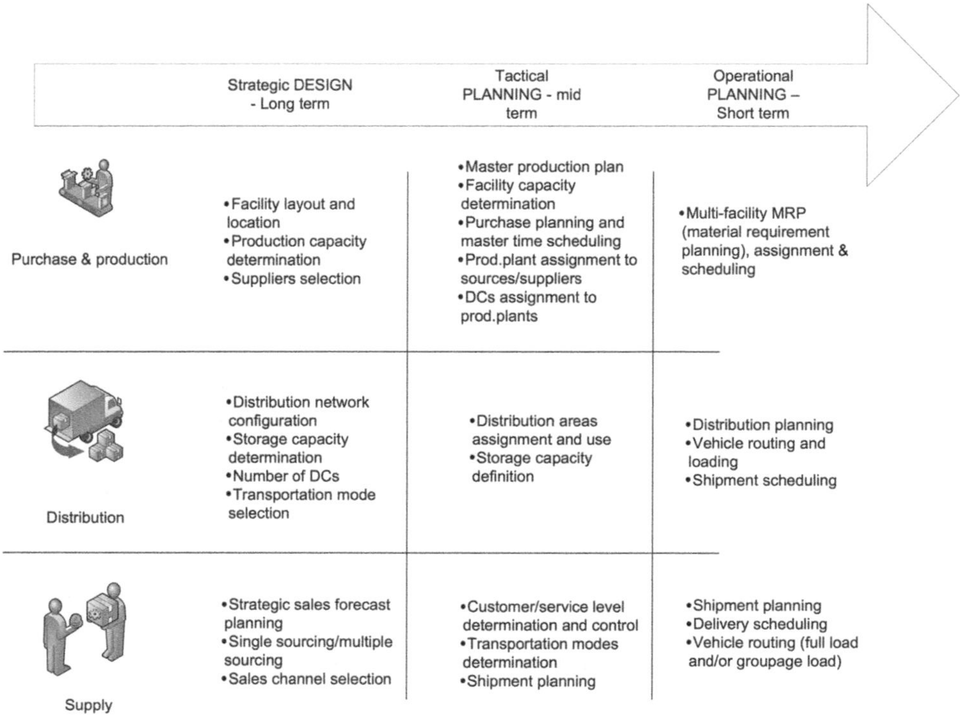

Figure 22.1 Conceptual framework on supply chain planning

Problems and Issues

Several problems and issues relate to logistics and/or SCM. These are familiar to practitioners and managers of industry, and compel taking decisions over strategic perspectives, usually referring to long-term issues, tactical perspectives, which refer to mid-term issues, and operational perspectives, which refer to short-term issues. These problems can significantly vary according to the level of complexity. Several of them belong to the NP, "non-polynomial in time", class of problems as defined by the Operations Research discipline.

For example, the common level of complexity of modern logistics systems can be associated with one or more of the following operating conditions:

- Multi-echelon production systems, made of several manufacturing plants, central and regional distribution centers, hubs, wholesalers, etc., and located worldwide in different locations. Typical problems and issues related to such a system are the coordination, the collaboration between different levels and partners of the system and the network design (Manzini et al. 2008; Manzini and Bindi 2009; Manzini 2012; Manzini et al. 2014).
- A large number of products, e.g. 200,000. A similar number of items is typical of spare-parts storage systems in the automotive industry (Accorsi et al. 2012; Accorsi et al. 2014a).
- Coexistence of different storage modes and equipments, as well as of retrieving activities, including the picking process, i.e. the less than unit loads retrieval (Bindi et al. 2009; Accorsi et al. 2014a).

- Perishable products subject to a shelf life, typical of the food industry and food supply chain (Manzini and Accorsi, 2013). Other related issues are the quality and safety of the supply chain at the place of consumption, e.g. on the table of the final consumer.
- Logistics outsourcing, i.e. presence of third- and fourth-party providers (Manzini et al., 2007).
- A combination of direct (i.e., forward) and indirect (i.e., reverse) flows of materials as typical of a distribution network of food products combined with the collection and pooling of pallets and reusable packages for food catering (Accorsi et al. 2014c). Special typologies of items are waste and by-products, the so-called "secondary products", which are generated by manufacturing processes and/or chemical reactions. A typical example is the management of waste and by-products of the food industry. Modeling the solid waste collection is a topic recently discussed by Mora et al. (2014). They include the environmental and economic assessment of logistics decisions. Gamberini et al. (2010) present a study on the WEEE transportation network.
- Different packaging systems and containment solutions with different levels of performance (Accorsi et al. 2014b), which eventually include track and trace solutions and traceability issues so critical in more and more supply chains (e.g. pharmaceutical and food industry) (Sarac et al. 2010; Regattieri et al. 2007).
- Multiple transportation modes, e.g. road, rail, sea and air, which eventually are combined with intermodal hubs and infrastructures.
- Assessment of the environmental impact due to logistics and SCM (Sundarakani et al. 2010; Sarkis et al. 2011). Nowadays, logistic decisions must be supported by life-cycle assessment, including the analysis of the environmental impacts and the evaluation of the generated externalities (Chaabane et al. 2012).

This list of operating conditions is not exhaustive, but represents a reference set of issues generating complexity and stimulating the theoretical and applied research.

Literature presents many supporting decisions models and methods, e.g. mixed integer linear programming (MILP), widely used in transportation problems, heuristic and meta-heuristic algorithms, etc. We decided not to treat these modeling issues, which is not the objective of this chapter. Conversely, in the next sections we depict the hotspots of logistics and operation issues with a focus on food supply chains, giving examples and potential research directions.

The Food Supply Chain

The food supply chain is one of the most interesting research arenas for logistics and operations. In particular, perishable products, e.g. cold chain products and fresh products, can be produced and consumed in locations significantly far away one from the other, affecting the cost of transportation, customer service level, the level of quality and safety of the items being transported, and the impact on the environment. These are fast-moving products because of their short shelf life. The design, management and control of such systems are significantly critical and strategic. For example:

- How to transport fresh fruit and vegetables from one country, e.g. South America, to another, e.g. Europe, to fulfill the demand for non-seasonal products there?
- Which is the logistic efficiency of such a distribution system and the environmental impact?

- Which is the best combination of packaging decisions (primary, secondary and tertiary), containment decisions (e.g. standard or refrigerated containment), transportation modes, etc., to control the quality of the products?

The following is a list of exemplifying and significant logistics and operational issues in the agro-food industry:

1. production planning and scheduling;
2. network design and "shelf life erosion";
3. storage allocation and assignment in a unit load warehousing system;
4. picking in less than unit load warehousing system.

All these issues involve material flows moving within a facility (this is the intra-facility material handling), or outside the single facility and within a network of nodes, e.g. sources, production plants, distribution centers, hubs, points of demand. This is the inter-facilities transportation.

Production Planning and Scheduling

The food supply chain deals with several issues in manufacturing and processing systems. A significant example is the bakery industry. This is usually a batch production system, made of a sequence of lots scheduled each day on a single continuous production line. The production scheduling accounts for set-up time to machines, and labor and set-up costs. Large batches generate great costs of storage and reduce the global set-up costs. Small batches generate large costs of set-up but reduce the inventory level. The planning and scheduling of different batches of products in a single production line and in agreement with capacity constraints is difficult, particularly in presence of strict demand due-dates and different storage modes and solutions (e.g. drive-in, block storage, automated systems, etc.), as typical of the food industry.

The food industry is not the only sector affected by large set-up costs, a large number of items and high production rates. For instance, in the tile industry a few hours of the continuous production of a single product will cover months of customer demand and generates many square kilometers of tiles to be handled and stored. Obviously, in the food industry high levels of inventories are not admissible, especially in view of shelf-life restrictions and the finite capacity of storage. This is a single machine scheduling problem with sequence-dependent set-up times (i.e. the set-up depends on the sequence of batches of products scheduled) and due dates (i.e. time to supply consumers/customers). The most valuable decision in these contexts is the determination of the production batches, i.e. the number of batches and the dimension of each of them scheduled in a period of time, e.g. a week or a month. How many batches for each item? Which is the number of products for each batch? Similar problems involve several issues: production, inventory, warehousing, cost accounting, etc.

In the bakery industry, the largest part of a product's shelf life (e.g. six months) is usually taken by the retailer and consequently it is possible that the consumer could find products that are close to their sell-by date (the retailer), due to the inefficiency of the logistics system. This erosion is a very critical issue in the food supply chain and logistics play a strategic role in controlling it and the product's quality.

Network Design and "Shelf Life Erosion"

Network design is another critical issue in the food supply chain. Moving perishable products with a short shelf life is a very complex and expensive activity, especially in terms of transportation and environmental costs. To achieve high-quality consumer service, frequent shipments of the product need to be transported in vehicles that are not fully loaded. The logistics manager must identify the shelf-life erosion within the system in order to control the most critical stages in terms of maintaining the remaining shelf life of the product from the place of production to the place of consumption, i.e. "from farm to fork".

Figure 22.2 exemplifies the evaluation of the distribution performance of fresh perishable products supply network of a large-scale European retail supplier. The network is made up of three levels:

1. the suppliers, corresponding to vendors located worldwide and producing products for the retailer;
2. the retailer's regional distribution centers (RDC), corresponding to different geographic areas or divisions;
3. the Pods, corresponding to grocery shops and malls.

The number of RDCs is about 20. The three levels correspond to two stages: the first is the suppliers—RDCs—the second the RDCs—Pods. This configuration is named AS-IS and is illustrated in Figure 22.2 (i.e. the left network).

Figure 22.2 (i.e. the right network) also illustrates the so-called TO-BE configuration, which differs from AS-IS with the addition of a central distribution system (CDC), named hub. This facility is located in Italy and collects products from about 100 suppliers. As a consequence, the TO-BE network counts for four levels and three stages.

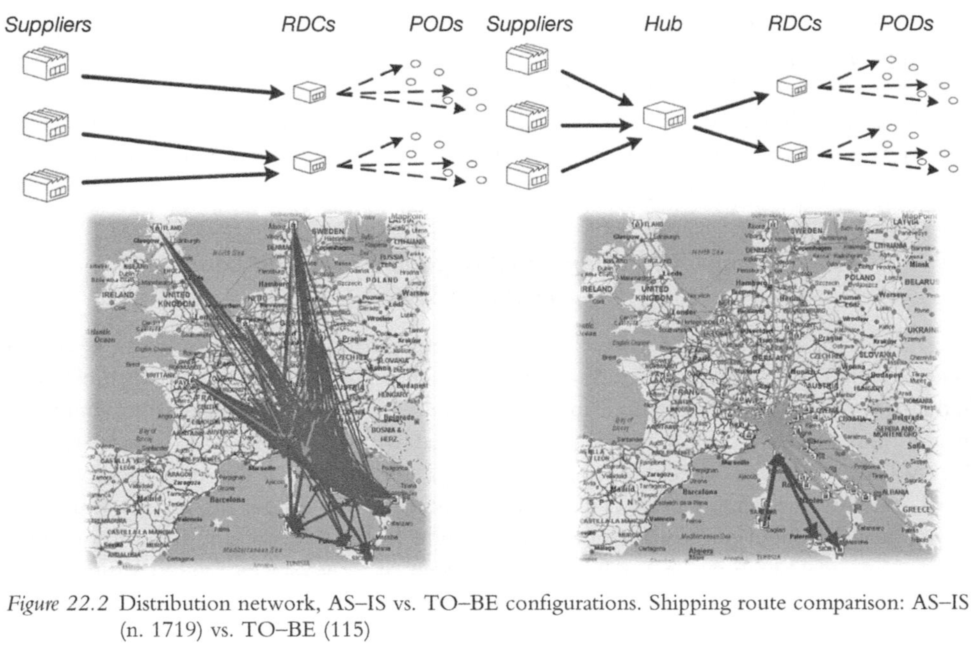

Figure 22.2 Distribution network, AS–IS vs. TO–BE configurations. Shipping route comparison: AS–IS (n. 1719) vs. TO–BE (115)

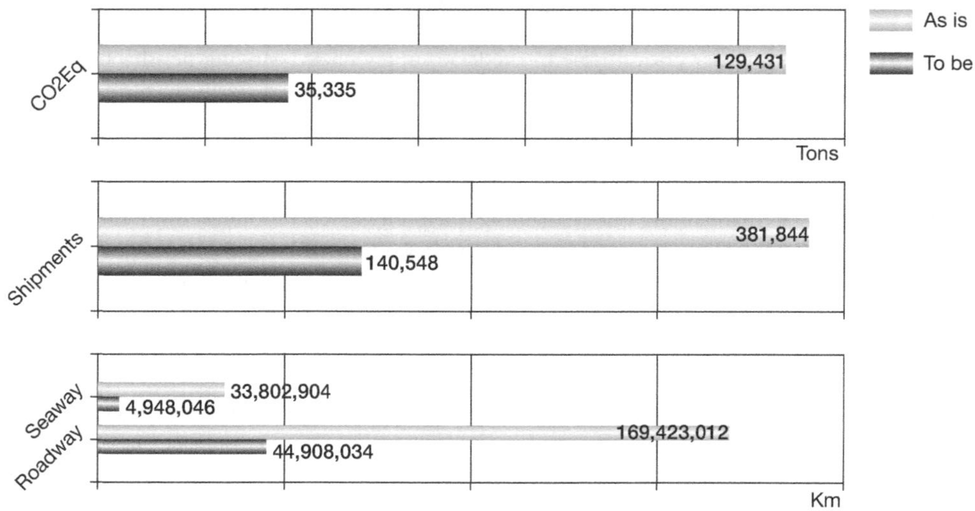

Figure 22.3 KPI comparing AS–IS and TO–BE configurations

Figure 22.2 illustrates the shipping routes at different stages respectively in the AS-IS and TO-BE network configurations. Obviously, the number of connections significantly reduces from AS-IS to TO-BE due to the establishment of the hub. These connections involve different transportation modes, including sea freight, which is mainly used for national and international shipments of fish. Figure 22.3 presents three different key performance indicators (KPI), evaluated in the AS-IS and TO-BE system configurations. Both networks are defined and compared, guaranteeing the same customer service level. In particular, the number of shipments reduces from about 381,000 to 140,500. In the network named TO-BE, shipments from the suppliers to the hub fill the capacity of vehicles and containers. The number of traveled kilometers significantly decreases both for the "road mode" and "sea mode". Finally, the corresponding environmental impact, measured in terms of CO_2 eq. tons, reduces by about 73 percent (from 129.431 to 35.335 CO_2 eq. tons).

This case study identifies the role of logistical decisions to determine the performances of the production/distribution system in terms of costs and efficiency, environmental impact, and consequently in terms of product quality.

Storage Allocation and Assignment in a Unit Load Warehousing System

A parallel issue in the food supply chain is the determination of the best layout of storage systems and material handling solutions to store and retrieve products. Which is the most useful infrastructure to store products unit loads in a warehousing system in order to guarantee the first in and first out (FIFO) policy usually adopted with perishable items subject to shelf life? What about the allocation of products, i.e. the determination of the space and volume to be allocated to the generic product stored in the system? What about the storage assignment of products, i.e. the determination of the best location within the system in order to minimize travel time and costs? A typical batch production system, e.g. the beverage industry (production of water, beer, etc.), is made up of parallel production and package lines operating simultaneously.

The unit load storage systems are typical of the food and beverage industry managing a limited number of items, usually no more than a few hundred. The suitable solution to store these products is the "lane", e.g. in drive-in, flow-rack, block storage solutions. The generic lane is made up of unit loads belonging to the same product and the same production batch. Material handling solutions to move such unit loads can be automatic. A few examples are the laser-guided vehicles (LGV) and the autonomous storage and retrieval solutions (Ekren and Heragu, 2009).

The new challenge of logistics and automation in the unit load warehousing system is the simultaneous design, management and control of the storage area and material handling vehicles by the adoption of integrated solutions, minimizing costs and increasing the customer service level.

Picking in Less Than the Unit Load Warehousing System

Order picking (OP) is defined as the process of retrieving products from a warehouse in order to fulfill customer orders. In general, OP processes are the most labor-intensive activities of all manual warehouse operations, especially for less-than-unit load order picking system (OPS) due to the variability of handled shapes and sizes. Indeed, on average the cost of OP is estimated to be 55 percent of the total warehouse operating costs (Tompkins et al., 2003). Furthermore, customer satisfaction is increasingly important in competition among companies in terms of both service levels and delivery time, which means OP is widely felt to be important in manufacturing and distribution environments.

Figure 22.4 presents a conceptual framework for the classification of storage systems and activities. In particular, unit load handling and picking operations are illustrated. The large number of involved stock-keeping units (SKUs), their variability of weights, volumes and shape increases significantly the complexity and performance of the order picking system.

Storage allocation determines the right inventory level per each stock-keeping unit (SKU) and the storage assignment deals with the assignment of SKU to the most convenient locations.

The forward or fast pick area in OPS is a zone within the storage system where the most popular SKUs are stored in rather small amounts so that the most part of picking operations can be carried out in a relatively small area. Therefore, all the related operations, e.g. stocking, traveling, searching, picking and restocking, become faster and more efficient.

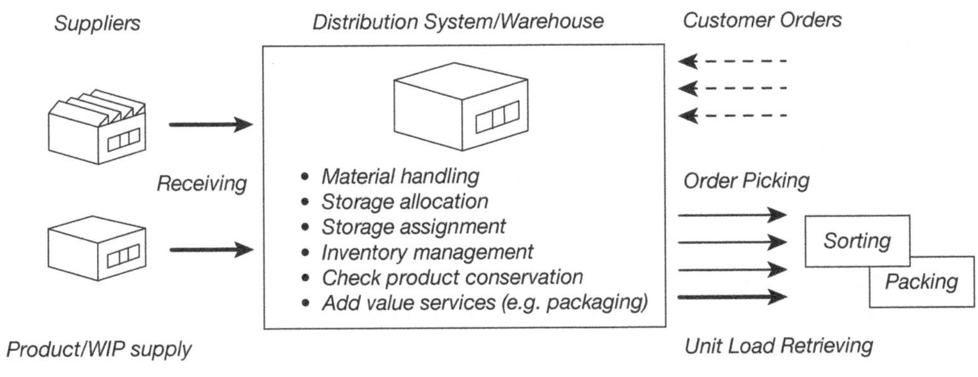

Figure 22.4 Storage system and activities classification in a distribution center (DC)

The main problems to be handled in optimizing fast pick areas are:

- Which items to store in the fast pick area?
- How much of each item to store?
- Which are the most suitable locations for each item where the traveling, searching, and picking times are minimized?

In particular, the storage assignment problem is the assignment of products to storage locations as introduced by Cahn (1948). It involves reducing both picking time and costs. The storage assignment problem has been classified as a non-polynomial (NP) hard problem. Hence, in order to obtain effective solutions to this problem in a reasonable time, various heuristic policies have been developed and applied in previous studies.

Picking activities are increasing more and more frequently in food supply chains and warehousing systems. For example, in the beverage industry the largest part of annual traveled kilometres within a facility is due to moving single-item unit loads, but the less than unit and multiple-item loads are significantly increasing because distribution centers directly supply small shops, bars and single consumers. This is the reason it is very important to support decisions on the design and management of order picking systems by the development of models and tools able to control and minimize costs and system efficiency (Accorsi et al. 2014a).

New Opportunities and Challenges

The new challenges in logistics, including freight transportation, deals with the following keywords: collaboration, integration, intermodality, sustainability. The aim of a modern supply chain is to guarantee quality, safety, efficiency and sustainability simultaneously; in particular, sustainability is economic, social and environmental.

Research is focusing on developing effective solutions to facilitate the collaboration between the actors and stakeholder of the supply chain in order to reduce time and cost on the one hand, and increase efficiency, product and process traceability, and the level of customer service on the other hand. ICT is the generator of innovation for logistics and operations.

Logistics and transportation are searching for integrated solutions, mainly models and tools (e.g. supporting decision systems (DSS)), to support decision making and system optimization. A few examples are:

- the strategic planning of distribution networks and logistics infrastructure;
- planning forward–reverse logistic networks;
- designing the closed-loop supply chain for the management of waste and by-products;
- the assessment of environmental KPIs of the reverse collection chain;
- the collection of fleet management and routing in waste management;
- planning intermodal freight infrastructure and networks.

Nowadays, literature presents several green solutions for logistics and freight transportation. All decisions have to be supported by the evaluation of the environmental impact according to a low carbon-driven and life-cycle perspective. A few examples are:

- measuring the environmental performance of alternative packaging materials;
- designing sustainable supply chain processes over economic and environmental goals;

- the assessment of environmental KPIs and impact categories throughout the supply chain and distribution systems (e.g. GHGs/shipped products);
- multi-scenario analysis of carbon trends in supply chain decisions, life-cycle assessment (LCA) of the supply chain, and distribution/storage/packaging processes.

Conclusions and Further Research

The aim of this chapter is to introduce the reader to the complexity of logistical and operational issues and problems. A few examples are illustrated. Research and industry are searching for integrated solutions supporting multiple choices of decision-making and involving different expertise. The food supply chain is a significant example of such a complexity, which generates new opportunities and challenges in modeling and developing supporting decision methods and tools. Green logistics properly supported by life-cycle assessment is a new perspective to face logistical and operational decisions.

References

Accorsi, R., Manzini, R. and Bortolini, M. (2012). A hierarchical procedure for the storage, allocation and assignment within an order picking system: a case study. *Internatinal Journal of Logistics*,15 (6): 351–364.

Accorsi, R., Manzini R. and Maranesi, F. (2014a). A decision-support system for the design and management of warehousing systems. *Computers in Industry*, 65: 175–186.

Accorsi, R., Manzini, R. and Ferrari, E. (2014b). A comparison of shipping containers from technical, economic and environmental perspectives. *Transportation Research: Part D, Transport and Environment*, 26: 52–59.

Accorsi, R., Cascini, A., Cholette, S., Manzini, R. and Mora, C. (2014c). Economic and environmental assessment of reusable plastic containers: a food catering supply chain case study. *International Journal of Production Economics*, 152: 88–101.

Bindi, F., Manzini, R., Pareschi, A. and Regattieri, A. (2009). Similarity-based storage allocation rules in an order picking system. An application to the food service industry. *International Journal of Logistics Research and Applications*, 12 (4): 233–247.

Cahn, A. S. (1948). The warehouse problem, *Bulletin of the American Mathematical Society*, 54: 1073.

Chaabane, A. A., Ramudhin, A. A., and Paquet, M. M. (2012). Design of sustainable supply chains under the emission trading scheme. *International Journal of Production Economics*, 135(1): 37–49.

Ekren, B. Y. and Heragu, S. S. (2009). Simulation based regression analysis for rack configuration of autonomous vehicle storage and retrieval system. Simulation Conference (WSC), Proceedings of the Winter, pp. 2405–2413.

Gamberini, R., Gebennini, E., Manzini, R. and Ziveri, A. (2010). On the integration of planning and environmental impact assessment for a WEEE transportation network: a case study. *Resources, Conservation and Recyling*, 54: 937–951.

Lambert, D. M, Cooper, M. C. and Pagh, J. D. (1998). Supply chain management: implementation issue and research opportunities. *International Journal of Logistics Management*, 9 (2): 1–18.

Manzini, R. (2012). A top-down approach and a decision support system for the design and management of logistic networks. *Transportation Research Part E: Logistics and Transportation Review*, 48 (6): 1185–1204.

Manzini, R. and Accorsi, R. (2013). The new conceptual framework for food supply chain assessment. *Journal of Food Engineering*, 115 (2): 251–263, March.

Manzini, R. and Bindi, F. (2009). Strategic design and operational management optimization of a multi stage physical distribution system. *Transportation Research Part E: Logistics and Transportation Review*.

Manzini, R., Accorsi, R., and Bortolini, M. (2014). Operational planning models for distribution networks. *International Journal of Production Research*, 52(1): 89–116.

Manzini, R., Pareschi, A., and Persona, A. (2007). Logistics outsourcing: an examination of third-party providers. *International Journal of Logistics Systems and Management*, 3 (2): 135–137.

Manzini, R., Gamberi, M., Gebennini, E., and Regattieri, A. (2008). An integrated approach to the design and management of a supply chain. *International Journal of Advanced Manufacturing Technology*, 37: 625–640.

Mora, C., Manzini, R., Gamberi, M. and Cascini, A. (2014). Environmental and economic assessment for the optimal configuration of a sustainable solid waste collection system: a 'kerbside' case study. *Production Planning & Control: The Management of Operations*, DOI:10.1080/09537287.2012.750386

Regattieri, A., Gamberi, M., Manzini, R. (2007) Traceability of food products: general framework and experimental evidence. *Journal of Food Engineering*, 81, 347–356.

Sahin, F. and Robinson, E. (2002). Flow coordination and information sharing in supply chains: review implications and directions for further research. *Decision Sciences*, 33 (4): 505–536.

Sarac, A. A., Absi, N. N. and Dauzre-Prs, S. S. (2010). A literature review on the impact of RFID technologies on supply chain management. *International Journal of Production Economics*, 128(1): 77–95.

Sarkis, J. A, Zhu, Q. B. and Lai, K.-H. C. (2011). An organizational theoretic review of green supply chain management literature. *International Journal of Production Economics*. 130 (1): 1–15.

Shen, Z.-J. M. A multi-commodity supply chain design problem (2005). *IIE Transaction*, 37: 753–762.

Sundarakani, B. B., De Souza, R., Goh, M. M., Wagner, S. and Manikandan, S. S. (2010). Modeling carbon footprints across the supply chain. *International Journal of Production Economics*,128 (1): 43–50.

Tompkins, J. A., White, J. A., Bozer, Y. A., Frazelle, E. H. and Tanchoco, J. M. A., 2003. *Facilities Planning*. Hoboken, NJ: John Wiley & Sons.

23

CITY LOGISTICS

Eiichi Taniguchi

KYOTO UNIVERSITY

Introduction

Urban freight transport issues are very important for the sustainable development of cities, but many urban planners and decision makers involved in urban transport have overlooked important freight-related issues in urban planning and urban transport planning. One of the main reasons for this is that they regard urban freight transport as something being dealt with by private companies and the public sector cannot address urban freight transport issues due to the lack of knowledge of logistics. However, in reality efficient and effective urban distribution systems are required in logistics and we face increasing levels of traffic congestion, negative impacts on the environment, crashes, and higher energy consumption caused by freight vehicle traffic in urban areas. These are complicated and difficult problems that should be solved for the sustainable development of cities by balancing smart economic growth with cleaner, quieter and safer environments in urban areas.

For solving these difficult urban freight transport issues, the concept of city logistics has been proposed (Taniguchi et al., 2001). City logistics is defined as:

> the process for totally optimising the logistics and transport activities by private companies with support of advanced information systems in urban areas considering the traffic environment, the traffic congestion, the traffic safety and the energy savings within the framework of a market economy.
>
> *(Taniguchi et al., 2001)*

City logistics provides approaches and measures towards mobility, sustainability and liveability. These three freight planning principles provide goals for enhancing the management of urban freight transport. Recently, resilience was added as the fourth principle of city logistics (Taniguchi et al. 2010), since risk management has become an important issue for the mitigation of damage and the quick recovery of logistics systems after disasters. There are a number of criteria for evaluating the policy measures of city logistics, including global competitiveness, efficiency, environmental friendliness, congestion alleviation, security, safety, energy consumption and labour force. The four principles comprise the pillars and multiple criteria provide bracing to form the structure of the visions for city logistics (Taniguchi et al., 2004).

There are four major stakeholders who are involved in city logistics: a) shippers, b) freight carriers, d) administrators, and e) residents (consumers). The unique characteristics of city logistics is that administrators take part in the activities relating to urban freight transport as a player. As the social and environmental concerns are dealt with in the procedure of the management of city logistics policy measures for urban freight transport, administrators play a vital role in coordinating the stakeholders who have different objectives and intentions. Public–private partnerships (Browne et al., 2004) among these stakeholders are a key element for successfully planning and implementing policy measures, including joint delivery systems with consolidation centres (Taniguchi and van der Heijden, 2000; Duin et al., 2010), off-hour delivery systems (Holguín-Veras et al., 2014), application of ITS (Intelligent Transport Systems) for optimizing vehicle routing and scheduling (Taniguchi and Thompson, 2002), and load factor controls (Teo et al., 2014). Public–private partnerships allow all stakeholders to participate in the discussion of policy measures from the initial stage of planning. Sharing data relating to urban goods movements and freight vehicle traffic among public and private entities is essential for identifying problems that stakeholders are facing and finding good solutions.

Challenging Issues

In the twentieth century most researchers and practitioners thought that there was a trade-off relationship between efficiency and environmental friendliness in urban freight transport. Namely, to improve logistics systems in terms of reducing NO_x emissions, there is a risk of increasing logistics costs. Also, to save the operating costs of logistics by using old trucks, higher levels of NO_x emissions will be caused in busy urban areas. However, fortunately in the twenty-first century, technological innovations including ICT (Information and Communication Technology) and ITS can be used. The development and deployment of ICT and ITS have substantially contributed towards establishing efficient and environmentally friendly urban freight transport systems. The main functions of ICT and ITS relating to urban freight transport are:

- collecting, processing and analysing data;
- storing data;
- sharing data, and
- communicating each other within a network of stakeholders.

These functions are fully used for rationalizing logistics operations. The real-time location of trucks can be identified using GPS (Global Positioning Systems) and these data can be used to optimize routes and schedules of trucks to meet the changing demands from customers and varying travel times on road network due to congestion. Mobile communication systems between truck drivers and control centres are also available, and are helpful to keep drivers informed of updates on customers' information and traffic conditions. In the logistics industry, historical data of truck operations provides a good source of improving the routing and scheduling of trucks, since the traffic conditions are always varying and the probabilistic treatment of vehicle operations can be achieved using historical data of travel times. An important function of ICT and ITS is sharing data between stakeholders. It is possible to share precise digital data of the logistics operation of private companies and traffic data from the public sector. This type of data sharing provides a good platform for public–private collaboration.

Another important element for promoting city logistics is the change in the mind-set of managers within transport and logistics companies. They are moving towards emphasizing the concerns about the environment, safety and security associated with logistics activities. A number of companies obtained ISO 9,001 (quality management) and ISO 12,001 (environment management) certificates to make them greener and more sustainable. They issue the annual report on the contribution to reduce environmental impacts. Since ISO systems are expensive for small and medium-size enterprises, a similar and simplified system of green management certificate is available from the Foundation for Promoting Personal Mobility and Ecological Transportation in Japan. These systems allow private companies to educate their employees on eco-driving and the safer driving of trucks. As a result, the companies can benefit from a reduction in fuel consumption, and traffic accident costs and society can also benefit from a decrease in CO_2, NO_x and SPM emissions.

Third, public–private partnerships play an important role in promoting city logistics. As there are four major stakeholders, namely shippers, freight carriers, administrators, and residents, who are involved in city logistics, the collaboration and coordination between the private and public sectors are needed for a successful implementation of city logistics policy measures. However, the results of planning and implementing city logistics measures should create win–win situations, namely all the stakeholders being satisfied to some extent, even if it is not ideal. Therefore, balancing economic efficiency and the reduction of negative environmental impacts is vital in planning city logistics policy measures. Although administrators in municipalities are ultimately responsible for the results of implementing policy measures for improving urban freight transport, they should listen carefully to shippers, freight carriers, residents, the chamber of commerce, developers of shopping malls, consumers' associations, and other related entities before making decisions regarding policy measures. Sharing data which private and public sectors have collected among these stakeholders is critical for better understanding the problems.

Management for Urban Freight Transport

To achieve the goals of city logistics, namely mobility, sustainability, liveability and resilience, good management of urban freight transport is needed, since multiple stakeholders are involved and problems include elements of complexity and uncertainty in terms of the economy, urban development, social interests and traffic conditions. Therefore, the methodology of adaptive management has been proposed (World Road Association, 2012), which encompasses PDCA (Plan, Do, Check and Act) cycle to incorporate different opinions and perspectives of stakeholders as well as to meet with dynamically changing social and economic requirements. Predicting and evaluating the results of implementing policy measures in advance and feedback of the results after the real execution is also essential. More details of the design, assessment, implementation, and evaluation stages for PDCA cycles will be given in the following sections.

Design Stage

A public–private partnership or consortium of public involvement should be established at the design stage to organize all stakeholders who are interested in urban freight transport. Discussing urban freight transport is important to identify problems, including delays in distribution to customers, difficulty of finding appropriate parking space on urban streets for loading/unloading goods, higher level of congestion, NO_x emissions, and crashes on some links of road network, improper route choice by heavy freight vehicles in residential areas. Business-related problems are sometimes recognized as critical for their business by the private sector, but they are not well perceived by the public sector and social problems in turn are recognized as critical for daily life of citizens by the public sector, but are not well understood by private companies. Intensive and in-depth discussions are needed to overcome the perception gap between private and public entities based on the data of real situations of urban freight transport operation. After problem identification, we can investigate the cause of problems. For example, illegal double parking of pickup-delivery trucks on streets may cause traffic congestion, and a large number of older trucks still used by small companies may cause more NO_x emissions. Then we can set up the goals of urban freight transport management. The goals should be simple enough to be fully accepted by all stakeholders and examined during the management procedure. The next step is to determine the desirable freight vehicle movements in urban areas from various viewpoints of land use, environment, and location of warehouses. Although drivers of freight vehicles hope to choose the shortest path to destinations, green routes that minimize NO_x emissions are a little different. Truck routes should not go through residential areas to maintain safe and quiet conditions for communities. Desirable freight vehicle movements can be determined based on the discussion within public–private partnerships following in-depth investigations regarding the reasons why so many trucks are moving on particular links of the road network.

Once the desirable freight vehicle movements are determined, we can consider approaches and policy measures for solving problems. The WRA report presents six approaches:

- infrastructure;
- regulatory;
- logistical;
- co-operative;
- technological, and
- behavioural.

It also describes a number of policy measures:

- traffic flow management;
- parking management;
- time management;
- vehicle management;
- joint delivery;
- intermodal transport;
- land use plan;
- improvement of vehicle movement, and
- organizational activities.

A combination of these approaches and policy measures is recommended for tackling challenging urban freight transport issues. Full utilization of innovative technology of ICT and ITS is also recommended.

Assessment Stage

During the assessment stage a pilot project is normally planned and evaluated, as the effects of policy measures are somewhat uncertain. In particular, the unexpected side effects of the policy measures should be carefully examined using traffic simulation models and other mathematical models. For example, bans of large trucks are often implemented to alleviate traffic congestion, but sometimes they fail because a larger number of small trucks were used by freight carriers instead of large trucks to carry the same amount of goods and the congestion level increased.

Sometimes subsidies from municipalities are provided for supporting a pilot project but we need to carefully examine if the project will become sustainable after the subsidies are ended. The financial feasibility of city logistics projects can be ensured based on the self-financing systems without subsidies.

Implementation Stage

During the implementation stage after the pilot project, we need to consider regulations relating to urban freight transport, provide the permanent facilities for city logistics solutions including consolidation centres, and create financial schemes to operate the systems without subsidies. During this stage an individual leader's intensive efforts can make progress in promoting collaboration among stakeholders. The innovative leadership of lead entities is a key factor for the successful implementation of city logistics initiatives.

Evaluation Stage

For evaluating the effects of implementing city logistics policy measures, monitoring the performance is required by measuring a set of indicators, including the travel times of trucks, emissions of hazardous gases and the number of crashes. Key Performance Indicators (KPIs) for urban freight transport provide performance measures for benchmarking a city comparing the effects with the ideal situation or other cities. The WRA report (2012) provides examples of KPIs, including "local air quality emissions," "complaints from residents or industry" for the quality of life, "number of visitors/amount of revenue generated per hour/day" for economic development, "number of vehicle-ton-kilometre," "travel time" for accessibility, "average load factor for vehicles" for transport efficiency. The results of evaluation can be fed back and if negative effects are observed, the process can be restarted from the problem identification stage in the PDCA cycle.

City Logistics Policy Measures

A number of city logistics policy measures have been proposed and implemented in cities. These policy measures can be grouped into four categories:

1. traffic management;
2. better transport method;
3. harmony with other urban planning;
4. others.

Traffic management measures include truck routes, access control to cities, low emission zones, traffic information using ITS, loading/unloading space on streets, off-hour delivery, time windows for entering cities, optimization of vehicle routing and scheduling, road pricing and load factor controls. Better transport methods include: urban consolidation centres, intermodal terminals as well as pick-up points for delivery. Harmony with other urban planning includes restrictions for locating logistics facilities. Finally, other schemes such as the recognition of green freight transport companies, underground freight transport systems, truck dimensions and building codes for parking as well as freight quality partnerships are also important.

These policy measures have been implemented in some cities and success factors have been identified from these case studies (World Road Association, 2012). A success factor is to fully understand the situation that is basically required to start discussion of urban freight transport. Then, establishing a cross-departmental organization provides an opportunity to take flexible action, since existing organizations in many cases are too rigid to adjust to the new situation. Talking with the right persons at the right time is essential for successful management. Partnerships among stakeholders and communication between each other are the most important factors affecting the results. The key is to discuss problems to find better solutions based on real data and build consensus among stakeholders. It may take a longer time to reach consensus, but spending enough time for understanding the costs and benefits of policy measures is needed. Also, advertising the importance of policy measures to individuals and the community, and obtaining public opinion enhances policy measures. After building consensus, vision/goal setting is the key issue. Qualitative or quantitative criteria should be defined but these should be simple and easy to understand by all stakeholders. Wise use of subsidies is also important to establish pilot project. However, as discussed in the previous section, it is important not to depend too much on subsidies, since subsidies will be stopped after a certain period.

Modelling

Mathematical modelling provides important techniques for evaluating the performance of city logistics schemes and policy measures. There are three types of models relating to city logistics: a) demand models, b) supply models, and c) impact models (Taniguchi et al., 2001). Demand models are basically developed for predicting the demand of goods relating to the generation, attraction and flows on the transport network (Tavasszy and De Jong, 2014). Supply models mainly deal with the capacity or the level of service of transport infrastructure. Impact models address the behaviour of stakeholders and response to city logistics policy measures to estimate impacts on the local economy, traffic conditions, environment and energy consumption.

Impact models are important and useful for evaluating policy measures and estimating the effects from the performance of freight carriers, including hazardous gas emissions and energy consumption. Recently, multi-agent models (Duin et al., 2007; Tamagawa et al., 2010; Teo et al., 2012) have been developed and used for improving understanding of the behaviour and interaction between stakeholders. Multi-agent models can mathematically represent the behaviour of shippers, freight carriers, administrators and residents who participate in urban freight transport. Reinforcement learning techniques are often incorporated in agent-based simulation to take into account the learning behaviour of agents. Multi-agent simulation cannot guarantee the conversion of changing values of costs, travel times, emissions of CO_2, NO_x, SPM in the time domain, but these can be compared with the results of game theory modelling. Multi-agent simulation allows decision makers to understand the satisfaction levels of each objective and to choose better policy measures to ensure balanced benefits for each stakeholder. Therefore, multi-agent simulation provides a scientific approach that can be used to identify better policy measures for meeting the goals of public–private partnerships.

The Vehicle Routing and scheduling Problem with Time Windows (VRPTW) models (Solomon, 1987; Taniguchi and Shimamoto, 2004; Braysy and Gendreau, 2005) have been used to represent the behaviour of freight carriers in urban distribution. In this model, pickup-delivery trucks visit multiple customers using a single route, and time windows are commonly set by customers in Just-In-Time transport situation. VRPTW models are suitable for mathematically representing the behaviour of freight carriers. Stochastic and dynamic VRPTW models are usually used in the real-time operation of vehicles to incorporate varying travel times and demands of customers. Stochastic and dynamic VRPTW models can be applied with historical and real-time travel times data on road networks using ITS. Multi-agent simulation models can incorporate VRPTW models as sub-models to represent the route choice behaviour of freight carriers.

Future Perspectives

Recently, there have been increasing concerns about emergency relief relating to the supply and distribution of water, food and daily commodities to people affected by disasters. Humanitarian logistics methods have been developed and implemented in major disasters such as the Tohoku disasters in Japan in 2011 (Taniguchi and Thompson, 2013), the Haiti earthquake in 2010, Hurricane Katharina in 2008 (Holguín-Veras et al., 2012). The principles of humanitarian logistics are similar to city logistics in terms of minimizing the total costs, which are the sum of the commercial logistics costs and social costs. However, the external costs of humanitarian logistics relate to the suffering of displaced people in disasters, whereas those of city logistics include environmental and safety costs in normal situations. In humanitarian logistics, the social capital of communities plays a very important role for facilitating last-mile deliveries in emergency cases. Good relationships among local community members assists greatly with the smooth distribution of relief goods even in cases where supplies are much less than demands. In the management of humanitarian logistics, collaboration between different public and private organizations is a key issue for the effective collection, transportation and storage of relief supplies.

Home health care (HHC) problems (e.g. Benzarti et al., 2013; Rasmussen et al., 2012) have gained more attention in ageing societies due to the shortage of hospitals and medical doctors and nurses compared with the increasing number of elderly patients. To effectively provide medical, nursing and healthcare services to patients at home, routing and scheduling problems have arisen. The HHC problem is similar to the conventional VRPTW, but includes additional objectives such as the inconvenience to patients due to the late arrival of health service providers. In this new area ICT and ITS can also contribute towards sharing patient information among different types of health service providers, and to reduce the inconvenience of patients as well as the costs of routing and scheduling of health service providers.

In the future, urbanization and the ageing population will be accelerated based on the predictions from the United Nations. Under such conditions urban freight transport will become more important, because it provides a fundamental framework for supporting better production of commodities and more efficient and environmentally friendly distribution of goods to people who need them. City logistics will contribute towards creating more mobile, sustainable and liveable cities using innovative ideas for improving urban freight transport systems based on public–private coordination.

References

Benzarti, E., Sahin, E. and Dallery, Y. (2013). Operations management applied to home care services: Analysis of the districting problem, *Decision Support Systems*, 55(2): 587–598.

Braysy, O. and Gendreau, M. (2005). Vehicle routing problem with time windows, part I: Route construction and local search algorithms, *Transportation Science*, 39(1): 104–118.

Browne, M., Nemoto, T., Visser, J. and Whiteing, T. (2004). Urban freight movements and public–private partnerships, in *Logistics Systems for Sustainable Cities*, E. Taniguchi and R.G. Thompson (Eds.), Elsevier, Oxford, 17–35.

Duin, J.H.R. van, Quak, H. and Muñuzuri, J. (2010). New challenges for urban consolidation centres: A case study in The Hague, *Procedia-Social and Behavioral Sciences* 2(3): 6177–6188.

Duin, J.H.R. van, Tavasszy, L.A. and Taniguchi, E. (2007). Real time simulation of auctioning and re-scheduling processes in hybrid freight markets, *Transportation Research, Part B*. B41(9): 1050–1066.

Holguín-Veras, J., Jaller, M., Wassenhove, L.N. van, Pérez, N. and Wachtendorf, T. (2012). On the unique features of post-disaster humanitarian logistics, *Journal of Operations Management*, 30(7–8): 494–506.

Holguín-Veras, J., Wang, C., Browne, M., Hodge, S.D. and Wojtowicz, J. (2014). The New York City Off-hour Delivery Project: Lessons for City Logistics, *Procedia-Social and Behavioral Sciences*, 125: 36–48.

Rasmussen, M.S., Justesen, T., Dohn, A. and Larsen, J. (2012). The home care crew scheduling problems: Preference-based visit clustering and temporal dependencies, *European Journal of Operational Research*, 219(3): 598–610.

Solomon, M. M. (1987). Algorithms for the vehicle routing and scheduling problems with time window constraints, *Operations Research*, 35: 254–265.

Tamagawa, D., Taniguchi, E. and Yamada, T. (2010). Evaluating city logistice measures using multi-agent model, *Procedia-Social and Behavioral Sciences* 2(3), 6002–6012.

Taniguchi, E. and Shimamoto, H. (2004). Intelligent transportation system based dynamic vehicle routing and scheduling with variable travel times, *Transportation Research Part C*, 12C(3–4): 235–250.

Taniguchi, E. and Thompson, R.G. (2002). Modelling city logistics, *Transportation Research Record*, 1790: 45–51.

Taniguchi, E. and Thompson, R.G. (2013). Humanitarian logistics in the Great Tohoku Disasters 2011, in Zeimpekis, V., Ichoua, S. and Minis, I. (Eds.) *Humanitarian and Relief Logistics*, Springer, 207–218.

Taniguchi, E. and van der Heijden, R.E.C.M. (2000). An evaluation methodology for city logistics. *Transport Reviews*, 20(1): 65–90.

Taniguchi, E., Thompson, R.G. and Yamada, T. (2004). Visions for city logistics, in *Logistics Systems for Sustainable Cities*, E. Taniguchi and R.G. Thompson (Eds.), Elsevier, Oxford, 1–16.

Taniguchi, E., Thompson, R.G. and Yamada, T. (2010). Incorporating risks in city logistics, *Procedia-Social and Behavioral Sciences*, 2(3): 5899–5910.

Taniguchi, E., Thompson, R. G., Yamada, T. and Duin, J.H.R. van. (2001). *City Logistics: Network Modelling and Intelligent Transport Systems*, Pergamon, Oxford.

Tavasszy, L. and De Jong, G. (2014). *Modelling Freight Transport*, Elsevier, Oxford.

Teo, J.S.E., Taniguchi, E. and Qureshi, A.G. (2012). Evaluation of distance-based and cordon-based urban freight road pricing on e-commerce environment with multi-agent model, *Transportation Research Record: Journal of the Transportation Research Board*, 2269, 127–134.

Teo, J.S.E., Taniguchi, E. and Qureshi, A.G. (2014). Evaluation of Load Factor Control and Urban Freight Road Pricing Joint Schemes with Multi-agent Systems Learning Models, *Procedia—Social and Behavioral Sciences*, 125: 62–74.

World Road Association (2012). *Public Sector Governance on Urban Freight Transport Management*, technical report by Technical Committee B4.

24

TRANSPORTATION, ENERGY, AND THE ENVIRONMENT

Konstantina Gkritza

PURDUE UNIVERSITY

1. Introduction

Transportation systems are crucial to the economy and quality of life. In 2012, transportation-related goods and services accounted for approximately 10 percent—over $1.5 trillion—of the United States (US) Gross Domestic Product (GDP) (Bureau of Transportation Statistics, 2014a). However, the construction, expansion, operation, and maintenance of the various transportation systems and modes have direct, indirect, and cumulative/synergistic impacts on energy use, air quality, water and soil quality, noise, land use, and nature conservation. The transportation sector accounted for 28 percent of US energy consumption, 71 percent of petroleum use, and 28 percent of US greenhouse gas (GHG) emissions (1.8 billion tons of CO_2 equivalent emissions annually) in 2012 (Oak Ridge National Laboratory, 2013a; US Environmental Protection Agency, 2014). Petroleum use is predominantly related to vehicle miles traveled (VMT), a widely used measure of vehicle travel demand in the US; almost 97 percent of the primary demand for transportation is met by petroleum (Bureau of Transportation Statistics, 2014b).

There is a consensus among leading climate scientists that the earth's climate is changing and that GHG emissions from human activities are major contributors to climate change. "Climate change refers to a statistically significant variation in either the mean state of the climate or its variability over an extended period, typically decades or longer, that can be attributed to either natural causes or human activity" (Transportation Research Board, 2008). Climate change effects such as intense rainfalls, heat waves, Arctic warming, storm surge and rising sea levels can have significant impacts on the state of good repair, reliability, safety, and environmental sustainability of transportation systems (Federal Transit Association, 2011; Schwartz et al., 2014). It is argued that as much as 80 percent reduction in worldwide emissions may be needed over the next four decades to stabilize atmospheric concentration of GHG (Transportation Research Board, 2011).

Sustainable development has become an essential component of international and domestic environmental policy. The US Department of Transportation (2012) includes environmental sustainability as one of the agency's strategic goals and promotes strategies to increase the use of environmentally sustainable practices in the transportation sector. Defining sustainability of transportation systems is challenging due to the inherently broad nature of the concept that contains three overlapping dimensions: environmental, economic, and social. A widely accepted definition of sustainable development is provided by the Brundtland Commission: "sustainable development seeks to meet the needs and aspirations of the present without compromising the ability to meet those of the future" (United Nations, 1987). This implies "the balancing of economic, social, environmental and technological considerations as well as the incorporation of a set of ethical values" (Council of Academies of Engineering and Technological Sciences, 1995).

This chapter first describes the impacts of transportation systems on energy (section 2) and natural (environmental) systems (section 3). The metrics, methodologies, and tools that are typically used to measure these impacts across a wide range of transportation modes are discussed. Then, section 4 presents potential solutions to mitigate the impacts of transportation on energy and environmental systems. The chapter concludes with some thoughts on moving forward towards an environmentally sustainable transportation system.

2. Transportation and Energy

Energy Consumption

The transportation sector in the US is the dominant consumer of energy since 1985 other than "energy input at electric utilities." Energy consumption is measured in British thermal units (Btu), which refers to the amount of energy required to raise the temperature of 1 pound of water by 1 degree Fahrenheit. The barrel of oil equivalent (BOE) is another energy unit that measures the energy released when one barrel (approximately 42 US gallons of diesel or gasoline) is burnt.

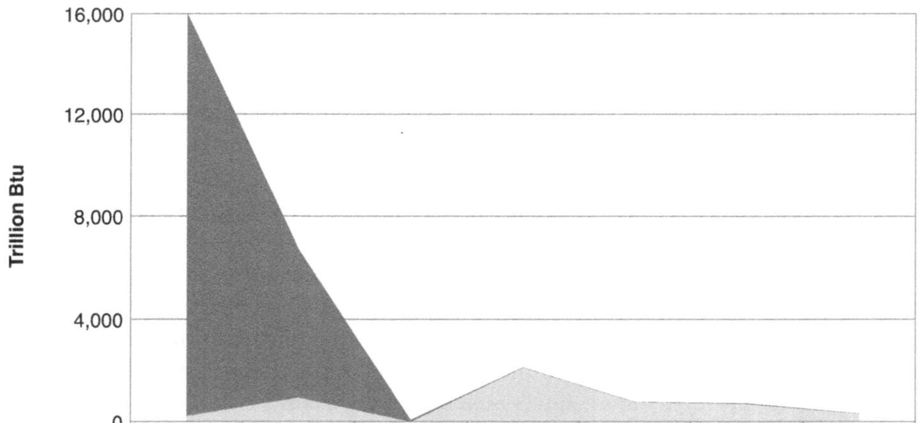

	Gasoline	Diesel fuel	Liquefied petroleum gas	Jet fuel	Residual fuel oil	Natural gas	Electricity
HIGHWAY	15,978.9	5,751.1	64.6	0.0	0.0	20.2	0.6
NONHIGHWAY	221.8	949.5	0.0	2,136.2	779.4	704.9	316.6

Figure 24.1 Domestic consumption of transportation energy by mode and fuel type, 2011
Source: Oak Ridge National Laboratory (2013a)

The sector's energy consumption increased from 23 percent in 1960 to 28.9 percent in 2006, and down to 28 percent in 2012, which accounts for 26.63 of the 95.10 quadrillion Btu energy consumption in the US (Bureau of Transportation Statistics, 2014b). The total transportation energy consumption decline over the last decade is also exhibited relative to GDP; it was increasing approximately 5 billion Btu per month, until mid 2007, when the cost of fuel rose dramatically (Young, 2009).

The domestic consumption of transportation energy varies by mode. Highway modes dominate transportation energy use with light-duty vehicles accounting for 72 percent of the highway total, and nearly 60 percent of the total sector's energy use. While freight modes use less energy than passenger modes in total, there is room for improvement in the energy efficiency of medium and heavy trucks that consume approximately 65 percent of the total energy use for freight movements. Petroleum is the dominant fuel for transportation, shown in Figure 24.1.

Energy Intensity

The concept of "energy intensity" is commonly used as a means to a comparison among transportation modes as far as their energy consumption of the (different) energy sources is concerned. It refers to the energy each mode requires to transport one passenger (or one ton of freight) one mile. The corresponding performance measures are: Btu per passenger-mile and Btu per freight ton-mile. Table 24.1 shows the evolution of the energy intensity by passenger mode in the US throughout the past decade (2002–2011). Light duty vehicles (long wheel base) are consistently the most energy intense (poorer in terms of fuel use). Air transportation presents a constant improvement (especially in domestic operations) as does Amtrak. Since 2009, transit motor-bus is on average less energy intense than light-duty vehicle, short wheelbase. Note that on a case-by-case comparison, a light-duty vehicle might be more intense than a transit bus depending on the occupancy rates (see Example 24.1). The load factor of transportation modes is critical to this comparison.

A similar comparison among freight modes is provided in Table 24.2. Freight modal intensities have been decreasing, but probably not as rapidly as for passenger modes. Significant improvements are evident in rail and water. Energy intensity is hard to estimate for trucks; instead, vehicle-mile intensities are provided.

Example 1

The table below shows the vehicle miles, passenger miles, and fuel consumed by light-duty vehicles (short wheelbase) retrieved from the 2011 Highway Statistics (Federal Highway Administration, 2013). Calculate the energy intensity and the average occupancy rate. Assume that the energy consumed is 125,000 Btu/gallon of fuel for light duty vehicles (see Table 24.1, Notes).

	Vehicle miles (millions)	*Passenger miles (millions)*	*Fuel consumed (million gallons)*	*Energy intensity (Btu/pas.-mile)*	*Average occupancy rate*
Light-duty vehicle, short wheelbase	2,043,409	2,839,083	88,537		

Table 24.1 Energy intensity of passenger modes (Btu per passenger mile)

	2002	*2003*	*2004*	*2005*	*2006*	*2007*	*2008*	*2009*	*2010*	*2011*
Air, certificated carrier										
Domestic operations	3,608	3,493	3,408	3,232	3,142	3,040	2,936	2,774	2,691	2,597
International operations	3,984	4,180	3,890	3,817	3,665	3,572	3,473	3,384	3,330	3,519
Highway[a]										
Light-duty vehicle, short wheelbase[b]	3,600	3,570	3,509	3,585	3,510	3,368	3,344	3,823	3,854	3,898
Light-duty vehicle, long wheelbase[b]	4,121	4,452	4,452	4,077	4,042	4,537	4,159	5,411	5,447	5,479
Motorcycle	1,969	1,969	1,969	1,784	1,805	2,185	2,315	2,688	2,675	2,669
Transit motor bus	3,766	3,778	3,886	3,657	3,702	3,596	3,511	3,472	3,350	3,343
Amtrak	2,537	2,145	2,068	2,025	1,948	1,824	1,745	1,773	1,668	1,628

Source: Adapted from Bureau of Transportation Statistics, 2014c.

Key: Btu = British thermal unit.

Notes: To calculate total Btu, multiply fuel consumed by 135,000 Btu/gallon for air carrier; 125,000 Btu/gallon for light duty vehicle, short wheelbase, light duty vehicle, long wheelbase, and motorcycle; 138,700 Btu/gallon for transit motor bus and Amtrak diesel consumption; and 3,412 Btu/KwH for Amtrak electric consumption. (a) Highway passenger miles were taken directly from highway statistics rather than derived from vehicle miles and average occupancy. (b) Data for 2007–11 were calculated using a new methodology developed by FHWA. Data for these years are based on new categories and are not comparable to previous years. The new category light-duty vehicle, short wheelbase includes passenger cars, light trucks, vans and sport utility vehicles with a wheelbase (WB) equal to or less than 121 inches. The new category of light-duty vehicle, long wheelbase includes large passenger cars, vans, pickup trucks, and sport/utility vehicles with wheelbases larger than 121 inches. The data from 1960–2006 are not comparable to the data from 2007–11.

Table 24.2 Energy intensity of freight modes, 2000–2011

Year	*Heavy single-unit and combination trucks (Btu per vehicle-mile)*	*Class I freight railroad*		*Waterborne commerce on taxable waterways (Btu per ton-mile)*
		(Btu per freight car-mile)	*(Btu per ton-mile)*	
2000	23,449	14,917	352	270
2001	23,024	15,108	346	253
2002	23,462	15,003	345	253
2003	22,461	15,016	344	251
2004	20,540	15,274	341	241
2005	22,866	15,152	337	241
2006	23,340[b]	14,990	330	235
2007	21,238	14,846	320	225
2008	21,008	14,573	305	252
2009	21,024	13,907	291	225
2010	21,499	13,733	289	217
2011	21,698	14,043	298	a
Average annual percentage change				
1970–2011	–0.3%	–0.6%	–2.0%	a
2001–2011	–0.6%	–0.7%	–1.5%	a

Source: Oak Ridge National Laboratory, 2013b

Notes: (a) Data are not available. (b) Due to changes in the FHWA fuel use methodology, truck data are not comparable with data before the year 2007.

a) The energy intensity is:

$$\frac{88{,}537 \text{ mill Gal} * 125{,}000 \text{ Btu/gallon}}{2{,}839{,}083 \text{ mil. pas.-miles}} = 3{,}898 \text{ Btu/pas.-mile}$$

b) The average occupancy rate is:

$$\frac{2{,}839{,}083}{2{,}043{,}409} = 1.39 \text{ passengers.}$$

c) The energy efficiency of transit bus has been estimated 3,343 Btu/pas. mile for the same year (2011). Estimate the occupancy rate of a light-duty vehicle, short wheelbase, needed to achieve equal energy intensity with a transit bus. Assume fixed vehicle miles and fuel consumed.

$$\frac{88{,}537 \text{ mill Gal} * 125{,}000 \text{ Btu/gallon}}{x \text{ mil. pas.-miles}} = 3{,}343 \text{ Btu/pas.-mile}$$

$$\Rightarrow x = \frac{88{,}537 \text{ mill Gal} * 125{,}000 \text{ Btu/gallon}}{3{,}343 \text{ mil. pas.-miles}} = 3{,}310{,}537 \text{ mil. pas.-miles}$$

$$\Rightarrow \frac{3{,}310{,}537}{2{,}043{,}409} = 1.62 \text{ passengers.}$$

Example 2

Assuming that for a transit bus the energy consumed is 138,700 Btu/gallon of fuel (see Table 24.1, Notes), what was the average occupancy for the achieved energy intensity in 2011 (3,343 Btu/pas. miles)? Assume that the average fuel efficiency of buses is 6.1 miles per gallon of diesel and all vehicles consumed diesel. (In practice other fuels, such as compressed and liquefied natural gas, and biodiesel are also used.)

$$\frac{3{,}343 \text{ Btu}}{1 \text{ passenger-mile}} = \frac{138{,}700 \text{ Btu}}{\text{gallon}} * \frac{1 \text{ gallon}}{6.1 \text{ vehicle-miles}} * \frac{1 \text{ vehicle}}{x \text{ passengers}}$$

$$\Rightarrow \text{occupancy } x = \frac{138{,}700}{6.1 * 3{,}343} = 6.8 \text{ passengers/bus.}$$

Estimate the average occupancy rate for transit bus required to achieve an equal energy intensity with Amtrak the same year (1,628 Btu/pas. miles, for 2011 data). (Assume fixed average fuel efficiency.)

$$\text{occupancy } x = \frac{138{,}700}{6.1 * 1{,}628} \approx 14 \text{ passengers/bus.}$$

Energy Systems Models

Energy systems models are mathematical tools that quantify the interdependent effects of energy and economic systems typically under constrained energy choices. The transportation sector is modeled in these tools as one of the sectors of energy end users (along with residential, commercial and industrial sectors). These optimization models have been applied to examine the energy, economic, environmental, and security impacts related to an oil crisis, alternative energy policies and new technologies, and in general, to assist in the decision-making and design of future energy systems (Nakata et al., 2011; Weijermars et al., 2012). The most common models are: the US NEMS (US Department of Energy, 2000), the European EFOM (van der Voort, 1982), the MARKAL model (Fishbone & Abilock, 1981), and the Integrated MARKAL-EFOM System TIMES (Loulou et al., 2005). A review of these models is provided in Fleiter et al. (2011).

3. Transportation and the Environment

Transportation and Air Quality Impacts

Transportation-Related Air Pollutants

Pollutant emission refers to the discharge of a pollutant from a source to the atmosphere and is commonly stated in terms of mass, such as grams, tons, or teragrams (Tg), which is equivalent to a million metric tons. Transportation sources are typically mobile (as opposed to stationary) sources of criteria air pollutants and greenhouse gases (GHG). They can be either on- or off-road mobile sources, or transport hub-related sources (airports, railyards, etc.).

Criteria Air Pollutants

Criteria air pollutants refer to the six most common air pollutants that can directly affect human health, the environment, and cause property damage. They are regulated by the US Environment Protection Agency (EPA) and include: Ozone (O_3), Particulate Matter (PM), Carbon Monoxide (CO), Nitrogen Oxides (NO_x), Sulfur Dioxide (SO_2), and Lead (Pb).

According to US EPA (2013) data, on-road transportation sources accounted for 39 percent of the NO_x emissions, 35 percent of the CO emissions, and 13 percent of the VOC emissions in 2012. If off-road vehicles such as railroads, construction equipment and boats are included, the contribution of transportation sources becomes even more significant (61 percent of the NO_x, 54 percent of the CO, and 24 percent of the VOC emissions).

Greenhouse Gases

GHG refers to gases that trap heat in the atmosphere. Common GHG are: Carbon dioxide (CO_2), methane (CH_4), nitrous oxide (N_2O), and fluorinated gases. Carbon dioxide emissions dominate US GHG emissions (82 percent in 2012); transportation sources account for a significant percentage (32 percent of the CO_2 emissions in 2012) (US Environment Protection Agency, 2014). Table 24.3 shows that highway vehicles have consistently the largest share of CO_2 and total (CO_2 equivalent) transportation-related emissions, with passenger cars having the largest share. For a number of modes, the GHG emission production peaked in 2006–2007 and has been declining since then.

Table 24.3 Transportation-related greenhouse gas emissions (Tg CO_2 equivalent)

Gas/vehicle	*2004*	*2005*	*2006*	*2007*	*2008*	*2009*	*2010*	*2011*	*2012*
Passenger cars	**693.1**	**712.6**	**688**	**855.4**	**817.9**	**811.5**	**805.8**	**798**	**793.8**
CO_2	641.6	662.3	639.1	804.4	769.3	766	763.7	760.1	759.8
CH_4	1.2	1.1	1	1.1	1	0.9	0.9	0.8	0.8
N_2O	19.5	17.8	15.7	17.3	14.7	12.4	10.9	9.4	8
HFCs	30.8	31.4	32.2	32.6	32.9	32.1	30.4	27.6	25.2
Light-duty trucks	**589.9**	**553.1**	**566.9**	**371.7**	**354.8**	**359.9**	**359.1**	**343.1**	**338.4**
CO_2	542.7	505.9	519.5	330.1	312.8	317.4	317.6	303.8	301.2
CH_4	0.7	0.7	0.7	0.3	0.3	0.3	0.3	0.3	0.3
N_2O	15.3	13.7	12.6	5.9	5.2	5.2	4.7	4.1	3.6
HFCs	31.2	32.8	34.2	35.4	36.4	37	36.5	34.9	33.3
Medium- and heavy-duty trucks	**377.7**	**408.4**	**418.6**	**444.7**	**427**	**389.2**	**402.9**	**402.4**	**403.4**
CO_2	365.6	396	406.1	431.6	413.9	376.3	390	389.6	390.6
CH_4	0.1	0.1	0.1	0.1	0.1	0.2	0.1	0.1	0.1
N_2O	1.2	1.1	1.1	1.4	1.4	1.1	1.1	1	0.9
HFCs	10.7	11.1	11.4	11.5	11.6	11.6	11.6	11.7	11.7
Buses	**15**	**12.1**	**12.3**	**18**	**17.4**	**16.5**	**16.3**	**17.5**	**18.6**
CO_2	14.8	11.8	12	17.6	17	16.1	15.9	17	18.2
CH_4	+	+	+	+	+	+	+	+	+
N_2O	+	+	+	+	+	+	+	+	+
HFCs	0.2	0.2	0.3	0.3	0.4	0.4	0.4	0.4	0.4

Motorcycles	**1.8**	**1.7**	**1.9**	**4.3**	**4.5**	**4.3**	**3.8**	**3.7**	**4.3**
CO_2	1.7	1.6	1.9	4.3	4.4	4.2	3.8	3.7	4.2
CH_4	+	+	+	+	+	+	+	+	+
N_2O	+	+	+	+	+	+	+	+	+
Commercial aircraft[a]	**126**	**134**	**138.3**	**141**	**128.5**	**120.7**	**114.4**	**115.7**	**114.4**
CO_2	124.8	132.7	137	139.7	127.3	119.5	113.3	114.6	113.3
CH_4	+	+	+	+	+	+	+	+	+
N_2O	1.2	1.3	1.3	1.3	1.2	1.1	1.1	1.1	1.1
Other aircraft[b]	**64.6**	**59.7**	**48.1**	**42.4**	**48.2**	**36.8**	**40.5**	**34.2**	**32.1**
CO_2	64	59.1	47.6	42	47.8	36.4	40.1	33.9	31.8
CH_4	+	+	+	+	+	+	+	+	+
N_2O	0.6	0.6	0.4	0.4	0.4	0.3	0.4	0.3	0.3
Ships and boats[c]	**40.1**	**45.2**	**48.4**	**55.2**	**45.9**	**39.3**	**45.3**	**47**	**40.8**
CO_2	39.6	44.5	47.7	54.4	45.2	38.7	44.6	46.3	40.1
CH_4	+	+	+	+	+	+	+	+	+
N_2O	0.5	0.6	0.7	0.8	0.6	0.5	0.6	0.7	0.6
HFCs	+	+	+	+	+	+	+	+	+
Rail	**52.3**	**53**	**55.1**	**54.4**	**50.7**	**43.4**	**46.3**	**48**	**46.9**
CO_2	49.7	50.3	52.4	51.6	47.9	40.7	43.5	45.3	44.1
CH_4	0.1	0.1	0.1	0.1	0.1	0.1	0.1	0.1	0.1
N_2O	0.4	0.4	0.4	0.4	0.4	0.3	0.3	0.4	0.3
HFCs	2.1	2.2	2.2	2.2	2.3	2.3	2.3	2.3	2.3
Other emissions from electricity generation[d]	0.1	0.1	0.1	0.1	+	+	+	+	+

Table 24.3 continued

Gas/vehicle	*2004*	*2005*	*2006*	*2007*	*2008*	*2009*	*2010*	*2011*	*2012*
Pipelines[e]	**31.1**	**32.2**	**32.3**	**34.2**	**35.6**	**36.7**	**37.1**	**37.8**	**40.1**
CO_2	31.1	32.2	32.3	34.2	35.6	36.7	37.1	37.8	40.1
Lubricants	**10.2**	**10.2**	**9.9**	**10.2**	**9.5**	**8.5**	**9.5**	**9**	**8.3**
CO_2	10.2	10.2	9.9	10.2	9.5	8.5	9.5	9	8.3
Total transportation	**2,002**	**2,022**	**2,019.8**	**2,031.5**	**1,939.9**	**1,866.9**	**1,880.9**	**1,856.4**	**1,841**
International bunker fuels[f]	*109.5*	*114.3*	*115.3*	*116.5*	*115.5*	*107.5*	*118.2*	*112.8*	*106.9*

Source: US Environment Protection Agency, 2014

Notes: + Does not exceed 0.05 Tg CO_2 Eq. (a) Consists of emissions from jet fuel consumed by domestic operations of commercial aircraft (no bunkers). (b) Consists of emissions from jet fuel and aviation gasoline consumption by general aviation and military aircraft. (c) Fluctuations in emission estimates are associated with fluctuations in reported fuel consumption, and may reflect data collection problems. (d) Other emissions from electricity generation are a result of waste incineration (as the majority of municipal solid waste is combusted in "trash-to-steam" electricity generation plants), electrical transmission and distribution, and a portion of other process uses of carbonates (from pollution-control equipment installed in electricity generation plants). (e) CO_2 estimates reflect natural gas used to power pipelines, but not electricity. While the operation of pipelines produces CH_4 and N_2O, these emissions are not directly attributed to pipelines in the US Inventory. (f) Emissions from international bunker fuels include emissions from both civilian and military activities; these emissions are not included in the transportation totals.

Metrics, Estimation Methods and Models

Emission levels are based on many factors, including levels of industrial activity, technological developments, fuel consumption, vehicle miles traveled, and other activities that cause air pollution. Specifically, transportation-related emission levels depend on a number of factors that involve the activity levels (e.g., vehicle miles traveled), the mode's energy intensity (or fuel economy), which is based on the type of vehicle and various vehicle characteristics (age, weight, etc.), the type of fuel consumed, the emission intensity of the fuel, and other factors (such as weather conditions and driver behavior). They can be estimated in a basic form as the product of the emission factor and the activity levels. As an example, in a simplified way, the CO_2 emissions of an automobile would be:

$$\text{Grams of CO}_2 = \text{vehicle miles traveled} \times \frac{\text{gallons}}{\text{mile}} \times \frac{\text{grams of CO}_2}{\text{gallon}}$$

Emission factor (or emission intensity) is an average estimated rate of a specific air pollutant as a result of a specific activity (such as transportation) per unit of activity (such as passenger miles traveled or vehicles miles traveled), and is used for any pollutant emission. Because of the CO_2 significance among the GHG emissions, a valuable metric is the *carbon intensity*. Similarly to energy intensity, carbon intensity can be defined as the amount of carbon emitted per passenger-mile or ton-mile to enable comparison among passenger and freight modes, respectively. As such, the carbon intensity of different vehicles and modes varies significantly depending on a number of factors including the vehicle's occupancy rate and energy efficiency. In general, the least to the most carbon intense modes are: intercity and transit bus, rail transportation, shared (two or more persons) automobile, air transportation, and non-shared automobile and truck (Bradley and Associates, 2014; Schipper et al., 2011).

The US EPA estimates nationwide emissions of ambient air pollutants and the pollutants from which they are formed (their precursors). These estimates are based on 1) actual monitored readings of pollutant concentrations in the ambient (outside) air at selected monitoring sites throughout the country (air concentrations) or 2) engineering calculations of the amounts and types of pollutants released into the air by vehicles, factories, and other emission sources. To quantify the impact of transportation systems (and changes on them) on air quality, both the resulting pollutant emissions and the change in pollutant concentrations on ambient air are required.

A large number of models to estimate the pollutant emissions from on-road transportation (automobile, trucks, and motorcycles) are currently available. Most models differ in the modeling approach, structure, and/or data requirements and can be broadly categorized into three categories: macroscopic (average speed, at a regional level; travel or fuel based), mesoscopic and microscopic (at a segment/facility and individual level respectively; instantaneous and integrated emission-traffic models) (Demir et al., 2014; Smit et al., 2010; Yu et al., 2009). The US EPA's current official on-road transportation emissions model is MOVES (which replaced MOBILE), designed at a macroscopic level. Other common macroscopic models in the US are: EMFAC (California Air Resources Board, CARB), and a GHG-focused model GREET (Argonne National Laboratory). Mesoscopic and microscopic models include: CMEM (Scora and Barth, 2006), PERE (Nam and Giannelli, 2005), and VT-Micro (Ahn et al., 2002). The US EPA also developed the NONROAD model, which estimated emissions for all non-road engines, equipment and vehicles except commercial marine, rail and air transportation. For air transportation related emissions EDMS by the Federal Aviation Administration is officially used.

For rail transportation, two modeling approaches are currently used that provide emission inventories at a county level: a fuel-based model by the US EPA, and a travel-time-based model by CARB (Gould & Niemeier, 2009).

To estimate the *concentration* (i.e., the amount in terms of mass or volume of a pollutant per unit volume of air) of pollutants in ambient air, dispersion models are used. A dispersion model uses input data that involve three main categories: data pertaining to the source of pollution (e.g., coordinates and emission rates), the meteorology (stability, wind speed/direction, mixing height, and ambient temperature), and each receptor, which can be either a point or an area (grid of points, e.g., coordinates and elevation). The sources of pollution can be considered point, line, area or volume, and the emission releases continuous or instantaneous. Transportation-related sources are near surface and the associated releases are instantaneous. On-road transportation sources are a typical example of line sources and are commonly modeled with the US EPA recommended models CALPUFF (for long-range emission transport and complex terrain), CALINE, and CAL3QHC. AERMOD is also recommended by the US EPA but it does not explicitly simulate line sources (US Environmental Protection Agency, 2015).

Other Environmental Impacts

Transportation Noise Pollution

Noise refers to any sound that becomes "unwanted or disturbing" disrupting humans' normal activities (e.g., sleep, working, learning, etc.) or diminishing humans' (or other living organisms) quality of life in any other way. Noise pollution has been proven to adversely affect human health, and a number of health issues have linked with it (such as noise-induced hearing loss (NIHL), stress-related illnesses, high blood pressure, speech interference, etc.) (US EPA). Transportation noise is created from highway transportation, rail and other mass transportation, marine noise (mainly affecting marine mammals), and aircraft operation related noise. From these sources, aircraft-related noise is the most difficult to mitigate. In the US, aircraft noise has been a major restraint of the aviation growth and, thus, a major incentive for aircraft technology improvement (Transportation Research Board, 2014). The most common source is highway transportation, and the most broadly employed noise mitigation methods for such sources are noise barriers, use of plants, and depressed highway design.

Water and Soil Quality

Transportation impacts on water quality are mainly caused by: marine transportation activities; transportation infrastructure that involves impervious surfaces (roadways, parking lots, airport aprons and runways, etc.); and pollutants such as fuel, oil, and deicing chemicals (US Environment Protection Agency, 1999). The most significant impact of marine transportation has been major oil spills from barges and onshore facilities. In general, the average volume of spills (and spill incidents) has been decreasing, although small-scale oil spills have been increasing (American Petroleum Institute, 2009). Furthermore, the dumping of solid waste and sewage dumping can have serious effects on the water and marine ecosystem. Runoffs from impervious areas might carry pollutants that can contaminate wetlands, lakes, and streams. In addition, impervious areas can affect the recharge rates of the groundwater and increase runoffs, contributing to the erosion of the soil. In the same manner, coastal infrastructures can contribute to waterfront soil erosion (US Environment Protection Agency, 1999).

Land Use and Nature Conservation

Transportation has both direct and indirect impacts on land use. Transportation land take refers to the area of land devoted to transportation infrastructures and other transportation-related facilities. Subsequently, this excludes the area for "greener" competing land uses (such as green or open spaces, wetlands, etc.) and transforms the landscape. Another impact pertaining to both land use and nature preservation is population fragmentation. Roadways and other transportation infrastructures create a barrier that might prevent groups of the same species from interacting (Litman, 2014). The degree of this impact depends on the type and size of the infrastructure, the traffic, the size of the integral land, and other factors (Swedish Environment Protection Agency, 2002). Other ecological impacts can include but are not limited to road kills, access to humans (e.g., hunters), habitat impacts (such as loss of habitat), and pollution and noise that might affect the quality of animals' lives (Litman, 2014). Finally, transportation can indirectly impact nature conservation and biodiversity, specifically in the context of natural resources consumption, and contribution to deforestation and other natural habitats.

4. Strategies

This section discusses a variety of environmental, energy, and transportation strategies and policies, and technological innovations that are promising for reducing energy consumption and emissions from the transportation sector. These range from market-based approaches, regulations and standards, to emerging vehicle and fuel technologies, and policies to encourage behavioral change. For information on climate change adaptation strategies to "reduce the vulnerability of natural and human systems or to increase system resiliency in light of expected climate change or extreme weather events", the reader is referred to National Cooperative Highway Research Program, 2014a; Wall and Meyer, 2013.

Environmental and Energy Policies, Standards, and Regulations

Standards

Air quality standards form the basis for the national air pollution control effort. The Clean Air Act (CAA) of 1970 established "national ambient air quality standards (NAAQS) for major air pollutants, hazardous air pollutants standards, state attainment plans, motor vehicle emissions standards, stationary source emissions standards and permits, acid rain control measures, stratospheric ozone protection, and enforcement provisions". These standards are applicable to all modes of transportation.

Corporate Average Fuel Economy (CAFE) requires car and light truck manufacturers to comply with the fuel economy standards set by the US Department of Transportation. First enacted by Congress in 1975, the purpose of CAFE is to reduce energy consumption by increasing fuel economy. The most recent standards are projected to require average fleet-wide performance levels of 35.5 mpg (250 grams/mile of CO_2) by model year 2016 and 54.5 mpg (163 grams/mile of CO_2) by 2025 (Schoettle and Sivak, 2013; US Environmental Protection Agency and the National Highway Traffic Safety Administration, 2012). Research shows that CAFE performance is meeting or exceeding the projected GHG reductions contained in the latest standards (Schoettle and Sivak, 2013). Research also cautions that increasing the fuel economy (and thus, decreasing the fuel expenditures) might induce additional travel, offsetting the savings resulting from increased fuel efficiency (known as the rebound effect) (Litman, 2013; Transportation Research Board, 2011).

Heavy-Duty National Program provides standards to reduce GHG emissions and improve fuel efficiency of medium- and heavy-duty vehicles. The rule to establish this program was adopted in September 2011 (US Environmental Protection Agency and the National Highway Traffic Safety Administration, 2011b). An impact analysis of this program suggested substantial fuel savings and emissions reductions (US Environmental Protection Agency and the National Highway Traffic Safety Administration, 2011a).

Renewable Fuel Standards (*RFS, RFS2*) are set for suppliers and refiners of transportation fuels annually by US EPA in a bid to reduce the carbon intensity of transportation fuels. These standards apply to the production volumes of cellulosic biofuels, biomass-based diesel, advanced biofuel, and renewable fuels to be used as transportation fuel, or blended with gasoline or diesel fuels (US Environmental Protection Agency, 2013). When projected volumes are lower than the mandatory volumes, US EPA is required to lower the standards (Kay et al., 2014). A supplemental program to US EPA's RFS program is the low carbon fuel standards (LCFS) program, adopted in California, which sets carbon-intensity performance targets for producers of petroleum-based fuels. LCFS also involves a credit trading system where "petroleum importers, refiners and wholesalers can either develop their own low carbon fuel products, or buy LCFS Credits from other companies" (Transportation Research Board, 2011).

Policies

Cap and trade policy involves a mandatory cap on emissions and issues allowances that allow a source (industry) to emit a specific amount. Sources can buy or sell allowances on a market-based price or "bank" them to use in future years. The government sets the goal that the industry must meet and monitors compliance. This policy is considered an effective and pragmatic solution to GHG emission reduction given that it is designed appropriately. Key challenges in its design and implementation involve identifying the sectors to be applied and the mechanism by which permits would be distributed (for free or through auctions) (US Environmental Protection Agency (Cap and Trade); RAND Corporation, 2009).

Feebates are a market-based policy for reducing emissions from passenger vehicles. These fees (per unit of energy consumption or unit of GHG emissions) are added to the purchase price of new vehicles with low fuel efficiency, and are used as rebates for lower-emitting vehicles. There is no existing feebate program in the US; the federal government imposes a fee on manufacturer sales of vehicles that test at 22.5 miles per gallon or lower, but this fee is not accompanied by a rebate program (Transportation Research Board, 2011). A study in California found that feebates can cost-effectively reduce fleet-average GHG emissions, and that a national feebate program would produce the greatest benefits (Bunch and Greene, 2010).

Taxes

Carbon tax involves a fixed fee per unit of GHG emitted. This policy can be applied across multiple sectors and is considered a more economically efficient approach compared to a cap and trade policy. While there is certainty about the price of emissions, the amount of emissions reductions is not certain (the opposite holds true under a cap and trade policy). A low carbon price might not result in the desirable reduction in GHG emissions, while a high carbon tax might result in more resources used than necessary (RAND Corporation, 2009).

Fuel tax (or gas tax) is an excise tax imposed on the sale of fuel. The first gas tax was enacted in Oregon in 1919; by 1929, every state had adopted such a tax (Black, 2010). Currently, the fuel tax consists of a federal excise tax at $0.184 per gallon of gasoline and $0.244 per gallon of diesel, plus state and local taxes. The intent of the tax is to raise funds for highway construction, operation, and maintenance (placed in the Highway Trust Fund (HTF)) but also to reduce emissions and energy use. While this tax promotes changes in driving intensity and provides an incentive for energy efficiency (RAND Corporation, 2009; Small, 2012), it is not sufficient to cover the financial needs of the surface transportation system; the federal tax has not been increased or indexed to inflation since 1993, vehicle fuel economy has improved, and the market penetration of alternative vehicles is increasing (Government Accountability Office, 2006). Supporters of a fuel tax raise argue that it will both boost the HTF and serve as an incentive for greener vehicles and reduction in travel (Litman, 2013; Small, 2012).

Transportation Policies and Strategies to Reduce Travel Demand and Intensity

Transportation Demand Management (TDM) is "a general term for strategies that result in a more efficient use of transportation resources" (Victoria Transportation Policy Institute, 2014). TDM strategies aim to affect travel behavior, reduce VMT and, as a result, reduce vehicle energy consumption and emissions. TDM strategies include traffic calming and speed reduction strategies, ridesharing (carpooling, vanpooling) and high-occupancy vehicle (HOV) lanes, congestion and other road pricing strategies (such as toll roads and VMT fees), parking management and pricing, transit improvements and other transit incentives, and telework or flexible work schedules (Black, 2010; Federal Highway Administration, 2012; Victoria Transportation Policy Institute, 2014).

Congestion pricing is widely viewed as an effective TDM strategy that charges drivers higher tolls to use certain roads at more congested times of day, and thus encourages drivers to use other routes, transport modes, or shift the time of their trips (Federal Highway Administration, 2008). Congestion pricing has been found effective in reducing VMT and emissions, raising funds for new construction, and promoting public transportation. However, gaining public acceptance remains a major barrier to implementing congestion pricing in urban areas (see discussion in Rentziou et al., 2011). The Federal Highway Administration (2008) describes four main types of road pricing strategies:

1. *Variably priced lanes* such as high-occupancy toll (HOT) lanes and express toll lanes within a highway. On HOT lanes, low-occupancy vehicles are charged a toll, whereas high-occupancy vehicles (HOVs), public transit buses, and emergency vehicles are allowed to use the lanes free of charge or at reduced rates. On express toll lanes all vehicles irrespective of the occupancy level are required to pay a toll.
2. *Variable tolls* refer to tolls on highway facilities or bridges that vary by time of day (higher during peak travel hours, lower during off-peak travel hours) to manage traffic flow.
3. *Cordon pricing (or zone-based charges)* involves charging a fee to enter or drive within a congested area, usually a city center. Cordon pricing was first established in Singapore in 1975 and was then followed by European cities facing traffic problems such as Oslo, London, and Stockholm. Zone-based charges can be flat or vary by time of day (such as in Stockholm).
4. *Area-wide or system-wide charges* refer to a per-mile charge that might involve higher charges during congested periods. This pricing strategy where users are charged a fee for externalities

(such as congestion and emissions), also known as VMT fee, is considered a potential replacement for the fuel tax per gallon and is viewed as a more equitable alternative to fuel taxes (Sorensen et al., 2013). It is also expected that VMT fees can change driver behavior and travel intensity. The application of this pricing strategy to passenger vehicles has been limited to only a few pilot studies in the US (Oregon, Washington, and Iowa).

Land use is considered one of the major factors influencing travel demand; various indicators, such as density, mixed use or diversity, urban form and development, have been found to have a significant impact on the amount and intensity of travel (see Rentziou et al., 2012 for a summary of past studies). Higher densities contribute to shorter trip lengths and make transit and non-motorized modes more viable. Smart growth initiatives, land-use controls, transit-oriented development, and better coordination between transportation and land-use planning can lead to changes in travel behavior and demand and, in the long run, can impact transportation energy and the environment (Black, 2010; RAND Corporation, 2009). Transit service improvements (such as new routes and services, increased frequencies and reliability) and other transit incentives (such as fare reduction) can further encourage mode shift and increase transit ridership. As discussed in sections 2 and 3, transit systems such as bus and passenger rail can transport passengers more efficiently and with fewer GHG emissions than automobiles (especially single-occupant vehicles). It has been estimated that changes in the US built environment have the potential to impact transportation energy and GHG emissions significantly over the long term, from less than 1 percent to as high as 10 percent by 2050 (Cambridge Systematics, 2013). These strategies are mostly effective in metropolitan areas with a well-developed transit network, and have limited applicability to travel in rural areas (Transportation Research Board, 2011). Transit or land-use policies would likely not affect freight emissions (Kay et al., 2014).

Information and communication technologies can facilitate telecommuting, e-commerce, distance-learning, and other activities that can be conducted without physical travel (personal or commercial trips). The impact of these technological innovations on travel demand has been conflicting since communications might serve as substitutes for only a small number of trips, and additional demand for other types of trips (such as non-work based) might be induced (Black, 2010).

Technology

Intelligent Transportation Systems

Intelligent transportation systems (ITS) can be defined as "the application of advanced information and communications technology to surface transportation in order to achieve enhanced safety and mobility while reducing the environmental impact of transportation" (US Department of Transportation, 2011). ITS technologies focus on creating intelligent vehicles, intelligent infrastructure, and an integrated system of these two components. ITS initiatives can improve both passenger and commercial vehicle operations and typically do not require a long time horizon for implementation. While these technologies do not directly address sustainability issues, they could be deployed to mitigate the negative impacts of the surface transportation sector on the environment, as well as reduce energy consumption (Black, 2010; Gkritza and Karlaftis, 2013).

For example, *advanced traffic management systems* that provide real-time traffic and traffic signal monitoring and control, coupled with *advanced traveler information systems* such as dynamic message signs and automated warning systems can reduce fuel consumption and emissions by reducing congestion, idling, and promoting "greener" routing (eco-routing) options (Gkritza and Karlaftis, 2013). Eco-routing in particular allows road users to change routes to avoid congestion and reduce their carbon footprint; the fuel and emissions savings have been found to be larger on a grid network compared to a freeway corridor network (Ahn and Rakha, 2013). *Advanced vehicle control systems* such as adaptive cruise control that senses the presence of other vehicles and slows if necessary to keep pace with the leading vehicle, and automated highway systems (AHS), where sensors and communication devices link the roadway and the vehicle to maximize driving performance, can offer benefits in safety, travel reliability, but also, emissions and fuel use savings (Bell, 2006; Congress, 1994). A special case of AHS are automated electric highway systems (AEHS) that are electrified roadways where the power is supplied continuously via inductive charging, or there is wireless transfer of electricity through electromagnetic induction. Electrified highways would allow a significant number of vehicles to shift from petroleum-based fuel sources, increasing energy independence and decreasing fuel price volatility. AHES can also provide significant benefits in terms of emissions and energy use savings (Lavrenz and Gkritza, 2013; MIT, 2010). One of the biggest outstanding questions is how infrastructure will be adapted to accommodate electrified highways, with the logical extension of who will pay for the infrastructure and how market penetration will occur (MIT, 2010).

Emerging Fuels and Vehicle Technologies

Natural gas is a naturally occurring form of fossil energy; chemically, it consists primarily of methane (CH_4). Over the past decade, the combination of horizontal drilling and hydraulic fracturing has greatly expanded access to large volumes of shale gas (natural gas that is trapped within shale formations) that make the production of natural gas economically feasible in the US. As such, natural gas is viewed as a cheap and abundant alternative fuel for transportation applications as well as feedstock for electric power generation. In addition, the US pipeline distribution network for natural gas is quite extensive—305,000 miles of interstate and intrastate transmission pipelines (US Department of Energy, 2007). Natural gas can be stored in a vehicle in either compressed form (CNG) or chilled and liquefied (LNG). Light- and medium-duty vehicle engines can be retrofitted with modest changes (but at a high cost) to run on CNG. Heavy-duty vehicles can use natural gas in CNG or LNG form (National Cooperative Highway Research Program, 2014b). While CNGs offer some reduction in per mile emissions (GHG and local air pollutants), the potential leakage of methane in the supply chain could more than offset this GHG emission reduction.

Biofuels refer to liquid fuels produced from biomass. First-generation biofuels include ethanol (mainly from corn or sugar cane) and biodiesel (from oil seeds such as soybeans); second-generation biofuels include cellulosic ethanol or butanol (from wood, switchgrass, or crop wastes. Another potential feedstock from biofuels is algae that require less land and water to produce. Second-generation biofuels are expected to achieve higher GHG reductions compared to gasoline. Blends of ethanol up to E10 (10 percent ethanol) and E15 (15 percent ethanol) can be used in older conventional and newer vehicles, respectively. Flex-fuel vehicles (FFVs) can use blends that range from 100 percent gasoline to 85 percent ethanol. The cost to purchase an FFV is comparable to that of conventional vehicles; however, the costs of biofuels are higher than that of conventional gasoline or diesel (National Cooperative Highway Research Program, 2014b).

Electricity is a low-cost transportation fuel that can be used in plug-in electric vehicles (PEVs). PEVs include battery electric vehicles (BEVs, such as the Nissan Leaf and Tesla Model S) and

Table 24.4 Factors influencing market potential and environmental benefits of various emerging fuels and vehicle technologies

Technology	*Production, distribution, and refueling*	*Vehicle cost and performance*
Natural gas CNG or LNG	• *Environmental concerns with fracking* • *Climate concerns from possible leakage of methane in the supply chain (which could more than offset GHG benefits of natural gas versus petroleum)* • Lack of refueling stations • Competition with other potential uses of natural gas	• High costs for onboard CNG or LNG storage, leading to higher vehicle costs • Perceived safety concerns for CNG • Limited vehicle range for CNG
Cellulosic E85 or drop-in biofuels FFV or ICE biofuels	• Insufficient feedstocks to fully displace petroleum and potential competition with food crops • *Uncertain ability of industry to meet RFS2 targets for advanced low-carbon biofuels such as cellulosic ethanol or, eventually, drop-in biofuels* • Limited distribution network, blending and storage capacity, and refueling stations for E85	• Reduced vehicle range with E85 given lower energy content per volume for ethanol versus gasoline
Electricity BEV or PHEV	• *High cost of increasing the share of renewable electricity on the grid* • Cost of upgrading local transformers to support at-home charging • Cost of upgrading residential electrical systems and installing home recharging equipment • Limited availability of fast recharging stations (most critical for BEVs)	• Unresolved battery performance concerns related to safety, longevity, and ability to handle fast recharging • Very high battery costs, leading to very high vehicle costs • Limited all-electric driving range and long required recharging time (most critical for BEVs)
Hydrogen FCV	• *High cost of renewable hydrogen in comparison to hydrogen produced from fossil fuels* • Lack of hydrogen distribution network and refueling stations	• Insufficient durability of fuel cells • High cost of onboard hydrogen storage and very high cost of fuel cells, leading to very high cost of vehicles

Source: Adopted from Table 3.2 in *National Cooperative Highway Research Program*, 2014b plug-in hybrid electric vehicles (PHEVs, such as the Chevy Volt). BEVs rely on the electricity contained in the battery system, while PHEVs include a combustion engine that provides additional driving range when the battery is depleted. Depending on the charging system used (Level 1 alternating current AC, Level 2 AC or direct current/fast charging DC), a PEV's battery can charge in less than 30 minutes to more than 24 hours. Production of PEVs targets passenger vehicles and light trucks such as delivery vans, small and large transit buses, utility service vehicles, and urban delivery trucks. There are limited production programs for medium- and heavy-duty commercial vehicles. Electricity also can be an alternative power source for non-road vehicles and equipment at ports, airports, and truck stops (Electric Power Research Institute, 2011). The impact of PEVs on GHG emissions depends on the electricity generation mix; a renewable and less-carbon intensive generation of electricity could result in GHG reduction benefits (National Cooperative Highway Research Program, 2014b).

Hydrogen (H_2) is a potentially emissions-free alternative fuel that exists in abundance in the US and could be used to power fuel cell electric vehicles (FCV) that are two to three times more efficient than internal combustion engines (ICE). However, hydrogen has a low volumetric energy density compared to gasoline, and as such, a larger tank for fuel storage is required compared to conventional vehicles. The environmental and health benefits related to FCVs depend on the method of hydrogen production; greater benefits would incur when hydrogen is produced from low- or zero-emission sources, such as solar, and wind. FCVs can refuel in as little as three minutes and can achieve a range of more than 300 miles on a single tank. As of 2014, there are only a handful hydrogen stations in the US, mostly in California (US DOE Office of Energy Efficiency and Renewable Energy).

Table 24.4 summarizes the main concerns related to the wide adoption of the aforementioned emerging fuels and vehicle technologies as alternatives to conventional fuels and petroleum-fueled vehicles.

High-Speed Rail Systems

Various countries in Europe and Asia have electrified part or their entire railroad network in an effort to reduce GHG emissions and provide improved service for passenger and freight transportation. France, Japan, and to a lesser extent China, Germany, Russia, and Great Britain have led railroad electrification worldwide. Japan has the most expanded railway system that serves mainly passenger transportation with high-speed and regional routes. In the US, the development of a nationwide high-speed rail (HSR) network has been suggested as a promising solution for reducing energy consumption and GHG emissions subject to sufficient ridership demand and continuous investment (Chester and Horvath, 2010; Sonnenberg, 2010). Candidate corridors for HSR range from 150 to 800 miles (with up to 500 miles being most attractive) where HSR would mainly substitute for air travel, and particularly among densely populated areas (Clewlow et al., 2014; Transportation Research Board, 2011). The lack of sufficient ridership has been one of the major barriers to embracing HSR technology in the US that can be attributed to the prevalence of an automobile culture or equivalently, the lack of a passenger railroad culture; the size of the airline service network, as well as the size of the country (Fricker and Whitford, 2004). The presence of sufficient demand is essential to ensure the financial viability of investments on HSR, but also to balance the emissions stemming from HSR-related infrastructure construction and power generation (Westin and Kageson, 2012). If the sufficient ridership goal is met, electric rail transportation could promote energy independence, transportation safety and efficiency, and better community interconnectivity (Federal Railroad

Administration, 2009). It is also anticipated that passenger rail growth will bring regional economic benefits such as business and urban development, increase mobility and liveability (Randolph et al., 2008).

Moving Forward

While the transportation sector as a whole has made significant progress in reducing its energy use and emissions per unit of transportation output, transportation activities are still energy- and emissions-intensive, particularly personal travel by cars and light trucks, and moving freight on medium- and heavy-duty trucks. Reducing GHG emissions by 50 percent to 80 percent by 2050 in order to limit global temperature increase to 2 °C would be challenging and require a multitude of environmental, energy and transportation policies, and technological innovations (Sperling and Gordon, 2009).

Technological innovations alone, while essential, will not be sufficient to reduce significantly the transportation sector's carbon footprint, at least in the near term (Skippon et al., 2012). Technology diffusion and adoption of these technologies would also not be immediate. While technological barriers continue to fall, societal barriers to the acceptance of electric vehicles (EVs) such as vehicle cost and payback period, range anxiety, and safety still exist (Al-Alawi and Bradley, 2013a; Electric Power Research Institute, 2011). The charging infrastructure is strongly related to EV adoption and expected to play a key role in public acceptability (Sierzchulaa et al., 2014). Moreover, the penetration rates of EVs would differ based on land use and urban sprawl, demographics, travel behavior, as well as market coordination, local policies and incentives (see Al-Alawi and Bradley, 2013b for a review of past studies). The current financial incentives alone do not ensure high EV adoption rates (Sierzchulaa et al., 2014). Feebate programs may motivate greater interest in EVs both from the demand (consumers) and supply side (vehicle manufacturers), as past research has concluded that upfront rebates on the vehicle purchase are more effective than energy use rebates (Mallette and Venkataramanan, 2010).

Fuel technology pathways are slow as research in underway for most emerging fuels. The production of first-generation biofuels and use in FFVs seem viable in the near term; however, concerns related to land availability, water use, and effects on food prices might hinder long-term production. The second generation of biofuels seems more promising for achieving reductions in GHG emissions. Environmental, climate, and safety concerns about the production and distribution of natural gas as well as the high cost of CNG vehicles might hinder the wide application of CNG as transportation fuel. Likewise, the future market penetration of FCVs would depend on the method of hydrogen production, the cost and durability of the fuel cells, and the availability of refueling stations. The pace of the technical progress for developing new fuels would also depend on LCFS programs that could encourage fuel and vehicle technological innovations.

Transport policies coupled with TDM strategies (pricing or other) could encourage changes in travel behavior (such as eco-driving, use of non-motorized modes of transportation, etc.) sooner. However, changes in the culture would be harder to achieve. The sociological literature suggests that "car use reflects much wider societal and cultural contextual factors, and as such, promoting alternatives to the car may need to be more creative in how those options are presented and the features that they offer" (Cairns et al., 2014). For example, marketing of electric vehicles is considered to be more persuasive if savings are framed in gallons, and could be more effective if marketing campaigns target environmentally concerned consumers rather than a general audience (Krupa et al., 2014). Further, promoting alternatives to car use for commuting trips should not only be based on time savings (Kent, 2014).

In summary, the pathway to a sustainable transportation system would involve policies to transform consumer and industry behavior to a less carbon-intensive behavior coupled with technological innovations in low-carbon fuels and vehicles.

Acknowledgment

The author gratefully acknowledges the generous assistance of Ms. V. Dimitra Pyrialakou, Ph.D. candidate at the Lyles School of Civil Engineering, in providing material for a portion of this chapter.

References

Ahn, K. and Rakha, H.A. (2013). Network-wide impacts of eco-routing strategies: A large-scale case study. *Transportation Research Part D: Transport and the Environment*, 25, 119–130.

Al-Alawi, B.M. and Bradley, T.H. (2013a). Total cost of ownership, payback, and consumer preference modeling of plug-in hybrid electric vehicles. *Applied Energy*, 103, 488–506.

Al-Alawi, B. M. and Bradley, T. H. (2013b). Review of hybrid, plug-in hybrid, and electric vehicle market modeling studies, *Renewable and Sustainable Energy Reviews*, 21 (0), 190–203.

American Petroleum Institute (2009). Oil spills in U.S. Navigable Waters, 1997–2006, Washington, DC. Available at: www.api.org/statistics/~/media/Files/Publications/2009_153_OIL_SPILLS_REPORT.pdf

Argonne National Laboratory, GREET. Available at: https://greet.es.anl.gov/

Bell, M.C. (2006). Environmental factors in intelligent transport systems. *Intelligent Transport Systems, IEE Proceedings*, 153 (2), 113–128.

Black, W.R. (2010). *Sustainable Transportation Planning and Solutions*, Guilford Publications, New York.

Bradley, M.J. and Associates (2014). Updated Comparison of Energy Use & CO_2 Emissions From Different Transportation Modes, submitted to American Bus Association. Available at: www.buses.org/files/green.pdf

Bureau of Transportation Statistics (2014a). National Transportation Statistics, Chapter 3, Table 3–3: U.S. Gross Domestic Product (GDP) Attributed to Transportation-Related Final Demand (Current billions of dollars). Available at: www.rita.dot.gov/bts/node/490986

Bureau of Transportation Statistics (2014b). National Transportation Statistics, Chapter 4, Table 4–2: U.S. Consumption of Energy from Primary Sources by Sector (Quadrillion Btu). Available at: www.rita.dot.gov/bts/sites/rita.dot.gov.bts/files/publications/national_transportation_statistics/html/table_04_02.html

Bureau of Transportation Statistics (2014c). National Transportation Statistics, Chapter 4, Table 4–20: Energy Intensity of Passenger Modes (Btu per passenger-mile). Available at: www.rita.dot.gov/bts/sites/rita.dot.gov.bts/files/publications/national_transportation_statistics/html/table_04_20.html

Bunch, D.S. and Greene, D.L. (2010). Potential design, implementation, and benefits of a feebate program for new passenger vehicles in California: interim statement of research findings, UCD-ITS-RR-10–13. Available at: www.its.ucdavis.edu/research/publications/publication-detail/?pub_id = 1400

Cairns, S., Harmer, C., Hopkin, J., and Skippon, S. (2014). Sociological perspectives on travel and mobilities: A review, *Transportation Research Part A: Policy and Practice*, 63, 107–117.

California Air Resources Board, EMFAC (2011). Available at: www.dot.ca.gov/hq/env/air/pages/emfac.htm

Cambridge Systematics (2013). Transportation Energy Futures Series. *Effects of the Built Environment on Transportation: Energy Use, Greenhouse Gas Emissions, and Other Factors*, A study sponsored by US Department of Energy, Office of Energy Efficiency and Renewable Energy. Available at: www.nrel.gov/docs/fy13osti/55634.pdf

Chester, M. and Horvath, A. (2010). Life-cycle assessment of high-speed rail: The case of California. *Environmental Research Letters*, 5(1).

Clean Air Act, February 24, 2004. Available at: www.epw.senate.gov/envlaws/cleanair.pdf

Clewlow, R.R., Sussman, J.M., and Balakrishnan, H. (2014). The impact of high-speed rail and low-cost carries on European air passenger traffic. *Transport Policy*, 33, 136–143.

Congress, N. (1994). The automated highway system: An idea whose time has come. *Public Roads*, 58 (1).

Council of Academies of Engineering and Technological Sciences (1995). *The Role of Technology in Environmentally Sustainable Development*. Report from the National Academies Press. Washington, DC.

Demir, E., Bektaş, T., and Laporte, G. (2014). A review of recent research on green road freight transportation. *European Journal of Operational Research*, 237 (3), 775–793.

Electric Power Research Institute (2011). *Transportation Electrification: A Technology Overview*. Palo Alto, CA. Available at: www.epri.com/abstracts/pages/productabstract.aspx?ProductID=000000000001021334

Federal Highway Administration (2008). Congestion pricing—A primer: Overview. Available at: ops.fhwa.dot.gov/publications/fhwahop08039/cp_prim1_00.htm

Federal Highway Administration (2012). Reference sourcebook for reducing greenhouse gas emissions from transportation sources, prepared under FHWA Project DTHF61-09-00117, RAND Corporation: Kalra, N., Ecola, L., Keefe, B., Weatherford, B., Wachs. B.; RSG, Inc.: Plumeau, P., Lawe, S., Smith, C. Available at: www.fhwa.dot.gov/environment/climate_change/mitigation/publications_and_tools/reference_sourcebook/

Federal Highway Administration (2013). 2011 Highway statistics (Washington, DC: annual issues), Table VM-1. Available at: www.fhwa.dot.gov/policyinformation/statistics/2011/vm1.cfm

Federal Railroad Administration (2009). Vision for High-Speed Rail in America, High-Speed Rail Strategic Plan, US Department of Transportation.

Federal Transit Administration (2011). Flooded Bus Barns and Buckled Rails: Public Transportation and Climate Change adaptation, FTA Repefomort No. 0001. Available at: www.fta.dot.gov/documents/FTA_0001_-_Flooded_Bus_Barns_and_Buckled_Rails.pdf

Fishbone, L.G. and Abilock, H. (1981). MARKet ALlocation (MARKAL), A linear-programming model for energy systems analysis: Technical description of the bnl version. *Energy Research* 5, 353–375.

Fleiter, T., Worrell, E. and Eichhammer, W. (2011). Barriers to energy efficiency in industrial bottom-up energy demand models—A review. *Renewable and Sustainable Energy Reviews*, 15 (6), 3099–3111.

Fricker, J.D. and Whitford, R.K. (2004). *Fundamentals of Transportation Engineering: A Multimodal Systems Approach*. Pearson Prentice Hall, Upper Saddle River, NJ.

Gkritza, K. and Karlaftis, M.G. (2013). Intelligent Transportation Systems Applications for the Environment and Energy Conservation (Part 1). Editorial in *Journal of Intelligent Transportation Systems*, 17 (1), 1–2.

Gould, G. and Niemeier, D. (2009). Review of Regional Locomotive Emission Modeling and the Constraints Posed by Activity Data, Transportation Research Record. *Journal of the Transportation Research Board* 2117, 24–32.

Government Accountability Office (2006). HIGHWAY TRUST FUND. Overview of Highway Trust Fund Estimates, GAO-06-572T. Available at: www.gao.gov/assets/120/113356.pdf

Kay, A.I., Noland, R.B., and Rodier, C.J. (2014). Transportation Futures: Policy Scenarios for Achieving Greenhouse Gas Reduction Targets. Mineta National Transit Research Consortium Report 12–11, Mineta Transportation Institute, San José State University, San José, CA.

Kent, J.L. (2014). Driving to save time or saving time to drive? The enduring appeal of the private car. *Transportation Research Part A: Policy and Practice*, 65, 103–115.

Krupa, J.S., Rizzo, D.M., Eppstein, M.J., Lanute, D. B., Gaalema, D.E., Lakkaraju, K., and Warrender, C.E. (2014). Analysis of a consumer survey on plug-in hybrid electric vehicles. *Transportation Research Part A: Policy and Practice*, 64, 14–31.

Lavrenz, S. and Gkritza, K. (2013). Environmental and Energy Impacts of Automated Electric Highway Systems. *Journal of Intelligent Transportation Systems*, 17 (2), 221–232.

Litman, T. (2013). Comprehensive evaluation of energy conservation and emission reduction policies. *Transportation Research Part A: Policy and Practice*, 47, 153–166.

Litman, T. (2014). Evaluating transportation land use impacts: Considering the impacts, benefits and costs of different land use development patterns. Victoria Transport Policy Institute. Available at: www.vtpi.org/landuse.pdf

Loulou, R., Remne, U., Kanudia, A., Lehtila, A., and Goldstein, G. (2005). Documentation for the TIMES Model—PART I 1–78.

Mallette, M. and Venkataramanan, G. (2010). Financial Incentives to Encourage Demand Response Participation by Plug-In Hybrid Electric Vehicle Owners. In Proceedings of the Energy Conversion Congress and Exposition (ECCE): 4278–4284, 2010, IEEE, September 12–16, Atlanta, GA.

MIT (2010). Electrification of the Transportation System. An MIT Energy Initiative Symposium. Massachusetts Institute of Technology. Available at: https://mitei.mit.edu/system/files/electrification-transportation-system.pdf

Nakata, T., Silva, D., and Rodionov, M. (2011). Application of energy system models for designing a low-carbon society. *Progress in Energy and Combustion Science*, 37 (4), 462–502.

Nam, E.K. and Giannelli, R.A. (2005). Fuel consumption modeling of conventional and advanced technology vehicles in the physical emission rate estimator (PERE). Technical report, Washington, DC. Available at: www.epa.gov/otaq/models/ngm/420p05001.pdf

National Cooperative Highway Research Program (NCHRP). (2014a). NCHRP 750 Strategic Issues Facing Transportation, Volume 2: Climate Change, Extreme Weather Events, and the Highway System: Practitioner's Guide and Research Report, Washington, DC.

National Cooperative Highway Research Program (NCHRP). (2014b). NCHRP 750 Strategic Issues Facing Transportation, Volume 5: Preparing State Transportation Agencies for an Uncertain Energy Future, Washington, DC.

Oak Ridge National Laboratory (2013a). Transportation Energy Data Book, Edition 32, US Department of Energy's Office of Energy Efficiency and Renewable Energy: Chapter 2, Table 2.5: Domestic Consumption of Transportation Energy by Mode and Fuel Type, 2011. Available at: http://cta.ornl.gov/data/chapter2.shtml

Oak Ridge National Laboratory (2013b). Transportation Energy Data Book, Edition 32, US Department of Energy's Office of Energy Efficiency and Renewable Energy: Chapter 2, Table 2.15: Energy Intensities of Freight Modes, 1970–2011. Available at: http://cta.ornl.gov/data/chapter2.shtml

RAND Corporation, 2009. Integrating U.S. Climate, Energy, and Transportation Policies: Proceedings of Three Workshops. Edited by L. Ecola, S. Hassell, M. Toman, and M. Wachs. Sponsored by the McCormick Foundation, Santa Monica, CA.

Randolph, S., Haveman, J., and Egan, T. (2008). California High-Speed Rail Economic Benefits and Impacts in the San Francisco Bay Area, report from the Bay Area Council Economic Institute, San Francisco, CA.

Rentziou, A., Milioti, C., Gkritza, K., and Karlaftis, M.G. (2011). Urban road pricing: Modeling public acceptability. *ASCE Journal of Urban Planning and Development*, 137 (1), 56–64.

Rentziou, A., Gkritza, K., and Souleyrette, R. (2012). VMT, Energy Consumption, and Greenhouse Gas Emissions Forecasting for Passenger Transportation. *Transportation Research Part A: Policy and Practice*, 46 (3), 487–500.

Schipper, L, Saenger, C., and Sudarshan, A. (2011). Transport and carbon emission in the United States: The long view, *Energies*, 4 (4), 563–581.

Schoettle, B. and Sivak, M. (2013). A Comparison of CAFE Standards and Actual CAFE Performance of New Light-Duty Vehicles, The University of Michigan Sustainable Worldwide Transportation, UMTRI-2013-35, October. Available at: http://deepblue.lib.umich.edu/bitstream/handle/2027.42/100179/102966.pdf

Schwartz, H.G., Meyer, M., Burbank, C.J., Kuby, M., Oster, C., Posey, J., and Rypinski, A. (2014). Ch. 5: Transportation. Climate Change Impacts in the United States: The Third National Climate Assessment, J.M. Melillo, Terese (T.C.) Richmond and G.W. Yohe (Eds.), U.S. Global Change Research Program, 130–149. Available at: doi:10.7930/J06Q1V53

Scora, G. and Barth, M. (2006). COMPREHENSIVE MODAL EMISSIONS MODEL (CMEM), version 3.01. User's Guide. University of California, Riverside, Center for Environmental Research and Technology. Available at: www.cert.ucr.edu/cmem/docs/CMEM_User_Guide_v3.01d.pdf

Sierzchulaa, W., Bakkerb, S., Maata, K. and van Weea, B. (2014). The influence of financial incentives and other socio-economic factors on electric vehicle adoption, *Energy Policy*, 68, 183–194.

Skippon, S.M., Veeraraghavan, S., Ma, H., Tait, N., and Gadd, P. (2012). Combining technology development and behaviour change to meet CO_2 cumulative emission budgets for road transport: Case studies for the USA and Europe. *Transportation Research Part A: Policy and Practice*, 46, 1405–1423.

Small, K.A. (2012). Energy policies for passenger motor vehicles. *Transportation Research Part A: Policy and Practice*, 46, 874–889.

Smit, R., Ntziachristos, L., and Boulter, P. (2010). Validation of road vehicle and traffic emission models—A review and meta-analysis. *Atmospheric Environment*, 44 (25), 2943–2953.

Sonnenberg, A. (2010). Transportation Energy and Carbon Footprints for U.S. Corridors, Ph.D. thesis. School of Civil & Environmental Engineering, Georgia Institute of Technology, Atlanta, Georgia.

Sorensen, P., Ecola, L., and Wachs, M. (2013). Mileage-based user fees for transportation funding: A primer for state and local decisionmakers. Available at: www.rand.org/content/dam/rand/pubs/tools/TL100/TL104/RAND_TL104.pdf

Sperling, D. and Gordon, D. (2009). *Two Billion Cars: Driving Toward Sustainability*, Oxford University Press, New York.

Swedish Environmental Protection Agency (SEPA) (2002). Environmental impact from different modes of transport-method of comparison, Report 5183. Available at: www.naturvardsverket.se/Documents/publikationer/620–5183–0.pdf?pid=2861

Transportation Research Board (TRB) (2008). TRB Special Report 290, Potential Impacts of CLIMATE CHANGE on U.S. Transportation, Committee on Climate Change and U.S. Transportation, Transportation Research Board, Division on Earth and Life Studies, National Academy of Sciences, Washington, DC.

Transportation Research Board (TRB) (2011). TRB Special Report 307, Policy Options for Reducing Energy Use and Greenhouse Gas Emissions from U.S. Transportation, Committee for a Study of Potential Energy Savings and Greenhouse Gas Reductions from Transportation, National Academy of Sciences, Washington, DC.

Transportation Research Board (2014), Transportation Research Circular, Number E-C184, Critical Issues in Aviation and the Environment; Chapter 1. Environmental impacts of aviation on human and natural resources, Section 1.1, Noise, by Eagan, M.E., Rolf, B., and Sizon, N. Available at: http://onlinepubs.trb.org/onlinepubs/circulars/ec184.pdf

US Department of Energy, Office of Energy Efficiency and Renewable Energy. Alternative Fuels Data Center. Available at: www.afdc.energy.gov/fuels/hydrogen_basics.html

US Department of Energy (2000). The National Energy Modeling System (NEMS): An Overview 2000. Energy Information Administration. Washington, DC.

US Department of Energy (2007). Transporting Natural Gas. Energy Information Administration. Available at: www.eia.gov/pub/oil_gas/natural_gas/analysis_publications/ngpipeline/fullversion.pdf

US Department of Transportation (2011). Intelligent Transportation Systems (ITS) Standards Program Strategic Plan for 2011–2014, DTFH61–05-D-00002, Intelligent Transportation Systems Joint Program Office, Research and Innovative Technology Administration. Available at: www.its.dot.gov/standards_strategic_plan/stds_strat_plan.pdf

US Department of Transportation (2012). Strategic Plan for Fiscal Years 2012–2016. Available at: www.dot.gov/sites/dot.dev/files/docs/990_355_DOT_StrategicPlan_508lowres.pdf

US Environmental Protection Agency. Air and Radiation, Noise Pollution. Available at: www.epa.gov/air/noise.html

US Environmental Protection Agency. Cap and Trade. Available at: www.epa.gov/captrade/index.html

US Environmental Protection Agency. MOVES (Motor Vehicle Emission Simulator). Available at: www.epa.gov/otaq/models/moves/index.htm

US Environmental Protection Agency. MOBILE (Mobile Source Emission Factor Model). Available at: www.epa.gov/otaq/mobile.htm

US Environmental Protection Agency. Renewable Fuel Standards-2013. Available at: www.epa.gov/OTAQ/fuels/renewablefuels/documents/420f13042.pdf

US Environmental Protection Agency (1999). Indicators of the Environmental Impacts of Transportation: Highway, Rail, Aviation and Marine Transport, 2nd Ed.. EPA 230-R-99–001. Available at: http://ntl.bts.gov/data/Indicators.pdf

US Environmental Protection Agency (2013). National Emission Inventory Air Pollutant Emission Trends website: www.epa.gov/ttn/chief/trends

US Environmental Protection Agency (2014). US Greenhouse Gas Emissions and Sinks (Inventory): 1990–2012, EPA 430-R-14–003, Chapter 2 Trends in Greenhouse Gas Emissions. Available at: www.epa.gov/climatechange/Downloads/ghgemissions/US-GHG-Inventory-2014-Main-Text.pdf

US Environmental Protection Agency (2015). 40 CFR Part 51, Part III, Revision to the Guideline on Air Quality Models: Adoption of a Preferred General Purpose (Flat and Complex Terrain) Dispersion Model and Other Revisions; Final Rule; Appendix W. Available at: www.epa.gov/ttn/scram/guidance/guide/appw_05.pdf

US Environmental Protection Agency and National Highway Traffic Safety Administration (2011a). Final Rulemaking to Establish Greenhouse Gas Emissions Standards and Fuel Efficiency Standards for Medium- and Heavy-Duty Engines and Vehicles, EPA-420-R-11–901. Available at: www.epa.gov/otaq/climate/documents/420r11901.pdf

US Environmental Protection Agency and National Highway Traffic Safety Administration (2011b). EPA and NHTSA Adopt First-Ever Program to Reduce Greenhouse Gas Emissions and Improve Fuel Efficiency of Medium- and Heavy-Duty Vehicles, EPA-420-F-11–031. Available at: www.epa.gov/otaq/climate/documents/420f11031.pdf

US Environmental Protection Agency and National Highway Traffic Safety Administration (2012). 2017 and Later Model Year Light-Duty Vehicle Greenhouse Gas Emissions and Corporate Average Fuel Economy Standards: Final Rule. Federal Register 77:199 (October 15), pp. 62623–63200. Available at: www.nhtsa.gov/staticfiles/rulemaking/pdf/cafe/2017–25_CAFE_Final_Rule.pdf

United Nations (1987). *G. H. Brundtland Report of the World Commission on Environment and Development: Our Common Future.* New York.

Van der Voort, E. (1982). The Energy Flow Optimization Model (EFOM) 12C energy supply model within the EC modelling system. *Omega*, 10 (5), 507–523.

Victoria Transportation Policy Institute (2014). Transportation Demand Management. Available at: www.vtpi.org/tdm/tdm59.htm#_Toc193865012

Wall, T.A. and Meyer, M.D. (2013). Risk-Based Adaptation Frameworks for Climate Change Planning in the Transportation Sector A Synthesis of Practice. Transportation Research Circular E-C181, prepared for the Climate Change, Energy, and Sustainability Impacts on the Transportation Infrastructure Subcommittee, National Academy of Sciences, Washington, DC.

Weijermars, R., Taylor, P., Bahn, O., Das, S.R., Wei, Y.M. (2012). Review of models and actors in energy mix optimization—can leader visions and decisions align with optimum model strategies for our future energy systems? *Energy Strategy Reviews*, 1(1), 5–18.

Westin, J. and Kageson, P. (2012). Can high speed rail offset its embedded emissions? *Transportation Research Part D: Transport and Environment*, 17, 1–7.

Young, P. (2009). A Time Series Analysis of Transportation Energy Use Per Dollar of Gross Domestic Product. Bureau of Transportation Statistics (BTS) Transportation Trends in Focus. Washington, DC.

Yu, L., Jia, S., and Shi, Q. (2009). Research on transportation-related emissions: Current status and future directions. *Journal of the Air & Waste Management Association*, 59(2), 183–195.

25

TRANSPORTATION ECONOMICS

Kara M. Kockelman and T. Donna Chen

UNIVERSITY OF TEXAS AT AUSTIN

Introduction

In the United States, transportation constitutes 17 percent of household expenditures (BLS 2012), 30 percent of greenhouse gas (GHG) emissions, and 28 percent of total domestic energy consumption (EIA 2014). Anywhere in the world, transportation planning, design, policy-making, and budgeting decisions have economic implications. From travel-time savings to job creation (both direct and indirect), income growth to property value changes, motor vehicle crashes to air quality and noise impacts, and microeconomic choices to macroeconomic shifts, transportation policies and investments carry great weight.

Consider this example: due to pavement aging and regular use, many rural state and county roads are in need of rehabilitation or reconstruction. Should transportation agencies install more expensive but longer lasting concrete pavements or rely on less expensive asphalt overlays? The rule of thumb is to go with asphalt, for a variety of reasons, but a definitive answer is not simple. If strict near-term budget constraints did not exist, the decision presumably would be based on a life-cycle cost analysis, used to reveal the solution that yields the lowest annual equivalent cost or maximum net present benefit over a long-term horizon, reflecting risk and uncertainty in flow volumes, materials prices, vehicle sizes, and other economic indicators. In the face of tight budgets, immediate tradeoffs loom. Asphalt pavements may be favored simply to ensure a consistent level of pavement quality across the district under limited funding conditions, while emphasizing equity in funds disbursement, thereby covering more funding requests in a given year. However, if certain rural roads carry significantly more truck traffic, and some are in areas with high levels of black clay (which causes premature distress on asphalt pavement and so requires higher maintenance costs), should these roads be candidates for concrete pavements? What if such a consideration requires some less used roads to be maintained less frequently? What is the cost passed onto the users of these other roads, who may experience slower travel times and increased vehicle repair and maintenance costs? This seemingly simple question is rife with economic considerations.

Relying on various ideas in Kockelman et al.'s (2014) *Economics of Transportation Systems* book, this chapter illustrates the fundamentals of transportation economics (costs, benefits, and pricing), describes key factors that affect these (movement and location choice), and introduces different methods of economic analysis for quantifying the trends in and interactions across these topics.

Costs and Benefits of Transportation

Transportation systems tend to provide many access benefits (to new markets, customers, healthcare, and education, for example), and a project's or policy's net benefits can be estimated by anticipating all its impacts. Project costs are generally easier to quantify, as listed in Table 25.1, with benefits regularly emerging via "cost savings", in the form of lowered travel times, fewer crashes, and reduced emissions.

Transportation costs can be broadly categorized as either internal (those borne by system operators and/or users) or external (generally borne by non-users). In general, internal costs include capital, operating, vehicle, travel time and travel time reliability, and safety costs. In contrast external costs include noise, air pollution, and ecological impact costs borne indirectly by members of the local and global community. To fully understand the nuances between types

Table 25.1 Typical costs of transportation projects and policies

Costs and benefits	*Description*	*Examples*
Capital	One-time design and construction costs	A new facility's capital costs include planning, preliminary engineering, project design, environmental impact analysis, right-of-way (ROW) acquisition, construction, equipment purchases, etc.
Operating	Recurring operations, maintenance, and rehabilitation costs	Typical highway operating costs include traffic management, crash- or weather-related repair and cleanup, equipment (vehicles, traffic signals, signs), utilities, resurfacing (but not reconstruction), etc.
Vehicle	Vehicle ownership and maintenance costs such as fuel, tire replacement, insurance, etc.	Pavement resurfacing improves road conditions and reduces vehicle wear and maintenance costs.
Travel time	Lost time and productivity	Implementation of signal timing coordination on an arterial street enables faster travel times and reduces delay.
Travel time reliability	Variance of schedule uncertainty	Dynamically priced high-occupancy/toll (HOT) lane keeps travel speeds close to free flow speed and reduces variability in travel time.
Safety	Number, severity, and cost of crashes	Addition of rumble strips reduces the number of crashes related to driver fatigue.
Emissions	Health and other costs of vehicle-produced pollution due to changes in travel speeds, distances, times of day, fuels, and vehicle types	Fleet conversion from diesel to compressed natural gas vehicles reduces emissions.
Noise	Discomfort and property value loss due to increased traffic noise	Construction of a sound wall between a freeway facility and an adjacent neighborhood reduces traffic noise.
Ecological impacts	Travel's impacts on wildlife habitat, water flow, and water quality	Planned roadway alignment runs through an endangered species habitat, impeding animal movements through the area.

Source: Kockelman et al. 2014.

of transportation costs, these costs can be further categorized into accounting and opportunity costs, capital and operating costs, and variable and fixed costs, as discussed below.

Internal Costs

To compute internal costs, a differentiation between real accounting costs and opportunity costs is necessary. The former are real, experienced, monetary costs (such as capital, operating, and vehicle costs), while the latter are the net benefits of opportunities foregone (from projects or policies not pursued).

Accounting costs refer to transactions when real monetary exchanges occur, including one-time capital costs (like design, right-of-way (ROW) acquisition, and construction costs) and reoccurring operating costs (including maintenance and rehabilitation costs). Categories of operating costs are essentially the same across all modes of transportation, whereas capital costs differ, as noted in Table 25.2.

Accounting costs can reflect marginal and average costs. Marginal cost (MC) refers to the change in total cost when the quantity produced is increased by one unit (or an infinitesimal unit), whereas average cost (AC) is simply the total cost divided by the total number of units produced (or by total vehicle miles served, for example). To get a sense of this difference, consider how the total cost of a new four-lane highway is unlikely to be twice as much as that of a two-lane highway in the same location. Essentially, marginal costs tend to fall with project size, since the added cost of the third and fourth lanes is likely to be less than that of the first and even second lanes. Such savings accrue from the economies of scale that emerge when bringing equipment and workers out to a job site, ordering materials, managing the inspections, and so forth.

Accounting costs also can be categorized as fixed or variable. Fixed costs *do not change* with the level of output, such as vehicle miles traveled (VMT) along a new highway. In other words, fixed costs are independent of output levels: they must be paid even if no output is being produced. Capital infrastructure, such as railroad tracks and highway lanes, is typically considered a fixed cost, although some components can be variable costs, due to variations in the number of tracks needed, runways used, gates at an airport, and/or lanes along a highway. Variable costs are costs that *do change* with the level of output, including fuel used, system operators paid, and vehicle maintenance fees. Lowered production means lower variable costs.

Researchers have modeled fleet operating costs as both fixed and variable, with Welter et al. (2009) defining fixed costs as vehicle depreciation, financing, and insurance, and variable costs as vehicle repair, maintenance, and fuel. Fixed and variable costs are often differentiated by time frame of reference. In the short run, capital costs (such as construction and design) are considered fixed, while operating costs (such as maintenance and traffic management) are

Table 25.2 Comparison of capital and operating costs for different transportation modes

Mode	*Capital costs*	*Operating costs*
Automobiles	Vehicles, roadways, traffic signals	Fuel, labor, maintenance, supplies, lighting power
Railroads	Rail cars, tracks, stations, rail yards, signal systems	
Airlines	Aircraft, airports, traffic control systems	
Ships	Ships, ports	

Source: Kockelman et al. 2014.

considered variable. However, in the long run, all transportation facilities will eventually need replacement or major rehabilitation, so capital costs can be considered variable.

Another component of internal costs is the opportunity cost associated with transportation expenditures. Unlike accounting costs, which involve real monetary exchange, opportunity costs represent the lost value/benefit of an activity's foregone alternative. For example, no physical monetary exchange occurs when a driver is stuck in traffic on a congested highway, yet that delay results in missed economic (and other) opportunities (Boyer 1998). The value of travel time (VOTT) is a type of implicit opportunity cost and represents the amount of money a traveler is willing to pay for time savings. In other words, VOTT is the marginal rate of substitution of money for travel time (while keeping a traveler's overall happiness or utility constant). VOTT can represent the monetary value a salesperson places on spending an extra hour pitching a product or a parent places on spending at his/her child's soccer game—as opposed to sitting in traffic.

Since VOTT estimates emerge from willingness-to-pay (WTP) considerations, higher wage and higher income individuals regularly exhibit higher VOTTs than lower income individuals. Such travelers may be more able and willing to pay a toll at congested times of day. In its guidance for economic analysis of federal transportation projects and policies, the US Department of Transportation (DOT) recommends VOTTs (in 2009 US $) of $12.42–$25.23 per person-hour for local surface modes of travel, $17.39–$24.44 for intercity travel, and $25.75 for truck driver travel, both local and long distance (USDOT 2014). Of course, a traveler's VOTT varies by other details of the trip as well. If one is late to an important meeting or the airport, VOTTs can run very high, even for low-income travelers.

In addition to people valuing the (average) time they spend traveling, they also value reliability, or the lack of uncertainty in travel times. For example, a truck driver en route to an important or time-constrained delivery can better anticipate and prepare for a consistent 30-minute travel time than one that averages 25 minutes, but regularly varies between 15 and 50 minutes. Transportation projects that improve travel-time reliability offer an important benefit; those that decrease reliability create a cost. Most theories explaining aversion to unreliable travel are based on the perceived cost of unexpected arrival times at work (Bates et al., 2001), which are greater for being late than for being early. Estimates of the value of reliability (VOR), vary from $10.20 to $32 per hour of late or early arrival (Brownstone and Small 2005). Both VOTT and VOR may be estimated with WTP values determined from revealed- or stated-preference surveys.

External Costs

Another critical, yet sometimes overlooked, project and policy impact is that of externalities, or external costs to society. These include the annoyance of highway noise, harm to adjacent properties and bystanders during crashes, and the visual and health impacts of air and water pollution. Reductions in these external costs (due to the provision of sound walls, low-noise pavements, cleaner vehicles, and safer transportation systems) can be considered external benefits.

Transportation projects or polices can either increase or decrease the number of crashes occurring, and their severity, resulting in net costs or benefits to travelers and society at large. Vehicle owners buy insurance to cover many crash costs, but auto insurance is typically a flat fee, rather than per mile driven, and many costs are not recovered from insurers (such as delays in lanes blocked by crashed vehicles, deployment of emergency personnel, and the pain and suffering of crash victims and their families). According to the USDOT (2003a, p. 98), just "about one-quarter of the cost of crashes is paid directly by those involved, while society in general pays the rest," suggesting that external costs are substantial. In 2000, driver-funded insurance companies paid about half the cost of all US highway crashes (USDOT 2003a).

Mobile source emissions of carbon dioxide (CO_2), oxides of nitrogen (NO_x), reactive organic gases or volatile organic compounds (ROG and VOC), particulate matter (PM2.5 and PM10), carbon monoxide (CO), and various other hazardous species constitute a major fraction of human-caused (anthropogenic) emissions and are responsible for air quality concerns in most major cities (Maibach et al. 2008; Litman 2009). Emissions costs arise from damages caused by these pollutants to human, plant, and animal life, as well as damage to buildings and ecosystems. Models like the US Environmental Protection Agency's (EPA's) MOBILE 6.2 anticipate emissions rates and totals, when projects and policies impact VMT, speeds, vehicle types, fuels used, and other relevant factors.

Sustained exposure to traffic noise can cause hearing impairment and sleep disturbance, increasing stress levels in those living and working nearby. In addition to human health impacts, studies indicate that traffic noise reduces the price of nearby homes (e.g., Lewis et al. 1997; Bowes and Ihlanfeldt 2001). Numerous factors influence levels of traffic noise, including vehicle type, engine type, traffic speed, pavement condition, and presence of noise barriers (Litman 2009). Passenger cars are generally quieter than buses, and motorcycle noise costs are about ten times those of passenger vehicles, per mile traveled (Litman 2009). Among passenger cars, electric and hybrid vehicles are generally quieter than gasoline and diesel vehicles.

External costs and benefits are reflected in comprehensive cost-benefit estimates, for the overall economic value and social welfare impacts of transportation projects and policies. Figure 25.1 presents a comprehensive, ranked list of costs associated with 1 mile of passenger vehicle travel. The highest costs are vehicle ownership, travel time, vehicle operation and crash risk borne by the travelers. When examined individually, external costs are much smaller in magnitude. However, once aggregated, external costs represent a significant portion of the per vehicle mile costs.

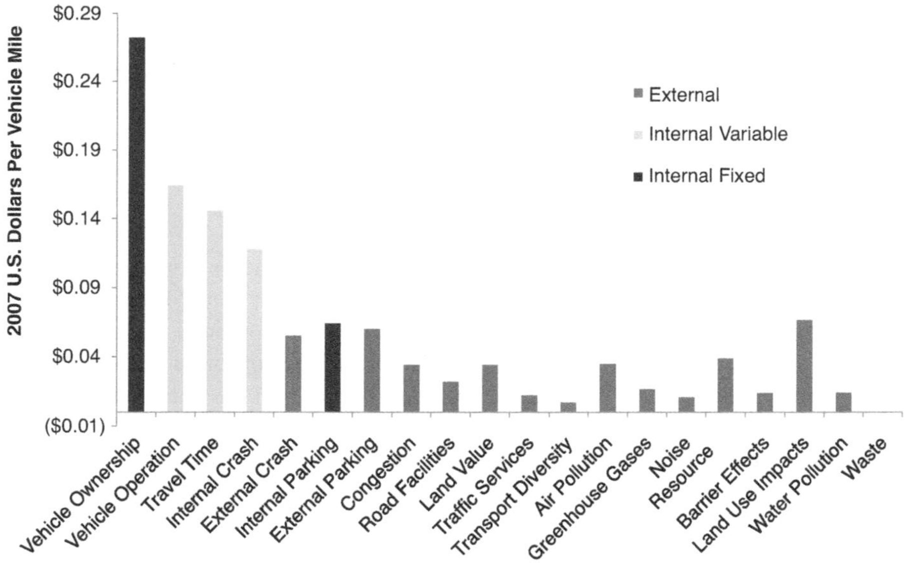

Figure 25.1 Average travel costs per passenger–vehicle mile
Source: Litman 2009.

Pricing of Transportation Services

Now that we have examined significant benefits and costs involved in providing and using transportation systems, the next step is to look at how these costs can best be allocated. In other words, who should pay for transportation services—and how? Transportation pricing refers to fees (and incentives) incurred by travelers, including transit fares, cargo fees, fuel taxes, tolls, parking fees, vehicle registration fees, and insurance payments. When the prices of vehicles, parking, fuel, and transit use change, travel activity patterns are affected.

A key concept here is consumer surplus, or the difference between the maximum prices travelers are willing to pay for a good or service (i.e., the value they actually derive) and the prices they pay. In markets with many consumers (travelers), this surplus can be represented by the area under the demand curve and above the market price, as shown in Figure 25.2. Producer surplus, on the other hand, is the benefit or profit that producers gain by selling at market prices above their cost. The sum of consumer surplus and producer surplus is called social welfare.

For transportation agencies, it is neither ideal to operate free of charge (since facilities not rationed by pricing are typically rationed by congestion) nor to maximize profits (since a public institution's objective should lie in serving the community's best interests). In theory, maximizing net social welfare ensures that transportation system resources are used optimally. Social welfare is maximized when the marginal cost of serving one more customer equates to the marginal social benefit derived from serving that customer. Application of marginal social cost pricing seeks to achieve this optimum by ensuring that users/consumers are made aware of and pay the equivalent of their added cost to the system (in terms of pavement damage and delays imposed on those who follow them, for example).

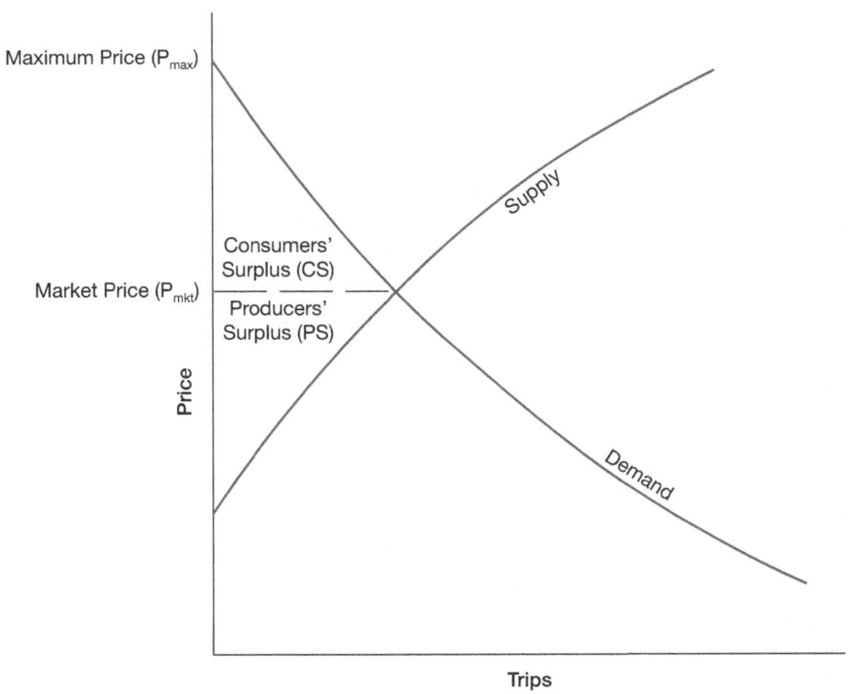

Figure 25.2 Supply, demand, and consumer surplus
Source: Kockelman et al. 2014.

Unlike most markets, where price-setting firms provide all the economic resources and bear all costs of production, transportation costs are substantially shared among system users, operators/owners, and non-users. For example, in short-run marginal cost (SRMC) pricing, where capital costs of facility provision are excluded, the marginal costs of roadway transport can be expressed as follows:

$$SRMC = MC_{\text{User}} + MT \cdot VOTT + MR \cdot VOR + MC_{\text{Gov}} + MSC_{\text{Crash}} + MSC_{\text{Env}} \quad (1)$$

where MC_{User} represents the marginal vehicle operation and ownership costs borne by users (such as tire wear and vehicle depreciation), $MT \cdot VOTT$ reflects travel time costs borne by the marginal traveler and any delays he or she imposes on others (with MT representing the change in total system travel time due to the addition of one more vehicle, which depends on existing congestion levels), $MR \cdot VOR$ represents schedule delay costs (the change in travel time reliability as a result of adding the marginal traveler), MC_p includes system maintenance and operation costs (like highway maintenance expenses, law enforcement, crash response and system administration), and MSC_{Crash} and MSC_{Env} are external costs resulting from marginal increases in crash risk and impacts on air, water, noise, and other ecological elements.

Due to culture, administrative costs, and other factors, road pricing is rare in practice—and truly variable or congestion-based marginal cost road pricing is very difficult to achieve. Moreover, traditional user fees (including vehicle registration fees) tend to be independent of road use and fuel use. Unfortunately, the flat-rate nature of federal and local or state fuel taxes in many countries and regions has not kept pace with increasing transportation capital and maintenance costs (Wachs 2006; Totty 2012). Furthermore, as natural gas and electric vehicles (including hybrid electric vehicles) increase their market shares in the US and abroad, traditional fuel taxes will reflect smaller shares of actual road use. As fuel tax revenues fall, a move toward alternate funding instruments, such as sales taxes and bonds (which are unrelated to roadway use), leads transportation finance farther away from marginal cost pricing, resulting in less efficient markets, where demand does not recognize the true costs of transport (Litman 2011).

Congestion pricing and distance-based tolls seek to allocate many travel costs based on usage. A flat-rate tolled road cannot moderate congestion as well as a dynamically/variably priced road can. Variable tolls are generally designed to fall when traffic is light (because external delay costs are negligible then) and rise when traffic is heavy (because added delays are quite high under near-capacity conditions). In theory, a dynamically tolled road can be priced so that forced flow (level of service F) never occurs. In reality, drivers have only incomplete information available to them, and unpredictable events (such as traffic crashes) and a lack of substitutable alternative routes and modes (and imperfect pricing of such alternatives) mean that even the most thoughtfully congestion-priced facilities will experience suboptimal flows.

Distance-based or VMT fees seek to more equitably charge for roadway usage (and infrastructure damages based on distances traveled), rather than fuel-consumption taxes (which may reflect fuel economy more than pavement damage and other costs) (Whitty 2007). Public services provided to vehicle users include policing, traffic lights, and emergency services, for which costs are estimated to run about 1–4¢ per vehicle mile in the US. VMT taxes can also be priced to approximately account for emissions or air-quality impacts, habitat loss, stormwater management, and heat-island effects (Litman 2011). Estimates of other external costs of light-duty vehicles run about 2–5¢ per VMT. In contrast, current US fuel taxes average about 2¢ per mile on a 20-mpg vehicle and 1¢ per mile on a 40-mpg hybrid electric vehicle (Litman 2011). At the other end of the spectrum, in Europe, fuel taxes run about 0.40–0.50€ per liter

(Hylen et al. 2013), or the equivalent of 9¢ per mile on a 20-mpg vehicle and 4.5¢ per mile on a 40-mpg hybrid electric vehicle.

Public and political opposition to roadway pricing policies often arise based on equity concerns (Schweitzer and Taylor 2008; Litman 2011). Because the traveler (and trip) population is heterogeneous, those with higher values of time tend to benefit the most from pricing (since their time-savings values are more likely to exceed tolls imposed). Even though the optimal price maximizes social welfare, some users will be "priced off" the toll road and take a less-preferred route or mode, or eliminate their trips altogether. Researchers have proposed various ways to increase equity in roadway pricing, including redistribution of toll revenues to transit (Eliasson and Mattson 2006) and credit-based congestion pricing (CBCP), which provides travelers with equal starting travel budgets (Kalmanje and Kockelman 2004). As budget constraints on public agencies grow, vehicle identification technologies become more cost-effective, and travelers become more familiar with flat-rate pricing, variable road pricing will become more likely, and creative welfare-enhancing policies, like CBCP, may allow for network-wide pricing applications.

Movement and Location Choices

Travel demand modelers and planners have long estimated the travel-time savings benefits of network improvements. However, travel is a derived demand, a by-product of the need to work, shop, visit with others, and access other opportunities. Other than the occasional joy ride or Formula One race, travel itself is not the desired activity. Getting from point A to point B only matters if point B is a quality destination that offers the traveler the satisfaction of a more direct demand (for labor, food, human interaction, etc.). While the quest for mobility looks solely at travel times, speeds, and distances, an accessibility objective also considers the quality and number or land-use intensity of the destinations.

Transportation infrastructure choices influence emissions by changing travel patterns and shaping land uses. Do business and household locations determine a transportation system's expansion? Or do the system's design and extent determine their location choices? Button (2010) likens the continual cycles of cause and effect between transportation and land use to the age-old "chicken or egg" dilemma. Given the longevity of transport infrastructure, system changes often have long-term effects on economic activity. Subsequent changes to residential and employment location patterns will, in turn, influence future transportation demand. Relative accessibility is the foundation for location choice theories, providing a basis for land values and wage gradients. With highway investments, the greatest land value increases tend to occur near infrastructure investments, even though relative accessibility changes occur throughout the region. Furthermore, investments in already extensive urban networks will have less pronounced impacts than those in rural areas, where access is relatively limited to begin with (Giuliano 2004). Unlike highway investment, rail transit investment generally affects accessibility only in the immediate vicinity of the rail corridor, specifically near access points, or rail stations (Cervero and Landis 1997; Giuliano 2004). When activity increases with rail station access, a corresponding increase in land values should reflect the uptick in activity (Giuliano 2004). The concept of an urban wage gradient describes the willingness of workers to accept lower-paying jobs closer to home, as long as commute cost savings cover the lost salary. Then, as is the case with land values, the highest wages should exist at the city core, with wages declining as jobs migrate outward (see, e.g., Zhang and Kockelman 2014).

The complex relationship between transportation and urban form will likely grow more complicated as regions and nations move towards an information-based economy that allows more flexibility in household and business location (Guiliano 2004). As job turnover rates increase, households tend to value access to overall job opportunities over access to any specific job. When flexible work relationships (e.g., telecommuting) are introduced, the existing location choice models become even more inadequate, given the complex relationship between transportation and land use in modern metropolitan areas. Ultimately, there is far more to the economic and other impacts of transportation policies and projects than simply travel time savings.

Methods of Economic Analysis

There are many tools for different types of economic analysis, including approaches that simply examine direct project costs and benefits, methods that anticipate wage and employment impacts, and statistical analyses that illuminate trends and decipher interactions by calibrating mathematical models for the associated data. Below is a brief overview of these different tools.

Engineering Economic Analysis Methods

Engineering decision-making often entails selecting the best option among many alternatives. Most alternatives require different levels of investment and yield distinctive benefits, so cost-benefit analysis (CBA) is regularly used to evaluate competing impacts in a relatively comprehensive fashion. CBA can be used to determine whether and when a project should be undertaken and to rank and prioritize projects. The following is a description of the CBA process (USDOT 2003b; Transport Canada 1994):

1. Identify project needs. Clearly state project needs, so that key relationships are identified and a wide range of alternatives can be examined.
2. Identify project constraints. Constraints include resource limitations and regulations governing process, and may require specific assumptions about the future (such as expected regional traffic growth).
3. Define the base case. This is also known as the "no action" (or "do nothing") case—the continued operation of the current system without any major investments or policy changes.
4. Identify alternatives. Identify project (and/or policy) alternatives, which may vary from major rehabilitation of existing facilities to new construction, full reconstruction, or replacement.
5. Define a time period. Set the analysis period over which the life-cycle costs and benefits of all of the alternatives will be measured. It should be long enough to reflect project (or policy) lifespan.
6. Define work scope. Define the level of effort for screening alternatives. A complete analysis of all options is neither achievable nor necessary. Screening alternatives allows a wide range of initial options to be considered with only a reasonable level of effort. The level of effort is typically proportional to the expense, complexity, and controversy of the project.
7. Analyze alternative effects. Analyze travel demand (and other) effects that each alternative will have on the future traffic patterns (wages, land values, etc.) to anticipate project costs and benefits.
8. Estimate benefits and costs. These estimates include investment costs, consumer welfare impacts, hours of delay, crash rates, and other effects of each alternative—relative to the base case.
9. Evaluate risk. Examine the risks associated with uncertainty in costs, traffic impacts, and other outcomes.

10. Conduct sensitivity analysis. Vary assumptions and inputs to find factors that have an important effect on outputs, or benefits and costs (and thus scores of competing alternatives).
11. Compute and compare benefit-cost ratios. Compare each project's net benefits to its net costs and rank alternatives based on their B-C ratios. These are the ratio of net discounted social benefits of any project (reflecting traveler welfare gains, emissions changes, and crash impacts, for example) to its total discounted accounting costs (reflecting construction and operations, for example). Generally, alternatives with higher B-C ratios rank higher, but one can also compare their internal rates of return (which discount net present value to zero (Kockelman et al. 2010)).
12. Make recommendations. Recommendations tend to target alternatives offering the highest B-C ratios. But net present values (NPVs), internal rates of return (IRRs), budget limitations, and other constraints (e.g., needing to spread a state's budget across multiple regions) also play a role in formulating final recommendations. For more details on carrying out engineering economic analyses, readers may wish to refer to Park (2007), Sullivan et al. (2012), or Kockelman et al. (2014).

Economic Impact Analysis

Transportation projects and policies can have significant short-term and long-term employment impacts, too. Economic impact analyses (EIAs) seek to anticipate such wage, job count, sales, and related impacts. Economic indicators such as spending by households and businesses, employment levels, wages, business sales, tax revenues, exports and imports, and capital investment expenditures can be the focus of an EIA, and are typically measured using data from local governments, counties, state agencies (such as the state controller's office), and federal agencies (such as the Department of Commerce, Bureau of Economic Analysis, and Census Bureau).

Transportation investments may affect a variety of economic indicators (via construction expenditures, traveler time savings, and materials price changes), and these impacts can be categorized as generative or redistributive. Generative impacts imply a net economic gain, while redistributive impacts essentially shift economic activity from one region or location to another, netting zero economic gains. Ideally, the economic goals of a project should be generative for as large a region as possible, but generative impacts at a local scale may result in redistributive impacts at a larger geographic scale. For instance, a region may wish to attract businesses from elsewhere via property tax incentives and transportation system improvements. In this case, the impact to the city or metro area would be generative, but the business relocations would not result in a net economic gain for the state or nation.

Comprehensive EIA can be carried out using different models, including input–output (IO) and computable general equilibrium (CGE) models. For more details on developing IO models, readers may refer to Horowitz and Planting (2009). For CGE model information, one can refer to Piermartini and Teh (2005) and Wing (2004).

Econometric Analysis

Econometric analysis is a departure from project-focused evaluation and impact analysis. It tackles the fundamentals of travel choices, land use, and other economic variables, characterizing mathematical relationships across data samples of observed values. Econometric analysis typically involves advanced statistical models to help practitioners understand a variety of relationships between distinct variables (in order to pursue more optimal policies and investments, while predicting future trends). For example, parametric estimates resulting from such models are used to predict ROW acquisition costs (Heiner and Kockelman 2005; Xiong and Kockelman 2012), mode and route choices in the presence of tolls (Arnott et al. 1994; Olszewski and Xie 2005), the impact of gas prices on VMT (Puller and Greening 1999; Bomberg and Kockelman 2005), the effects of speed limits and vehicle characteristics on crash counts and injury severities (Chen and Kockelman 2012; Kockelman et al. 2006), the impact of household characteristics and income on vehicle ownership choices (Dargay 2001; Fang 2008), and much more.

The standard econometric investigative approach is to:

1. Specify questions (or hypotheses) of interest (e.g., what is the response of businesses and households to new highway capacity?).
2. Acquire data sets of interest (e.g., sales data from a tax office, household travel survey data from a planning organization, and/or property values from an appraisal district).
3. Specify the mathematical model that makes the most sense for the data (e.g., linear equations to predict home prices using ordinary least squares regression and multinomial logit equations for mode choice probabilities).
4. Select statistical software to estimate the unknown parameters (βs) that fuse the predictive variables (Xs) in estimating the response variables (Ys). One can then forecast the response of other actors (e.g., other travelers, businesses, or regions) in new, related settings (e.g., to anticipate the demand for another highway corridor in a different city or a different part of the same region).

Econometrics is used extensively in transportation economics, and econometric tools are instrumental in anticipating travel demands, emissions, and crash counts; forecasts of project costs and benefits; other economic impacts; and behavioral responses to road pricing. For more information on the variety of econometric methods, readers may wish to refer to Greene (2008) and Gujarati (1995).

All analytical methods come with some caveats. For example, while mathematical equations are meaningful tools for understanding behavior and anticipating future outcomes, they are all abstractions of reality. Human behavior is highly variable, and many outside-the-model factors intervene (e.g., economic recessions, technological advances, and restrictive work schedules). As statistician George Box (Box and Cox 1964) is credited for saying, "essentially, all models are wrong, but some are useful." We hope the methods outlined in this short chapter on a massive subject (transport economics) are a useful starting point for those seeking to understand and/or ultimately apply such tools.

Summary

This chapter overviews many core concepts in transportation economics. They include the primary costs and benefits of transportation projects and policies, supply and demand under system pricing, transportation and land-use interactions, engineering project evaluation tools, economic impact analysis, and statistical (econometric) methods for data investigations. Comprehensive assessment of the costs and benefits of project and policy alternatives (including opportunity costs) leads to better selection, design and management of transportation services. These markets are unusual in that travel is a derived demand, and an understanding of movement patterns and location choice theories fundamentally distinguishes transportation from traditional markets. This chapter also discusses three types of evaluation and analysis tools specific to transportation's many decision-making settings. These tools can be employed to address issues ranging from budget allocation between construction and maintenance projects, to the pricing of a dynamically tolled facility, to anticipating trends in VMT in the face of rising fuel prices. Fundamentally, travel is an economic activity and good economic judgment is vital to transportation planning and engineering.

Acknowledgments

Much of the content from this chapter comes from Kockelman et al. (2014), a product emerging from a two-year Texas Department of Transportation (TxDOT) research project (#0–6628). We would like to acknowledge the contributions work of our co-authors on that reference, Dr. Kathryn Larsen and Brice Nichols, and the administrative support of Annette Perrone, without whom this work would not be possible.

References

Arnott, R., De Palma, A., and Lindsey, R. (1996) The Welfare Effects of Congestion Tolls with Heterogeneous Commuters. *Journal of Transport Economics and Policy*, 28(2), 139–161.

Bates, J., Polak, J., Jones, P., and Cook, A. (2001) The Valuation of Reliability for Personal Travel. *Transportation Research*, 191–229.

Bomberg, M. and Kockelman, K.M. (2011) Traveler Response to the 2005 Gas Price Spike, in *Household Energy: Economics, Consumption and Efficiency*. B. Mendez and J. Pena (Eds.), NOVA Science Publishers.

Bowes, D. R. and Ihlanfeldt, K. R. (2001) Identifying the Impacts of Rail Transit Stations on Residential Property Values. *Journal of Urban Economics*, 50, 1–25.

Box, G.E.P. and Cox, D.R. (1964) An Analysis of Transformations. *Journal of the Royal Statistical Society, Series B*, 26, pp. 211–252.

Boyer, K.D. (1998). *Principles of Transportation Economics*. New York: Addison Wesley Longman.

Brownstone, D. and Small, K.A. (2005) Valuing Time and Reliability: Assessing the Evidence from Road Pricing Demonstrations. Department of Economics, University of California, Irvine, CA.

Bureau of Labor Statistics (2014). *Consumer Expenditures in 2012*. Bureau of Labor Statistics. Available at: www.bls.gov/cex/csxann12.pdf

Button, K. (2010) *Transport Economics*. Northampton, MA: Edward Elgar Publishing.

Cervero, R. and Landis, J. (1997) Twenty Years of the Bay Area Rapid Transit System: Land Use and Development Impacts. *Transportation Research*, 31A(4), 309–333.

Chen, T.D. and Kockelman, K.M. (2012) Roles of Vehicle Footprint, Height, and Weight in Crash Outcomes: Application of a Heteroscedastic Ordered Probit Model. Transportation Research Record 2280, pp. 89–99.

Dargay, J.M. (2001) The Effect of Income on Car Ownership: Evidence of Asymmetry. *Transportation Research Part A: Policy and Practice*, 35(9), 807–821.

Eliasson, J. and Mattsson, L. (2006) Equity Effects of Congestion Pricing: Quantitative Methodology and a Case Study for Stockholm. *Transportation Research*, 40A, 602–620.

Energy Information Administration (2014) Monthly Energy Review, April. Available at: www.eia.gov/totalenergy/data/monthly/pdf/mer.pdf

Fang, H.A. (2008) A Discrete-Continuous Model of Households' Vehicle Choice and Usage, with an Application to the Effects of Residential Density, *Transportation Research Part B: Methodological*, 43(9), pp. 736–758.

Greene, W.H. (2008) *Econometric Analysis* (6th ed.). Upper Saddle River, NJ: Prentice Hall.

Greene, D.L. and Plotkin, S.E. (2011). *Reducing Greenhouse Gas Emissions from U.S. Transportation*. Pew Center on Climate Change. Available online at: www.c2es.org/docUploads/reducing-transportation-ghg.pdf

Guiliano, G. (2004) Land Use Impacts of Transportation Investments: Highway and Transit. Chapter 9 in *The Geography of Urban Transportation*. New York: The Guilford Press.

Gujarati, D.N. (1995) *Basic Econometrics* (3rd ed.). New York: McGraw-Hill.

Heiner, J. and Kockelman, K.M. (2005) The Costs of Right of Way Acquisition: Methods and Models for Estimation. *Journal of Transportation Engineering*, 131(3).

Horowitz, K.J. and Planting, M.A. (2009). *Concepts and Methods of the U.S. Input–Output Accounts*. Bureau of Economic Analysis. Available at: www.bea.gov/papers/pdf/IOmanual_092906.pdf

Hylen, B., Kauppila, J. and Chong, E. (2013) *Road Haulage Charges and Taxes: Summary Analysis and Data Tables 1998–2012*. International Transport Forum Discussion Paper 2013–8. Available at: www.internationaltransportforum.org/jtrc/DiscussionPapers/DP201308.pdf

Kalmanje, S. and Kockelman, K. (2004) Credit-Based Congestion Pricing: Travel, Land Value and Welfare Impacts. *Transportation Research Record*, 1864, 45–53.

Kockelman, K.M., CRA International (2006) *Safety Impacts and Other Implications of Raised Speed Limits on High-Speed Roads*. NCHRP Project 17–23, Transportation Research Board. Available at: http://onlinepubs.trb.org/onlinepubs/nchrp/nchrp_w90.pdf

Kockelman, K., Fagnant, D., and Xie, C. (2010) Documentation of Project Evaluation Toolkit and Appendix D of Final Project report.

Kockelman, K.M., Chen, T.D., Larsen, K.A., and Nichols, B.G. (2014) *The Economics of Transportation Systems: A Reference for Practitioners*. University of Texas at Austin, TX. Amazon's CreateSpace.

Lewis, C.A., Buffington, J.L., Vadali, S.R., and Goodwin, R.E. (1997) Land Value and Land Use Effects of Elevated, Depressed, and At-Grade Level Freeways in Texas. TxDOT Research Report 1327–2.

Litman, T. (2009) Transportation Cost and Benefit Analysis (2nd ed.). Victoria Transport Policy Institute. Available at: www.vtpi.org/tca/

Litman, T. (2011) Socially Optimal Transport Prices and Markets: Principles, Strategies, and Impacts. Victoria Policy Institute. Available at: www.vtpi.org/sotpm.pdf

Maibach, M., Schreter, C., Sutter, D., van Essen, H.P., Boon, B.H., Smokers, R., Schroten, A., Doll, C., Pawlowska, B., and Bak, M. (2008) *Handbook on Estimation of External Costs in the Transport Sector*, Delft, the Netherlands. Available at: http://ec.europa.eu/transport/sustainable/doc/2008_costs_handbook.pdf

Moore, T.G. (1986) U.S. Airline Deregulation: Its Effects on Passengers, Capital, and Labor. *Journal of Law and Economics*, 29(1), 1–28.

Olszewski, P. and Xie, L. (2005) Modelling the Effects of Road Pricing on Traffic in Singapore, *Transport Research Part A, Policy and Practice*, 39(7–9), 755–772.

Park, C.S. (2007) *Contemporary Engineering Economics*. Upper Saddle River, NJ: Pearson Education.

Piermartini, R. and Teh, R. (2005) *WTO Discussion Papers No. 10: Demystifying Modelling Methods for Trade Policy*. Geneva: World Trade Organization.

Puller, S.L. and Greening, L.A. (1999) Household Adjustment to Gasoline Price Change: An Analysis Using 9 Years of US Survey Data. *Energy Economics*, 21(1), 37–52.

Schweitzer, L. and Taylor, B. (2008) Just Pricing: The Distributional Effects of Congestion Pricing and Sales Taxes. *Transportation*, 35(6), 797–812.

Small, K. and Verhoef, E. (2007) *The Economics of Urban Transportation*. New York, Routledge.

Sullivan, W.G., Wicks, E.M., and Koelling, C.P. (2012) *Engineering Economy* (15th ed.). Upper Saddle River, NJ: Pearson Higher Education.

Totty, M. (2012, September 17). The Gas Tax is Running Low. But What Should Replace It? *The Wall Street Journal*. Available at: http://online.wsj.com/news/articles/SB10000872396390443864204577619082194372886

Transport Canada (1994) Guide to Benefit–Cost Analysis in Transport Canada. Available online at: www.tc.gc.ca/media/documents/corporate-services/bca.pdf

US Department of Transportation (DOT), Bureau of Transportation Statistics (BTS) (2003a) *Transportation Statistics Annual Report*. Available at: www.bts.gov/publications/transportation_statistics_annual_report/2003/html/chapter_02/economic_impacts_of_motor_vehicle_crashes.html

US Department of Transportation (DOT), Federal Highway Administration (FHWA) (2003b) Asset Management—Economic Analysis Primer. Available at: www.fhwa.dot.gov/infrastructure/asstmgmt/primer.cfm.

US Department of Transportation (DOT) (2014) *TIGER Benefit-Cost Analysis (BCA) Resource Guide*. Available at: www.dot.gov/sites/dot.gov/files/docs/TIGER_BCARG_2014.pdf

Wachs, M. (2006) *A Quiet Crisis in Transportation Finance: Options for Texas*. Testimony presented before the Texas Study Commission on Transportation Finance on April 19. Available at: www.senate.state.tx.us/75r/Senate/commit/c865/assets/c865.quietcrisis.pdf

Welter, D., Ates, M., Loftus-Otway, L., Matthews, R., and Harrison, R. (2009) Estimating Texas Motor Vehicle Operating Costs. Center for Transportation Research, University of Texas at Austin, TX. Available online at: www.utexas.edu/research/ctr/pdf_reports/0_5974_1.pdf

Whitty, J.M. (2007) Oregon's Mileage Fee Concept and Road User Fee Pilot Program: Final Report. Oregon Department of Transportation. Available at: www.oregon.gov/ODOT/HWY/RUFPP/docs/RUFPP_finalreport.pdf

Wing, I.S. (2004) *Computable General Equilibrium Models and Their Use in Economy-Wide Policy Analysis*. Massachusetts Institute of Technology, Cambridge, MA: MIT Joint Program on the Science and Policy of Global Change.

Xiong, X. and Kockelman, K. (2014) Cost of Right-of-Way Acquisition: Recognizing the Impact of Condemnation via a Switching Regression Model. *Journal of Infrastructure Systems*, 10.1061.

Zhang, W. and Kockelman, K. (2014) Urban Sprawl, Job Decentralization, and Congestion: The Welfare Effects of Congestion Tolls and Urban Growth Boundaries. Proceedings of the 93rd Annual Meeting of the Transportation Research Board (2014), and under review for publication in *Journal of Urban Economics*.

26

QUEUEING IN TRANSPORTATION SYSTEMS

Dušan Teodorović and Ranko Nedeljković

UNIVERSITY OF BELGRADE

Introduction

Queueing is a part of our everyday life. Pedestrians usually wait before crossing the street until traffic has stopped from both approaches or the street is clear. When the airport is busy, the aircraft is accommodated into a holding pattern with other aircrafts waiting to land. Queueing time for security passenger search has become very high at many airports. Vessel bunching causes significant port congestion. Trucks wait in line in ports to pick up loaded containers or drop off empty containers. When entering a tall building we press a button and we wait for an elevator. Our waiting time depends on the number of other requests for the elevator, as well as on the number of elevators. Every day thousands of vehicles are delayed for a signal. Callers call specific telephone numbers expecting that there will be an operator to answer their call. In many cases they have to wait before they can talk with an operator. Why do we wait? The answer is simple: Demand that during certain time periods exceeds capacity is what results in queues. A lot of customers (pedestrians, drivers, cars, aircraft, vessels) request various services on a daily basis (crossing the street, passing through the intersection, landing on the runway, unloading at the dock, etc.). Service demand is characterized by the customer arrival rate. Arrival rates significantly vary depending on a time of a day, day in a week, or month in a year. Service rate indicates the number of customers that could be served in a time unit. For example, the average service time for check-in is about 2–3 minutes on many domestic flights. In the case of group arrival (whole family) this service time is significantly longer. Obviously, there are fluctuations in arrival rates and service times in many queueing systems. These fluctuations create queues and decrease the level of service offered to the passengers.

Queueing theory is the mathematical analysis of queues. The roots of the queueing theory are related to the Danish engineer Agner Krarup Erlang who studied telephone traffic problems and published in 1909 the first paper on queueing theory. Queueing theory has been applied in urban and road traffic control, elevator traffic control, airport operations, air traffic control, crowd dynamics, emergency egress analysis, railway, telephone, and Internet traffic. This theory has been also extensively used to study and plan operations in various areas like bank services, supermarkets, health services, manufacturing, amusement parks, or mining.

Queueing theory helps us to evaluate levels of service and operational performance of the systems. The average waiting time a customer spends in a queue, and the average number of customers in a queue, represent the usual metrics for the level of service. Utilization of the service facility has been frequently used as a metric for the system's operational performance. Planners and designers use queueing theory techniques in various design stages of the future service facility (calculation of the number of lanes at intersection, the estimation of the length of left-turning bays, calculation of the size of the check-in area at the airport, calculation of the required number of parking spaces, etc.). In essence, the analysts use queueing theory techniques to answer the following questions: What is the operational efficiency of the observed queueing system? What is the level of service offered to the clients? Should transportation capacity be increased in response to expected demand?

Elements of Queueing Systems

There are many different queueing systems. The following are the basic characteristics of every queueing system:

a. arrival process type;
b. service process;
c. number of servers;
d. queue discipline;
e. queue capacity.

In some queueing systems arrivals are deterministic. In others, the arrivals are random. In the case of deterministic arrivals, the arrival rate is constant or varies over time. Bottleneck analysis in transportation systems is based on deterministic queueing. Service process could be characterized by deterministic service time, or by service time that is random variable. In the case of stochastic queueing, clients' arrivals and service times are described by probability distributions known as arrival and service time distributions. Queues in this case are defined by probabilistic metrics such as the expected number of entities in the system, expected waiting time per client, percentage of time when the server is busy, etc.

The number of servers equals one, for example, in the case of an airport with one runway. The number of servers equals, for example, five when five toll booths are open on the highway. In the majority of queueing systems queue discipline is FIFO (First In, First Out). The queue discipline FIFO is also frequently called FCFS (First Come, First Served). There are also the other queue disciplines like queue discipline Last In, First Out (LIFO), and Service In Random Order (SIRO).

In some queueing systems there are no special constraints related to the allowed queue capacity. In such cases, queue capacity is considered to be equal to infinity. In many other cases the queue has a certain capacity. For example, shock waves happen when left-turning vehicles are required to slow down in the through lanes, in that way causing through traffic to also slow down. Left-turn bays could significantly decrease the negative shock-effect. The lengths of the left-turn bay (queue capacity) must be appropriate to meet left-turners' storage requirements. (The length of the left-turn bay depends on many factors. The most important are traffic volumes, cycle length, number of approach lanes, and the probability of the event that the queue exceeds storage).

The following standardized format has been used to denote the main queueing system characteristics (arrival process type, service process type, the number of servers, etc.):

(A/B/C):(D/E/F)

where:

A = arrivals distribution;
B = service time distribution;
C = number of servers;
D = service discipline;
E = maximum number of clients allowed to be in the queueing system;
F = size of calling source.

Most of the time we will use only the notation A/B/C to describe arrivals distribution, service time distribution and the number of servers. The uniform, deterministic distribution of arrivals or departures is denoted by D, while the exponential distribution is denoted by M. A "general distribution" is denoted by G. For example, the notation M/M/1 denotes the queueing system that is characterized by exponentially distributed interarrival times, exponentially distributed service times and the existence of one server, while notation D/D/1 describes queueing system with deterministic arrivals, deterministic departures and one server.

All queueing systems could be graphically represented in the following way (Figure 26.1).

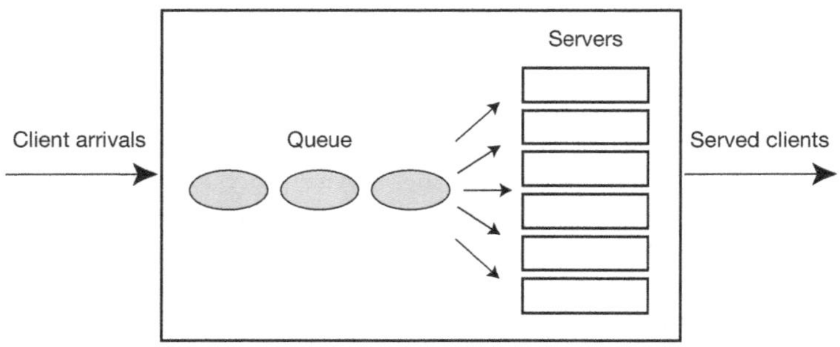

Figure 26.1 Queueing system

Example

Let us consider signalized "T" intersection (Figure 26.2). We assume that there is only one lane in both approaches.

Let us note vehicles coming from approach 1. The shown approach could be treated as a queueing system. There is only one server in the system. The server is a traffic light. The server is busy (red light for approach 1) when vehicles from approach 2 are passing through the intersection. Customers are vehicles that show up from approach 1. The service is a pass through the intersection. The service time (time necessary to pass through the intersection) could be treated as a deterministic quantity. In the majority of cases, vehicle arrivals are random. In other words, vehicles show up from approach 1 in random time points. Vehicles that request service (pass through the intersection) are immediately accepted if there is a green light assigned to approach 1 and if there is already no queue of vehicles from approach 1. If there is a queue, the newly arrived vehicle will join the queue. The queue discipline is the FIFO discipline.

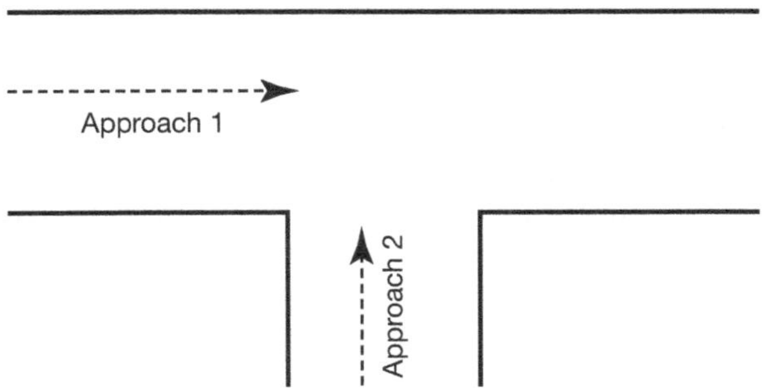

Figure 26.2 "T" intersection

D/D/1 Queueing

There is no randomness in the case of deterministic arrivals. This means that the time points of appearance of the first, second, third, etc. customer are precisely known (Figure 26.3).

The total number of customers that will show up during a certain time period is known in the case of constant arrival rate, as well as in the case of deterministic arrival rate that varies over time. Deterministic queueing is a technique that has been used, among other applications, for bottleneck analysis in transportation systems. In the case of D/D/1 queueing systems, the service process is also characterized by the deterministic service time. In the case shown in Figure 26.3, the demand rate (clients/hour) is known.

This rate increases from the beginning of the observation until 7.00 a.m. The rate is constant between 7.00 a.m. and 10.00 a.m. The demand rate decreases between 10.00 a.m. and 12.00.

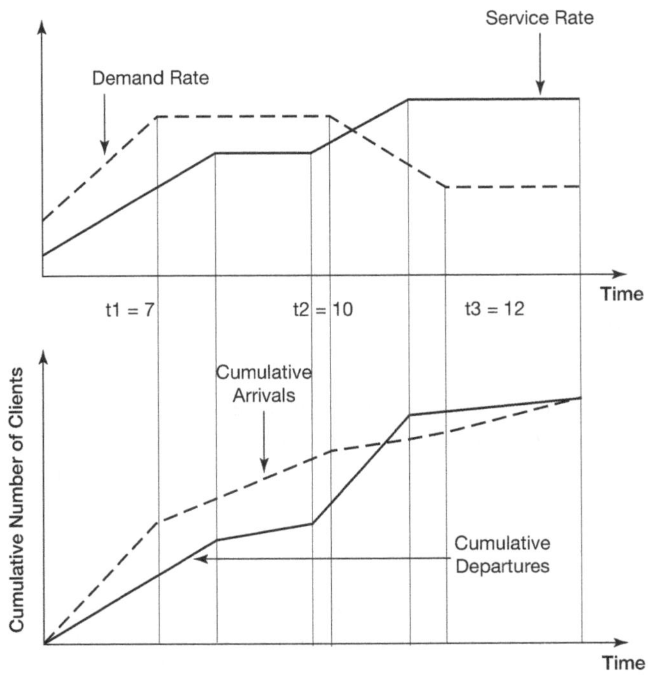

Figure 26.3 Arrival and departure rates

Finally, after 12.00 the rate is constant. The service rate (clients/hour) could be described in a similar way. It is relatively easy to generate the cumulative number of arrivals and the cumulative number of departures for known arrival and departure times. The cumulatives provide the information about the total number of arrived customers and the total number of departed customers until a certain time point. When we say that the cumulative number of arrived vehicles at the toll booth at 7.00 a.m. equals 150, it means that by 7.00 a.m., a total of 150 vehicles have arrived (Figure 26.3).

The basic queueing concepts can be easily understood after studying the simple D/D/1 queueing system. Deterministic queueing is analogous to a continuous flow of entities passing a point over time. We use continuous lines to represent cumulatives in the D/D/1 queueing systems. It has been shown that these lines are a very good approximation for the cumulative stepped lines that represent the cumulative number of arrivals.

Example

Work zones generate a very high non-recurring delay. Every year thousands of work zones cause significant loss in capacity, poor traffic flow and many frustrations to drivers. One should also take into account the delay cost when estimating the total cost of future work zones. The working hours in a work zone start at 7.00 a.m. The cumulative arrivals and the cumulative departures are shown in Figure 26.4.

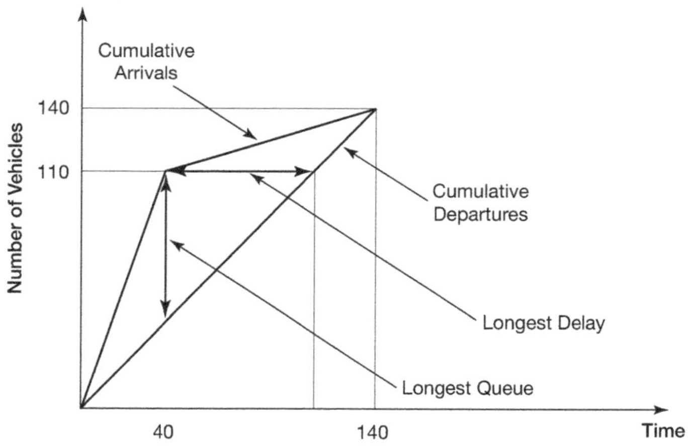

Figure 26.4 Client arrivals modeling: Arrivals are deterministic

The queue length equals the ordinate distance between the cumulative arrivals and cumulative departures. The longest queue happens 40 minutes after 7.00 a.m. Customers entering the system 40 minutes after 7.00 a.m. suffer the longest delay. The waiting time (delay) represents the horizontal distance between the two cumulative curves.

Example

Vehicles travel through the work zone. The time period whose length equals 100 minutes was analyzed. After analyzing the statistical data, it has been concluded that both arrival and service rates were deterministic. Service rate was constant. Arrival rate changes over time. These rates (vehicle/min) are given by the following functions:

$$\lambda(t) = -\frac{1}{5}t + 20$$

$$\mu = 10$$

where t is in minutes after beginning the observation of the queueing system. Cumulative number of arrivals $A(t)$ and cumulative number of departures $D(t)$ are respectively equal:

$$A(t) = \int_0^t \lambda(t)\mathrm{d}t = \int_0^t \left(-\frac{1}{5}t + 20\right)\mathrm{d}t$$

$$= \left(-\frac{1}{5}\frac{t^2}{2} + 20t\right)\Bigg|_0^t = -\frac{1}{10}t^2 + 20t$$

$$D(t) = \int_0^t \mu(t)\mathrm{d}t = \int_0^t 10\mathrm{d}t = 10t$$

The cumulative number of arrivals and the cumulative number of departures are shown in Figure 26.5.

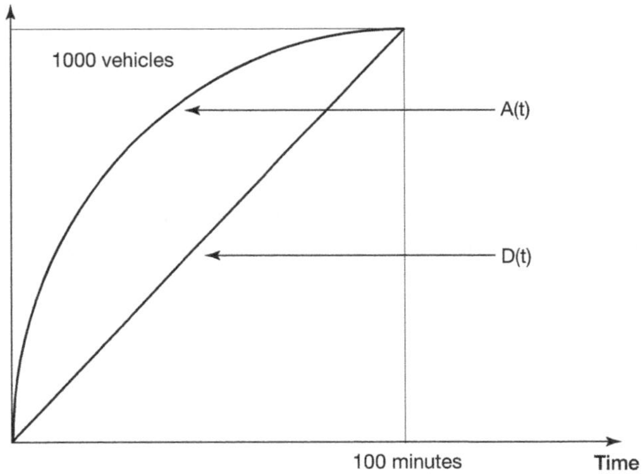

Figure 26.5 Cumulative arrivals and cumulative departures

The queue will dissipate in the time point when the total number of cumulative arrivals equals the total number of cumulative departures, i.e.:

$$A(t) = D(t)$$

After substitution, we get the following equation:

$$-\frac{1}{10}t^2 + 20t = 10t$$

$$-\frac{1}{10}t^2 + 10t = 0$$

The solutions of the equation are $t_1 = 0_t$ and $t_2 = 100$. The cumulative number of arrivals $A(t)$ equals the cumulative number of departures $D(t)$ for the first time at the beginning of our observation ($t = 0$). These cumulatives are equal for the second time when the queue dissipates. We conclude that the queue will dissipate after 100 minutes.

The total number of arrivals during 100 minutes equals:

$$A(100) = D(100) - 1{,}000$$

The total delay D of all vehicles is represented by the area between cumulative number of arrivals and the cumulative number of departures. In other words:

$$D = \int_0^{100} A(t)\mathrm{d}t - \int_0^{100} D(t)\mathrm{d}t$$

$$D = \int_0^{100} \left(-\frac{1}{10}t^2 + 20t\right)\mathrm{d}t - \int_0^{100} 10t\mathrm{d}t$$

$$D = \left(-\frac{1}{10}\frac{t^3}{3} + 20\frac{t^2}{2} - 10\frac{t^2}{2}\right)\Bigg|_0^{100}$$

$$D = 16{,}700$$

The average delay d per one vehicle equals:

$$d = \frac{16{,}700}{1{,}000} = 16.7$$

Because of the work zone, average delay per vehicle equals 16.7 minutes. Queue length $L(t)$ in any moment t represents the difference between cumulative number of arrivals $A(t)$ by moment t, and cumulative number of departures $D(t)$ by moment t:

$$L(t) = A(t) - D(t)$$

$$L(t) = -\frac{1}{10}t^2 + 20t - 10t$$

$$= -\frac{1}{10}t^2 + 10t$$

We determine the maximum queue length as follows:

$$\frac{\mathrm{d}\left(L(t)\right)}{\mathrm{d}t} = 0$$

$$\frac{\mathrm{d}\left(-\frac{1}{10}t^2 + 10t\right)}{\mathrm{d}t} = 0$$

$$-\frac{t}{5} + 10 = 0$$

$$t = 50$$

We conclude that the maximum queue length happens 50 minutes after the beginning of observation. The maximal queue length equals:

$$L_{\max} = -\frac{1}{10}50^2 + 10 \cdot 50$$
$$L_{\max} = 250 \text{ vehicles.}$$

Example
(by courtesy of Professor Antonio Trani)

In the planning program for renovating an airport terminal facility it is important to estimate the requirements for the ground access area. It has been estimated that an hourly capacity of 1,500 passengers can be adequately served with the existing facilities at a medium-size regional airport. Due to future expansion plans for the terminal, one-third of the ground service area will be closed for 2 hours in order to perform inspection checks to ensure expansion compatibility. A recent passenger count reveals an arrival rate as shown below

$$\lambda = \begin{cases} 500, & t > 1 \\ 1{,}500, & 0 \le t \le 1 \end{cases}$$

where t is in hours after the beginning of the observation.

Estimate the following parameters:

- The total waiting time of all passengers.
- The average waiting time per passenger.
- The waiting time of a passenger arriving 30 minutes after terminal closure.

Let us first analyze the cumulative number of arrivals $A(t)$ and the cumulative number of passengers served $D(t)$. The cumulative number of arrivals $A(t)$ during the first hour equals $(0 \le t \le 1)$:

$$A(t) = \int_0^t \lambda(t)\mathrm{d}t = \int_0^t 1{,}500 \cdot \mathrm{d}t = 1{,}500 \cdot t$$

The total number of arrivals during the first hour equals:

$$A(1) = 1{,}500 \cdot 1 = 1{,}500$$

The cumulative number of arrivals $A(t)$ during the second hour $(1 \le t \le 2)$ equals:

$$A(t) = \int_0^1 \lambda(t)\mathrm{d}t + \int_1^t \lambda(t)\mathrm{d}t$$
$$= \int_0^1 1{,}500 \cdot \mathrm{d}t + \int_1^t 500 \cdot \mathrm{d}t$$
$$A(t) = 1{,}500 + 500 \cdot (t-1)$$
$$= 500 \cdot t + 1{,}000$$

Regular hourly capacity is 1,500 passengers. Since one-third of the ground service area will be closed, this means that the new hourly capacity equals 1,000 passengers (2/3 of the regular capacity). The service rate μ equals:

$$\mu = 1{,}000 \left(\frac{\text{passengers}}{\text{hour}}\right)$$

The cumulative number of passengers served $D(t)$ is:

$$D(t) = \int_0^t 1{,}000 \, \mathrm{d}t = 1{,}000 \cdot t$$

Figure 26.6 shows passengers in and passengers served at the airport terminal facility.

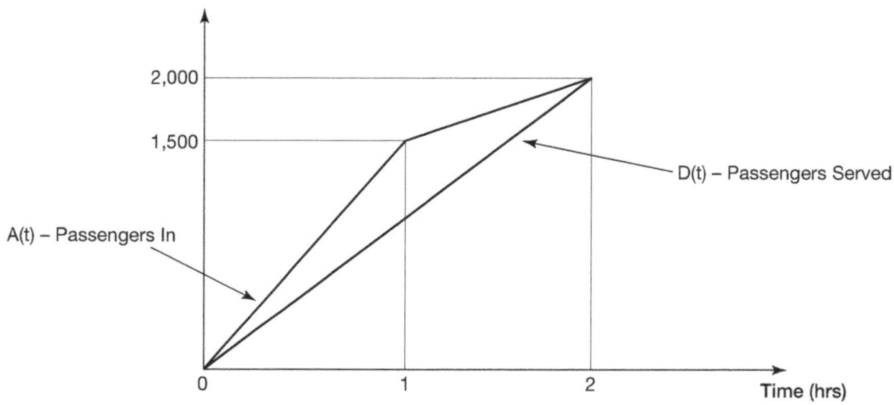

Figure 26.6 Passengers in and passengers served at the airport terminal facility

The total delay D represents the area between two cumulative curves, i.e.:

$$D = \int_0^2 A(t)\mathrm{d}t - \int_0^2 D(t)\mathrm{d}t$$

$$D = 2 \cdot \frac{1}{2} \cdot (1{,}500 - 1{,}000) \cdot \; = 500\left(\frac{\text{passengers}}{\text{hour}}\right)$$

The total number of passengers arrived N equals:

$$N = \int_0^1 \lambda(t)\mathrm{d}t + \int_1^2 \lambda(t)\mathrm{d}t$$
$$= \int_0^1 1{,}500 \cdot \mathrm{d}t + \int_1^2 500 \cdot \mathrm{d}t$$
$$= 2{,}000$$

The average delay per passenger W equals:

$$W = \frac{D}{N} = \frac{500}{2{,}000} = \frac{1}{4}$$

The average delay per passenger delayed during the queueing incident equals 1/4 hours = 15 minutes. Now, we can find the delay for a passenger entering the terminal 30 minutes after the partial terminal closure occurs. Note that at t = 30 minutes = 0.5 hours, 750 passengers have entered the terminal before the passenger in question. Thus, we need to find the time when the supply function, $D(t)$, achieves a value of 750 so that the passenger "gets serviced". This occurs at

$$D(t+\Delta t) = A(t) = 750$$

therefore, Δt is just 15 minutes (the passenger actually leaves the terminal at a time equal to 3/4 hours). This can be shown in the following diagram.

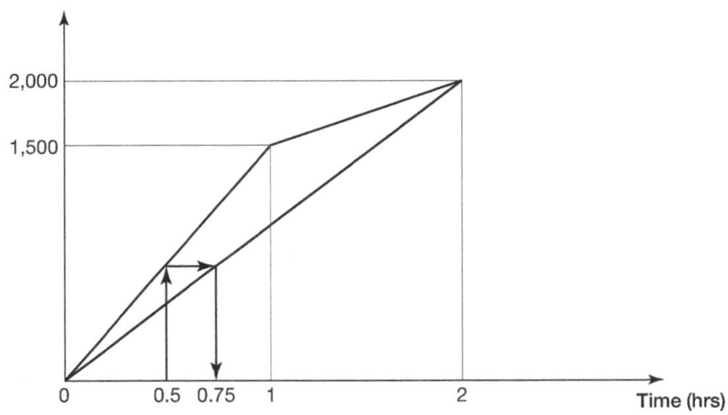

Figure 26.7 Waiting time of a passenger arriving 30 minutes after the terminal closure

M/M/1 Queueing

Many queueing systems in transportation are M/M/1 queueing systems. The examples could be one open toll booth at the highway, one bridge toll booth, one open check-in counter at the airport, vehicle inspection facility with one repairman working, etc. The M/M/1/ queueing system has the following characteristics: a) Poisson arrivals (exponential interarrival times); b) exponential service times; c) one server; d) FIFO queue discipline.

Let us introduce the following notation:

λ: mean arrival rate
μ: mean service rate

The following relations describe M/M/1 queueing.

The probability of having no customers in the queueing system equals:

$$p_0 = 1 - \frac{\lambda}{\mu}$$

The probability of having n customers in the queueing system equals:

$$p_n = \left(\frac{\lambda}{\mu}\right)^n \cdot p_0$$

The average number of customers in the queue:

$$L_q = \frac{\lambda^2}{\mu \cdot (\mu - \lambda)}$$

The average number of customers in the queueing system:

$$L = L_q + \frac{\lambda}{\mu} = \frac{\lambda^2}{\mu \cdot (\mu - \lambda)} + \frac{\lambda}{\mu}$$
$$= \frac{\lambda}{\mu - \lambda}$$

The average waiting time a customer spends in the queue:

$$W_q = \frac{L_q}{\lambda} = \frac{\lambda^2 / \left(\mu \cdot (\mu - \lambda)\right)}{\lambda} = \frac{\lambda}{\mu \cdot (\mu - \lambda)}$$

The average waiting time a customer spends in the queueing system:

$$W = W_q + \frac{1}{\mu} = \frac{\lambda}{\mu \cdot (\mu - \lambda)} + \frac{1}{\mu}$$
$$= \frac{1}{\mu - \lambda}$$

Example

Consider the bridge shown in Figure 26.8. Drivers pay in cash when entering the bridge. Service time varies depending on the availability of the cash and coins, but on average, the bridge attendant needs 20 seconds to collect the money. Average vehicles arrival rate equals 150 [veh/hr].

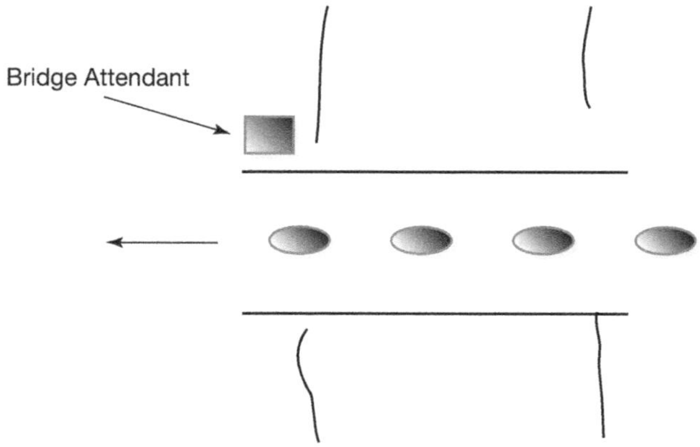

Figure 26.8 Bridge as M/M/1 queueing system

Treat the bridge as M/M/1 queueing system and calculate the following:

- The average number of vehicles in the queue.
- The average number of clients in the queueing system.
- The average waiting time a client spends in the queue.
- The average waiting time a client spends in the queueing system.

The average arrival rate λ equals 150 [veh/h], i.e.

$$\lambda = \frac{150 \text{ vehicles}}{3{,}600 \text{ sec}} = 0.0417\left(\frac{\text{veh}}{\text{sec}}\right).$$

The average service rate equals:

$$\mu = \frac{1}{20} = 0.05\left(\frac{\text{veh}}{\text{sec}}\right).$$

The average number of clients in the queue equals:

$$L_q = \frac{\lambda^2}{\mu \cdot (\mu - \lambda)} = \frac{0.0417^2}{0.05 \cdot (0.05 - 0.0417)} = 4.19 \text{ vehicles}$$

The average number of clients in the queueing system equals:

$$L_q = \frac{\lambda}{\mu - \lambda}$$

The average waiting time a client spends in the queue equals:

$$W_q = \frac{\lambda}{\mu \cdot (\mu - \lambda)} = \frac{0.0417}{0.05 \cdot (0.05 - 0.0417)} = 100.5 \text{ sec}$$

The average waiting time a client spends in the queueing system equals:

$$W = \frac{1}{(\mu - \lambda)} = \frac{1}{0.05 - 0.0417} = 120.5 \text{ sec}$$

M/M/*s* Queueing

In this type of queueing system both arrivals and departures happen according to Poisson distribution. The total number of servers is equal to s. This means that maximum s customers can be served at once. Obviously, the system's service rate is much higher than in the case of one server. The examples of the M/M/s queueing system are: a parking lot, where each parking place represents one server; airport operations in the case of multiple runways, or a few toll booths on the highway. Utilization factor ρ of the facilities is defined in the following way:

$$\rho = \frac{\lambda}{s \cdot \mu}$$

The following relations describe M/M/s queueing:

$$p_0 = 1 \Big/ \sum_{k=0}^{s-1} \left(\frac{(\lambda/\mu)^k}{k!} + \frac{(\lambda/\mu)^s}{s!} \cdot \frac{1}{1-(\lambda/(s \cdot \mu))} \right)$$

$$p_k = \frac{(\lambda/\mu)^k}{k!} \cdot p_0 \quad \text{for } 0 \le k \le s$$

$$p_k = \frac{(\lambda/\mu)^k}{s! s^{k-s}} \cdot p_0 \quad \text{for } k > s$$

$$L_q = \frac{p_0 \cdot (\lambda/\mu)^s \cdot \rho}{s! \cdot (1-\rho)^2}$$

$$W_q = \frac{L_q}{\lambda}$$

$$L = \lambda \cdot W = \lambda \cdot \left(W_q + \frac{1}{\mu} \right)$$

$$= L_q + \frac{\lambda}{\mu}$$

Example

The airport terminal shown in the next figure has two security checkpoints for all passengers boarding aircraft. Each security checkpoint has two X-ray machines. A survey reveals that on average a passenger takes 45 seconds to go through the system (exponential distribution service time).

The arrival rate is known to be random (this equates to a Poisson distribution) with a mean arrival rate of one passenger every 25 seconds. In the design year the demand for services is expected to grow by 60 percent compared to the present one.

In order to properly plan further airport development, transportation engineers and planners are looking for answers to the following questions:

a. What is the current utilization of the queueing system (i.e., two X-ray machines)?
b. What should be the number of X-ray machines for the design year of this terminal if the maximum tolerable waiting time in the queue is 2 minutes?
c. What is the expected number of passengers at the checkpoint area on a typical day in the design year?
d. What is the new utilization of the future facility?
e. What is the probability that more than four passengers wait for service in the design year?

Security Checkpoint Solutions

a) Utilization of the facility: Note that this is a multiple server case with infinite source.

The mean arrival rate equals one passenger on every 25 seconds, i.e.

$$\lambda = \frac{1}{25/3{,}600}\left(\frac{\text{pass.}}{\text{hour}}\right) = 144\left(\frac{\text{pass.}}{\text{hour}}\right)$$

On average it takes 45 seconds for a passenger to go through the system. The service rate equals:

$$\mu = \frac{1}{45/3{,}600}\left(\frac{\text{pass.}}{\text{hour}}\right) = 80\left(\frac{\text{pass.}}{\text{hour}}\right)$$

The utilization of the facility equals:

$$\rho = \frac{\lambda}{s \cdot \mu}$$

$$\rho = \frac{144}{2 \cdot 80} = 0.90$$

Other queueing parameters for a multi-server queueing system with infinite population are:
Idle probability:

$$p_0 = 1 \Big/ \sum_{k=0}^{s-1}\left(\frac{(\lambda/\mu)^k}{k!} + \frac{(\lambda/\mu)^s}{s!} \cdot \frac{1}{1-\left(\lambda/(s \cdot \mu)\right)}\right)$$

$$p_0 = 1 \Big/ \sum_{k=0}^{2-1}\left(\frac{(144/80)^k}{k!} + \frac{(144/80)^s}{2!} \cdot \frac{1}{1-\left(144/(2 \cdot 80)\right)}\right)$$

$$p_0 = 0.052632$$

Expected number of clients in the queue:

$$L_q = \frac{p_0 \cdot (\lambda/\mu)^s \cdot \rho}{s! \cdot (1-\rho)^2}$$

$$L_q = \frac{0.052632 \cdot (144/80)^2 \cdot 0.90}{2! \cdot (1-0.90)^2}$$

$$L_q = 7.6737$$

Expected number of customers in the system:

$$L = L_q + \frac{\lambda}{\mu}$$

$$L = 7.6737 + \frac{144}{80}$$

$$L = 9.4737$$

Average waiting time in the queue:

$$W_q + \frac{L_q}{\lambda}$$

$$W_q = \frac{7.6737}{140} = 0.055 \text{ (hr)} = 197 \text{ (sec)}$$

Average waiting time in the system:

$$W = W_q + \frac{1}{\mu}$$

$$W = 197 + 45 = 242 \text{ (sec)}$$

b) The solution to this part is done by trial and error. As a first trial, let us assume that the number of X-ray machines is 3 ($s = 3$).

In the design year, the demand for services is expected to grow by 60 percent compared to that today. The average arrival rate in the design year will be:

$$\lambda = 144 \cdot 1.6 = 230 \left(\frac{\text{pass.}}{\text{hour}}\right)$$

Finding p_0 for the design year:

$p_0 = 0.0097$, or less than 1 percent of the time the facility is idle.

Finding the average waiting time in the queue:

$W_q = 332$ (sec)

Since this waiting time violates the desired 2-minute maximum, it is suggested that we try a higher number of X-ray machines to expedite service (at the expense of cost). The following results show that four X-ray machines are needed to satisfy the 2-minute operational design constraint:

$$p_0 = 0.045$$

$$L_q = 1.16$$

$$W_q = 18 \text{ (sec)}$$

c) The expected number of passengers in the system is (with $s = 4$):

$L = 4.04$ passengers in the system on the average design year day

d) The utilization of the improved facility (i.e., four X-ray machines) is:

$$\rho = \frac{\lambda}{s \cdot \mu}$$

$$\rho = \frac{230}{4 \cdot 80} = 0.72$$

e) The probability that more than four passengers wait for service is just the probability that more than eight passengers are in the queueing system, since four are being served and more than four wait.

$$P(n > 8) = 1 - \sum_{k=0}^{8} p_k$$

where

$$p_k = \frac{(\lambda/\mu)^k}{k!} \cdot p_0 \quad \text{for } 0 \le k \le s$$

$$p_k = \frac{(\lambda/\mu)^k}{s! \cdot s^{k-s}} \cdot p_0 \quad \text{for } k > s$$

from where, $P(n > 8) = 0.0879$.

Note that this probability is low and therefore the facility seems properly designed to handle the majority of the expected traffic within the 2-minute waiting-time constraint.

Little's Law

Little's Law is the most important result of the Queueing Theory. This law is valid for *any* queueing system that is in stable conditions (stable conditions do not assume, for example, the beginning of operations in the system). In case of stochastic queueing, client arrivals and service times are described by probability distributions known as arrival and service-time distributions. The relationship described by Little's Law requires no assumptions about probability distributions of the interarrival and service times. Some queueing systems are composed of queueing subsystems. Little's Law is valid to queueing subsystems, as well as for a whole system.

The following is an interpretation of the Little's Law. Assume that N customers arrive in the queueing system during the time interval $(0, T)$. Figure 26.9 shows cumulative arrivals and cumulative departures.

The queue length in any time point corresponds to the maximum ordinate distance between the cumulative departure and cumulative arrival curves. For example, let us assume that the queueing system started with operations at 7 a.m. The total number of arrived customers by 9 a.m. equals 1,000. The total number of served customers by 9 a.m. equals 800. Obviously, the queue length at 9 a.m. equals 200. It is also very easy to "read" from the figure, by visual

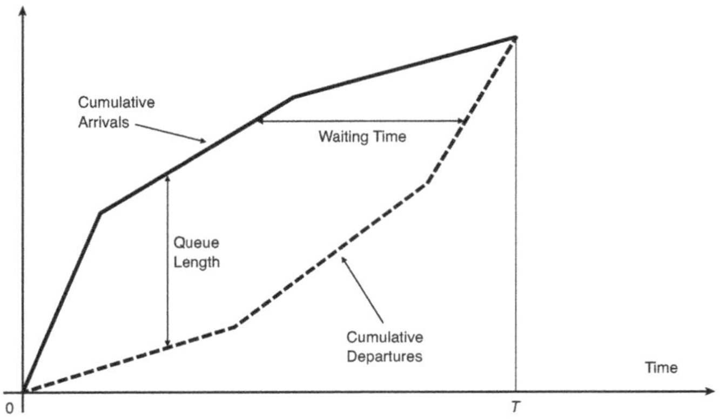

Figure 26.9 Graphical illustration of the queue length and waiting time

inspection, the waiting time of every customer. The area shown in Figure 26.9 represents the total waiting time (the total delay) of all customers. The average delay time W is the quotient of the total delay and the number of clients, i.e.:

$$W = \frac{\text{Area}}{N}$$

The area also represents the total length of all queues in all time points. The average queue length L equals:

$$L = \frac{\text{Area}}{T}$$

We conclude the following:

$$W \cdot N = L \cdot T$$

i.e.:

$$L = \frac{N}{T} \cdot W$$

The ratio N/T actually represents the arrival rate λ.

Finally, we can write:

$$L = \lambda \cdot W$$

The relation above has been known as Little's Law. This law offers the following conclusion: The average number of clients in a queueing system (over some time interval) is equal to their average arrival rate multiplied by their average time in the system. In the same manner, we can also write the following:

$$L_q = \lambda \cdot W_q$$

$$W = W_q + \frac{1}{\mu}$$

Traffic Congestion and Delays

Due to the expansion of metropolitan areas, increase in population, increase in the number of licensed drivers, the choices people make about where they live, where they work and how they travel to work, the number of trips by private cars has significantly increased in recent decades in many cities, and on many highways. At the same time, road network capacities have not kept up with this increase in travel demand. Urban road networks in many countries are severely congested, resulting in increased travel times, increased number of stops, unexpected delays, greater travel costs, inconvenience to drivers and passengers, increased air pollution and noise level, and increased number of traffic accidents. Government estimates show that people living in the US spend a few billion hours per year in traffic.

To alleviate traffic congestion and its associated costs, investments have been made by governments and transportation organizations. Expanding traffic network capacities by building more roads is extremely costly as well as environmentally damaging. Since traffic flows continue to outpace the current resources available to improve the infrastructure and facilities, researchers and policy makers have begun to consider strategies that focus on demand management. One of the most important demand management strategies is congestion pricing. Congestion pricing introduces different fees for road usage. The different fees or tolls may vary with the location in the network, the time of a day and/or the level of traffic congestion. Drivers should pay for using a specific road, corridor, bridge, or for entering a particular area during some high-volume time period. The other demand management strategies are: High Occupancy Vehicle (HOV) and HOT (High Occupancy Toll) lanes; carpooling (ride sharing), which assumes the participation of a few drivers who all together share their privately owned vehicles when traveling from a few places of origin to a few destinations), and administrative measures such as not allowing vehicles to enter certain areas.

Demand-management projects that are well planned and successfully implemented could also result in significant toll revenues, a decreased total number of vehicle trips, decreased total number of vehicle trips during peak periods, increased number of vehicle trips during off-peak periods, increase in ridesharing, greater number of passengers in public transit, and in some cases increased cycling and walking.

A similar situation occurs in air transportation. Airspace systems in many countries are overloaded. The number of passengers in air transportation and the total number of flights have significantly increased in recent decades. At the same time, airport capacities have not kept up with this increase in demand. This situation results in canceled flights, delayed flights (departure delays and aircraft delays in queues before landing), greater airline and airport operating costs, inconvenience to passengers, decrease in the quality of air traffic service and potential decrease in airspace safety. Every day, hundreds of flights in the USA are delayed for at least 15 minutes. There are estimates that the current ground delays in the USA are more than 2,000 hours/day. Yearly airline delay cost (primarily airline fuel and crew operating costs) totals several billion dollars. Since all aircraft fly several legs every day, in many cases flights are delayed because the arriving flight is delayed. Because of the "hub-and-spoke" type of most airlines' network, the delay of one flight could cause the delay of many different flights. Approximately one-fifth of all European flights are delayed, which on average equals 21 minutes. In years to come, the demand for airport use will continue to increase. In the USA, airport capacity is considered the main problem, while in Europe, en route capacity is the critical point in air traffic control.

Queueing Theory and Congestion Pricing

Congestion pricing already exists in many different sectors. Airlines have been offered off-peak discounts for many years. Traditionally, in many countries, hotel rooms cost more during peak tourist seasons. Telephone companies have been applying congestion pricing concepts for many years. In some countries, road-use charges vary with the level of congestion (Singapore, USA, Europe). In this way, it is possible to shift some trips to off-peak times, less-congested routes, or alternative modes of transportation.

The expectation is that the congestion pricing concept, once fully applied, could decrease the number of delayed trips, mitigate the frustration of drivers and passengers, decrease queues at intersections and airports, and decrease the average trip time, energy consumption, and air pollution. Congestion pricing practically represents basic economic concepts of supply and demand. The basic idea of congestion pricing in transportation is to introduce peak-period pricing and other possible strategies as a powerful tool that will be capable of modifing users' use of existing roads, intersections, streets, airways and airports, and changing drivers' and passengers' behavior. In other words, the main role of any congestion pricing would be in shifting some users to alternative routes, alternative departure times, alternative airports or alternative time slots. The basic goal of congestion pricing (to distribute transportation demand more evenly over time and over space) is graphically illustrated in Figure 26.10.

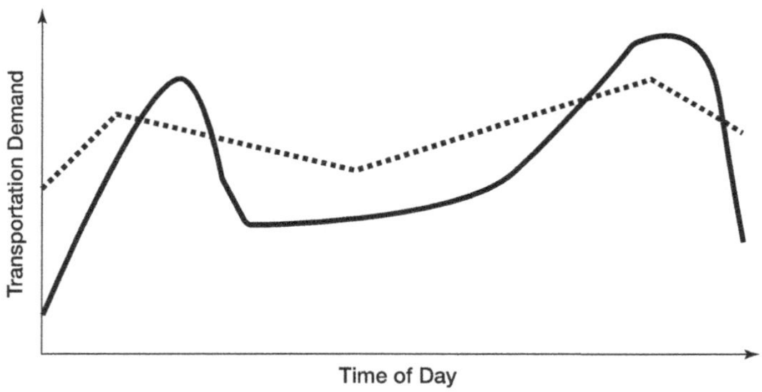

Figure 26.10 More evenly distributed passenger demand

It is known that drivers and passengers perceive only their own travel time costs. In the same way, bus operators and airlines perceive primarily their own operating costs. They do not take into account the additional delay their trips impose on other passengers. Noticing these facts, Vickrey (1955, 1963, 1969) claimed that "Charges should reflect as closely as possible the marginal social cost of each trip in terms of the impacts on others. There is no excuse for charges below marginal social cost." In other words, it seems very logical that users should pay a price equivalent to the delay cost they impose on others. In this way, travel patterns could probably be changed. The likely consequence could be efficiency improvement of transportation capacity utilization, and probably the reduction of future transportation capacity requirements. It is fair to say that there are many opponents of the congestion pricing idea. It seems that there are no easy solutions to traffic congestion problems. Implementation of any political and/or economic traffic congestion easing strategies requires the previous development of the appropriate models that can help us to properly understand the complex problems of traffic congestion.

Let us denote by c the delay cost per unit time per client. The total cost of delay per unit time equals:

$$C = c \cdot L_q$$

where L_q is the average number of clients in the queue.

We conclude, based on the Little's Law, that the total cost of delay per unit time equals:

$$C = c \cdot \lambda \cdot W_q$$

where λ is the average arrival rate and W_q the average waiting time a client spends in the queue.

The more vehicles there are on a highway, the higher is the congestion level and the higher are vehicle delay costs. The more aircraft there are in the air, the higher is the aircraft waiting time for landing, and the higher are aircraft operating costs. The marginal cost indicates the additional increase in queueing cost with every additional customer. The marginal delay cost MC imposed by an additional client equals:

$$MC = \frac{\mathrm{d}C}{\mathrm{d}\lambda}$$

i.e.:

$$MC = c \cdot W_q + c \cdot \lambda \cdot \frac{\mathrm{d}W_q}{\mathrm{d}\lambda}$$

The last equation indicates that the marginal delay cost MC has two terms. The term $c \cdot W_q$ represents the internal cost experienced by the additional customer. The second term $c \cdot \lambda \cdot (\mathrm{d}W_q/\mathrm{d}\lambda)$ represents the "external" cost imposed. The majority of customers in queueing systems in transportation do not realize the existence of these "external" costs. In other words, when we enter a congested highway, we are exposed to significant delay and an increase in travel cost, but at the same time we must realize that, by entering the congested highway, we further increase the delay costs of all other drivers and passengers. Aircraft trying to land during high airport congestion are sent to the holding pattern. In this way, their operating costs are increased. At the same time, these aircraft also increase the operating costs of all the other aircraft that are trying to land.

The marginal delay cost in the case of M/M/1 queueing system equals:

$$MC = c \cdot W_q + c \cdot \lambda \cdot \frac{\mathrm{d}W_q}{\mathrm{d}\lambda}$$

In the case of M/M/1 queueing system, the average waiting time W_q equals:

$$W_q = \frac{\lambda}{\mu \cdot (\mu - \lambda)}$$

The marginal delay cost MC equals:

$$MC = c \cdot \frac{\lambda}{\mu \cdot (\mu - \lambda)} + c \cdot \lambda \cdot \frac{\mathrm{d}\left(\lambda / \left(\mu \cdot (\mu - \lambda)\right)\right)}{\mathrm{d}\lambda}$$

$$MC = \frac{c \cdot \lambda \cdot (2 \cdot \mu - \lambda)}{\mu \cdot (\mu - \lambda)^2}$$

Example

There are two alternative routes between point A and point B. The work zone is located on the shorter route (Figure 26.11). Because of the work zone, this route is characterized by the high congestion level.

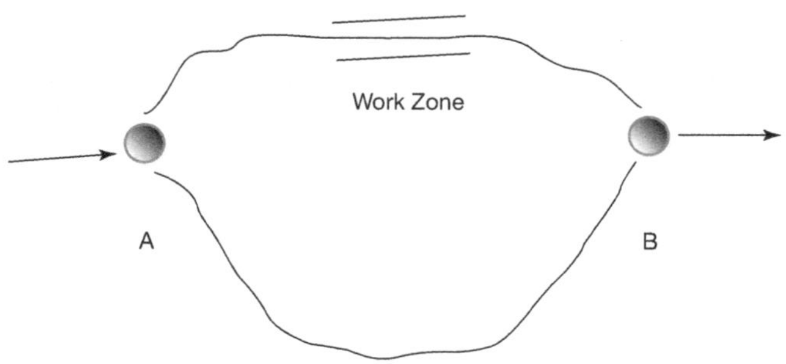

Figure 26.11 Congestion in the work zone neighborhood

By analyzing traffic conditions, traffic engineers concluded that the route with the work zone could be treated as the M/M/1 queueing system. The average arrival rate equals 150 [veh/h], i.e.

$$\lambda = \frac{150 \text{ vehicles}}{3{,}600 \text{ sec}} = 0.0417 \left(\frac{\text{veh}}{\text{sec}}\right).$$

The average service rate equals

$$\mu = \frac{1}{20} = 0.05 \left(\frac{\text{veh}}{\text{sec}}\right).$$

The traffic authorities and the local government can introduce congestion pricing in order to decrease congestion level. Calculate the congestion price (marginal delay cost) that should be charged to drivers who use the congested route.

The marginal delay cost MC imposed by an additional vehicle equals:

$$\begin{aligned} MC &= \frac{c \cdot \lambda \cdot (2 \cdot \mu - \lambda)}{\mu \cdot (\mu - \lambda)^2} \\ &= \frac{0.0417 \cdot (2 \cdot 0.05 - 0.0417)}{0.05 \cdot (0.05 - 0.0417)^2} \cdot c \\ &= 706 \cdot c \\ MC &= 706 \cdot c \end{aligned}$$

Table 26.1 shows the congestion price MC (\$) (marginal delay cost) as a function of the delay cost per unit time per driver c.

Table 26.1 Congestion price *MC* ($) (marginal delay cost) as a function of the delay cost per unit time per driver *c*

c ($/hr)	*c ($/sec)*	*MC ($)*
18	0.005	3.53
15	0.00417	2.942
12	0.00333	2.353
9	0.0025	1.765

Queueing Theory and Capital Investments in Transportation

By using queueing theory techniques we try to measure operational efficiency of the observed queueing system, as well as the level of service offered to customers. At the same time, we try to understand the queueing system operations, and to increase transportation capacity in response to demand expected in the future.

By varying the number of servers in the system, we explore the sensitivity of the service facility idle time, the sensitivity of the queue length, and the average waiting time as the number of servers increases. In many cases, we can significantly improve the queueing system operations by adding more servers (more through-lanes on a highway, a new airport runway, expanding the dock in a harbor). Expanding traffic network capacities is extremely costly, as well as environmentally damaging. Frequently, citizens are not ready to accept new projects that increase capacity. Transportation capacity-increase projects are also highly correlated with complex land-use policy issues. To mitigate traffic congestion, traffic engineers, planners, and authorities should combine the expansion of existing facilities, and the construction of new transportation facilities with the use of various demand management strategies (congestion pricing, HOV lanes) and various advanced technologies (Intelligent Transportation Systems). Queueing theory techniques should always be used in the analysis of the potential transportation capacity increase.

Imagine that we are at the stage of a new building design. We have to make a decision about the number of elevators in the building (Figure 26.12).

It is intuitively clear that the queue length during the rush hour will decrease with the increase in the number of elevators. How many elevators do we need in the building? Using queueing theory techniques, we can calculate average queue length during the rush hour, average waiting time per client, percentage of elevator idle time, etc. Figure 26.13 shows the expected user waiting cost, and the total elevator construction and maintenance cost.

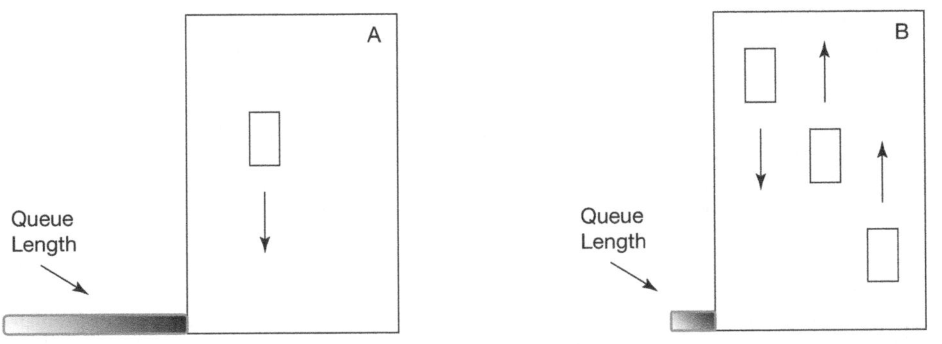

Figure 26.12 Queue length as a function of the number of elevators

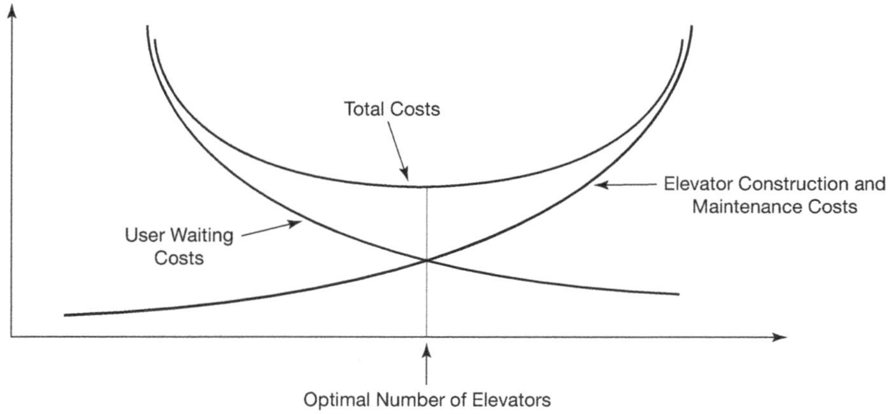

Figure 26.13 User waiting cost, construction cost, and the total cost in the queueing system

The higher the number of elevators, the lower the user waiting cost, and the higher the elevator construction and maintenance cost. This is also true for any other transportation facility (highway, airport, railway station, port). The curves shown in Figure 26.13 characterize operations of all transportation facilities. A high level of service (short waiting times, short queue lengths) is expensive, but customer waiting costs are very low, and vice versa. It is clear that the "optimal" number of servers represents a compromise between queue lengths, user waiting times and construction and maintenance costs.

Queueing theory enables us to perform detailed analysis of the queueing phenomenon, and to explore the trade-off between the various service costs and the costs of waiting for the service.

Bibliography

Andreatta, G. and Odoni, A.R. (2003). Analysis of Market-Based Demand Management Strategies for Airports and En Route Airspace, in *Operations Research in Space and Air*, Ciriani, T. (Ed.) Kluwer Academic Publishers, Boston, MA.

Hillier, F.S. and Lieberman, G.J. (2010). *Introduction to Operations Research*, McGraw Hill, New York.

Larson, R.C. and Odoni, A.R. (1981). *Urban Operations Research*. Prentice Hall, NJ.

Larson, R.C. and Sasanuma, K. (2010). Congestion Pricing: A Parking Queue Model. *Journal of Industrial and Systems Engineering*, 4, 1–17.

Odoni, A.R. and Larson, R.C. (2006). Lecture Notes in Logistical & Transportation Planning Methods, MIT, Cambridge, MA.

Taha, H. (1976). *Operations Research: An Introduction*, Prentice Hall, Upper Saddle River, NJ.

Teodorović, D. and Edara, P. (2007). A Real-Time Road Pricing System: The Case of a Two-link Parallel Network. *Computers & Operations Research*, 34, 2–27.

Trani, A.A. (2011). Lecture Notes in Introduction to Transportation Engineering—Introduction to Queueing Theory, Virginia Polytechnic Institute and State University, Blacksburg, VA.

Vickrey, W. (1955). A Proposal for Revising New York's Subway Fare Structure. *Journal of the Operations Research Society of America*, 3, 38–68.

Vickrey, W. (1963). Pricing in Urban and Suburban transport. *The American Economic Review* (papers and proceedings), 53, 452–465.

Vickrey, W. (1969). Congestion Theory and Transport Investment. *American Economic Review* (papers and proceedings), 59, 251–261.

Vickrey, W. (1994). Statement to the Joint Committee on Washington, D.C., Metropolitan Problems, The Economizing of Curb Parking Space—A Suggestion for a New Approach to Parking Meters. *Journal of Urban Economics*, 36, 42–65 (republished).

27

LOCATION ANALYSIS IN TRANSPORTATION

Dušan Teodorović, Branka Dimitrijević, and Milica Šelmić

UNIVERSITY OF BELGRADE

Introduction to Location Analysis

Bus stops, metro stations, hubs, vehicle depots, airports, etc., are examples of facilities whose locations in transportation networks, number and customer allocation schemes highly influence the quality of services provided and a system's total costs. The engineers and analysts optimize an explicit or implicit spatially dependent objective(s) when deciding the locations of these facilities. Typical objective functions for such decisions include: average distance between demands and servers that should be minimized (median problem), maximal distance that should be minimized (center problem), minimum distance that should be maximized (anti-center problem), a cost function of travel that should be minimized (simple plant location problem), etc. Optimization criteria are often used to classify location problems. Bearing in mind the nature of traffic control and transportation planning problems, one can claim that location analysis and transportation are inextricably linked.

In the text to follow we will first focus on basic location problems, their formulations and possible applications in transportation. Our attention will then be directed to particular transportation problems that belong to the location theory.

Brief History of Location Analysis

Throughout the history of human civilization, the problem of selecting one or more locations for specific purposes has been solved continuously. The fact that a written record of a location problem can be found in the Bible (Wilamowsky et al., 1995) confirms the existence of awareness of the importance of these issues in the distant past.

The first mathematically expressed problem that can be viewed as a location problem is recorded by Pierre de Fermat (1601–1665). It can be formulated in the following way: "For a given three points in a plane, find a fourth point such that the sum of its distances to the three given points is a minimum". The criterion used in this problem represents the root of those that belong to an important class of location problems called *minisum* or *median problems*. In the following centuries other scientists, Evangelista Toricelli (1608–1647), Bonaventura Francesko

Cavalieri (1598–1647), Thomas Simpson (1710–1761), and Giovanni Francesco dé Toschi Fagnano (1715–1797), studied that problem and contributed to its generalization by increasing the number of points in a plane.

In 1857, James Joseph Sylvester (1814–1897) recorded a new problem that reads: "For a given set of points in a plane, find a circle with a minimal radius that covers all points." The spirit of this objective appears in and characterizes another important class of location problems called *covering problems*.

Alfred Weber (1868–1958) is considered to be the pioneer of modern location analysis. Weber studied the locations of industrial facilities and warehouses (Weber, 1929). Hakimi (Hakimi, 1964) investigated the locations of switching centers in communication networks, as well as locations of police stations in highway networks. He also considered median and center problems in general. Since the mid-1960s, the study of location theory has boomed. A large number of researchers contributed significantly to the location theory, either from a theoretical point of view, or by solving real-life problems.

Numerous papers have been published dealing with location problems in transportation networks. Warehouses, freight terminals, bus stops, airports, public garages, police stations, firefighters' brigades, ambulances, hubs, schools, hospitals, restaurants, gasoline stations, swimming pools, and undesirable facilities are some of the facilities whose locations were analyzed and determined by various location analysis models. There are many books and review papers which are very useful in understanding the diversity of location analysis and state-of-the-art achievements. Some of these publications are: Love et al. 1988; Handler and Mirchandani 1979; Larson and Odoni 1981; Current et al. 2002; Brandeau and Chiu 1989; Mirchandani and Francis 1990; Drezner 1995; Daskin 1995; Drezner and Hamacher 2002; Eiselt and Marianov 2011.

Characteristics and Classification of Location Problems

The space in which facilities are located is primarily used to distinguish between classes of location problems. In that sense, there is a difference between location problems in *n*-dimensional real space (most commonly in a plane) and network location problems, each of which can be further divided into continuous and discrete location problems. In continuous problems, facilities can generally be located at any point of the plane or at any point in the network, while in discrete problems they can be placed only at a limited number of acceptable points selected in a pre-processing phase of modeling.

Distances in *n*-dimensional real space are most often derived from Minkowski distances (l_p, distances with a parameter p, $1 \leq p \leq \infty$), which are defined in the plane (between point $I(x_i, y_i)$ and point $J(x_j, y_j)$ with $i \neq j$) by:

$$l_p(I,J) = \left(|x_i - x_j|^p + |y_i - y_j|^p\right)^{1/p} \tag{1}$$

In the literature, most commonly used distance measures are the Manhattan (or rectilinear or rectangular or l_1) distance and the Euclidean (or straight line or l_2) distance. Love et al. (1988) recommended a weighted l_p distance function when solving real-life problems that use approximations of road distances between points purely based on the coordinates of the endpoints. They have shown that the road distances are usually 10–30 percent greater than the corresponding Euclidean distances.

When solving network location problems, distances are measured on the network itself and defined by the shortest paths between any two points (nodes) (Dijkstra, 1959).

Typically, continuous location problems, that are mostly planar problems, are defined as non-linear optimization problems, while discrete location problems, that are most often network problems, are zero-one programming/combinatorial optimization problems. The types of the algorithms used to solve location problems are exact and heuristic. The algorithm used depends on the size of a problem considered, acceptable CPU time, as well as on the analyst and decision-maker's readiness to accept non-optimal solutions.

The facility location also depends on the type of service that facility provides. For example, it is clear that an airport must be located as far as possible from the city center for environmental reasons. On the other hand airports should not be too far away because of the increase of total travel time and decrease of the level of service. In that sense we can distinguish *desirable* facilities (schools, hospitals, restaurants, bus stops, etc.), that do not have any significant detrimental effects to any individual, from *undesirable* (polluting plants, waste storage sites, noise, odor or heat emitters, etc.) that have considerable adverse effects on their environment or their customers.

When solving some location problems, it is important to take into account more than one objective function. Two, or more objective functions are usually in conflict. In such a situation the problem considered should be solved by a multi-objective approach. The following are the other characteristics of the location problem that could be used in its classification: facility capacities (uncapacitated vs. capacitated), type of demand (deterministic vs. stochastic), type of facility service (unique vs. various), demand allocation (to the nearest facility vs. demand is split between facilities), etc.

It is obvious that various characteristics are present in the description of location problems and that standardized classification does not exist. Various authors have proposed different classifications that attempt to describe all location models (Hamacher and Nickel (1998); Brandeau and Chiu (1989)).

In the following sections we will present, in more detail, both discrete and network location problems, since the majority of transportation terminals can be located only in pre-specified nodes of the transportation network due to various geographic, urban, engineering, legal, organizational, and economic constraints.

Basic Location Problems on (Transportation) Networks

In the following text, some of the most studied problems in the facility location literature, not only from a theoretical point of view, but also from its complexity in finding a solution, will be presented based on the criteria they use. All of them are *NP*-hard and could be solved by various optimization and heuristic procedures.

Median Problems

Median problems are particularly important for transportation activity since they appear when designing different distribution systems. In those problems the effectiveness of a facility location is measured by determining the average travel distance by those (customers) who visit it. Thus, the objective of the *p-median problem* is to locate p facilities to minimize the total demand-weighted travel distance between demand points and facilities. This problem was introduced by Hakimi (1964). To formulate it as a zero-one programming problem, the following notation is necessary:

I = set of demand nodes;
i = index of demand node;

J = set of nodes-candidates for facility sites;
j = index of potential facility site;
d_{ij} = distance between node i and node j;
w_i = demand at node i;
p = number of facilities to be located.

Decision variables are:

$$x_j = \begin{cases} 1 & \text{if a node } i \text{ is chosen to be a facility location site } j \\ 0 & \text{otherwise} \end{cases}$$

$$y_j = \begin{cases} 1 & \text{if demands at node } i \text{ are served by a facility at node } j \\ 0 & \text{otherwise} \end{cases}$$

The p-median formulation follows:

$$\min Z = \sum_{i \in I} \sum_{j \in J} w_i d_{ij} y_{ij} \quad (2)$$

subject to:

$$\sum_{j \in J} y_{ij} = 1 \quad \forall i \in I \quad (3)$$

$$y_{ij} \leq 1 \quad \forall i \in I, j \in J \quad (4)$$

$$\sum_{j \in J} x_j = p \quad (5)$$

$$y_{ij} \in \{0,1\} \quad \forall i \in I, j \in J \quad (6)$$

$$x_j \in \{0,1\} \quad \forall j \in J \quad (7)$$

The objective (2), as mentioned above, is to minimize the total demand-weighted distances between customers and facilities. Constraint (3) requires that exactly p facilities be located. Constraint (4) ensures that every demand is assigned to some facility site, while constraint (5) allows assignment only to sites at which facilities have been located. Constraints (6) and (7) are binary requirements for the problem variables.

There are plenty of works presenting generalizations and extensions to the original p-median problem (Mirchandani and Francis, 1990). Some of them are related to different amounts of demand for different products or commodities, as well as different routes for each product (multi-commodity p-median). Various p-median problems could include specific constraints related to the facility capacity, arc capacity, distance constraints and implementation constraints, probabilistic travel distances and probabilistic demands, oriented and non-oriented networks, non-linear transportation costs, etc.

The p-median problem and its extensions are useful to model many real-world situations, such as the location of industrial plants, warehouses, shops and public facilities (schools, hospitals, public transit stops, post offices, etc.). Mladenović et al. (2007) survey different approaches to the solution of the p-median.

Covering Problems

In many cases, when locating facilities in networks, analysts try to cover as much as possible of the clients' demand. "Coverage" of the node assumes the existence of a facility within a pre-specified coverage radius or service distance (D_p). D_p is the maximum distance that any client would have to travel to reach a facility. It defines the so-called catchment area. Airports, freight terminals, bus stops and other transportation facilities have corresponding catchment areas. If clients are located within a defined service distance ("covered"), they receive the service, otherwise they do not (Figure 27.1). The covering concept can be very useful in locating emergency services. Two major covering problems are the location set covering problem and the maximal covering problem.

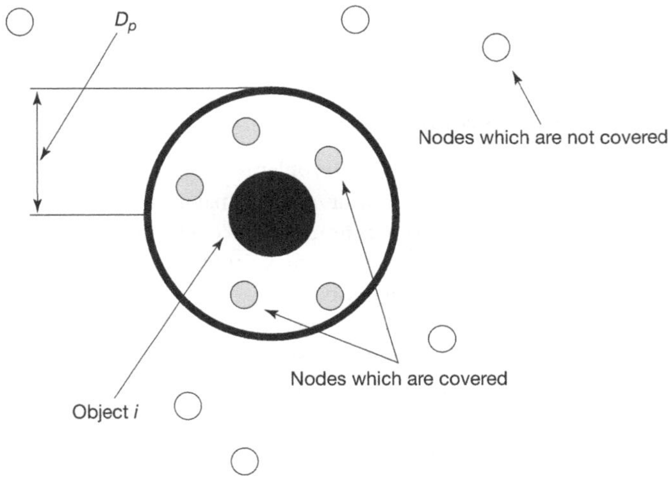

Figure 27.1 "Covering" of nodes located within a defined maximal service distance D_p

In the set-covering problem, the objective is to minimize the cost of facility locations opening such that all demand nodes are covered in a way that every demand node has at least one facility located within the maximal service distance. If the costs of all the facility locations are the same, then the objective is simply to minimize the number of sited facilities and to cover all demands.

Notation as in median problems will be used in this problem formulation. In addition, let N_i be a set of facility sites j that "cover" demand node i (i.e., $N_i = \{j \mid d_{ij} \leq D_p\}$), and c_j be a fixed cost of sitting a facility at node j.

The location set-covering problem has the following mathematical formulation (Toregas et al. (1971); ReVelle et al. (1976); Current et al. (2002)):

$$\text{Minimize} \sum_{j \in J} c_j x_j \,, \tag{8}$$

subject to:

$$\sum_{j \in N} x_j \geq 1 \;\; \forall i \in I, \text{ and} \tag{9}$$

$$x_j \in \{0,1\} \;\; \forall j \in J \tag{10}$$

When considering the location set-covering problem, we try to discover the smallest number of candidates for facility locations necessary to cover all demand nodes. On the other hand, in some cases (due to budget constraints) it is not possible or not necessary to cover all demand nodes. In such cases, it is logical to maximize the coverage (population covered) within a prescribed service distance by locating the given number of service facilities. This problem is known as the maximal covering location problem (Church and ReVelle, 1974).

Both covering models presume that if demand is covered by a facility, then that facility will be available to serve the demand. An emergency service vehicle, already responding to a call for service, will not be available to answer additional demands. In that case, the concept of "multiple coverage" must be applied. Another practically interesting extension of covering problems is the variable coverage radius. The signal strength of a radio station determines the coverage area. Longer runways at an airport allow it to service larger planes, thus allowing arrivals and departures of airplanes that fly large distances. A recent review of the variable radius-covering problems is provided by Berman et al. (2009).

Center Problems

Another class of location problems is the class of p-center problems. The objective of the p-center problem is to locate p facilities on a network to minimize the maximum weighted or unweighted distance that separates any demand node from its nearest facility (Hakimi 1964; Kariv and Hakimi 1979). The p-center problem is also called the minimax problem.

If facility locations are restricted to the nodes of the network, the problem is a vertex center problem. Problems that allow facilities to be located along edges of the network are called absolute center problems. The center problems in location analysis are the problems related to the location of emergency-type facilities in the transportation networks (fire station, police station, ambulances depot, etc.).

For some cases of the p-center problem, exact algorithms have been proposed (Daskin (1995); Chen and Chen (2009)). Still, the majority of p-center problems are solved by heuristics and procedures based on meta-heuristics (Mladenović et al. (2003); Davidović et al. (2011)).

Additional Location Problems

The fixed charge facility location problem (or simple plant location problem) includes the fixed costs associated with establishing a facility at each potential facility site. The uncapacitated fixed-charge facility location problem is similar to the p-median problem. It is formulated by adding fixed costs to the p-median objective function and removing the constraints that dictate the number of facilities to be located. In this way, the problem is transformed in the problem that endogenously determines the number of facilities to locate so as to minimize the total (opening facility plus travel) costs (Balinski 1965; Current et al. 2002). The formulation of the capacitated version of the problem assumes additional facility capacity constraints.

The hub location problem is one of the most important problems in location theory. This problem appears in transportation systems that have a large number of origin-destination pairs. Usually, it is cheaper for the carrier to avoid direct routing between all pairs of nodes between which flow occurs. Instead, hub objects perform the transfer of passengers (from one plane/bus line to another) or goods reloading (reloading letter mail and packages from small to larger aircraft, reloading of goods from trucks to ships, etc.) with a higher frequency of operations and lower costs per unit. Passengers or goods travel longer and change the means/modes of transport in the hubs (Figure 27.2) but travel much more frequently to their destinations. This area of research was introduced to the literature by O'Kelly (1986, 1987). The reviews by Bryan and O'Kelly (1999) and Campbell et al. (2002) are comprehensive.

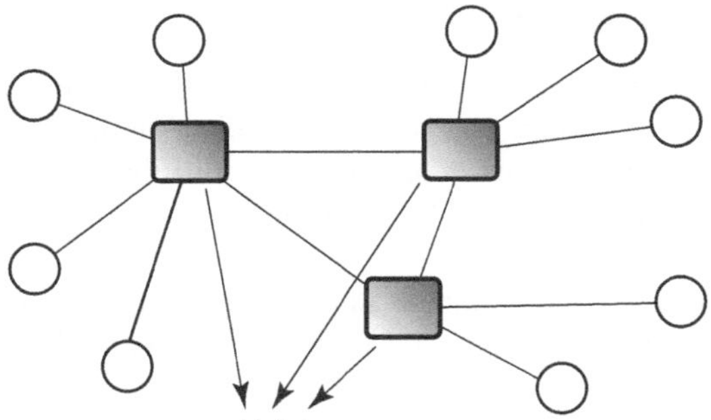

Figure 27.2 Transport network with hubs

Unlike other basic location problems, the *p*-dispersion problem takes into account only the distances between objects and tries to maximize the minimum distance between any two objects in the network (Kuby (1987); Daskin (1995)). A similar problem is the anti-covering location problem. When solving this problem, an analyst tries to find the maximally weighted set of location sites in such a way that no two selected sites are within a pre-specified distance (Murray and Church, 1997). Those problems are related to the franchise's locations distribution, as well as an undesirable facility location. Most of the earlier approaches for an undesirable facility location address the problem of locating a single facility under a maximin criterion. There has been little focus on the single facility maxisum problem. An undesirable facility location problem with covering concept is called the minimum covering location problem. Berman et al. (1996) studied this problem on a network and presented an algorithm to solve it. Combining it with the anti-covering location problem Berman and Huang (2008) formulated the minimum covering location problem with distance constraints. Other concepts to solve an undesirable facility location has focused on the multi-objective models (Erkut and Neuman, 1989) and fuzzy models (Rakas et al., 2004).

A problem closely linked to undesirable facility location problems is that of transporting hazardous materials or wastes and locating undesirable facilities to treat or dispose them. Selecting good disposal sites for waste and hazardous materials, and effectively transporting them with optimal routing, minimizes population exposure and reduces further pollution of the environment. Location-routing models are frequently used when solving desirable facility location problems. For example, Bouliane and Laporte (1992) considered the simultaneous location of postal relay boxes and the routing of postal trucks for deliveries and pickups.

Traffic and Transportation Problems that Belong to Location Theory

Traffic authorities are responsible for mitigating traffic congestion in cities and on the freeways, to help drivers find a way through the network, to properly organize checks on truck weight limits, to prevent drink driving, etc. The following are some of the traffic engineering problems that should be considered within the location theory context:

- Location of sensors in transport network.
- Location of inspection stations in transport network.
- Location of automated vehicle identification equipment locations.
- Location of parking guidance signs.

Location of Sensors in Transport Network

Sensors or detectors are deployed on roadways to collect traffic data including volume, occupancy and speed. Sensors may be classified as "passive" or "active" depending on the level of information that a vehicle provides en route from its point of origin to its destination. Passive sensors are simpler than active, and they have a role to detect vehicles, while active sensors have the ability to record more data about vehicles passing by.

In general, whether a sensor is active or passive, a sensor obtains an image of traffic flow and processes it to identify characteristics of the vehicle (car, bus, truck, etc.) or the traffic flow. Recognition of a bus or a truck could provide information on its path. Recognition of a vehicle license plate could provide its origin and/or destination. For both active and passive sensors, a new set of location problems arises on where to locate such sensors to monitor or manage the particular class of traffic detected.

A few papers from the relevant literature deal with the problem of locating sensors and detectors in the transport network. Gentili and Mirchandani (2005) refer to an active sensor when it reads identifications of vehicles, including their routes in the network, that the vehicles actively provide when they use the network. They consider some type of generic sensors and their locations that address a large number of sensor location problems.

Edara et al. (2008) analyzed the location of the conventional inductance loop detectors, which are passive sensors that mostly count vehicles at points in a network to obtain traffic volumes (e.g., vehicles per hour) on a lane or road of the network. The authors developed a method to indicate the most appropriate locations for detector deployment such that the travel time estimate error is minimized within the budgetary constraints. The problem of optimal placement of point detectors is studied in the case of Northern Virginia region in Virginia. The authors used a relatively new meta-heuristic Bee Colony Optimization, as a tool for solving the considered problem. The obtained results showed good performances and high accuracy.

The problem of optimal traffic count locations is studied, among others, in Yang and Zhou (2005). The authors proposed an integer linear programming model and a heuristic greedy algorithm to discover the optimal traffic count locations in a network for the purpose of providing more accurate origin-destination trip matrices. The same problem was studied by Ehlert et al. (2006). They also propose a mixed integer programming formulation to optimize traffic count locations.

Different aspects of sensor or detector location problems are also defined and solved in Chan and Lam (2002), Fei et al. (2007), and Bianco et al. (2001).

Location of Inspection Stations in the Transport Network

There are various risks related to transportation networks. High-level risks are primarily associated with potential terrorist attacks, natural catastrophes, and the transportation of some hazardous materials. The sudden degradation of transport systems opens a lot of research questions associated with the transport network's reliability, emergency response, evacuation strategies, crowd dynamics, and urban, highway and air traffic congestion mitigation.

There are also significant risks related to driving under influence, overweight truck operations or the operation of motor vehicles that do not satisfy safety inspection standards. According to the World Health Organization, drink driving is a permanent problem in many countries in the world and one of the most contributing factors in fatal traffic accidents.

At weigh stations, traffic authorities check weight limits, hours and service regulations, vehicle equipment safety and collect road-use taxes. Depending on the state and country, there are various scales at different weigh stations. Simple, portable scales enable a weigh station to be set up practically at any node along a highway.

Several billion tons of hazardous materials are shipped yearly in the world. Authorities in some countries established a system of checkpoints to detect leaks from trucks transporting hazardous materials. At these checkpoints vehicles in violation of safety limits are placed out of service.

In order to properly organize the checkup of truck weight limits, hours and service regulations, and vehicle equipment safety, as well as the prevention of drink driving, several authors tried to find the answers to the following questions: a) What should be the total number of inspection stations in the transport network? b) Where should these facilities be located?

Inspection stations (''Driving under the influence" checkpoints, weigh stations, routine police checkups, hazardous material check points) belong to the class of flow-intercepting facilities. In the case of flow-intercepting facilities, clients (drivers) obtain service if they pass through one of the facilities. Every client (driver) who passes by such a facility is treated as a "captured" or "intercepted" client.

Different aspects of the problem of determining the locations of the flow-capturing facilities in the transportation network were treated by Hodgson (1981, 1990); Hodgson et al. (1996); Berman et al. (1992); Berman et al. (1995); Berman and Krass (1998); Gendreau et al. (2000); Wu and Lin (2003); Kuby and Lim (2005); and Jun and Min (2006).

Šelmić et al. (2010) developed two models to determine the locations of the uncapacitated inspection stations in the traffic network. The first model formulation is related to a single-objective optimization problem. The objective function, to be maximized, represents the largest possible risk decrease on the traffic network. The multi-objective approach to the problem of inspection stations deployment in traffic networks is elaborated within the second version of the model. The first objective function tends to maximize the reduction of risk in the traffic network. The second objective minimizes the total number of deployed facilities. The inspection stations location problem is solved by the combination of the Bee Colony Optimization meta-heuristic and compromise programming.

Location of Automated Vehicle Identification Equipment Locations

The automated vehicle identification (AVI) equipment location problem entails determination of the best locations for the automated vehicle identification equipment. AVI technology allows vehicles equipped with special tags to be detected at specific points in the transportation network without any action by the driver as they pass under a reading station. The potential benefits of this technology are found in the real-time measurement of traffic parameters, traffic operations, reduction and control of traffic congestion in transportation facilities (e.g., toll plaza), transportation planning studies, fleet management, information and control, electronic toll collection, vehicle weighting and classification, etc.

The quality of the AVI coverage and the total costs of an AVI system depend primarily on the number and positions for the AVI receiving antennas. The antenna location problem is also combinatorial by nature. First, the quality of a location is highly dependent on the measure of effectiveness being used. Then, every possible deployment of AVI readers within a network is characterized by an appropriate number of vehicle "readings". The readers must be located in such a way that they could "catch" a maximum number of trips within the network.

Teodorović et al. (2002) solved the AVI equipment location problem as a multi-objective optimization problem, thus determining the best locations on the basis of several criteria. They adopt the fact that when determining the quality of certain locations it is necessary to take into account the total number of readings, as well as the total number of "covered" pairs of nodes by readings. The problem considered by Teodorović et al. (2002) can be formulated in the following way: For a given number of AVI readers, find the optimal reader locations along the network links that would maximize or minimize a given objective function. The develop model is based on the genetic algorithms.

Chen et al. (2007) considered the same problem. They assumed that there are three possible objectives when deciding locations for AVI readers: (a) a minimum number of AVI readers, (b) maximum O-D coverage, and (c) a maximum number of trips (or AVI readings). To satisfy all three objectives as much as possible, the authors presented a multi-objective integer-optimization problem formulation. A genetic algorithm is applied to solve this multi-objective AVI reader-location problem by explicitly generating the non-dominated solutions. Numerical results are presented to demonstrate the feasibility of the proposed multi-objective model.

Location of Parking Guidance Signs

Drivers searching for a parking space considerably contribute to the increase of traffic congestion. Every day, a significant number of drivers in single-occupancy vehicles search for a parking space. In addition, the least knowledgeable drivers or out-of-towners further contribute to the increase in traffic congestion.

During the past two decades, traffic authorities in many cities (Helsinki, Cologne, Mainz, Stuttgart, Wiesbaden, Aalborg, Hague) implemented various parking guidance and information. These systems provide service to users by modern technology related to traffic monitoring, communication, and variable message signs. The parking guidance and information systems inform and guide drivers to parking facilities with real-time variable message signs (directional arrows, the names of the parking facilities, status (full, not full, the number of available parking spaces, etc.)). Information about the number of available parking spaces could be displayed on major roads, streets and intersections, or it could be distributed through the Internet.

Parking guidance systems notably increase the probability of finding a vacant parking space, and alleviate frustration of drivers–visitors unfamiliar with city centers. At the same time, these systems decrease queues in front of parking garages, decrease the total amount of vehicle miles travelled (particularly in city centers), and decrease the average trip time, energy consumption, and air pollution.

When analyzing the problem of locating parking guidance signs in transportation networks, numerous research questions appear. The information concerning the capacity of each parking facility, the location of the parking guidance sign, and the decision of drivers to select a specific parking location are all interconnected. The research up to date has provided answers to some of these concerns (Teodorović and Šelmić (2009); Wardman et al. (1997); Waterson et al. (2001); Chou et al. (2008)).

Teodorović and Šelmić (2009) analyzed the problem of the optimal placement of parking guidance signs. They assume that optimal decisions related to the locations of the parking guidance signs should be made in the presence of trade-offs between two or more objectives. The problem considered in this chapter could be defined in the following way: Locate the given number of parking guidance signs in the network in such a way to maximize the total number of informed drivers, as well as to maximize the number of possible diversions for the drivers to react according to the obtained information. A multi-objective formulation of the parking guidance signs location problem was proposed. The considered location problem is solved by a combination of Bee Colony Optimization meta-heuristic and compromise programming. The model is supported by numerical examples.

Waterson et al. (2001) suggested driver parking choice models (both during the journey and pre-trip). The authors also analyzed potential parking guidance sign locations (placing signs close to car park entrances, or placing signs at major decision points in the network far from the car park entrances).

References

Balinski, M.L. (1965). Integer programming: methods, uses, computation. *Management Science*, 12, 253–313.

Berman, O. and Huang, R. (2008). The minimum weighted covering location problem with distance constraints. *Computers & Operations Research*, 35, 356–372.

Berman, O. and Krass, D. (1998). Flow intercepting spatial interaction model: a new approach to optimal location of competitive facilities. *Location Science*, 6, 41–65.

Berman, O., Drezner, Z. and Wesolowsky, G.O. (1996). Minimum covering criterion for obnoxious facility location on a network. *Networks*, 28, 1–5.

Berman, O., Drezner, Z., Krass, D. and Wesolowsky, G.O. (2009). The variable radius covering problem. *European Journal of Operational Research*, 196, 516–525.

Berman, O., Hodgson, M.J. and Krass, D. (1995). Flow-interception problems, facility location. In Z. Drezner (Eds.) *A Survey of Applications and Methods*, New York, Springer-Verlag, 389–426.

Berman, O., Larson, R.C. and Fouska, N. (1992). Optimal location of discretionary service facilities. *Transportation Science*, 26, 201–211.

Bianco, L., Confessore, G. and Reverberi, P. (2001). A network based model for traffic sensor location with implications on O-D matrix estimates. *Transportation Science*, 35(1), 50–60.

Bouliane, J. and Laporte, G. (1992). Locating postal relay boxes using a set covering algorithm. *American Journal of Mathematical and Management Sciences*, 12, 65–74.

Brandeau, M.L. and Chiu, S.S. (1989). An overview of representative problems in location research. *Management Science*, 35 (6), 645–674.

Bryan, D. and O'Kelly, M. (1999). Hub-and-spoke networks in air transportation: an analytical review. *Journal of Regional Science*, 39, 275–295.

Campbell, J.F., Ernst, A.T. and Krishnamoorthy, M. (2002). Hub location problems. In Drezner, Z. and Hamacher, H.W. (Eds.) *Facility Location: Applications and Theory*, 373–407, Springer, Berlin.

Chan, S. and Lam, H.K. (2002). Optimal speed detector density for the network with travel time information. *Transportation Research*, 36, 203–223.

Chen, D. and Chen, R. (2009). New relaxation-based algorithms for the optimal solution of the continuous and discrete p-center problems. *Computers & Operations Research*, 36(5), 1646–1655.

Chen, A., Chootinan, P. and Pravinvongvuth, S. (2004). Multiobjective model for locating automatic vehicle identification readers. *Transportation Research Record*, 1886, 49–58.

Choua, S., Linb, S.W. and Lia, C.C. (2008) Dynamic parking negotiation and guidance using an agent-based platform. *Expert Systems with Applications*, 35, 805–817.

Church, R.L. and ReVelle, C. (1974). The maximal covering location problem. *Papers of the Regional Science Association*, 32, 101–118.

Current, J., Daskin, M. and Schilling, D. (2002). Discrete network location models. In Drezner, Z. and Hamacher, H. (Eds.) *Facility Location: Applications and Theory*, Springer, Berlin, 81–118.

Daskin, M.S. (1995). *Network and Discrete Location: Models, Algorithms, and Applications*, Wiley & Sons, New York.

Davidović, T., Ramljak, D., Šelmić, M. and Teodorović, D. (2011). Bee colony optimization for *p*-center problem, *Computers & Operations Research*, 38 (10), 1367–1376.

Dijkstra, E.W. (1959). A note on two problems in connexion with graphs. *Numerische Mathematik*, 1, 269–271.

Drezner, Z. (1995). *Facility Location: A Survey of Applications and Methods*, Springer Verlag, Berlin.

Drezner, Z. and Hamacher, H.W. (Eds.) (2002). *Facility Location: Applications and Theory*, Springer, Berlin.

Edara, P., Šelmić, M. and Teodorović, D. (2008). Heuristic solution algorithms for a traffic sensor optimization problem. *INFORMS*, Washington, DC.

Ehlert, A., Bell, M. and Grosso, S. (2006). The optimization of traffic count locations in road networks. *Transportation Research B*, 40, 460–479.

Eiselt, H.A. and Marianov, V. (Eds.) (2011). *Foundations of Location Analysis*, Springer, Berlin.

Erkut, E. and Neuman, S. (1989). Analytical models for locating undesirable facilities. *European Journal of Operational Research*, 40 (3), 275–291.

Fei, X., Mahmassani, H.S. and Eisenman, S.M. (2007). Sensor coverage and location for real-time traffic prediction in large-scale networks. *Transportation Research Record*, 2039, 1–15.

Gendreau, M., Laporte, G. and Parent, I. (2000). Heuristics for the location of inspection stations on a network. *Naval Research Logistics*, 47, 287–304.

Gentili, M. and Mirchandani, P. (2005). Locating active sensors on traffic networks. *Annals OR*, 136, 229–257.

Hakimi, S.L. (1964). Optimal locations of switching centers and the absolute centers and medians of a graph. *Operations Research*, 12, 450–459.

Hamacher, H.W. and Nickel, S. (1998). Classification of location models. *Location Science*, 6, 229–242.

Handler, G.Y. and Mirchandani, P.B. (1979). *Location on Networks: Theory and Algorithms*, MIT Press, Cambridge, MA.

Hodgson, M.J. (1981). The location of public facilities intermediate to the journey to work. *European Journal of Operational Research*, 6, 199–204.

Hodgson, M.J. (1990). A flow-capturing location-allocation model. *Geographical Analysis*, 22, 270–279.

Hodgson, M.J., Rosing, K.E. and Zhang, J. (1996). Locating vehicle inspection stations to protect a transportation network. *Geographical Analysis*, 28, 299–314.

Jun, Y. and Min, Z. (2006). Flow capturing location-allocation problem with piecewise linear value-time function based on max-min ant colony optimization. In Proceedings of the IEEE International Conference on Computational Intelligence and Security, Volume 2, Guangzhou, China, 1172–1175.

Kariv, O. and Hakimi, S.L. (1979). An algorithmic approach to network location problems: Part I: The *p*-centers. *SIAM Journal of Applied Mathematics*, 37, 513–538.

Kuby, M. (1987). The *p*-dispersion and maximum dispersion problems. *Geographical Analysis*, 19, 315–329.

Kuby, M. and Lim, S. (2005). The flow-refueling location problem for alternative-fuel vehicles. *Socio-Economic Planning Sciences*, 39, 125–145.

Larson, R. and Odoni, A. (1981). *Urban Operations Research*, Prentice Hall, Upper Saddle River, NJ.

Love, R.F., Morris, J.G. and Wesolowsky, G.O. (1988). *Facilities Location: Models and Methods*. North Holland, New York.

Mirchandani, P. and Francis, R. (Eds.) (1990). *Discrete Location Theory*, John Wiley & Sons, New York.

Mladenović, N., Brimberg, J., Hansen, P. and Moreno-Perez, J. (2007). The *p*-median problem: A survey of metaheuristic approaches. *European Journal of Operational Research*, 179, 927–939.

Mladenović, N., Labbe, M. and Hansen, P. (2003) Solving the p-center problem with tabu search and variable neighborhood search. *Networks*, 42(1), 48–64.

Murray, A.T. and Church, R.L. (1997). Solving the anti-covering location problem using Lagrangian relaxation. *Computers & Operations Research*, 24, 127–140.

O'Kelly, M. (1986). The location of interacting hub facilities. *Transportation Science*, 20, 92–106.

O'Kelly, M. (1987). A quadratic integer program for the location of interacting hub facilities. *European Journal of Operational Research*, 32, 393–404.

Rakas, J., Teodorović, D., Kim, T. (2004). Multi-objective modeling for determining location of undesirable facilities. *Transportation Research D*, 9, 125–138.

ReVelle, C., Toregas, C. and Falkson, L. (1976). Applications of the location set covering problem, *Geographical Analysis*, 8, 67–76.

Šelmić, M., Teodorović, D. and Vukadinović, K. (2010). Locating inspection facilities in traffic networks: an artificial intelligence approach. *Transportation Planning and Technology*, 33, 481–493.

Teodorović, D. and Šelmić, M. (2009). Parking guidance signs location in transport networks. In *Symposium on Operation Research- SYMOPIS*, Tara, Serbia, 663–666 (in Serbian).

Teodorović, D., Van Aerde, M., Zhu, F. and Dion, F. (2002). Genetic algorithms approach to the problem of the automated vehicle identification equipment locations. *Journal of Advanced Transportation* 36, 1–21.

Toregas, C., Swain, R., ReVelle, C. and Bergman, L. (1971). The location of emergency service facilities. *Operations Research*, 9, 1363–1373.

Wardman, M., Bonsall, P.W. and Shires, J.D. (1997). Driver response to variable message signs: a stated preference investigation. *Transportation Research Part C: Emerging Technologies*, 5, 389–405.

Waterson, B.J., Hounsell, N.B. and Chatterjee, K. (2001). Quantifying the potential savings in travel time resulting from parking guidance systems—a simulation case study. *Journal of the Operational Research Society*, 52, 1067–1077.

Weber, A. (1929) *Theory of the Location of Industries*. Chicago, The University of Chicago Press.

Wilamowsky, Y., Epstein, S. and Dickman, B. (1995). How the oldest recorded multiple facility location problem was solved. *Location Science*, 3 (1), 50–60.
Wu, T.H. and Lin, J.N. (2003). Solving the competitive discretionary service facility location problem. *European Journal of Operational Research*, 144, 366–378.
Yang, H. and Zhou, J. (2005). Optimal traffic counting locations for origin-destination matrix estimation. *Transportation Research B*, 32, 109–126.

28

EFFICIENCY PERFORMANCE MEASUREMENT FOR TRANSPORTATION SYSTEMS

A Primer Using an Empirical Frontier Perspective

Konstantinos Triantis

VIRGINIA TECH

Introduction and Initial Considerations

The purpose of this chapter is to introduce the reader to performance evaluation of transportation systems using an empirical frontier perspective. In essence, we will explore the notion of an empirical frontier as a means to assess the performance of transportation systems. In order to do this we will introduce concepts and methods that have originated in the efficiency measurement literature and have been applied to transportation systems and processes (Triantis, 2011).

The first fundamental question is why is performance evaluation important? Two issues come immediately to mind. The first is to assess how well the transportation system has met specific goals in the past (ex-post analysis). The second is to understand how policies and specific interventions can improve performance in the future (meta-analysis or ex-ante analysis).

Underlying these two issues is the definition and understanding of what constitutes system efficiency performance. One can begin by suggesting that efficiency performance is the attainment or not of specific desirable goals. For example, if a certain level of service is important, then a deviation from the desirable level of service is potentially a measure of performance. This can be labelled as service effectiveness. If the consumption of scarce resources (labor, fuel, materials, etc.) is an important consideration, then the deviation from a desirable level of resource consumption is a measure of performance. This could be labelled resource or input efficiency. Therefore, one of the initial important challenges is to define which goals are of interest and how we ascribe desirable levels for these goals.

A second core challenge is the in-depth understanding of the internal transformation processes of the system. For example, in a manufacturing context this notion is easily understood. In a paper manufacturing environment, energy, materials, labor and equipment are all used to physically transform the resources (wood) into physical paper. The capital and process technologies are well understood. For transportation systems, however, the transformation processes are more difficult to understand and represent. The physics associated with traffic flows provide an understanding of the physical flows that are part of transportation networks, but what remains a challenge is the representation of the outcomes of these networks as an expression of the services provided.

Thus, at the core of efficiency performance measurement is how we represent the transformations that take place in all engineering systems including transportation systems. There are two approaches. The first relies on functional and/or mathematical and/or descriptive representations of the physical, engineering, informational, behavioral laws that precisely describe how the transformations in any system take place. This is akin to being able to provide chemical reactions in chemistry where material transformations are the norm. However, in transportation systems, as is the case with the majority of engineering systems, these descriptions are very hard to obtain, in part because of the interactions among materials, information and people in the context of organizational realities. For example, when studying a highway system, one needs to consider the physical pavement, signs, shoulders, etc. along with the users (drivers) and decision makers (within a transportation agency) that decide appropriate maintenance strategies. Taking all of these perspectives into account is extremely difficult. Therefore, it is very challenging, if not impossible, to come up with accurate representations of these transformation processes.

What we are left with is an indirect approach that suggests that if we can effectively represent the resources used (inputs) and the outputs/outcomes achieved, then this constitutes a representation of the underlying transformation processes of any engineering system. For example, Sheth, Triantis, and Teodorović (2007) in describing bus routes as a transportation system, suggest headway, cost, service duration, # of intersections, # of priority lanes as inputs (resources), and average traveling time, vehicle miles, schedule reliability and passenger miles as outputs.

This input/output representation, while intuitively appealing, is hard to implement. What we mean by input/output representation or specification is the definition of all input (resources that are used in the transformation process) variables and the output/outcome (resources that are generated from the transformation process) variables along with their units of measurement. Typically, in most efficiency measurement applications, all variables are represented as interval variables.

There is no agreement in the literature as to the best way to achieve a representative input/output specification for any transportation or engineering system. Initially, the researcher/modeler needs to have access to the transportation system that will be assessed. By access, we mean access to the key stakeholders involved with the system (decision makers, users, others), an in-depth understanding of the technologies and the environment in which the system operates, and access to operational data/information. This type of access is not easy to obtain. Furthermore, the time required to obtain an in-depth understanding and to collect the requisite data can easily take months, if not years, to accomplish.

There are an additional two challenges. First, depending on how the transportation system is defined in terms of its boundaries, the number of observations that can be reasonably collected is typically relatively small (for example, in the teens as opposed to the hundreds or thousands). This presents an issue when attempting to obtain statistical inferences using either optimization or econometric approaches (see p. 433). This is what is typically referred to in the literature as the curse of dimensionality. When this is the case, the researcher/modeler is faced with the difficult task of altering the input/output specification by reducing the number of variables while at the same time not compromising the fidelity and appropriateness of the input/output specification. In other words, you do not wish to reduce the number of variables where the indirect representation of the transformation process is lost.

Second, the data and information that are available are typically never in the format that is required for an operational/performance analysis. This means that we may not have representations (definitions) that are consistent with the required notion of resources used (inputs) or resources produced (outputs/outcomes). Additionally, there is a certain degree of imprecision associated with the actual numerical values that are collected. This imprecision is due to the appropriateness or not of the measurement instruments that are used to obtain the data. Thus, this imprecision is a type of uncertainty that needs to be captured, understood and incorporated in the modeling. Therefore, it is not uncommon to initially complete an outlier/influential (Seaver and Triantis, 1992, 1995) observation analysis to account for this type of uncertainty.

To summarize, the prerequisite to conducting an empirical frontier analysis is understanding the transformation process, understanding the key performance goals and the respective levels that need to be reached, obtaining an appropriate input/output specification for the transportation system, and collecting the requisite data and information to conduct the analysis. What follows next is a brief introduction of an empirical frontier analysis. For an in-depth discussion of this topic, please refer to Cooper, Seiford and Tone (2006).

Frontier Analysis

Preliminaries

We begin after identifying the inputs and outputs associated with a transportation system and have collected data (observations) for each input and output. The observations can be with respect to a transportation system of a similar type (bus systems for various metropolitan areas: cross-sectional analysis), with respect to a transportation system over time (transportation network observed over time: temporal analysis), or with respect to transportation systems of a similar type and over time (panel data approach). It is desirable to have more observations rather than less for statistical inference reasons. However, the one caveat that needs to be added is that the basic assumption is that the observations that have been collected are comparable in terms of the technologies and environments in which they operate. If not, then one needs to complete a different type of analysis where the distinction between operational (efficiency change) and technical change is made.

All the observations that have been collected constitute the observed production possibility set (PPS). Each observation in the PPS represents an instantiation of the production (or service and, in our case, of transportation) technology that we are observing. All observations are critical in the sense that they define the boundaries and the possibilities of the production (or service and, in our case, the transportation) technology. Typically, this PPS assumes the interface between the variable inputs and outputs where people, materials, energy and assets come together. There is a necessary distinction that needs to be made between the variable inputs and outputs, and those that are fixed (for example, equipment, infrastructure, etc.). All fixed inputs (for example, equipment) and outputs are variable over the long term. This is why it is necessary to also determine the time horizon of the analysis and the distinction between a variable, and a fixed variable becomes a function of this determination.

In the frontier terminology, each observation is labelled a "decision-making unit" (DMU). The general assumption is that someone is making a decision within or about the observation. For example, when we are considering bus routes as the unit of analysis, the question of who is making decisions about the bus routes comes to mind. In this example, we can assume that individuals in the transportation agency are making decisions about resource utilization and the level of service that is to be provided, and that these individuals are concerned with the efficiency (resource utilization) and effectiveness (level of service) performance of the bus routes. Alternatively, one could also stipulate that the bus riders make their own decisions in terms of which bus routes to take. They are also concerned with resources (for example, cost) and the level of service. What this discussion points to is that in performance evaluation the linkage to decisions and decision science is extremely important.

Given that we have reasonable representations of the DMUs and of the production possibility set, we then move to the notion of an empirical frontier. The question is why it is desirable to identify the empirical frontier. The answer is that once we have done so, then we can provide best practice units, provide efficient reference set (peers) for the inefficient (under-performing) units, identify performance targets, and finally we can provide an aggregate performance (efficiency) score, taking into account all the input and output variables. From a decision-making point of view, the most useful information is what constitutes the best practice units, the peers, and the performance targets (see examples in Medina-Borja, Pasupathy and Triantis (2007) and Jain, Triantis, and Liu (2011)). There are two approaches to computing/estimating the empirical frontier. The first is based on linear programming and optimization, and is referred to as data envelopment analysis (DEA), and the second is based on statistical estimation, and is referred to as stochastic frontier analysis (SFA). There are multiple commercial software packages available that can be used to estimate the empirical frontier using optimization (DEA) or statistical estimation (SFA).

DEA Empirical Frontier

Identification of the empirical frontier has its roots in the definition of efficiency/inefficiency. When should a DMU be considered inefficient? When another unit produces more with less and/or when another unit has produced at least as much output and consumed at most as much input. One defines a simple engineering benefit/cost ratio where Y is the output and X is the input as:

$$\frac{\text{Output}}{\text{Input}} = \frac{Y}{X} \tag{1}$$

The larger the value of the ratio, the better is the DMU's performance. This occurs when you have more output, Y and/or less input, X.

In almost all cases, the DMU uses multiple inputs (m) and produces multiple outputs (t). The original formulation by Charnes, Cooper and Rhodes (1978) for DMU j^0 (n DMUs) whose efficiency we wish to compute is as follows (this is typically called the CCR model in the literature):

$$\max h_0 \frac{\sum_{r=1}^{t} u_r y_{rj}^0}{\sum_{i=1}^{m} v_i x_{ij}^0} \tag{2}$$

subject to:

$$\frac{\sum_{r=1}^{t} u_r y_{rj}^0}{\sum_{i=1}^{m} v_i x_{ij}^0} \leq 1, \quad j = 1,2,\ldots,n \tag{3}$$

$$u_r, v_i \geq \varepsilon \tag{4}$$

$$\forall \; u_r, v_i \tag{5}$$

For an inefficient DMU, the solution to the math programming model 2–5 identifies corresponding efficient DMUs (i.e. efficient with the same weights) which are said to form a peer group for the inefficient unit. Efficiencies for every DMU relative to all DMUs in the data set can be found by solving 2–5 for each DMU. The values of the weights u_r and v_i are different from DMU to DMU. There is a great deal of flexibility in the choice of the weights u_r and v_i, so the following question arises: If the DMU is designated as efficient, is it because it is truly technically efficient or is it because there is flexibility in choosing the weights?

The dual formulation (envelopment formulation) is as follows:

$$\min 100z_0 - \varepsilon \sum_{r=1}^{t} s_r^+ - \varepsilon \sum_{i=1}^{m} s_i^+ \tag{6}$$

$$x_{ij_0} z_0 - s_i^+ - \sum_{j=1}^{n} x_{ij}\mu_j = 0, \;\; i = 1,2,\ldots,m \tag{7}$$

$$-s_i^- + \sum_{j=1}^{n} y_{ij}\mu_j = y_{rj_0}, \;\; r = 1,2,\ldots,t \tag{8}$$

$$\mu_j, s_i^+, s_r^- \geq 0 \;\; \forall j, r \text{ and } i \tag{9}$$

$$z_0 \text{ unconstrained} \tag{10}$$

The envelopment formulation 6–10 tries to find values of μ_j so as to construct a composite unit with outputs

$$\sum_{j=1}^{n} y_{rj}\mu_j, \;\; r = 1,2,\ldots,t$$

and inputs

$$\sum_{j=1}^{n} x_{ij}\mu_j, \;\; i = 1,2,\ldots,m$$

that outperforms the decision-making unit j^0. The DMU j^0 will be efficient if

$$z_0 = 1,$$
$$\sum_{r=1}^{t} s_r^+ = 0,$$
$$\sum_{i=1}^{m} s_i^- = 0.$$

This indicates that it is impossible to construct a composite unit that outperforms j^0. If j^0 is inefficient, z_0 will be smaller than one and/or its slacks will be positive. The optimal values of μ_j form a composite unit outperforming j^0 and providing targets for j^0. z_0 represents the maximum proportion of its input levels that unit j^0 should be expending to secure at least its current output levels (input reducing measure of technical efficiency).

Cooper, Seiford and Tone (2006) provide a whole host of alternative DEA empirical frontier formulations. These formulations are contingent upon the specific assumptions of the modeler and researcher. In this chapter we focus only on the most basic notions and formulations. Additionally, even though when the empirical frontier is derived using the mathematical programming approach there is no consideration of uncertainty, there has been a considerable effort in the literature the last decade to do so using alternative methods such as bootstrapping.

SFA Empirical Frontier

One of the key criticisms of DEA is that it does not consider uncertainty/noise and from that point of view one cannot draw statistical inferences. Noise could be measurement error in either an input or an output, improper construction of an input or an output, shocks to producers that are out of their control, etc.

The Stochastic Frontier Analysis (SFA) is an alternative method to DEA for estimating efficiency of an operating unit. At the end, while both SFA and DEA will produce metrics of interest, the issue is how the analyst will arrive at the final destination.

Let us consider the following relationship:

$$y_j = f\left(x_j,\beta\right)\ \exp\varepsilon_j \quad j = 1,2,\ldots n \tag{11}$$

One could consider that in the single equation production function specification as (11) $\varepsilon_j = u_j - v_j$. u_j is the part of the error process representing random noise or factors that are not controllable by the organization/firm/system. u_j is usually assumed to be normally distributed with zero mean and finite variance σ^2. v_j represents technical efficiency and is truncated, i.e., it is usually represented as a one-sided distribution.

Typically, equation 11 is estimated using a maximum likelihood approach. This makes the composed error approach computationally difficult for the practitioner who is not well versed in statistics. The technical efficiency for each observation j requires that one computes $E(v_j/u_j)$ for each observation *j*. Again, there are software and commercial packages that estimate the empirical frontier using the SFA approach.

Transportation Systems: Opening the Black Box

Even though efficiency measurement has been a fruitful area of scholarly research, engineers and decision-makers have not extensively used the efficiency measurement paradigm to evaluate and design the performance of engineered/transportation systems. Consequently, the challenge remains on how to focus on processes at the system and subsystem levels, i.e., how to study in detail the "input/output transformation box" that takes place in transportation systems. Nevertheless, in order to accomplish this task, this entails understanding the technologies used, the processes employed, the organizational structures, the decision-making rules and procedures, and equally important how various stakeholders behave. Another important modeling and research issue is to decide the appropriate level of aggregation (or the necessary level of detailed complexity) of the efficiency performance measurement analysis.

One of the main reasons for the lack of efficiency performance studies for transportation systems at the disaggregated process level is that one needs to invest a considerable amount of time within organizations studying their systems and transformation processes, interviewing both decision-makers and employees, and collecting useful data and information. Once in the organization, the researcher may not have the data to support a complete model where all the input and output/outcome variables are included.

A modeling approach that has been proposed in the literature to measure and design a system at a subsystem level is the network DEA model (Färe and Grosskopf, 2000). This approach identifies inefficiency sources embedded in the interactions among various components (subsystems) of the evaluated unit. It allows one to further investigate the structure and processes inside the decision-making unit, to identify the misallocation of inputs among subprocesses (subsystems) and generate insights about the sources of inefficiency within the DMU. Consequently, this approach allows one to investigate what is transpiring within the "black box" input/output representation of the traditional DEA analysis.

Considerations of Temporal Changes and Transportation System Dynamics

The majority of the efficiency measurement literature focuses on static performance measurement representations that assume that inputs are completely used for producing outputs or achieving outcomes in the same period. Thus, there is no time interdependence between the input utilization/output or outcome realization in subsequent time periods (Fallah-Fini, Triantis, and Johnson 2014). These static representations usually ignore the effects of engineering design decisions in one period on output or outcome levels over several subsequent time periods. Moreover, they do not take into account the dynamic nature of engineered systems where delays, non-linearities and feedback mechanisms are the norm.

In the context of the dynamic characteristics of engineering systems, this is an issue that is of importance both to the engineering literature as well as the efficiency measurement literature. There is a growing body of literature that addresses the non-parametric representation of dynamic efficiency (Fallah-Fini, et al. 2014). Additionally, in the engineering literature dynamic systems and modeling continue as a vibrant field of inquiry. One of the issues that remain open is the extent to which the efficiency literature can provide performance considerations that are part of the initial design decisions. This is extremely important for the concepts of resilience and sustainability, especially in the context of transportation systems where both temporal and life-cycle considerations are key.

One possible solution to deal with temporal and dynamic issues is to perform a system simulation to evaluate the various engineering design solutions before committing to a final subset of design solutions. In this context, one could adopt the Dynamic Productive Efficiency Model (DPEM) introduced by Vaneman and Triantis (2007) as a mechanism to track resilience, sustainability and other transportation system performance measures over time.

While the static network DEA performance models for engineering systems provide an initial understanding of the determinants of efficiency performance within these systems, they do not consider the dynamic characteristics of these systems. In terms of dynamic network DEA models, the efficiency research community has chosen to approach this issue using two distinct and separate directions.

The first simply extends network DEA formulations to include time (see for example, Tone and Tsutsui, 2013). An alternative approach is to model the system's dynamic behavior explicitly (for example, using either system dynamics (Vaneman and Triantis, 2007) or agent based modeling (Dougherty, Ambler, and Triantis, 2014)) and then include efficiency concepts as a way to assess system performance. This latter direction allows for the explicit consideration of causation, feedback mechanisms, delays, and non-linear relationships, whereas the former approach introduces temporal variations of efficiency measures explicitly. The structure of network DEA model does not suggest anything in terms of the resulting system behavior, whereas the structure of system dynamics or agent-based models results in various forms of system behaviors once one executes the simulations. On the other hand, the ways to measure efficiency performance with the system dynamics (Fallah-Fini, Triantis, Rahmandad, and de la Garza 2015) or agent-based models (Dougherty et al. 2014) are not straightforward. Consequently, one can view both directions as complementary since they address alternative representations of dynamics and efficiency measurement.

Conclusions

The purpose of this chapter is to introduce the reader to the efficiency performance evaluation of transportation systems using an empirical frontier analysis. The term "empirical" implies that there is a possibility of collecting relevant data that characterizes the key determinants of performance, that the modeler has access to the organization, technologies and decision-makers that can inform the performance representations and validate the modeling results.

There are two methods that are at the core of the estimation/computation of the frontier. The first, i.e., DEA, is based on mathematical programming representations that do not assume statistical noise but can accommodate a large number of input and output variables and have flexibility in terms of the representation of the transformation processes. While the other approach, i.e., SFA is based on statistical estimation, includes statistical noise, but requires assumptions about the functional form of the transformation processes and the distribution of the error terms.

In terms of transportation systems, it is important to open the black box, i.e., study in detail how the various transportation processes manifest themselves. An approach to accomplish this is to use a network approach. We also emphasize the importance of taking into account the dynamics that are associated with the transportation systems and their network representations. A few approaches to accomplish this are discussed.

References

Borja-Medina, A. and Triantis, K., 2011,"Modeling Social Services Performance: A Four-Stage DEA Approach to Evaluate Fundraising Efficiency, Capacity Building, Service Quality, and Effectiveness in the Nonprofit Sector," Annals of OR, DOI: 10.1007/s10479-011-0917-0.

Charnes, A., Cooper, W. W. and Rhodes, E., 1978, "Measuring the Efficiency of Decision Making Units," *European Journal of Operational Research*, 2, 429–444.

Cooper, W. W., Seiford, L., and Tone, K., 2006, Data Envelopment Analysis: A Comprehensive Text with Models, Applications, References and DEA-Solver Software, Kluwer Academic Publishers.

Dougherty, F., Ambler, N. and Triantis, K., 2014, "A Complex Adaptive Systems Approach for Productive Efficiency Analysis: Building Blocks and Associative Inferences" under review, Annals of OR.

Fallah-Fini, S., Triantis, K. and Johnson, A., 2014, "Efficiency Performance: State-of-the-Art," *Journal of Productivity Analysis*, Volume 41(1), 51–67, February.

Fallah-Fini, S., Triantis, K., Rahmandad, H., and de la Garza, J., 2015, "Measuring Dynamic Efficiency of Highway Maintenance Operations," *Omega*, 50, 18–28.

Färe, R. and Grosskopf, S., 2000, "Network DEA," *Socio-Economic Planning Sciences*, 34, 35–49, 2000.

Jain, S., Triantis, K., and Liu, S., 2011, "Manufacturing Performance Measurement and Target Setting," *European Journal of Operational Research*, November 2011, 144(2), 412–428, DOI: 10.1016/j.ejor.2011.05.028.

Seaver, B. and Triantis, K., 1992, "A Fuzzy Clustering Approach Used in Evaluating Technical Efficiency Measures in Manufacturing," *Journal of Productivity Analysis*, Volume 3, 337–363.

Seaver, B. and Triantis, K., 1995, "The Impact of Outliers and Leverage Points for Technical Efficiency Measurement Using High Breakdown Procedures," *Management Science*, 41(6), 937–956.

Sheth, C., Triantis, K., and Teodorović, D., 2007, "The Measurement and Evaluation of Performance of Urban Transit Systems: The Provision of Bus Service along Different Routes," *Transportation Research: Part E*, 43, 453–478.

Tone, K. and Tsutsui, M., 2013, "Dynamic DEA with Network Structure—A slacks-based measure approach. Workshop on DNDEA." Tokyo, Japan: Workshop on DNDEA.

Triantis, K., 2014, "Network Representations of Efficiency Analysis for Engineering Systems: Examples, Issues and Research Opportunities," in W.D. Cook and J. Zhu (Eds.), *Data Envelopment Analysis, International Series in Operations Research & Management Science* 208, DOI: 10.1007/978-1-4899-8068-7_22, © Springer Science+Business Media, New York.

Triantis, K., 2011, "Engineering Applications of DEA," (Chapter 14, pp. 363–402) *Handbook of Data Envelopment Analysis*, Kluwer Publishers, DOI: 10.1007/978–1-4419–6151–8.

Vaneman, W. and Triantis, K., 2007, "Evaluating the Productive Efficiency of Dynamical Systems," *IEEE Transactions on Engineering Management*, 54 (3), 600–612.

29

GIS AND TRANSPORTATION

Alan T. Murray

DREXEL UNIVERSITY

Introduction

Good transportation planning supported by insightful analysis is essential for the efficient operation of any town, city, metropolitan area, state and/or country, but also local, national and global companies/firms/corporations. Geographic information systems (GIS) have become an important part of transportation planning because they enable detailed spatial analysis to be carried out, taking into account data associated with demographic, environmental, social, economic and other factors.

In most areas of planning and analysis, GIS is so central to daily and long-term operations that separating it in some distinct manner is virtually impossible. Transportation planning is a little different in this respect given the range of issues and associated methodologies relied upon. Nevertheless, there have been a number of overviews of GIS and transportation in the literature (see Waters 1999; Goodchild 2000; Thill 2000; Miller and Shaw 2001; Murray 2001).

This chapter provides an introduction to GIS and discusses its connection to transportation planning and analysis. Basic principles and components of GIS are reviewed. This is followed by a discussion of geo-database design issues for transportation. Next, data and geo-spatial technologies are detailed. GIS to support various transportation planning contexts is then considered. Finally, concluding comments are given.

GIS

A geographic information system (GIS) is broadly conceived to be a collection of hardware, software and procedures that support spatial data creation, management, manipulation, analysis and display (Longley et al. 2011; Church and Murray 2009). Of course, developing components and bringing this collection together has required considerable research across a range of academic disciplines. What has made GIS unique and successful is viewing the world in a coordinated fashion, focusing on layers of information that can be integrated and combined in meaningful ways. This layer centric view is depicted in Figure 29.1.

GIS would be of little use without spatial data, so there is an entire industry focused on geographic data collection and processing. While GIS software sales are thought to exceed \$1 billion annually, the GIS data industry is estimated to be substantially higher in total annual

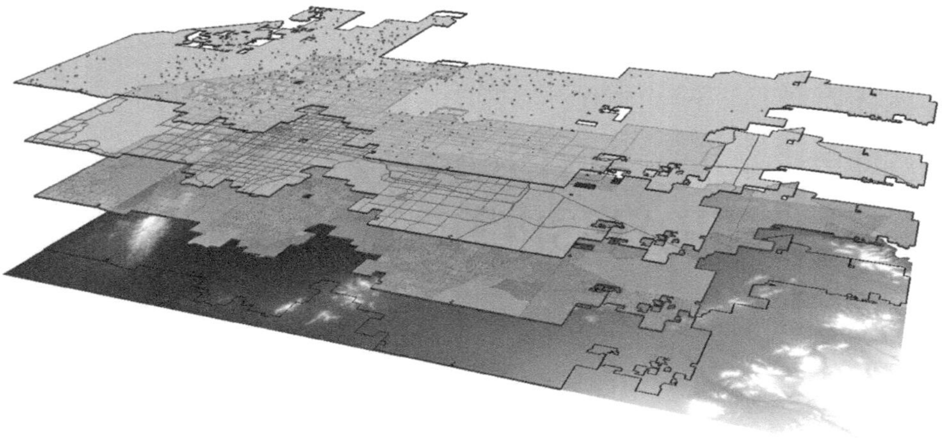

Figure 29.1 Layer-based view and integration in GIS

revenue (Longley et al. 2011). Of course, there is also a sizeable GIS services sector. In total, then, GIS is a multi-billion dollar industry.

Many descriptions of GIS can be found based on the classic definition of GIS given above (see Murray 2010). In this chapter a somewhat different type of overview is provided. What follows is a discussion of information systems, geo-referencing, spatial data models and analysis to provide a characterization of GIS that will be meaningful in the context of transportation planning and analysis.

Information Systems

In many ways, GIS is a derivative of a database management system (DBMS) that has been tailored to deal with the representation of the earth in a digital environment and capable of summary through visual methods, often map-based displays. Given data recording attributes and relationships of things, a database then is an assemblage of data organized in such a way that (efficient) digital storage and retrieval is possible. A DBMS is hardware, software and procedures that support creation, use and maintenance of a database. A detailed discussion of DBMS can be found in Silberschatz et al. (2006) and Shekhar and Chawla (2003).

A number of different DBMS approaches are relied upon in developed systems. The most popular and successful commercial systems are relational, network or object-oriented, but are also hierarchical and variations/combinations of these. The relational DBMS approach structures the database as a series of tables linked in various ways. The simplest, of course, is a single table with rows corresponding to observations, and columns being the associated attributes of each observation. In contrast, the network DBMS is centered around nodes and arcs that create a graph. Nodes and arcs may each have associated attributes, with additional information corresponding to paths, travel, flow, etc. through the network. This clearly is of significance for transportation planning and analysis. The object-oriented DBMS is a byproduct of the emergence of the object-oriented programming paradigm. The database then recognizes objects but also facilitates integration of unique operations that apply to particular object types.

Irrespective of the approach, a DBMS deals with the storage of data, management and manipulation as well as basic query. Doing this efficiently is particularly important for the success and operational performance of any system, and will be dependent on context and intended usage. As a result, different approaches have emerged for DBMS design because some are more amenable to being tailored to particular needs, making them better accessible and more efficient.

Geo-Referencing

What makes GIS special and important is that the data is associated with things and observed conditions on the surface of the earth. This goes beyond the generic observations recorded in a DBMS. As a result, it necessarily means that data being managed is geo-referenced in some way, and requires care and attention in terms of creation/maintenance, query, analysis and display.

Geodetics is a term generally used to refer to the earth, its shape and/or other defining characteristics. What we know is that representing the earth in a digital environment requires assumptions about its shape. A common assumption is that the earth is a sphere; however, it is not a sphere. At the equator the earth is some 67 km wider than it is between the North and South Poles. This means that it is more of an ellipsoid, but that is not exactly correct either because it is lumpy and irregularly formed in places. A more accurate representation is theoretically a geoide, but this remains an imperfect approximation as well. All assumed digital representations of the earth are therefore inaccurate in some way. For many types of analysis, approximation of the earth as a sphere, as an example, may be satisfactory. However, for some applications this may be a major complicating issue, with associated error and uncertainty impacting the validity and findings of a study. Care and attention must always be given to the fundamental assumptions of spatial representation when GIS is utilized.

Given an assumed shape, a referencing mechanism is used to describe relative position on the surface of the earth. The prevailing approach is the use of latitude and longitude defining relative position from an established origin and is measured in degrees (0–360), with a degree subdivided into minutes (0–60) and a minute subdivided into seconds (0–60). Latitude and longitude are then a reference pair, (ϕ, λ). Latitude reflects degrees north or south of the equator (zero degrees), with a particular degree representing a parallel. Longitude is an imaginary line connecting the North Pole to the South Pole known as a meridian, representing degrees east or west of the Greenwich meridian (prime meridian, or zero degrees that goes through the town of Greenwich, England).

With an assumed shape of the earth and a referencing system, GIS facilitates the integration of spatial information layers in a common coordinate system. Of course, each layer of spatial information must use the same referencing system for this to work. If not, then data layers must be projected and/or transformed so that each layer is in the same system. Projection involves assigning data referenced on the three-dimensional earth to a corresponding two-dimensional coordinate frame. This can be done in general terms as:

$$x = f\left(\phi,\lambda\right)$$
$$y = \hat{f}\left(\phi,\lambda\right)$$

where $f(\,)$ and $\hat{f}(\,)$ are functions of a specific mathematical form depending on the type of projection sought. There are, in fact, many different possible projections, such as Mercator, Robinson, Albers conic, Tobler hyperelliptical, etc. All projections introduce some sort of distortion and/or error because it is impossible to represent the 3-D earth in a 2-D environment. Thus, different projections exist because some decrease distortion/error in certain ways at certain locations. Depending on the intent of the analysis and area of study, one projection may be more appropriate than others.

A transformation involves changing the current referencing system for a data layer(s) to a different coordinate system. In 3-D, a transformation between different assumed shapes of the earth may be necessary. For example, this would correspond to moving between a sphere, ellipsoid and/or geoide (also referred to as a datum transformation). This is also possible in 2-D, often carried out using a complex affine or curvilinear transformation.

Spatial Data Models

Similar to DBMS application and development, a major concern in GIS has always been the efficient storage of geographic data and fast processing in order to ensure quick system response as part of analytical and exploratory inquiry. Two fundamental spatial data models have emerged for representing space, serving to organize the storage of spatial data (and attributes) in a manner that facilitates access and analytical processing. The two models are raster and vector.

The raster data model accounts for a continuously varying attribute or surface, and typically regular (square) cells are used to completely cover an area. Examples of spatial tessellations are depicted in Figure 29.2. For a regular raster (Figure 29.2a), a cell then has a spatial footprint or boundary on the surface of the earth with an attribute that records the observed characteristics of this cell (elevation in this case). Other tessellations are possible, such as the triangulated irregular network (TIN) shown in Figure 29.2b and the hierarchical raster (quadtree) shown in Figure 29.2c. Attributes recorded for an individual raster cell include, but are not limited to, temperature, rainfall, soil type, vegetation, air quality, elevation, etc. As suggested previously, it is often the case that GIS would manage many layers of data. For the raster model, each layer would correspond to a particular attribute type. Beyond the more obvious usefulness of recording any attribute type and value for any location, the raster data model has proven particularly adept for supporting analysis of water runoff, air quality and visibility analysis.

The technical challenges for raster GIS have included efficient storage and processing of both geographic location and associated attribute information. Various schemes have been devised to deal with this. Scan orders, run length encoding and hierarchical as well as irregular representations have all proven to be beneficial in certain application contexts. Commercial systems that are predominantly raster based include IDRISI, ERDAS IMAGINE and GRASS GIS, among others.

Of more direct relevance to transportation is the vector data model. The distinction of vector GIS is that discrete objects are of interest. These objects are typically in the form of points, lines and polygons, representing a non-exhaustive sample of geographic features or entities. An example of a vector GIS data layer is illustrated in Figure 29.3 showing a transportation network. This layer can be combined with other information layers in GIS to support display and analysis.

There have also been technical challenges for the vector GIS data model. The fundamental distinction of the vector data model is the object primitives (points, lines and polygons). Formally, these object primitives defined in 2-D are:

- point: $\{(x_1, y_1)\}$
- line: $\{(x_1, y_1), (x_2, y_2)\}$
- polyline: $\{(x_1, y_1), (x_2, y_2), (x_3, y_3), \ldots, (x_n, y_n)\}$
- polygon: $\{(x_1, y_1), (x_2, y_2), (x_3, y_3), \ldots, (x_m, y_m)\}$

where (x_i, y_i) corresponds to a point reference and $(x_1, y_1) = (x_m, y_m)$. A common assumption is that two consecutive points are connected by a straight line, but this need not be the case in theory as a mathematical function could be used. Relational DBMS approaches have historically been relied upon for the storage and structuring of data. Geographical data have presented a range of challenges because there are obvious/intuitive ways to display information, but also processing efficiencies that are possible. Relational and object-oriented DBMS approaches

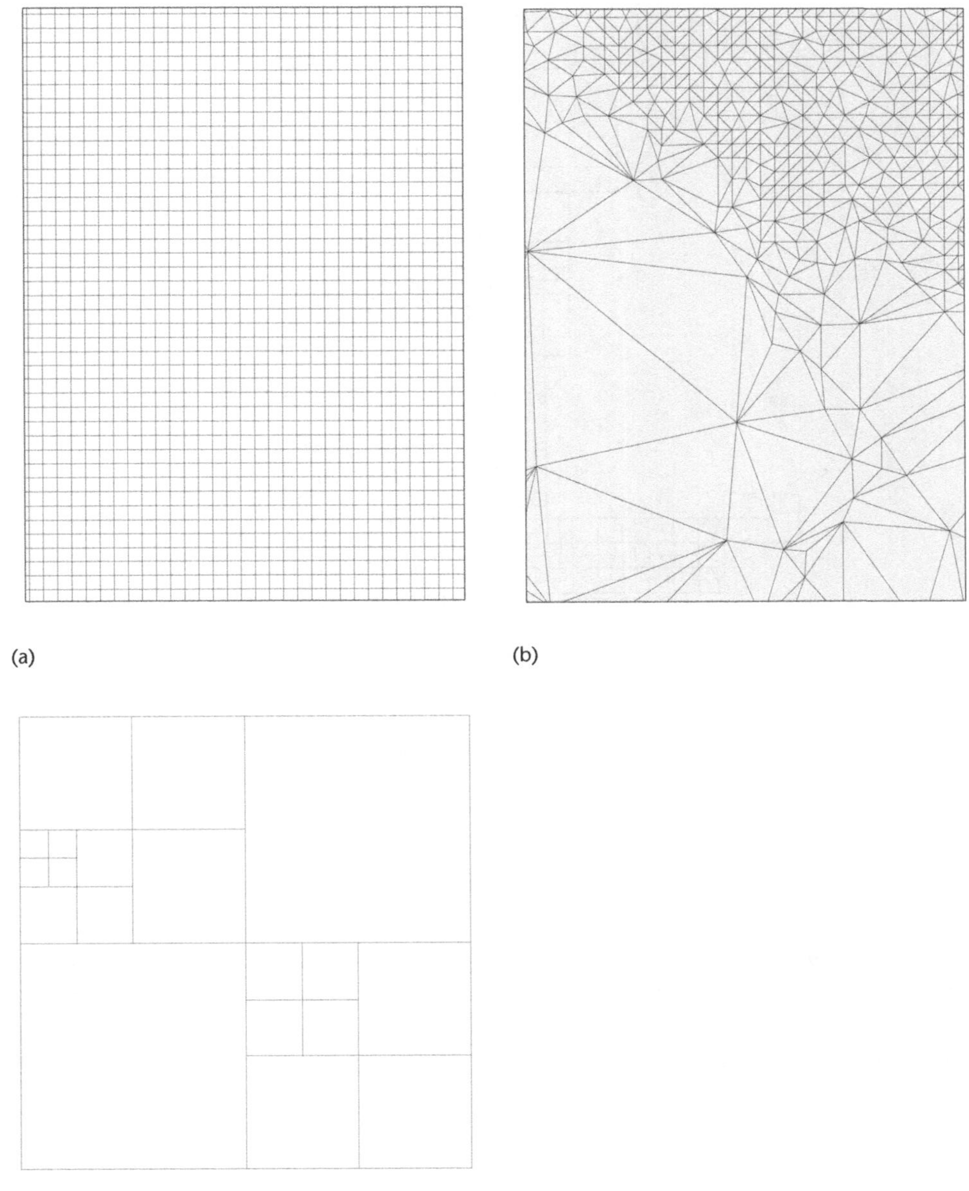

(a) (b) (c)

Figure 29.2 Tessellations of geographic space

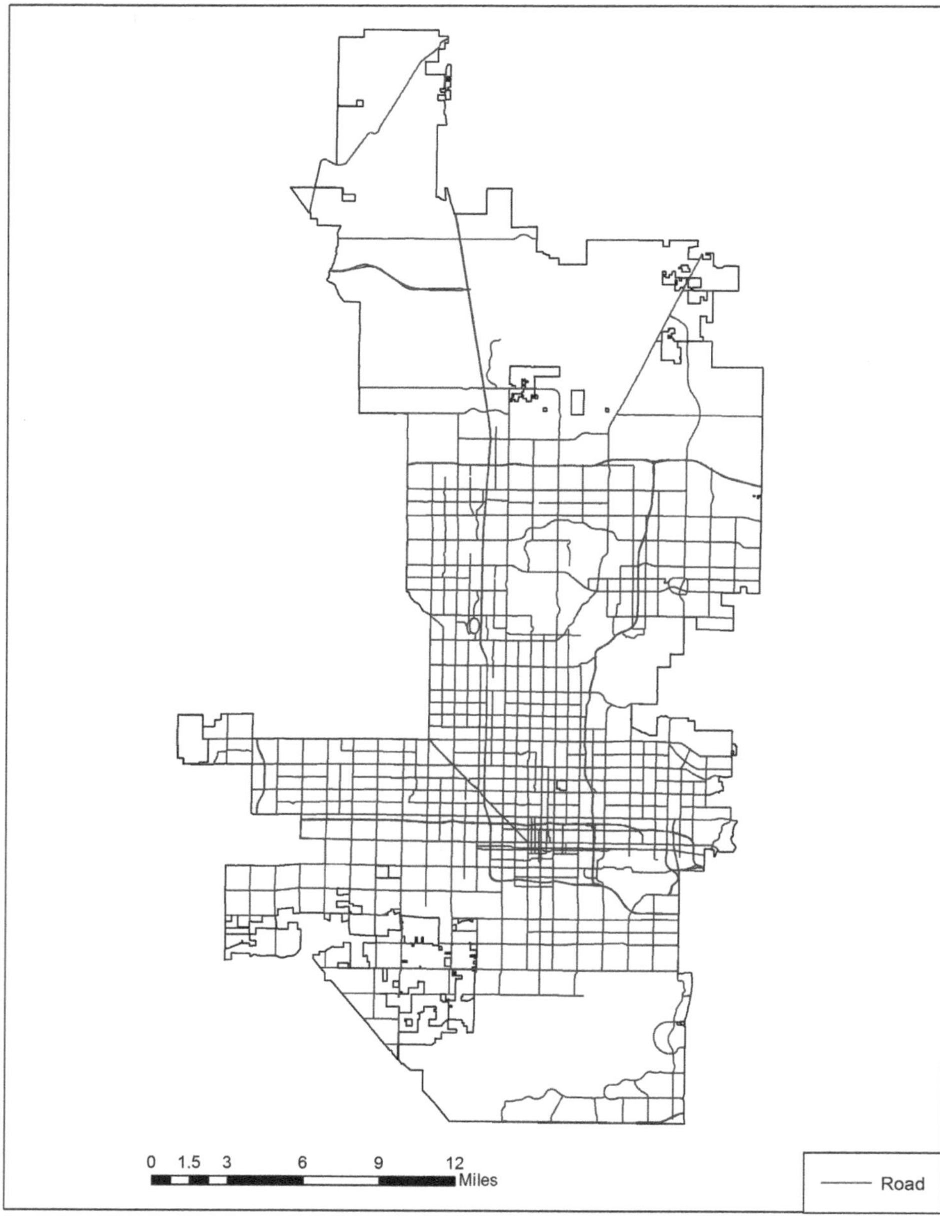

Figure 29.3 Vector data layer

dominate the commercial vector GIS market. Commercial systems that are predominantly vector based include ArcGIS, MapInfo and QGIS, among others.

Analysis

The ability to carry out analysis has been an important asset of GIS. Analytical methods range greatly, extending structured/standard query language (SQL) based query found in DBMS. In particular, spatial information gives rise to a need for geographic query. Thus, GIS supports identifying objects in a layer that are near or far away from features or other objects, or in a user defined area(s). As information is spatially based, mapping and geovisualization are essential, so GIS supports this in various ways. Of course, these methods and more must be supported by spatial analytical operations of all sorts. Given this, standard GIS functionality includes analysis approaches such as proximity, buffer, overlay, and map algebra, among others. Proximity analysis includes identification of neighbors as well as deriving distance between objects. Buffer analysis is the ability to derive proximity zones/areas around objects. Overlay has two interpretations, depending on the spatial data model. For vector GIS, overlay refers to the spatial combination of two or more vector object layers, creating a new layer that is all of the smallest unique object that results from geometric unification of multiple layers (note as well that resolution of the attributes for newly created objects is also necessary, often through the use of an areal interpolation process). Finally, map algebra (also equivalent to raster overlay) is the process of taking one or more input layers and creating an output layer of the same spatial resolution with cell attribute values that are a function of the input layers. Details of these approaches (and more) can be found in Longley et al. (2011) and Church and Murray (2009), among others.

Historically, analysis capabilities in GIS such as those detailed above were thought to be rather basic, lacking quantitative rigor and specification (or statistical significance). The reality is that these approaches are, in fact, more sophisticated than they might appear. Nevertheless, advanced methods continue to be developed, extended and incorporated as standard GIS analysis tools. Examples include various interpolation techniques as well as statistical and geo-statistical/spatial-statistical methods. Some approaches/methods can be found in the commercial GIS packages mentioned above. A number of GIS-based analysis packages exist that provide access to certain analytical approaches. Examples include the following GIS based software: GeoDa, SANET, CrimeStat, SaTScan, Fragstats, GWR, etc.

Data and Geo-Spatial Technologies

An incredibly important element in the application and use of GIS is spatial data. Obtaining high quality data may be both difficult and costly, depending on the needs and purpose of analysis. As noted previously, GIS data provision is big business. There are, in fact, many sources for obtaining existing geographic information, including governmental agencies and private companies. Longley et al. (2011) as well as Church and Murray (2009), among others, provide details on data sources. If geographic data does not exist in a digital form capable of being used in a commercial GIS, then data creation and acquisition as well as supporting geo-spatial technologies must be pursued.

The creation of digital geographic information has been a major concern since the inception of GIS. Manual and automated digitizing/scanning approaches have long been relied upon for the generation of supporting spatial data. To ensure quality and accuracy it is often the case that substantial time and effort are invested. Digitizing and scanning approaches make use of existing paper-based maps or areal images, and can be thought of as a technology for extracting from a map/image into a digital format for use in GIS. Airplanes, helicopters and drones are commonly used to acquire areal images and video. Considerable data used in GIS comes from such a process, though it may have been cleaned, updated or refined in various ways. Another major form of spatial data creation is through the use of geocoding approaches. Specifically, this involves taking a local reference, such as a street address, postal code or place name, and associating it to a latitude and longitude on the surface of the earth. Often one has a spreadsheet of observation addresses, then must convert these local references to geographic coordinates for use in GIS. In support of transportation planning and analysis, this might be gas stations in a metropolitan area or transit stops. Software exists for taking local references and producing geographic coordinates, relying on spatial databases and interpolation approaches for inferring latitude and longitude. Geocoded data too has much potential for error and inaccuracy (see Murray and Grubesic 2012), so care must be taken when such data is used.

Various forms of observation-generated data are heavily relied upon in transportation studies. Examples include traffic counts, driver behavior, etc. In the case of traffic counts, people are stationed at various locations to count passing vehicles or modal split, but this is also done through video-/camera-based observation as well. Not unrelated is so-called volunteered geographic information (VGI). This is where individuals throughout an area or region contribute unsolicited information associated with a particular topic. For example, current gas price information can be found for a region through sources like GasBuddy.com, but the information is collected/reported by users, not the actual service providers. This is also the case for other items of interest, such as road blocks, accidents, congested roadways, police check points and speed traps, etc., through TomTom Map Share and other similarly conceived products/sites. There are increasingly available sources of VGI-based information that can readily be used in GIS to support transportation planning.

On the more automated spatial data-creation side, acquisition of geographic information is increasingly relying on remote sensing and GPS. Remote sensing is both hardware and software to support semi-automated spatial data collection. Often this is specific to satellite-based sensors that detect electromagnetic radiation. However, more generally this could include LIDAR and other areal data collection approaches. Irrespective of the particular technology, remote sensing approaches make it possible to derive land cover composition, elevation, temperature, etc. There are numerous government and commercial remote-sensing options, including Landsat (US) and SPOT (French). Global position system (GPS) refers to a constellation of satellites orbiting the earth, base stations (in the case of differential correction) and individual receivers. Collectively, the satellites, base stations and receivers enable position and time on the surface of the earth to be recorded digitally. NAVSTAR GPS is operated by the US Department of Defense, but similar constellations exist, operated by other agencies. GPS enables vector-based data collection for almost anyone in any context, with little cost and with reasonable spatial accuracy. Further details may be found in Thurston et al. (2003) and Longley et al. (2011).

GIS-T

Commercial GIS has evolved to be a general-purpose environment capable of facilitating analysis for a range of concerns across a variety of application contexts. This means that GIS is not necessarily tailored to address the needs of transportation planning and analysis. In recognition of this, GIS-T (GIS for Transportation) has evolved as a term that acknowledges the uniqueness of transportation and the potential shortcomings of off-the-shelf GIS packages to provide sufficient functionality to address the uniqueness of transportation contexts (Waters 1999; Shaw 2010).

Transportation planning and analysis may focus on problems, issues and/or concerns that include but are not limited to infrastructure inventory, modal splits, traffic conditions and congestion, routing, navigation, emergency response, equity, safety, intermodal operations, and environmental impact, among others. As noted previously, GIS is generally viewed (and operationalized) in terms of raster/vector spatial data models. A particular transportation context may or may not be suited to either raster or vector. While the vector spatial data model theoretically enables a transportation network to be encoded and represented as an example, this may not be sufficient. The vector model has the ability to deal with points (nodes) and lines (arcs), and nodes and arcs are the fundamental building blocks of a network data structure. However, planar enforcement is generally assumed (and imposed through topological cleaning processes) in the vector model, meaning that an arc could not cross another arc without going through a node. Yet, if an arc segment represents a road overpass, there is no node reflective of an intersection. Similarly, GIS-based databases of road networks often simplify road infrastructure, using only the street centerline. This then omits important features of the infrastructure, such as individual lanes, turn lanes and HOV (high occupancy vehicle) lanes. Routing and navigation is highly dependent on this sort of detailed information, yet is not necessarily supported by the vector data model only capable of dealing with points and lines. Alternatively, some transportation contexts may well be suited for raster/vector data models. If a study is interested in examining air quality in and around transportation corridors or roadways, then a raster data model is important for representing continuous variation using raster cells with attributes of air quality indicators, such as carbon dioxide, nitrogen oxides, particulates, etc. Further, relating this to people likely would require census-oriented information represented in a vector data model, where polygons give recorded counts of people living there.

Shaw (2010) discusses networks, features and data unique to transportation contexts. There are many types of transportation networks, such as residential, freeway, interstate, highway, transit, railway, airline, ferry, and others. Further, there are a range of transportation-oriented features, including signage, signals, lighting, safety equipment, refueling stations, maintenance facilities, etc. Traffic data would therefore depend on whether it applies to a network or features, and may include vehicle miles travelled (VMT), counts, accidents, level of service, demographic and socioeconomic characteristics, pavement type, etc. While seemingly straightforward, technically this is particularly important and unique for transportation. VMT and flows between origins and destinations, as an example, are a dyad or locational pair attribute. This requires storage of the dyad, as well as storage of the associated attribute information, and is typically accomplished using matrices. This goes beyond the traditional vector data model. As a result, the transportation arena abounds with unique information systems designed to address concerns for transit, congestion, travel patterns, etc. A primary commercial transportation oriented GIS is TransCAD, produced by Caliper (the more generic GIS product being Maptitude). There are, in fact, many transportation GIS-oriented products, and a number of them will be mentioned in the next section.

Beyond the issues of origin-destination flows and routes, there are primary areas of focus unique to GIS-T: linear referencing, dynamic segmentation and intelligent transportation systems. Linear referencing represents storage and retrieval in GIS of objects and information associated with a linear feature. Examples of objects/attributes are mile markers, crash sites and signs/signals along a linear route. With greater access to and familiarity with GPS, there is a move towards storage of such information as a unique vector-based object. Dynamic segmentation is a storage and operational process for dealing with attributes of a linear feature without modifying the object to break it down to its smallest unique segments/parts. For example, consider volume of traffic along a roadway, pavement quality, level of service, etc. With multiple attributes of the linear segment, they may or may not overlap in coincidental ways, so the challenge is encoding and storage of this information without manipulating the object. This is the essence of dynamic segmentation. Intelligent transportation systems constitute a range of technologies brought together to provide operational benefits, safety improvements and enhanced user information. The technologies include sensor-based systems like radar and LIDAR combined with GPS and computer vision/processing. This is used in fleet management, collision avoidance, and signal control, but also is the foundation for emerging autonomous vehicle operation.

Transportation Areas

The diversity of areas in transportation is rather impressive. GIS has been applied and/or become a part of the many different areas, including aviation, cycling, evacuation, highways, intermodal, logistics, pedestrian access and accessibility, port operations, transit and ride sharing, among others. Generic GIS software packages like ArcGIS and TransCAD have been and continue to be applied in most areas/contexts in one way or another. However, the unique demands of transportation have necessarily meant that general commercial systems often are either customized or specialized packages are developed to address particular concerns or issues for a transportation area. Major GIS software providers such as Esri have even developed products along these lines, such as ArcGIS for Transportation Analytics, ArcGIS for Aviation, ArcGIS for Maritime and Esri Roads and Highways, and firms have developed extensions and standalone software, such as RouteSmart, Agile Route, etc.

Transit is arguably an area that integrates GIS most broadly. Table 29.1 attempts to reflect the diversity and widespread usage of GIS, where commercial and custom/tailored GIS packages are used to support planning, operations and customer information in many different ways. Of course, transit is not alone in recognizing the importance and significance of GIS. The above noted all make fairly extensive use of GIS in one way or another. Reviews highlighting the use of GIS can be found in Murray et al. (1998), Southworth and Peterson (2000), Grubesic et al. (2013), and Hong and Murray (2013), among others.

Table 29.1 GIS-based transit planning software

Task	*Software*
Route planning	TomTom, MyTracks
Route scheduling	Hastus, Trapeze, Master Scheduling, Init
Paratransit	Access, Trapeze, RouteMatch, Stratagan
Dispatch, automatic vehicle location and navigation	Siemens, Avail, Clever Devices, Init, Trapeze, RouteMatch, Indra, Xerox
Automated passenger counting	Urban Transportation Associates, Clever Devices, IRIS, Init, Eurotech
Customer trip planning and real-time arrival	HastInfo, Trapeze, Siemens, Init, RouteMatch, Avail, Google Transit
Transportation demand modeling, network and corridor design	Cube, Visum, EMME, TransCAD
Traffic, signaling analysis and simulation	Vissim, Corsim, Synchro Studio, Sidra
Connectivity analysis	Via City

Conclusions

While transportation is a broad and encompassing field, GIS has become a methodological approach that has proven to be invaluable for planning, management and policy tasks. GIS facilitates operational aspects of transportation systems to be considered, but also demographic, environmental, economic and other factors. GIS enables data to be created, managed, manipulated and analyzed in many different ways, making it central to daily and long-term transportation planning processes. However, the unique considerations and concerns addressed in transportation contexts have meant that analytical capabilities beyond GIS have invariably been necessary in supporting analysis and decision-making.

References

Church, R.L. and A.T. Murray (2009). *Business Site Selection, Location Analysis and GIS* (New York: Wiley).

Goodchild, M. (2000). "GIS and transportation: status and challenges." *GeoInformatica*, 4, 127–139.

Grubesic, T.H., A.T. Murray and T.C. Matisziw (2013). "Putting a price on politics as usual: rural air transport in the United States." *Transport Policy*, 30, 117–124.

Hong, I. and A.T. Murray (2013). "Efficient measurement of continuous space shortest distance around barriers." *International Journal of Geographical Information Science*, 27, 2302–2318.

Longley, P.A., M.F. Goodchild, D.J. Maquire and D.W. Rhind (2011). *Geographic Information Systems and Science*, 3rd ed. (New York: Wiley).

Miller, H. and S.-L. Shaw (2001). *Geographic Information Systems for Transportation: Principles and Applications* (New York: Oxford University Press).

Murray, A.T. (2001). "Strategic analysis of public transport coverage." *Socio-Economic Planning Sciences*, 35, 175–188.

Murray, A.T. (2010). "Quantitative geography." *Journal of Regional Science*, 50: 143–163.

Murray, A.T., R. Davis, R.J. Stimson and L. Ferreira (1998). "Public transportation access." *Transportation Research D*, 3, 319–328.

Murray, A.T. and T.H. Grubesic (2012). "Spatial optimization and geographic uncertainty: implications for sex offender management strategies." In M. Johnson (Ed.) *Community-Based Operations Research: Decision Modeling for Local Impact and Diverse Populations*, 121–142 (Berlin: Springer).

Shaw, S.-L. (2010). "Geographic information systems for transportation: from a static past to a dynamic future." *Annals of GIS*, 16, 129–140.

Shekhar, S. and S. Chawla (2003). *Spatial Databases: A Tour* (Upper Saddle River, NJ: Prentice Hall).

Silberschatz, A., H.F. Korth and S. Sudarshan (2006). *Database System Concepts*, 5th ed. (New York: McGraw-Hill).

Southworth, F. and B.E. Peterson (2000). "Intermodal and international freight network modeling." *Transportation Research C*, 8, 147–166.

Thill, J.-C. (2000). "Geographic information systems for transportation in perspective." *Transportation Research C*, 8, 3–12.

Thurston, J., T.K. Poiker and J.P. Moore (2003). *Integrated Geospatial Technologies* (New York: Wiley).

Waters, N. (1999). "Transportation GIS: GIS-T." In P. Longley, M. Goodchild, D. Maquire, and D. Rhind (Eds.) *Geographical Information Systems: Principles, Techniques, Management and Applications*, 827–844 (New York: Wiley).

INDEX